METALLURGICAL EFFECTS AT HIGH STRAIN RATES

The Metallurgical Society of AIME Proceedings published by Plenum Press

1968 – Refractory Metal Alloys: Metallurgy and Technology
    Edited by I. Machlin, R. T. Begley, and E. D. Weisert

1969 – Research in Dental and Medical Materials
    Edited by Edward Korostoff

1969 – Developments in the Structural Chemistry of Alloy Phases
    Edited by B. C. Giessen

1970 – Corrosion by Liquid Metals
    Edited by J. E. Draley and J. R. Weeks

1971 – Metal Forming: Interaction Between Theory and Practice
    Edited by A. L. Hoffmanner

1973 – Titanium Science and Technology (4 Volumes)
    Edited by R. I. Jaffee and H. M. Burte

1973 – Metallurgical Effects at High Strain Rates
    Edited by R. W. Rohde, B. M. Butcher, J. R. Holland, and C. H. Karnes

A Publication of The Metallurgical Society of AIME

METALLURGICAL EFFECTS AT HIGH STRAIN RATES

Edited by

R. W. ROHDE
B. M. BUTCHER
J. R. HOLLAND
C. H. KARNES
Sandia Laboratories
Albuquerque, New Mexico

PLENUM PRESS • NEW YORK–LONDON • 1973

Library of Congress Cataloging in Publication Data

Main entry under title:
Metallurgical effects at high strain rates.

"Sponsored by: Sandia Laboratories and the Physical Metallurgy Committee of the Metallurgical Society of AIME. [Held in] Albuquerque, New Mexico, February 5-8, 1973."
   Includes bibliographical references.
   1. Physical metallurgy—Congresses. 2. Dislocations in metals—Congresses. 3. Strains and stresses—Congresses. 4. Shock (Mechanics)—Congresses. I. Rohde, R. W., ed. II. Sandia Laboratories. III. Metallurgical Society of AIME. Institute of Metals Division. Physical Metallurgy Committee.

TN689.2.M45   620.1'6'3   73-10497
ISBN 0-306-30754-5

Proceedings of a technical conference sponsored by Sandia Laboratories and The Physical Metallurgy Committee of The Metallurgical Society of AIME and held in Albuquerque, New Mexico, February 5-8, 1973

© 1973 Plenum Press, New York
A Division of Plenum Publishing Corporation
227 West 17th Street, New York, N.Y. 10011

United Kingdom edition published by Plenum Press, London
A Division of Plenum Publishing Company, Ltd.
Davis House (4th Floor), 8 Scrubs Lane, Harlesden, London, NW10 6SE, England

All rights reserved

No part of this publication may be reproduced in any form without written permission from the publisher

Printed in the United States of America

PREFACE

A conference on <u>Metallurgical Effects at High Strain Rates</u> was held at Albuquerque, New Mexico, February 5 through 8, 1973, under joint sponsorship of Sandia Laboratories and the Physical Metallurgy Committee of The Metallurgical Society of AIME. This book presents the written proceedings of the meeting.

The purpose of the conference was to gather scientists from diverse disciplines and stimulate interdisciplinary discussions on key areas of materials response at high strain rates. In this spirit, it was similar to one of the first highly successful conferences on this subject held in 1960, in Estes Park, Colorado, on <u>The Response of Metals to High Velocity Deformation</u>. The 1973 conference was able to demonstrate rather directly the increased understanding of high strain rate effects in metals that has evolved over a period of roughly 12 years.

In keeping with the interdisciplinary nature of the meeting, the first day was devoted to a tutorial session of invited papers to provide attendees of diverse backgrounds with a common basis of understanding. Sessions were then held with themes centered around key areas of the high strain rate behavior of metals. Most sessions were introduced by an invited lecturer with subsequent papers accepted from submitted abstracts. Invited papers are designated in the Table of Contents. Conferees were encouraged to submit written discussion of the papers, and a few discussions were formally presented at the meeting. These discussions and the author's replys are included in the volume and are so designated in the Table of Contents.

The papers in the conference proceedings have been reproduced from camera ready copy prepared by the authors. As is customary in such cases, the manuscripts underwent no further editing after acceptance for these proceedings. The editors have relied primarily upon submitted discussions to point out any ambiguities in the manuscripts.

Thanks are due to the session chairman, the authors, invited lecturers and participants for their contributions to the conference. Acknowledgment is due to the many people somewhat behind the scenes who contributed in innumerable ways to the success of this conference. The efforts of G. A. Alers, North American Rockwell, J. C. Swearengen and T. V. Nordstrom of Sandia Laboratories, and L. E. Pope formerly of Sandia Laboratories are particularly appreciated. We also extend our grateful acknowledgment to Mrs. Wanda Maes who served as conference secretary.

The Organizing Committee

R. W. Rohde

B. M. Butcher

J. R. Holland

C. H. Karnes

CONTENTS

Introduction, G. E. Duvall (Invited) 1

SESSION A

TUTORIAL PAPERS

Chairman: G. E. Duvall

Shock Wave Physics 15
    G. R. FOWLES (Invited)

Shock Wave Mechanics 33
    O. E. JONES (Invited)

Numerical Analysis Methods 57
    W. HERRMANN AND D. L. HICKS (Invited)

Continuum Plasticity in Relation To Microscale
    Deformation Mechanisms 93
    J. R. RICE (Invited)

Experimental Methods in Shock Wave Physics 107
    J. W. TAYLOR (Invited)

Metallurgical Effects of High Energy Rate Forming 129
    R. N. ORAVA (Invited)

A Practical Guide To Accurate Grüneisen Equations
    of State 157
    D. J. O'KEEFFE AND D. J. PASTINE

SESSION B

PHASE TRANSFORMATIONS

Chairman: R. A. Graham

Relation Between Dynamic and Static Phase
    Transformation Studies 171
    W. J. CARTER (Invited)

Characteristics of the Shock Induced Transformation
  in BaF_2 185
    D. P. DANDEKAR AND G. E. DUVALL

A Theory of the $\alpha \rightarrow \epsilon$ Transition in Fe and of Possible
  Higher Pressure Transitions in Fe and in the Lighter
  Elements of the First Transition Series 201
    D. J. PASTINE

  Presented Discussion 223
    L. M. BARKER

Remanent Magnetization and Structural Effects Due to
  Shock In Natural and Man-Made Iron-Nickel Alloys 225
    P. J. WASILEWSKI AND A. S. DOAN, JR.

  Presented Discussion: Shock Induced Melt
    Transitions in Aluminum and Bismuth 247
    J. R. ASAY

  Discussion: Rapid Melting of Aluminum Induced
    by Pulsed Electron Beam Exposure 251
    F. B. MCLEAN, R. B. OSWALD, JR.,
    D. R. SCHALLHORN AND T. R. OLDHAM

SESSION C

DISLOCATION DRAG MECHANISMS

Chairman: G. A. Alers

Microscopic Mechanisms of Dislocation Drag 255
    A. V. GRANATO (Invited)

  Discussion 274
    F. A. MCCLINTOCK

  Discussion 274
    T. VREELAND, JR. AND K. M. JASSBY

  Author's Reply 275

Experimental Measurement of the Drag Coefficient 277
    T. VREELAND, JR. AND K. M. JASSBY (Invited)

  Discussion 285
    F. A. MCCLINTOCK

  Author's Reply 285

CONTENTS

The Role of Dislocation Drag in Shock Waves 287
    P. P. GILLIS AND J. M. KELLY (Invited)

    Discussion 309
    T. VREELAND, JR. AND K. M. JASSBY

    Discussion 309
    F. A. MCCLINTOCK

    Author's Reply 310

Plastic Deformation at Low Temperatures and Drag
    Mechanisms 311
    J. M. GALLIGAN AND M. SUENAGA (Invited)

SESSION D

DISLOCATION DYNAMICS

Chairman: T. Vreeland, Jr.

Dislocation Mechanics at High Strain Rates 319
    J. WEERTMAN (Invited)

    Discussion 333
    T. VREELAND, JR. AND K. M. JASSBY

A Microdynamical Approach to Constitutive Modeling of
    Shock-Induced Deformation 335
    H. E. READ

Wave Propagation in Beryllium Single Crystals 349
    L. E. POPE AND A. L. STEVENS

Shock-Induced Dynamic Yielding in Lithium Fluoride
    Single Crystals 367
    Y. M. GUPTA AND G. R FOWLES

Effects of Microstructure and Temperature on Dynamic
    Deformation of Single Crystal Zinc 379
    P. L. STUDT, E. NIDICK, F. URIBE
    AND A. K. MUKHERJEE

    Discussion 399
    C. H. KARNES

    Author's Reply 400

A Constitutive Relation for Deformation Twinning in
   Body Centered Cubic Metals 401
   R. W. ARMSTRONG AND P. J. WORTHINGTON

SESSION E

FRACTURE PROCESSES

Chairman: G. T. Hahn

Models of Spall Fracture by Hole Growth 415
   F. A. MCCLINTOCK (Invited)

   Discussion 424
   F. A. MCCLINTOCK

Observations of Spallation and Attenuation Effects in
   Aluminium and Beryllium From Free-Surface Velocity
   Measurements 429
   C. S. SPEIGHT, P. F. TAYLOR AND A. A. WALLACE

Effects of Metallurgical Parameters on Dynamic Fracture . . 443
   W. B. JONES AND H. I. DAWSON

Wave Propagation and Spallation in Textured Beryllium . . . 459
   A. L. STEVENS AND L. E. POPE

The Influence of Microstructural Features on Dynamic
   Fracture 473
   D. A. SHOCKEY, L. SEAMAN AND D. R. CURRAN

SESSION F

DEFORMATION MECHANICS

Chairman: M. Wilkins

Work Softening of Ti-6Al-4V Due to Adiabatic Heating 501
   A. U. SULIJOADIKUSUMO AND O. W. DILLON, JR.

Thermal Instability Strain in Dynamic Plastic
   Deformation 519
   R. S. CULVER

The Propagation of Adiabatic Shear 531
   M. E. BACKMAN AND S. A. FINNEGAN

CONTENTS

Shear Strength of Impact Loaded X-Cut Quartz as Indicated
  By Electrical Response Measurements (Abstract Only) . . . 545
    R. A. GRAHAM

Metallurgical Effects at High Strain Rates in the
  Secondary Shear Zone of the Machining Operation 547
    P. K. WRIGHT

Miniature Explosive Bonding With a Primary Explosive 559
    J. L. EDWARDS, B. H. CRANSTON AND G. KRAUSS

SESSION G

MICROSTRUCTURAL EFFECTS

Co-Chairmen: L. E. Murr and U. Lindholm

Microstructural Effects of High Strain Rate Deformation . . 571
    W. C. LESLIE (Invited)

The Substructure and Properties of Explosively Loaded
  Cu-8.6% Ge 587
    D. J. BORICH AND D. E. MIKKOLA

Fragmentation, Structure and Mechanical Properties of
  Some Steels and Pure Aluminum After Shock Loading 605
    T. ARVIDSSON AND L. ERIKSSON

Twinning in Shock Loaded Fe-Al Alloys 619
    M. BOUCHARD AND F. CLAISSE

    Discussion 630
    R. J. WASILEWSKI

Annealing of Shock-Deformed Copper 631
    E. A. CHOJNOWSKI AND R. W. CAHN

Energy Absorption and Substructure in Shock Loaded
  Copper Single Crystals 645
    A. M. DIETRICH AND V. A. GREENHUT

X-Ray Topography of Shock Loaded Copper Crystals 659
    P. W. KINGMAN

The Effect of Heat Treatment on the Mechanical Properties
  and Microstructure of Explosion Welded 6061 Aluminum
  Alloy 669
    R. H. WITTMAN

Discussion 687
J. LIPKIN

Author's Reply 688

Index 689

PROBLEMS IN SHOCK WAVE RESEARCH

George E. Duvall

Shock Dynamics Laboratory, Physics Department

Washington State University, Pullman, Wa.

I. HISTORICAL REVIEW

Theoretical studies of shock waves, their structure and their propagation date well back into the 19th century. Poisson, Stokes and Earnshaw were early pioneers; Rankine, the great Scottish engineer, and a less reknowned French scientist, H. Hugoniot, established foundations which were later elaborated by Rayleigh,[1] G. I. Taylor,[2] and P. Duhem.[3] Their work, along with more recent contributions by Bethe,[4] von Neuman,[5] Gilbarg,[6] R. Courant and K. O. Friedrichs[7] and others, has been adapted to solids in recent years[8] and serves us well today for most purposes. Recent developments by G. R. Fowles and R. Williams[9] promise a new dimension in the interpretation of experiments in solids, but fulfillment of their promise may await new measuring techniques.

Early experiments on the shock waves produced by projectiles in air were done by Ernst Mach,[10] who established a tradition for the use of high speed optics which has been carried on by Cranz and Schardin,[11] Walsh,[12] Fowles[13] and others. Consideration of the problems of supersonic aircraft gave impetus to the study of air shocks before and after WWII (Howarth[14]), and since 1950 there has been detailed and extensive study of shocks in gaseous plasmas.[15,16]

Modern developments in the study of shock waves in solids really arise from the Manhattan Project during World War II. Details of this period are lost in files of the Atomic Energy Commission and in the memories of various individuals. However, we do know that in this time and place it was realized that the jump conditions could be used to obtain pressure-volume relations, that experimental techniques for producing and measuring plane shock waves from explosives were developed, and that in the years following World War II a series of pioneering papers on this subject came out of Los Alamos.[12,17-19] Work done there also provided a frame and foundation for studies of elastic precursors and phase transitions, which have occupied so much of our energies during the last fifteen to eighteen years.

Progress in shock wave physics has been strongly tied to developments of experimental techniques. Los Alamos studies were initially largely made with pins used to record free surface motion. The use of these was highly developed by Stanley Minshall,[17] but they have been supplanted by flash gaps, initially developed by Walsh,[12] which are still widely used for pressure-volume measurements above 100 kbars. As interest developed in the detailed structure of shock waves, it also turned toward lower pressures and more refined recording methods. Optical level techniques developed by Fowles[13] and Doran[20] at Stanford Research Institute provided sensitivity for measurements at low pressures and quasi-continuous records of free surface motion. A condensor microphone method developed by Taylor and Rice of Los Alamos[71] offered significant improvement in time resolution, and an electromagnetic procedure used by Fritz and Morgan[72] has recently produced records of high resolution. Major steps forward were provided by Sandia Laboratories: first in Lundergan's development of the gas gun for impact studies[21,22] and then in development of the quartz gauge by Neilson, Benedick, Brooks, Graham and Anderson[23] and the laser interferometer by Lynn Barker.[24,25] These combined developments have led to resolution times of one to five nanoseconds in shock structure measurements below forty kilobars and to sharply enhanced abilities to evaluate theoretical models of material behavior. In a somewhat different class are the electromagnetic velocity gauge invented by E. K. Zavoiskii of the USSR[26] and the manganin gauge first developed by Keough and Bernstein at Stanford Research Institute.[27] These are gauges to be imbedded in a sample. They will probably never compete with quartz gauge and laser interferometer for time resolution, but they can be used to much higher pressures and can reduce problems of impedance mismatch. The potential of neither, nor of their various offspring, has yet been realized.

It has turned out that mechanical measurements yielding pressure-volume relations, precursor structure and phase transitions have been relatively easy to do. Electric, magnetic and optical measurements are much harder, though many have been done and some have been done well.[28] Still, the possibilities for research in this area are great, and, as mechanical measurements become harder, more attention will probably be directed toward these problems.

High speed computing machines play a particularly significant role in shock wave research. Without them one is constrained to consider shocks as discontinuities and to give minimum attention to details of shock structure between the discontinuities. Shock problems are relatively easy to solve numerically, and with high speed machines there is no barrier except cost to the most detailed comparison of shock structure with the predictions of various models. In this way extremely critical tests of theories of constitutive relations are possible.[29] Some use has been made of this capability, but its use is still limited--perhaps primarily by the scarcity of good physical models.

II. ACHIEVEMENTS

In 1963 Fowles and I attempted to collect references to all Hugoniot data that had been published and we found measurements on about eighty substances, not counting minor variations in composition of steel and aluminum.[30] In 1967 the Lawrence Livermore Laboratory issued a three volume, looseleaf compendium of shock wave data which contains entries for about 160 materials, with the same restrictions.[31] I doubt that the pace of data production has slackened; linear extrapolation from these two points suggests that the number of substances for which data are available today is about 300. Collection and publication of such data provides a real service to the technical community. The data are expensive to obtain and not easy to duplicate without special facilities. They should be made available to the general user.

In spite of the amount of data available, it turns out that few substances are well characterized over a large range of pressure. From the jump conditions one finds that the r.m.s. errors in pressure and compression in terms of particle velocity u and shock velocity D are

$$\delta p/p = [(\delta u/u)^2 + (\delta D/D)^2]^{1/2} = \delta(\Delta V/V_0)/(\Delta V/V_0)$$

Variations in arrival times of the shock over a free surface in the average experiment is probably not less than 50 nanosec over a 3 cm diameter specimen. If total travel time through the specimen is two microsec., the uncertainty in D is $\delta D/D \sim .05/2 = 0.025$. Measurements of u are probably better than this on the average, so the uncertainty in p and $\Delta V/V_0$ in the average published data point is probably 2.5 to 3%. It can be much more unless the work is done carefully. It can be appreciably less if the work is painstaking. As more measurements are published for a given material, one may expect the error in the mean Hugoniot curve to diminish.

The existence of good Hugoniot data on many materials has prompted much study of theoretical equations of state with the result that keener understanding of the compression process now exists, particularly for the rare earths and rare gases.[32]

In 1968 Jones and Graham published a table of elastic precursor measurements.[33] There were a hundred and thirty published measurements at that time, including duplicates and measurements on twenty different iron and steel alloys. The total has increased substantially since then, and it includes extensive series of precursor measurements in LiF made by J. Asay[34] and Y. Gupta.[35] As it presently stands, it is established that elastic precursors are indeed elastic waves. Their amplitude is directly related to the resolved shear stress which the material is supporting at the instant of measurement, and this amplitude decays as the wave propagates into the sample. The rate of decay is related to the dynamic failure of the material, and, in ductile materials, it can probably be related to the velocity and rate of generation of dislocations in the material, though this last statement must be labelled speculation at present. With some adjustment of parameters, a reasonable dislocation model can be used to fit most, but not all, of the measured shock profiles. Precursor decay measurements and the associated dislocation analysis have been made in lithium fluoride, tungsten, iron and aluminum, but not in other materials. Measurements at three different crystal orientations in tungsten strongly suggest that the slip mechanisms operating in shock loading are the same as those operating in quasistatic slip.[36] Electron transmission micrographs from recovered metal specimens suggest that the details of dislocation behavior in shocked materials may be quite different from those found in thin bar experiments, perhaps because of the very short distances travelled by dislocations during the shock process.[37]

In 1954 Stanley Minshall reported a 130 kbar "plastic wave" in iron which he tentatively identified as being due to a polymorphic phase transition induced by shock waves.[38] He tentatively identified this as the α-γ transition, but in a brilliant series of experiments which traced out the phase diagram in iron it was determined in 1961 to be a new phase,[39] later identified as hcp.[40] Since 1954 quite a number of solids have been found to undergo phase transitions under the influence of shock waves.[41] Shock transition pressure does not usually exceed the static pressure of transition, where static values are known. This is curious because the time available for transition is small and, since transitions are sometimes slow in occurring under static conditions, it might reasonably be assumed that they might not occur at all in a very short time, or that they might occur at higher pressures. This suggests that a study of the kinetics of phase transition under shock conditions may be fruitful. Calculations indicated that a finite transition rate produces a decaying wave similar to the elastic precursor resulting from dynamic failure[42] and that, if transition time is between 10^{-8} and 10^{-5} seconds, it can be detected in a shock experiment.

In a recent series of experiments on potassium chloride, D. B. Hayes has obtained results which tend to heighten the puzzle, rather than resolve it.[43] He has found that the kinetic behavior depends on crystal orientation. When the shock propagates along the <100> direction, the material transforms to a new and metastable phase or partially transforms to the CsCl structure in less than 10^{-8} seconds. He finds some evidence of slower decay from this intermediate state to some undefined state. When the shock propagates in the <111> direction, transition is slower, the transition time being 10 to 40 nanosec, depending on driving pressure, but the final state reached after this time is the well known CsCl state. There is no evidence of a transition state as found for the <100> orientation. The transition pressure determined from his experiments may be higher than the static pressure by about a kilobar, but this difference may be due to uncertainties in both static and shock experiments.

Effects of shock waves on magnetic materials has been of both practical and theoretical interest. Three processes have been identified as being responsible for producing demagnetization of magnetic materials by passage of shock waves. One is depression of the Curie temperature by compression. This occurs in iron-nickel alloys with nickel content greater than 30%. A second is transformation from a ferromagnetic to a non-magnetic state through a first-order phase transition. This is observed in iron when it changes from fcc to hcp at 130 kilobars. The third is anisotropic demagnetization, a kind of inverse magnetostriction resulting from rotation of the magnetic momentic vector when elastic strain is imposed on the lattice. This last effect occurs in nickel ferrite, yttrium-iron-garnet, manganese-zinc ferrite and other ceramic materials. It turned out to be rather complicated and has been resolved by elegant theoretical and experimental developments.[44-47]

Electrical measurements to determine the effects of shock compression on resistivity have been made for a number of materials.[28] The combination of geometric requirements, shock reflection problems and electronic response times make such measurements very difficult. Quite good measurements have been made in xenon, argon, carbon tetrachloride, germanium, iron, manganin and copper. Those in liquids were helpful in elucidating certain anomalies in the equations of state.[48] Measurements on germanium in the range of elastic compression were the basis for a detailed evaluation of band structure parameters, indicating that uniaxial compression is a valuable adjunct to hydrostatic compression for such studies.[49] Resistivity of iron shows some curious anomalies below 100 kilobars which have yet to be explained, and that of copper is anomalously high when compared with static compression measurements.[48,50] A substantial number of resistivity measurements have been made on alkali halides for the purpose of studying the collapse of the electron energy band gap under compression. The results are ambiguous, but enough information is obtained to show that shock resistivity experiments can provide valuable information in this area.[28,51]

An isolated but striking result which has dramatic implications for future research is the production of x-ray diffraction patterns in the vicinity of or behind the shock front. This technique was developed with LiF as specimen material. Its recent application to boron nitride[52] suggests that it may become an effective tool for structural studies.

III. PROBLEMS FOR FUTURE STUDY

Much is to be done, of course, in digesting past work, making it available in a synthesized form for others, and developing its physical implications. This is particularly true in measurements of pressure-volume relations. There are at least two approaches to the problem of determining an equation of state from shock experiments. One is strictly thermodynamic. Shock experiments provide data on a single curve in p,V,E space. Supplemental experiments are then required to provide off-Hugoniot and thermal data. Various methods for doing this have been tried and have not been very successful.[53-56] The situation will be improved if bulk sound velocities and temperatures can be measured in the shocked state, but it is unlikely that a complete thermodynamic characterization of any material will be achieved without reference to physical models. A second approach, and the one most used to date, is to assume a rough physical model for the substance, derive the equation of state, including undetermined parameters, and use shock wave data to determine the parameters. This procedure can be improved upon by combining thermodynamics and model in such a way that all thermodynamic data can be used in determining parameters of the equation of state.[57] This procedure is useful but is, in a sense, a stopgap. At the present time what is required is precise model development for restricted classes of materials based on elementary principles. These can then be combined with Hugoniot and/or other thermodynamic data to produce equations of state in which one can have reasonable confidence. The success of this procedure for special materials has been demonstrated by Ross, Pastine and others.[58-60] It is a demanding process, but it yields valuable results.

Careful study of dynamic failure is just beginning. Measurements of elastic precursor decay and shock structure for simple, pure, well-characterized and well-controlled single crystals are required. This must be coupled with the best micro-mechanical models available in order to determine the role played by various imperfections and atomic processes in dynamic failure under impact. When this is established, we may be in a position to predict dynamic failure in a material from ordinary laboratory measurements of yield stress, hardness, impurity content, etc. One thing that is needed rather badly is satisfactory reconciliation of shock experiments and ordinary thin bar experiments. The latter are used to measure failure stresses at strain rates up to about 10^3/sec. The former are essentially stress relaxation experiments. If data from both kinds of experiments are reduced to a common form, we may gain significantly in understanding of the underlying processes of dynamic failure.

The above remarks are directed primarily toward failure of ductile solids by the yield process. Fracture is much less understood, but concepts of fracture in ductile materials developed by conventional metallurgical techniques[61] and shock wave methods[62] are converging on what seems to be a reasonable understanding. The failure of brittle materials under shock conditions is not at all understood. Two questions are outstanding, and their investigation will lead to some insight into the total process. One concerns the transition from brittle to ductile behavior, which apparently occurs in some materials under pressure, and the role it plays in failure under impact. The other concerns the apparent total collapse of the stress deviator in some brittle materials, of which quartz and sapphire are notable examples.[63] Inasmuch as ceramic materials are coming to play an increasing role in our society, brittle failure will be of increasing future importance. If we understand it under the extreme conditions of impact, we may come to understand it otherwise.

Geometric aspects of fracture and failure in shock experiments have been largely neglected. It is reasonable during the formulation of concepts to concentrate on plane geometry, but an important test of concepts so developed lies in their extension to other geometries. It is not too early to start designing and planning experiments with other than uniaxial strain.

Insofar as phase transitions are concerned, we know essentially nothing about the kinetics of transition under shock conditions. Comprehensive and searching experiments on well-defined materials of various classes are needed before we can even state the problems clearly. A very critical question here, of deep meaning for physical theory, is whether or not this fast transition can be understood by application of quasi-equilibrium statistics. The only feasible alternative seems to be large scale machine simulation of particle dynamics.

It is unlikely that electron behavior is significantly influenced by the dynamics of shock compression. Electrons in solids move too rapidly for that. But we can't be sure without further experiments and interpretations of experiments. Anomalies exist, as indicated earlier, and until they are resolved we don't know whether electrical effects are understood or not. The anomalous thermoelectric effect reported by several workers[64] is a good example of a large effect beyond that expected from static experiments. It has been suggested that this is an essentially dynamic effect, but the argument is not conclusive. Alkali halides deserve more study under optimum conditions. Independent variations of temperature and pressure have been attempted, but more work along such lines is required.

Absorption spectroscopy is a powerful tool for studying the internal structure of solids under static conditions. Time resolved spectroscopy is possible in shock experiments, but it has been little used. Experimental problems are formidable, but not apparently insoluble.[51] Used in conjunction with resistivity or shock polarization experiments, it may tell us a great deal about the internal states of shocked materials.

Almost all insulating materials produce electrical signals on being shocked. This is commonly called "shock polarization" or "charge release," depending on the nature of the material. The effects are significant theoretically and practically. Practically, because these signals are often unwanted in experimental systems and they can obscure or confuse the nature of wanted signals. Their theoretical significance follows from the inference that they indicate the occurrence of dramatic changes in electrical structure of the solid in the vicinity of the shock front. These effects are well-documented[65] and have been characterized phenomenologically, but little progress has been made toward developing atomic models.

Problems of yield, flow and fracture are probably of greatest interest to the group assembled here this week. Such problems can usually be expressed in terms of behavior of stress deviators in the field. Because the amount of energy that can be stored in elastic deformation is limited, these deviators stop increasing at some point in the loading process and we call this failure. Failure of this kind is associated with fracture or flow of the material. But at least one other situation appears to exist which can produce collapse of the stress deviators, at least in uniaxial strain. If a first order phase transition occurs as a consequence of shock compression, it seems plausible that the new phase will form so as to reduce the energy of deformation as well as that of compression. Looked at macroscopically, one would say that the stress deviators had collapsed as a consequence of the transition. If the material had been on the verge of yield or fracture before transforming, no such failure is imminent after transition. It may then be possible for it to absorb additional deformation. An effect like this has been observed in CdS[66] and InSb,[33] so the speculation is not pointless. It is then reasonable to inquire what the material behavior is on being cycled through the transition and whether or not its ultimate strength is substantially modified. These are interesting questions because they may have significant applications in addition to their scientific implications.

IV. PROBLEMS OF APPLICATION

Most of the preceding remarks related to scientific questions having to do with shock waves. Many problems of application remain to be resolved. Technologists seem inclined to respect the principle that improved understanding of fundamental processes leads to better technology, but to ignore it in practice. This is done with good reason because technology has gone very far with little understanding and the road to better technology through better understanding is a long and tortuous one.

This seems to have been less true in shock wave problems than others, perhaps because of the precedent set in the Manhattan Project. Perhaps also because of the difficulty of a "cut and try" approach. So problems of application and science are not always far apart. There are, of course, continuing problems of major importance in weapons design and military defense, with which many of you are familiar. Progress is being slowly made in these areas and efforts along present lines will undoubtedly be continued.

There are other important applications. Explosive or impact welding is not understood, despite the fact that it is an important commercial enterprise. There is no continuum mechanical model which will predict the gross features of the bond. The first light of mechanical understanding may exist, but more is required.[67] Some of the qualitative metallurgical features can be rationalized, but there is, for example, no theory which tells us why apparent diffusion coefficients are so large. This feature is reminiscent of some early, rather poorly documented, observations which suggested that under some conditions carbon can be driven freely through an iron lattice. Is it possible for shock waves to differentially accelerate dissimilar atoms so that the usual barriers to diffusion are lowered?

Diamonds are being commercially produced by shock compression. They are not very large and the business may not be very profitable, but it exists and might be better if the transition process were understood. Some ideas exist,[68] but a great deal of work will be required to develop them. There may be other products sufficiently valuable for manufacture by shock methods, but the question has not been thoroughly explored.[70]

The hardening effects of shock waves on metals are still not understood, though they are frequently used. Understanding is intimately related to questions of dynamic failure and deformation, and therefore to the motion and creation of dislocations and other defects.[69] Commercial applications of these effects may provide additional motivation for understanding them.

Explosive or shock-actuated devices are frequently suggested and sometimes developed for engineering applications. They might include such items as one-shot electrical generators, timing devices and fast-acting valves. They may depend on changes in conductivity or interaction of waves with associated fracture and flow. Their development is usually very costly. Development of a quantitative engineering discipline soundly based on the known behavior of materials under shock conditions would accelerate such applications.

V. CLOSING REMARKS

Problems of shock wave propagation in solids involve continuum mechanics, thermodynamics and materials or solid state science, all interacting in a very intimate way. A great deal of progress has been made in sketching a framework of theoretical and experimental techniques within which it is possible to do meaningful, perhaps even revolutionary, experiments in solid state science. Within this framework many significant experiments have been done relating to mechanical, thermodynamic, electrical and magnetic properties of solids.

But in a deep sense the real science of shock waves in solids has hardly been touched. When nothing had been done, exploratory experiments were appropriate. Now what is needed is intensive study of problems chosen primarily for their scientific import, by specialists in materials and solid state science, using, where possible, established and reliable experimental procedures. When this becomes common, we shall begin to see the real significance of shock wave research.

ACKNOWLEDGMENT

This work was supported by the United States Air Force Office of Scientific Research Contract 71-2037A and National Science Foundation Grant Number GH 34650.

REFERENCES

The following bibliography is in no sense complete. It should, however, provide entry to the literature of various problems mentioned in the text.

1. Lord Rayleigh, Proc. Roy. Soc. (London) 84, 247 (1910). This paper provides a summary of and references to earlier work.
2. G. I. Taylor, Proc. Roy. Soc. (London) 84, 371 (1910).
3. P. Duhem, Zeits. f. Physik. Chemie, 69, 169 (1909).
4. H. Bethe, "The Theory of Shock Waves for an Arbitrary Equation of State," OSRD Report No. 545 (1942).
5. J. von Neumann, "Oblique Reflection of Shocks," U. S. Navy Bu. Ord. Explosives Research Report No. 12 (1943).
6. D. Gilbarg, Am. J. Math. 73, 256 (1951).
7. R. Courant and K. O. Friedrichs, Supersonic Flow and Shock Waves, Interscience, 1948. This contains an extensive bibliography.
8. W. Band, J. Geophys. Res. 65, 695 (1960).
9. G. R. Fowles and R. F. Williams, J. Appl. Phys. 41, 360 (1970).
10. E. Mach, Akad. Wiss. Wien. 77, 819 (1878).
11. C. Cranz and H. Schardin, Zeits. f. Phys. 56, 147 (1929).
12. J. M. Walsh and R. H. Christian, Phys. Rev. 97, 1544 (1955).
13. G. R. Fowles, J. Appl. Phys. 32, 1475 (1961).
14. L. Howarth, Modern Developments in Fluid Dynamics. High Speed Flow, Vols. I and II, Oxford, 1953.
15. Physics of High Energy Density, Proc. of the International School of Physics "Enrico Fermi," Course 48, Varenna, 14-26 July, 1969. Published by Academic Press, 1971, P. Caldirola and H. Knoepfel, Eds.
16. R. G. Fowler, "Electrically Energized Shock Tubes," U. Oklahoma, Norman, Okla. (1963).
17. F. S. Minshall, J. Appl. Phys. 26, 463 (1955).
18. R. W. Goranson, D. Bancroft, B. L. Burton, T. Blechar, E. E. Houston, E. F. Gittings, S. A. Landeen, J. Appl. Phys. 26, 1472 (1955).
19. M. H. Rice, R. G. McQueen and J. M. Walsh, "Compression of Solids by Strong Shock Waves," Solid State Physics, Vol. 6, pp. 1-63. Academic Press, 1958. F. Seitz and D. Turnbull, Eds.
20. D. G. Doran in Proc. of High Pressure Measurement Symposium, ASME, Nov. 1962. Published by Butterworth's, 1962. A. A. Giardini and E. C. Lloyd, Eds.
21. C. D. Lundergan and W. Herrmann, J. Appl. Phys. 34, 2046 (1963).
22. R. A. Graham, "Impact Techniques for the Study of Physical Properties of Solids under Shock Wave Loading," Paper 66-WA/PT-2, Presented at the Winter Annual Meeting and Energy Systems Exposition, N.Y. (Nov. 27-Dec. 1, 1966). ASME.

23. F. W. Neilson, W. B. Benedick, W. P. Brooks, R. A. Graham and G. W. Anderson, "Electrical and Optical Effects of Shock Waves in Crystalline Quartz," in Les Ondes de Detonation, No. 109, Editions du Centre National de la Recherche Scientifique, 15, Quai Anatole-France-Paris (VIIe) (1962).
24. L. M. Barker and R. E. Hollenbach, Rev. Sci. Instr. 36, 1617 (1965).
25. L. M. Barker and R. E. Hollenbach, J. Appl. Phys. 43, 4669 (1972).
26. A. N. Dremin, S. V. Pershin and V. F. Pogurelov, "Structure of Shock Waves in KCl and KBr under Dynamic Compression to 200,000 Atm.," Combustion, Explosion and Shock Waves 1, 1 (1965).
27. D. Bernstein and D. Keough, J. Appl. Phys. 35, 1471 (1964).
28. D. L. Styris and G. E. Duvall, High Temperatures-High Pressures, 2, 477 (1970).
29. J. J. Gilman, Micromechanics of Flow in Solids, McGraw-Hill, New York, 1969. p. 222 ff.
30. G. E. Duvall and G. R. Fowles, "Shock Waves," High Pressure Physics and Chemistry, Vol. 2, Academic Press, 1963. R. S. Bradley, Ed.
31. M. Van Thiel, A. S. Kusubov, and A. C. Mitchell, "Compendium of Shock Wave Data," UCRL 50108 (TID-4500) (1967).
32. E. B. Royce, "High Pressure Equations of State from Shock Wave Data," Proc. Int. School of Physics, "Enrico Fermi," op. cit. pp. 80-95.
33. O. E. Jones and R. A. Graham, "Shear Strength Effects on Phase Transition 'Pressures' Determined from Shock-Compression Experiments," Accurate Characterization of the High Pressure Environment, NBS Special Publication 326, Supt. Doc., U.S. Govt. Printing Office, March 1971.
34. J. R. Asay, G. R. Fowles, G. E. Duvall. M. H. Miles and R. F. Tinder, J. Appl. Phys. 43, 2132 (1972).
35. Y. M. Gupta, "Stress Relaxation in Shock-Loaded LiF Single Crystals," Ph.D. Thesis, Washington State University (1973).
36. T. E. Michaels, "Orientation Dependence of Elastic Precursor Decay in Single Crystal Tungsten," Ph.D. Thesis, Washington State University (1972).
37. J. W. Edington in Behavior of Metals Under Dynamic Load, p. 191. Springer-Verlag (1968). U. S. Lindholm, Ed.
38. F. S. Minshall, Bull. APS 29, 23 (12/28/54).
39. P. C. Johnson, B. A. Stein and R. S. Davis, J. Appl. Phys. 33, 557 (1962).
40. J. C. Jamieson and A. W. Lawson, J. Appl. Phys. 33, 776 (1962).
41. G. E. Duvall and G. R. Fowles, "Shock Waves," op. cit. p. 271.
42. Y. Horie and G. E. Duvall, "Shock Waves and the Kinetics of Solid-Solid Transitions," Proc. Army Symposium on Solid Mechanics, Sept. 1969, AMMRC MS 68-09.
43. D. B. Hayes, "Experimental Determination of Phase Transformation Rates in Shocked Potassium Chloride," Ph.D. Thesis, Washington State University (1972).

44. E. B. Royce, J. Appl. Phys. $\underline{37}$, 4066 (1966).
45. L. C. Bartel, J. Appl. Phys. $\underline{40}$, 3988 (1969).
46. R. C. Wayne, G. A. Samara and R. A. Lefever, J. Appl. Phys. $\underline{41}$, 633 (1970).
47. D. E. Grady, G. E. Duvall, and E. B. Royce, J. Appl. Phys. $\underline{43}$, 1948 (1972).
48. R. N. Keeler, "Electrical Conductivity of Condensed Media at High Pressures," Proc. Int. School of Physics, "Enrico Fermi," op. cit. pp. 106-122.
49. R. A. Graham, O. E. Jones and J. R. Holland, J. Phys. Chem. Solids $\underline{27}$, 1519 (1966).
50. J. Y. Wong, R. K. Linde and P. S. DeCarli, Nature $\underline{219}$, 713 (1968).
51. M. Van Thiel and A. C. Mitchell, Physics Dept. Progress Report, June-Sept. 1965. UCRL-14538, pp. 50-52.
52. Q. Johnson and A. C. Mitchell, Phys. Rev. Letters $\underline{29}$, 1369 (1972).
53. J. M. Walsh and M. H. Rice, J. Chem. Phys. $\underline{26}$, 815 (1957).
54. L. V. Al'tshuler, "Use of Shock Waves in High-Pressure Physics," Soviet Physics - USPEKHI $\underline{8}$, No. 1, (July-Aug. 1965).
55. J. W. Forbes and N. L. Coleburn, J. Appl. Phys. $\underline{40}$, 4624 (1969).
56. R. W. Rohde, J. Appl. Phys. $\underline{40}$, 2988 (1969).
57. D. J. Andrews, J. Phys. Chem. Solids. To be published.
58. M. Ross, J. Phys. Chem. Solids 33, 1105 (1972).
59. M. Ross, Phys. Rev. $\underline{171}$, 777 (1968).
60. D. J. Pastine, J. Phys. Chem. Solids $\underline{28}$, 522 (1966).
61. A. H. Cottrell, "Theoretical Aspects of Fracture," Swampscott Conference, 1959. H. Paxton, Ed.
62. T. Barbee, L. Seaman, R. C. Crewdson, Bull. APS II, $\underline{15}$, 1607 (1970).
63. R. A. Graham and W. P. Brooks, J. Phys. Chem. Solids $\underline{32}$, 2311 (1971).
64. A. Migault et J. Jacquesson, Le J. de Physique $\underline{33}$, 599 (1972).
65. G. E. Hauver, Bull. APS II, $\underline{14}$, 1163 (1969).
66. J. D. Kennedy and W. B. Benedick, J. Phys. Chem. Solids $\underline{27}$, 125 (1966).
67. G. R. Cowan and A. H. Holtzman, J. Appl. Phys. $\underline{34}$, 928 (1968).
68. P. S. DeCarli, Stanford Research Institute, Private Communication.
69. L. F. Trueb, J. Appl. Phys. $\underline{40}$, 2976 (1969).
70. S. S. Batsanov, "Physics and Chemistry of High Dynamic Pressure," Behaviour of Dense Media Under High Dynamic Pressures, Symposium H.D.P., I.U.T.A.M., Paris, Sept. 1967. Gordon and Breach, N.Y., (1968).
71. John W. Taylor and Melvin H. Rice, J. Appl. Phys. $\underline{34}$, 364 (1963).
72. J. N. Fritz and J. A. Morgan, "An Electromagnetic Technique for Measuring Material Velocity," Los Alamos Scientific Laboratory, LA-DC-72-815, Aug. 1972.

SHOCK WAVE PHYSICS

G. R. Fowles

Shock Dynamics Laboratory, Physics Department

Washington State University, Pullman, Washington

I. INTRODUCTION

Stress waves can be produced in solids by impact with another object, by gas pressure of adjacent detonating explosive, or by sudden deposition of radiation. These waves are in general very complicated, even for simple geometries, because of the strong interaction that occurs between the wave propagation behavior and the material response behavior when the stresses exceed the elastic limit. In this lecture we will limit our attention to plane longitudinal waves, non-linear in general, and emphasize the relation between the thermomechanical behavior of the material and the wave propagation behavior. Many interesting physical phenomena are not included in such a continuum approach. In part this elision is due to lack of space and in part because many microscopic phenomena such as conductivity, magnetic effects, dislocation mechanics, etc., are still in the exploration stage; many of the papers of this symposium deal with several of those aspects of shock wave physics. The thermomechanical behavior, however, is not only of intrinsic interest but occupies a central and essential role in the study of all shock related phenomena. To display the interrelationship between material properties and wave propagation it is customary to begin with the well-developed theory of high speed flow of fluids. Our approach will be similar except that we will delay somewhat the introduction of any assumptions concerning material properties. In this way we hope to avoid some confusion about the applicability of the results to solids.

Figure 1 shows a drawing of a typical stress wave, similar to some that have been observed, that illustrates most of the major

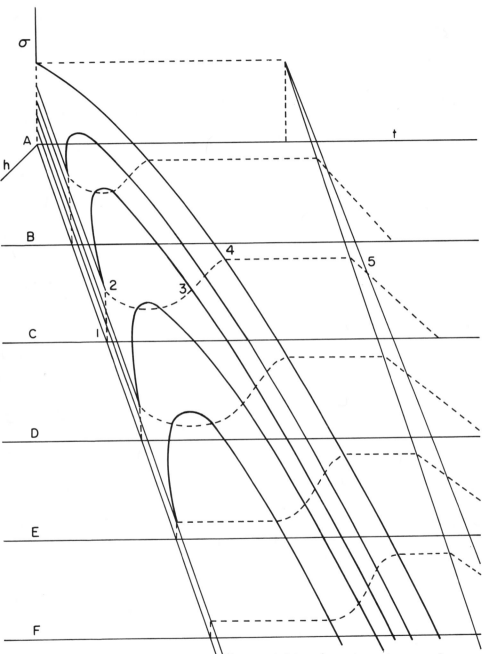

Figure 1--Stress-time profiles (dashed lines) and contours of constant stress (solid lines) for typical plane stress wave.

features of plane compression waves in solids. The figure represents a surface in distance-time-stress space. Stress-time curves are shown at a series of locations, A-F, fixed with respect to material coordinates; also shown are contours of constant stress amplitude. A square stress pulse is presumed to be applied to the boundary, A, and a wave propagates into the medium, changing shape as it goes. The change of shape is due to the non-linearity and rate dependence of the response of the material; in a linearly elastic medium no change of shape would occur. The associated stress-volume curves are shown in Fig. 2.

The leading part of the wave is shown as a discontinuity in stress that decays in amplitude up to location E, and is steady beyond that point. In the profile at C this jump is indicated by 1-2. At that same location the stress drops behind the front from 2-3 and then increases. A slow-rising compression wave is indicated between 3 and 4; state 4 is frequently assumed to be a thermodynamic equilibrium state. The transition from 4 to 5 is shown as a discontinuity in stress rate. Discontinuities of this type are called "acceleration waves." The acceleration wave, in this case, is the leading part of the "rarefaction" wave, in which unloading occurs.

With increasing propagation distance not only the first discontinuous front, 1-2, but also the second compression wave, 3-4, become steady with time. Thus, beyond point E the wave consists of two steady compression fronts (one of them a discontinuity) and an unsteady rarefaction wave, each travelling with different velocities. It is common to refer to the entire pulse as a shock wave; however, only fronts representing either discontinuities in the dependent variables or steady transitions are properly termed "shocks." The remaining ambiguity is then not very serious because the same conservation relations, or jump conditions, apply to either. Moreover, although they may be mathematically convenient, it is probable that true discontinuities do not occur in nature. Whether a shock is considered a steady transition or a discontinuity then depends on the time resolution of the experiment or some other characteristic time of the problem. Similar remarks apply to acceleration waves.

The pulse illustrated in Fig. 1 is typical of relatively low amplitude waves. With higher stress applied to the boundary the second compression front, after the initial transient behavior (before E in Fig. 1), normally travels somewhat faster and exhibits an increasingly smaller rise time. At still higher stresses the second shock may overtake the first (or never separate from it) and the stress pulse then resembles a square pulse except for a gradual fall off in the rarefaction tail. The stability conditions under which a single front will develop into two shock fronts, as in Fig. 1, are derived later in this chapter.

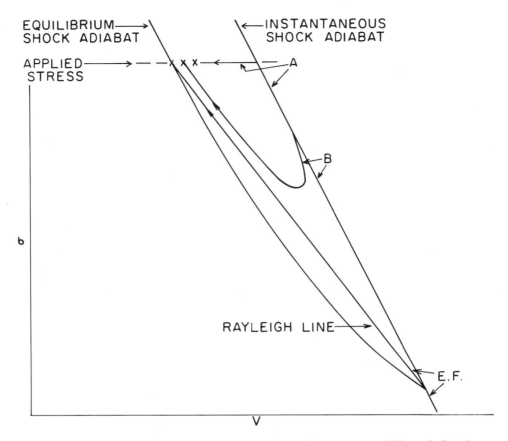

Figure 2--Stress-volume curves corresponding to profiles A-F of Fig. 1.

Of the various features of plane stress waves, shock transitions hold a certain fascination because of the extreme conditions that can occur. At high impact velocities strain rates are often so large as to be unmeasurable with current nanosecond instrumentation and the pressures and temperatures can reach millions of bars and tens of thousands of degrees.

Surfaces similar to that depicted in Fig. 1 can be drawn for each of the other dependent variables, such as density or particle velocity. Under certain conditions these all have the same shape, differing only by a scale factor that makes the dimensions commensurable. In the general case, however, the surfaces do not have similar shapes and there may thus be different wave velocities associated with each dependent variable. The scaling factors that relate the different surfaces are the slopes of the projections on the time-distance plane of the contours of constant amplitude, shown as constant stress contours in Fig. 1. The relations are derived from the equations of continuity and momentum conservation and therefore applicable to all materials.

II. CONSERVATION RELATIONS

In material (Lagrangian) coordinates each particle is specified by the initial value of its spatial (Eulerian) coordinate. Denoting the material coordinate by "h" and the spatial coordinate by "x," the transformation is given by

$$x(h,t) = x(h,0) + \int_0^t u(h,t)dt$$

where $u = (\partial x/\partial t)_h$, is the particle velocity.

Moreover, since either (x,t) or (h,t) can be taken to be independent variables, the transformation for time derivatives is,

$$(\partial/\partial t)_h = (\partial/\partial t)_x + u(\partial/\partial x)_t.$$

Material coordinates are frequently more convenient, especially for comparisons with experimental measurements. In these coordinates the mechanical flow equations representing conservation of mass and momentum are,

$$\rho_o(\partial V/\partial t)_h - (\partial u/\partial h)_t = 0 \qquad (1)$$

$$\rho_o(\partial u/\partial t)_h - (\partial \sigma/\partial h)_t = 0 \qquad (2)$$

where ρ_o is the initial density, V is the specific volume, and σ is the stress component in the direction of propagation, measured positive in tension.

Before a solution can be obtained to a specific problem these equations must be supplemented by a relation between σ and V and, perhaps their derivatives or past history. This specifying equation, or constitutive relation, is not well known for most materials and one must usually rely for this information on data from stress wave experiments themselves. Certain general results can be derived, however, that are independent of the properties of the medium and that are helpful for the interpretation of experiments.

We introduce "velocities" defined for each dependent variable by the slopes of the projections in the (h,t) plane of the contour lines of constant amplitude. Thus,

$$C_u \equiv (\partial h/\partial t)_u = -(\partial u/\partial t)_h (\partial u/\partial h)_t^{-1} \qquad (3a)$$

$$C_\sigma \equiv (\partial h/\partial t)_\sigma = -(\partial \sigma/\partial t)_h (\partial \sigma/\partial h)_t^{-1} \quad (3b)$$

$$C_V \equiv (\partial h/\partial t)_V = -(\partial V/\partial t)_h (\partial V/\partial h)_t^{-1} \quad (3c)$$

Substitution into Eqs. 1 and 2 then gives,

$$\rho_0 (\partial V/\partial t)_h + C_u^{-1} (\partial u/\partial t)_h = 0$$

$$\rho_0 (\partial u/\partial t)_h + C_\sigma^{-1} (\partial \sigma/\partial t)_h = 0$$

or, alternatively,

$$\rho_0 C_V (\partial V/\partial h)_t + (\partial u/\partial h)_t = 0$$

$$\rho_0 C_u (\partial u/\partial h)_t + (\partial \sigma/\partial h)_t = 0$$

Note that in each of these last four equations the differentiation is with respect to only a single independent variable. Consequently, we have,

$$\rho_0 C_u = -(\partial u/\partial V)_h \quad (4a); \qquad \rho_0 C_u = -(\partial \sigma/\partial u)_t \quad (4c)$$

$$\rho_0 C_\sigma = -(\partial \sigma/\partial u)_h \quad (4b); \qquad \rho_0 C_V = -(\partial u/\partial V)_t \quad (4d)$$

These equations can be used to relate the various wave surfaces of the dependent variables, σ, V and u. For example, experimental measurements commonly give stress as a function of time. From a series of such measurements at different locations the surface $\sigma(h,t)$, as in Fig. 1., can be constructed. Numerical integration of Eq. 4b then gives $u(h,t)$; finally, integration of Eq. 4a provides $V(h,t)$. Thus, in principle at least, measurements of a single dependent variable as a function of time at several locations when combined with Eqs. 4 are sufficient to provide a complete solution.

Several interesting and informative relations can be derived among the velocities defined by Eq. 3. For example, combining Eqs. 4a and 4b gives,

$$(\partial \sigma/\partial V)_h = \rho_0^2 C_\sigma C_u. \quad (5)$$

This derivative is the local instantaneous modulus for the material and is the direct link between the constitutive relation and the wave surfaces.

In high speed gas dynamics it is customary to assume that the stress is everywhere the equilibrium pressure, ($\sigma = -P$) and that entropy is conserved on particle paths. Under these conditions, although not generally, the sound speed in material coordinates is given by,

$$\rho_0 C^2 = (-\partial P/\partial V)_h = (-\partial P/\partial V)_s$$

and therefore, $C^2 = C_\sigma C_u$. Only under special conditions however are the "contour" velocities, C_σ and C_u, each individually equal to the sound speed, C.

Relations among the various contour velocities are easily derived. Thus, a well known partial differentiation formula gives,

$$C_\sigma - C_u = (\partial h/\partial t)_\sigma - (\partial h/\partial t)_u$$

$$= (\partial h/\partial t)_\sigma - [(\partial h/\partial t)_\sigma + (\partial h/\partial \sigma)_t (\partial \sigma/\partial t)_u].$$

Hence, $\quad C_u/C_\sigma = 1 - (\partial \sigma/\partial t)_u (\partial \sigma/\partial t)_h^{-1}.$ \hfill (6a)

Similarly, $\quad C_v/C_\sigma = 1 - (\partial \sigma/\partial t)_v (\partial \sigma/\partial t)_h^{-1}$ \hfill (6b)

and, $\quad C_u/C_v = 1 - (\partial V/\partial t)_u (\partial V/\partial t)_h^{-1}.$ \hfill (6c)

It is clear from Eqs. 6 that the contour velocities will be equal whenever the second term vanishes, i.e., whenever $\sigma = f(u) = g(V)$, or whenever there is a discontinuity in stress and volume so that,

$$(\partial \sigma/\partial t)_h^{-1} = (\partial V/\partial t)_h^{-1} = 0.$$

Special interest attaches to waves propagating into an undisturbed medium, as in Fig. 1, since most experiments are performed under these conditions. For such "simple" waves some other useful relations can be derived among the contour velocities. Thus, beginning with Eq. 4c and considering u and t to be the independent variables,

$$\rho_0 C_u = -(\partial \sigma/\partial u)_t$$

$$\rho_0 (\partial C_u/\partial t)_u = -\partial^2 \sigma/\partial u \partial t = -\partial/\partial u [(\partial \sigma/\partial t)_u]_t.$$

Or, substituting from Eqs. 6a and Eq. 2,

$$(\partial C_u/\partial t)_u = \partial/\partial u [(C_\sigma - C_u)(\partial u/\partial t)_h]. \tag{7a}$$

In integrated form,

$$[C_\sigma(u,t) - C_u(u,t)][(\frac{\partial u}{\partial t})_h(u,t)] = \int_0^u [(\frac{\partial C_u}{\partial t})_u(u,t)]du.$$

Other equations analogous to Eq. 7a can be derived by starting with each of the other expressions 4a, 4b, and 4d. The results are,

$$(\partial C_u^{-1}/\partial h)_u = \partial/\partial u [(C_v^{-1} - C_u^{-1})(\partial u/\partial h)_t] \tag{7b}$$

$$(\partial C_\sigma^{-1}/\partial h)_\sigma = \partial/\partial \sigma [(C_u^{-1} - C_\sigma^{-1})(\partial \sigma/\partial h)_t] \tag{7c}$$

$$(\partial Cv/\partial t)_V = \partial/\partial V [(Cu-Cv)(\partial V/\partial t)_h]. \quad (7d)$$

Consider now the case when all the contour lines of one kind, say u, are straight lines. Then the only non-trivial solution of Eq. 7a is $C\sigma$ = Cu. Further, from Eq. 7b, Cu = Cv. Finally from Eqs. 6 we conclude that $\sigma = f(u) = g(V)$. This result is of some importance since it shows that curvature of the contour lines is related to rate or history or entropy dependence of the constitutive relation. Experiments that measure only an impact time and a single stress or velocity time profile do not provide any measure of curvature and therefore cannot directly give more than a relation of the form $\sigma = \sigma(V)$.

This conclusion can be expressed alternatively by combining Eqs. 5 and 7a,

$$-V_o^2(\partial\sigma/\partial V)_h = Cu^2 - Cu(\partial u/\partial t)_h^{-1} \int_0^u (\partial Cu/\partial t)_u \, du.$$

Measurements of u and Cu at a single location do not permit the integral to be evaluated and hence cannot determine the modulus, $(\partial\sigma/\partial V)_h$.

We are now in a position to derive some of the more familiar and widely used formulas of shock wave physics. Consider the discontinuous leading front of Fig. 1. For this shock the stress-contour lines are clearly all superimposed. Moreover, $C\sigma$ = Cu = Cv from Eqs. 6. Hence, denoting the shock velocity in spatial coordinates by U, we can integrate Eqs. 4 to get the Rankine-Hugoniot relations,

$$u - u_o = \rho_o(U - u_o)(V_o - V) \quad (8a)$$

and,

$$-(\sigma - \sigma_o) = \rho_o(U - u_o)(u - u_o) \quad (8b)$$

where σ_o and u_o are the values ahead of the shock. Note that these relations also apply to the steady continuous shock transition beyond point E since there the contour lines are all parallel and the contour velocities are therefore independent of position, time, or amplitude.

Another relation is usually added to these expressions. It is derived by assuming that changes in the internal energy of the equilibrium end states are due only to mechanical work. Thus, we supplement Eqs. 1 and 2 with,

$$(\partial E/\partial V)_h = \sigma$$

or, from Eqs. 8b and 8a: $dE = +[\sigma_o + \rho_o^2 (U - u_o)^2(V_o - V)] \, dV$.

Since $(U - u_0)$ is constant throughout the wavefront, integration is immediate giving,

$$E - E_0 = -1/2\,(\sigma + \sigma_0)(V_0 - V). \qquad (9)$$

This equation, involving only thermodynamic quantities, is called the Hugoniot relation. It represents a surface in σ, V, E space whose intersection with the equilibrium equation of state for a material gives the locus of σ, V, E, equilibrium states attainable by means of a shock transition. This curve is usually referred to as the "Hugoniot equation of state" or "shock adiabat" for the material. Even for an ideal fluid it is incomplete, of course, since it is only a single curve and, moreover, does not explicitly include entropy. Nevertheless, even a single curve is very valuable when no other information is available and a very large number of shock adiabats have been measured for a variety of materials. The measurements typically yield directly corresponding values of shock and particle velocity which are combined with Eqs. 8 and 9 to give the thermodynamic quantities of Eq. 9.

Two other useful equations can be obtained from Eqs. 8 by eliminating either the shock or the particle velocity. Thus,

$$\begin{aligned}-(\sigma - \sigma_0) &= \rho_0^{\,2} U^2 (V_0 - V)\\ &= u^2/(V_0 - V).\end{aligned} \qquad (10)$$

From the first of these formulas it is clear that, since the shock velocity is constant throughout the shock transition, the σ-V states through which the material passes within the transition layer must lie on the straight line joining the equilibrium end states. This line is called the Rayleigh line, as indicated in Fig. 2, for example. Except at the end points, it does not intersect the equilibrium equation of state surface.

The difference in stress between the Rayleigh line and the equilibrium surface at a given volume is due to irreversible strain rate effects. This difference is the principal cause of dissipation and the concomitant entropy increase of the shocked state. This difference can also be related to the shape of the shock transition region if the rate dependence of the constitutive relation is known. Note that for a given material and applied stress the stress-rate within the transition region adjusts itself to just those values that cause the states to lie on the Rayleigh line. Thus, the shape of the shock transition is a material property, for a given stress, and is otherwise independent of the initial and boundary conditions.

The second expression of Eq. 10 gives a σ-u relation for a material once the shock adiabat, $\sigma(V)$, is known. This relation is very important for analyzing interactions of shock waves with boundaries with other materials, because reflections at interfaces must preserve continuity of stress and particle velocity.

It is also possible to derive jump conditions for acceleration waves like that at the head of the rarefaction wave of Fig. 1. They are of the same form as the shock jump conditions except that they involve derivatives of the dependent variables rather than the dependent variables themselves. Acceleration waves have not been widely used in the measurement of material properties, but are of interest theoretically partly because the characteristic speeds of gas dynamics are the same as acceleration wave speeds.

III. EQUATIONS OF STATE

For many purposes it is sufficient to neglect stress anisotropy and to treat solids as if they were fluids. The shock adiabat then yields information about the equation of state of the material in the form, $P_H(V, E_H)$, where $P_H = -\sigma$ is the pressure and subscripts H refer to shock adiabat states. This simplification breaks down at stresses comparable to the yield stress but even in that region it is useful as a first approximation.

Numerous empirical fits to shock adiabat data have been proposed. One of the most widely used is the relation,

$$U = C_o + au + bu^2 \tag{11}$$

where C_o, a, and b, are constants, U is shock velocity and u is particle velocity. In many cases the quadratic term is negligible and the corresponding pressure-volume relation, obtained by combining the above with the jump conditions, is,

$$P_H = \rho_o C_o^2 n / (1 - an)^2$$

where, $n = 1 - V/V_o$ is the compression. A compliation of nearly, if not all, data of this type has been made by Van Thiel.

A large amount of data has been obtained at pressures of about 100 to 2000 kilobars.* The highest reported measurements are at a pressure of 10 Mb, produced by symmetric impact of metal plates at an initial relative velocity of 14 Km/sec. Under these extreme conditions the shock temperatures may exceed 50×10^{3}°K, the density may increase by a factor of two to three, and the internal energy may be an order of magnitude larger than that of conventional high explosive.

*1 bar = 10^{-3} Kb = 10^{-6} Mb = 14.504 psi

Of greater interest to metallurgy, however, are pressures below about 100 Kb where strength effects are usually not negligible. Much research is being carried out in this area but the material behavior is complex and progress in understanding correspondingly slow.

Given data on the principal shock adiabat, it is of interest to estimate the other state parameters, specifically temperature and entropy, and to attempt to extend the equation of state in the vicinity of the measured curve. For example, if a reference curve such as the zero-degree isotherm and the specific heat at constant volume, C_v, are known, the temperature and entropy can be calculated at each volume by means of the formulas,

$$E_H(V,T_H) - E_0(V,T_0) = \int_{T_0}^{T_H} C_v(V,T) dT$$

and,

$$S_H(V,T_H) - S_0(V,T_0) = \int_{T_0}^{T_H} \frac{C_v(V,T)}{T} dT$$

where E_0, S_0, T_0 are the values on the reference curve.

The most common approach is to assume a form for the interatomic potential and thence to construct the zero-degree isotherm. The problem is thereby reduced to determining the function $C_v(T,V)$. The Debye theory is then invoked to provide the temperature dependence of the specific heat in the form

$$C_v = f(T_D(V)/T)$$

where $T_D(V)$ is the Debye temperature and the function f is known.

The volume dependence of C_v, contained in $T_D(V)$, is then found by integrating the Gruneisen relation

$$\gamma(V) = -d\ln T_D/d\ln V$$

where $\gamma(V)$ is the Gruneisen parameter.

Three theories exist for the volume dependence of γ: the Slater, Dugdale-McDonald, and the Zubarev-Vashchenko models, none of which have been demonstrated to be either correct or incorrect. Calculated temperatures are therefore uncertain at present, especially at high pressures.

Note that if $\gamma(V)$ is indeed independent of temperature, as assumed in the Gruneisen model, then it is the same as the thermodynamic parameter,

$$\Gamma = V\left(\frac{\partial P}{\partial E}\right)_V$$

and, moreover,

$$P(E_H, V) - P(E_0, V) = V^{-1} \Gamma(V) [E_H - E_0] \qquad (12)$$

The volume and temperature, if any, dependence of Γ is thus of paramount importance to equation of state determinations using shock waves. Experiments are possible in principle that can provide this information. Note that the Hugoniot relation, Eq. 9, includes the initial volume and internal energy as parameters. Hence, varying these (usually V_0) permits other shock adiabats to be determined and values for the thermodynamic Γ follow from differences in these shock adiabats, according to Eq. 12. The initial state is varied by changing the temperature or by using initially porous samples. The former method does not provide much variation, however, and the latter method suffers from difficulties of precision and questions of equilibrium. Consequently, although the measured values are not inconsistent with the various theoretical values, neither do they permit discrimination among them or, indeed, verification of any of them.

Figure 3 shows some representative curves obtained from shock wave experiments. The temperatures are increasingly uncertain at higher pressures as discussed above. At higher temperatures the above treatment is modified to account for the specific heat of the electrons, but this need not concern us here.

IV. STABILITY OF SHOCK FRONTS

Under certain conditions a single shock front is unstable and splits into two or more fronts travelling with different speeds. Figure 1 illustrates such a case, in which a single front develops into two shocks. These instabilities are of particular importance experimentally because they provide a very pronounced indication of the onset of critical type phenomena such as yielding at the elastic limit or phase transformations.

A criterion for the stability of a shock can be derived by considering the behavior of small amplitude sound waves following or leading the front. If the shock is not to spread as it travels, its speed must be greater than the speed of sound waves in the medium ahead of the front and, conversely, must be slower than the speed of sound waves in the compressed medium behind the front. Thus the shock is supersonic with respect to the medium ahead and subsonic with respect to the medium behind. This criterion can be expressed mathematically as follows.

Define the Mach number of the shock front as $M = (U-u)/c$, where c is the speed of sound, $c = (\partial P/\partial \rho)_s^{1/2}$,

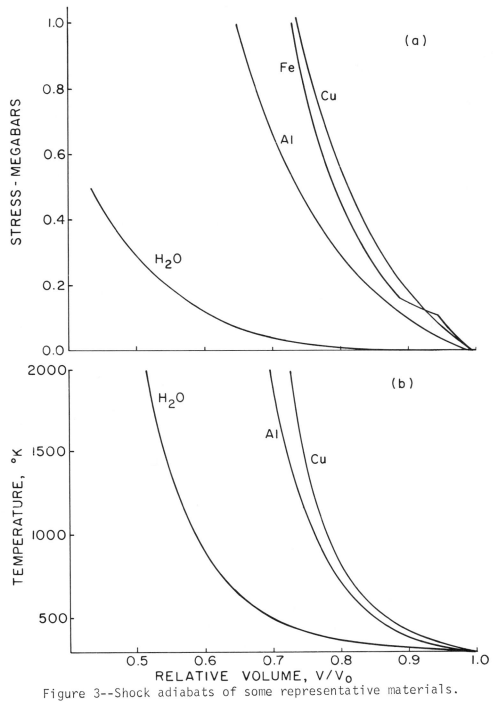

Figure 3--Shock adiabats of some representative materials.

then, if the slope of the Rayleigh line is denoted by

$$j^2 = (P - P_o) / (V_o - V) = \rho_o^2 U^2$$

the supersonic-subsonic condition can be written,

$$0 < M^2 < 1, \quad -1 < j^2(\partial V/\partial P)_s \quad \text{(behind)}$$
$$1 < M^2, \quad -1 > j^2(\partial V/\partial P)_s \quad \text{(ahead)}.$$
(13)

That is, the Rayleigh line must be steeper than the isentrope through the initial state at the foot of the shock adiabat and shallower than the isentrope passing through the compressed equilibrium state. Moreover, it can be shown that as a consequence the shock adiabat cannot intersect the equilibrium equation of state surface at intermediate points between the initial and the final states. For normally behaved material the relation, $(\partial^2 P/\partial V^2)_s > 0$ is everywhere valid and the stability conditions are automatically satisfied.

At a point of instability the slopes of the shock adiabat, the Rayleigh line, and the isentrope are all equal and the flow behind the front is just sonic. That all three slopes are simultaneously equal can be demonstrated as follows.

Standard partial differentiation formulas give,

$$(\partial P/\partial V)_s = (\partial P/\partial V)_E + (\partial P/\partial E)_V (\partial E/\partial V)_s$$

and,

$$(\partial P/\partial V)_H = (\partial P/\partial V)_E + (\partial P/\partial E)_V (\partial E/\partial V)_H$$

where the subscript H means differentiation along the shock adiabat.

Differentiating Eq. 9, and using Eq. 12, these relations can be combined to give,

$$(\partial V/\partial P)_H = (\partial V/\partial P)_s \left[\frac{1-a}{1-M^2 a} \right]$$

where $a = \Gamma(V_o - V)/2V$. Clearly the derivatives are equal whenever $M^2 = 1$, i.e. whenever

$$-j^2 \left(\frac{\partial V}{\partial P} \right)_s = 1.$$

The two principal types of material behavior that lead to shock instabilities are yielding at the elastic limit and phase changes. Other sources of instability include anomalous compressibility, (increasing compressibility with increasing pressure), exhibited by some glasses, and detonation. By way of illustration of instability phenomena we briefly consider some of the equilibrium thermodynamic aspects of phase transformations.

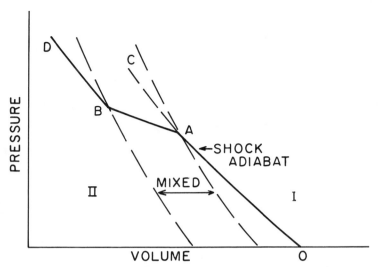

Figure 4--Shock adiabat for a material exhibiting a first-order phase change.

Figure 4 shows the projection on the P-V plane of the phase boundary lines and the mixed phase region for a material for which the slope, dP/dT, is positive; the derivations to follow apply equally to phase changes with negative dP/dT. Also shown is a shock adiabat, O-A-B-D, and its metastable projection in phase I, A-C. The curve O-A-B-D is seen to be similar to that determined experimentally for the α-ε phase change in iron in the vicinity of 130 Kb (Fig. 3). Since the slope of the shock adiabat A-B is shallower than the Rayleigh line from O to A, a single shock wave with pressure greater than A is unstable; a compressive pulse therefore forms two fronts, the first compressing the material to A and the second to the final state. At A the stability criterion, Eq. 13, is satisfied for both of these shocks. Thus, at point A we must have

$$j^2 \left(\frac{\partial V}{\partial P} \right)_s > -1 \text{ (low pressure shock)}$$
$$< -1 \text{ (high pressure shock)}.$$

Consequently, the slope of the isentrope at A must be discontinuous. The jump in slopes is derived as follows.

Let v and s be the specific volume and entropy of the mixed phase, of composition α. Then,

$$v = \alpha v_1 + (1 - \alpha) v_2; \quad s = \alpha s_1 + (1 - \alpha) s_2$$

where subscripts 1 and 2 refer to the appropriate quantities in the pure phases at the phase boundary.

Differentiating and setting $ds = 0$ gives,

$$d\alpha = \frac{ds_1 + (1-\alpha)ds_2}{s_2 - s_1}$$

and

$$\left(\frac{\partial v}{\partial P}\right)_s = \left(\frac{v_1 - v_2}{s_2 - s_1}\right)\left[\alpha\left(\frac{ds_1}{dP}\right) + (1-\alpha)\frac{ds_2}{dP}\right] + \alpha\frac{dv_1}{dP} + (1-\alpha)\frac{dv_2}{dP}$$

where the total derivatives are taken along the phase boundary.

At point A, $\alpha = 1$; moreover,

$$dP/dT = \frac{s_2 - s_1}{v_2 - v_1}$$

and,

$$\frac{dv_1}{dP} = \left(\frac{\partial v_1}{\partial P}\right)_{s_1} + \left(\frac{\partial v_1}{\partial s_1}\right)_P \frac{ds_1}{dP} ,$$

combining these last three equations gives, finally,

$$\left(\frac{\partial v}{\partial P}\right)_s - \left(\frac{\partial v_1}{\partial P}\right)_{s_1} = -\left(\frac{ds_1}{dP}\right)^2 \frac{T}{C_{P1}}$$

Since the right hand side is always negative we conclude that the isentrope in the mixed phase region is always shallower than the isentropes in either of the pure phase regions at the phase boundaries. This condition is clearly necessary for instability to occur but is not sufficient since the slope of the Rayleigh line, j^2, is not determined. In fact a shock instability due to a phase change can be suppressed by increasing the initial volume by means of porosity until the Rayleigh line is less steep than the isentrope in the mixed phase region.

If the transformation rate of the transition is not negligible, the development of a stable two wave structure may take place qualitatively as illustrated in Fig. 1. The initial applied pressure may overdrive the transition momentarily and the initial shock then carries the material into the metastable region, A-C. The material relaxes as the transformation proceeds, eventually forming two steady fronts. The reaction rates can be determined from experiments in which the transient behavior as the steady waves form is observed.

ACKNOWLEDGMENT

This work was supported by the National Science Foundation Grant No. GA 35064.

REFERENCES

Al'tshuler, L. V., "Use of Shock Waves in High Pressure Physics," Soviet Physics, Uspekhi $\underline{8}$, 52 (1965).

Duvall, G. E. and G. R. Fowles, "Shock Waves," in High Pressure Physics and Chemistry, Chapter 9, R. S. Bradley, Editor, Academic Press (1963).

Duvall, G. E. and Y. Horie, "Shock Induced Phase Transitions in Iron," in Behavior of Dense Media under High Dynamic Pressures, Gordon and Breach (1968).

Kinslow, R., Editor, High Velocity Impact Phenomena, Academic Press (1970).

Landau, L. and E. M. Lifshitz, Fluid Mechanics, Pergamon Press (1959).

Miklowitz, J., Editor, Wave Propagation in Solids, Proceedings of the ASME Winter Meeting, Los Angeles, Ca., November 1969.

Royce, E. B., "High-Pressure Equations of State from Shock-Wave Data," Lawrence Livermore Laboratory. Presented at the International School of Physics "Enrico Fermi," Varenna, Italy, July 1969.

Van Thiel, M., Editor, Compendium of Shock Wave Data, UCRL 50108, Lawrence Livermore Laboratory, Livermore, California.

SHOCK WAVE MECHANICS*

O. E. Jones

Sandia Laboratories, Albuquerque, New Mexico

I. INTRODUCTION

Shock stresses in solids typically range from tens of kilobars to megabars. Such stress levels are produced in a solid for durations of the order of several microseconds when an explosive is detonated in contact with it, when a projectile traveling at high velocity impacts on it, or when energy is deposited in it at very high power levels. Most solids deform irreversibly or fracture at stresses typically of the order of a few kilobars; thus, the generation and propagation of shock waves are intrinsically violent and destructive processes, and this accounts for their traditional importance in military and mining technology.

The destructive nature of shock waves makes it important to be able to understand and predict the dynamic response and failure of materials and structures subjected to shock loading. Predicting what actually occurs on a submicrosecond time scale is obviously an enormously complex problem whose solution requires both detailed knowledge of the dynamic constitutive relations representing the specific mechanical and physical properties of the materials, and sophisticated techniques for calculation of nonlinear wave propagation and interaction. Much of current shock-wave research is centered on these problems.[1,2]

Conversely, during the past twenty years engineers have begun to constructively exploit the energy released by detonation of an explosive charge to working and fabrication of materials, primarily metals. High-energy explosive forming and compaction, explosive hardening, and explosive welding have all been commercially exploited. In turn, metallurgists have undertaken extensive investigations of the terminal changes in the microstructure and mechanical properties of metals recovered after shock loading.[3,4,5]

Several topics of metallurgical interest which bear significantly on these two areas of activity are reviewed in this paper. In Section II the mechanics of shock-wave propagation in solids are briefly summarized with emphasis being given to shear strength effects. Recent work on the relationship between dislocation dynamics and shock-induced dynamic yielding of single crystals is described in Section III. Section IV deals with the effect of shock pulse duration, as well as amplitude, on the residual hardness of recovered metal specimens. Finally, in Section V, numerical experiments are discussed which are aimed at establishing whether observed residual effects are generated primarily during the shock compression cycle, or during the following stress release and recovery process.

II. BASIC CONCEPTS

Longitudinal plane-wave shock compression produces a state of uniaxial strain in which, as shown in Fig. 1, deformation occurs only in the direction of wave propagation, say the x-direction. Such a state may be produced experimentally by simultaneously applying a uniform normal stress σ_x over the entire face of a homogeneous specimen in the form of a flat, circular disk whose thickness is much less than its diameter. For a short time period,

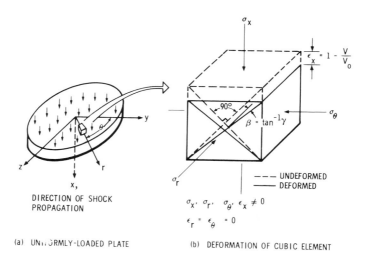

Fig. 1 Uniaxial-strain deformation.

before the arrival of radial unloading waves from the disk circumference, symmetry requires that each macroscopic volume element must deform in the same manner as its neighbors. There is no net lateral deformation and the strains ϵ_r and ϵ_θ in the plane of the disk are maintained equal to zero by the principal lateral stresses σ_r and σ_θ. Thus, only the principal strain component ϵ_x in the wave propagation direction is nonzero. The resulting one-dimensionality greatly aids in separating material properties from the mechanics of nonlinear wave propagation. However, volume elements at the circumference of the disk are not constrained against lateral motion and they accelerate outward as the plate is compressed. The resulting radial stress-release wave converges inwardly on the disk axis and progressively erodes the uniaxial-strain state. Eventually multiple reverberations and interactions of the longitudinal and radial wave systems in conjunction with dissipative processes relieve the stresses in the disk to zero, after which it may be recovered with suitable care. This unloading process will be discussed in Section V; only the uniaxial-strain part of the deformation will be considered in the remainder of this section.

It is evident from the deformed unit cube of Fig. 1(b) that longitudinal plane-wave deformation involves both a volume change and a geometric shape change. The engineering strain ϵ_x and the specific volume V are simply related since the mass of the element is conserved. Thus, $\epsilon_x = 1 - (V/V_o)$, where V_o ($\equiv 1/\rho_o$) is the initial specific volume of the material. (Compressive stresses and strains are taken to be positive.) Decreasing the volume of the cube by shock compression decreases interatomic distances and results in the generation of a pressure P. This pressure is elastic in nature and can reach values as great as megabars. From Fig. 1(b) it is clear that the geometric shape change causes a shear strain. The angular change β in the original right angle between the face diagonals of the cube is equal to $\tan^{-1}\gamma$, where γ is the engineering shear strain. The resistance of a solid to such a shape change gives rise to shear stresses τ on planes inclined to the wavefront. The pressure P is defined to be

$$P \equiv \frac{1}{3}(\sigma_x + \sigma_r + \sigma_\theta) = \sigma_x - \frac{2}{3}(\sigma_x - \sigma_r) \quad , \tag{1}$$

where, an isotropic solid has been assumed so that $\sigma_r = \sigma_\theta$. The maximum resolved shear stress τ for uniaxial strain is $\tau = (\sigma_x - \sigma_r)/2$; hence Eq. (1) may be rewritten as

$$\sigma_x = P + \frac{4}{3}\tau \tag{2a}$$

$$\sigma_r = \sigma_\theta = P - \frac{2}{3}\tau \tag{2b}$$

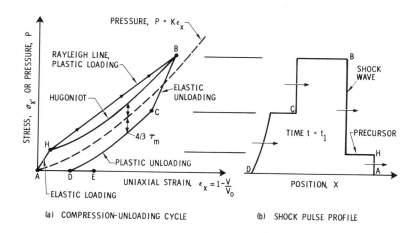

Fig. 2 Uniaxial-strain shock deformation of a rate-independent, elastic-plastic material.

Thus, each stress component consists of a spherical, or pressure, contribution P and a shear contribution τ. This is illustrated for σ_x in Fig. 2(a) which shows a typical stress-strain cycle AHBCD for longitudinal plane-wave shock loading of a ductile solid which is insensitive to rate of deformation (strain-rate independent).[1] When σ_x reaches some limiting value at state H, often referred to as the Hugoniot elastic limit σ_{xH}, the solid loses its resistance to further shear deformation and, if a metal, yields plastically when σ_x is typically of the order of five to ten kilobars. For either a Tresca or von Mises yield criterion, yielding occurs when

$$|\sigma_x - \sigma_r| = 2\tau_m(W_p) = Y(W_p) , \quad (3)$$

where isotropic work-hardening may be incorporated by permitting τ_m, the limiting shear stress the material can withstand, and Y, the yield stress determined in a conventional uniaxial stress compression test, to depend on the plastic work W_p. For strain-rate independent materials the relation between σ_{xH} and Y is[6]

$$\sigma_{xH} = \frac{(1-\nu)}{(1-2\nu)} Y , \quad (4)$$

where ν is Poisson's ratio. For an elastic-perfectly plastic material, Y is a constant in Eq.(3) so that the stresses increase during plastic flow according to the bulk modulus K, as shown in Fig. 2(a), since $P = K(\epsilon_x)\epsilon_x$ in Eq. (2).

The shock pulse profile corresponding to Fig. 2(a) defines the equilibrium states which may be achieved by shock compression. Note that the Hugoniot curve does not describe the state points through which the material passes during shock deformation; rather, it defines only the locus of possible final equilibrium states.[1,6] This is due to the fact that the material undergoes irreversible, nonequilibrium processes as it is engulfed by the shock front. The Hugoniot curve (Fig. 2(a)) and measurable wave propagation parameters characterizing the loading phase of the shock pulse (Fig. 2(b)) are related by the Hugoniot jump relations[6] which express conservation of mass, momentum, and energy across a shock front.

Mass: $\quad \epsilon_x - \epsilon_{xi} = (1 - \epsilon_{xi})(u - u_i)/(U - u_i)$, (5)

Momentum: $\quad \sigma_x - \sigma_{xi} = \rho_0 (U - u_i)(u - u_i)/(1 - \epsilon_{xi})$, (6)

Energy: $\quad E - E_i = (\sigma_x + \sigma_{xi})(\epsilon_x - \epsilon_{xi})/2\rho_0$, (7)

where U is the laboratory shock velocity, u is the particle velocity, E is the specific internal energy, and the subscript i denotes the state of the material ahead of the particular shock front under consideration. Only the axial stress component σ_x appears in Eqs. (5)-(7); thus, the lateral in-plane stresses can only be determined by inference from the dynamic constitutive relation for the material. Since knowledge of the constitutive relation is often the goal, it is clear that progress cannot be direct, but requires successive iterations between assumed theoretical constitutive models and experimental measurements.

Simultaneously solving Eqs. (5) and (6) yields two particularly useful formulae for the shock velocity ($U - u_i$) relative to the material in advance of the shock, and for the particle velocity jump ($u - u_i$):

Shock velocity: $\quad U - u_i = \dfrac{(1 - \epsilon_{xi})}{\sqrt{\rho_0}} \sqrt{\dfrac{\sigma_x - \sigma_{xi}}{\epsilon_x - \epsilon_{xi}}}$, (8)

Particle velocity: $\quad u - u_i = \sqrt{[(\sigma_x - \sigma_{xi})(\epsilon_x - \epsilon_{xi})]/\rho_0}$. (9)

From Eq. (8) it is evident that the chord of slope $\Delta\sigma_x/\Delta\epsilon_x$ which connects the initial state to some final shock state B on the Hugoniot curve determines the shock velocity. This chord is called the Rayleigh line (see Fig. 2(a)) and, since the shock front is steady in this theory, it defines the successive states of the material passing through the shock. If we attempt to connect the initial state A to some state B with a single Rayleigh line, i.e., with a single shock front, then the resulting shock velocity will be less than the elastic wave velocity C corresponding to the slope of a Rayleigh line from A to H. Hence, as shown in Fig. 2(b), a single shock is unstable and splits into a leading elastic precursor and a following shock (or plastic) wave whose velocity will correspond to the Rayleigh line drawn from point H to the final state B.[6] As the stress σ_{xB} is increased, it is obvious that the velocity of the shock wave increases. Eventually the shock velocity equals the elastic precursor velocity at a stress called the overdrive stress. At higher stresses a single shock front is stable.

The path BCD in Fig. 2(a) describes the states through which the material passes upon adiabatic unloading from state B. Unloading first occurs elastically along BC and then, after reverse yielding occurs at C (cf. Eq. (3)), along the path CD. The unloading phase of the shock pulse consists of a faster traveling elastic wave and a slower rarefaction wave, as shown in Fig. 2(b). Hysteresis occurs and a residual strain ϵ_x remains in the material after σ_x has been reduced to zero at state D. The waste energy left behind immediately after passage of the shock pulse corresponds to the enclosed shaded area. At state D it is apparent from Eq. (3) that the in-plane stresses σ_r and σ_θ have nonzero compressive values which may be as large as the yield stress Y. Release of the compressive in-plane stresses by the radial release process described earlier allows the disk thickness to decrease further, corresponding to state E, as a result of Poisson's effect.

It should be recalled that the longitudinal plane-wave loading and unloading behavior we have been describing is based on the assumption that rate-dependent phenomena are negligible. There now exists a body of experimental evidence which clearly indicates that this assumption may often be unjustified, except as a first approximation. Much of current shock-wave research on metallurgical effects at high strain rates is aimed at understanding how the basic behavior described above is modified by rate-dependent effects. In the following sections, attention will be focused on rate-dependent phenomena associated with points H (Section III) and D (Section IV), and with the transition to E (Section V) in Fig. 2(a).

III. SHOCK-INDUCED DYNAMIC YIELDING OF SINGLE CRYSTALS

The relationship between the macroscopic response of solids to shock loading and microscopic dislocation dynamics is a subject of increasing experimental and theoretical interest. This is because constitutive relations developed in terms of dislocation dynamics contain slip-plane parameters, such as dislocation mobility, multiplication, etc., which are identifiable with physical deformation processes. The effects on these parameters of changing processing variables or material are fairly well understood; thus, it is possible, at least qualitatively, to predict dynamic behavior with some confidence.

Experimentally, shock-induced dynamic yielding of metals having low initial dislocation densities can show very significant rate-dependent effects, and the concept of a Hugoniot elastic limit σ_{xH}, denoted by state H in Fig. 2(a), has little meaning. In particular, as shown qualitatively in Fig. 3, it has been found that the peak amplitude $\hat{\sigma}_x$ of the elastic precursor may not have a unique value, but instead may decrease with increasing propagation distance and

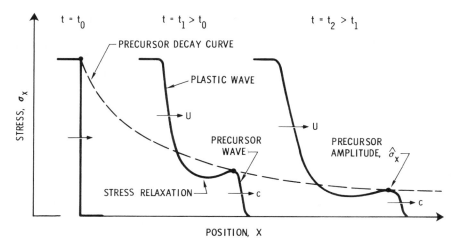

Fig. 3 Stress-position wave profiles illustrating typical rate-dependent, longitudinal plane-wave, dynamic yielding of metals.

may be followed by stress relaxation. Such behavior has been observed in a variety of polycrystalline materials, including commerically-pure iron,[7-10] mild steel,[8] hardened[8] and annealed[11] SAE 4340 steel, 1060 aluminum,[11] and annealed tantalum.[12] Taylor[13] showed that dislocation dynamics theory could qualitatively, and to a great extent quantitatively, account for the observed elastic precursor decay in Armco iron. Subsequently this theory has been extended and elaborated upon by other investigators[14-20] and comprehensively reviewed.[21] Although the use of polycrystalline materials simplifies experimental procedure, it greatly complicates analysis because of the random distribution of dislocation glide planes and directions, and because of grain boundaries which may act as dislocation sources and barriers. Most of these difficulties may be avoided by performing experiments on single crystals, and this constitutes the thrust of recent effort.

Jones and Mote[22] performed the first systematic transient studies of shock-induced dynamic yielding in metal single crystals. Precursor amplitudes observed at 5-mm propagation distances in copper single crystals were found to correspond to critical resolved shear stresses about twenty times the critical resolved shear stress for quasi-static macroscopic flow. Compared to the prediction of Eq. (4), this was dramatic evidence of strain-rate dependent phenomena. Similar results have also been observed for shock-loaded tungsten,[23] sodium chloride,[24] lithium fluoride,[25], and beryllium.[26] Additionally, precursor amplitudes have been found to differ by orders of magnitude, depending on whether the orientation enhances or suppresses dislocation glide on the primary slip systems.[24,26] In terms of dislocation dynamics, Jones[22] derived single-crystal constitutive relations for analyzing such behavior in face-centered cubic crystals. The analysis was later extended[27] and relations applicable to body-centered cubic, rocksalt, and hexagonal-close-packed crystals were derived.

For small strains, the application of elementary dislocation dynamics theory to shock-induced dynamic yielding of single crystals is straightforward, and is illustrated schematically in Fig. 4: As discussed in Section II, longitudinal plane-wave compression causes a mass element to undergo both a change in its volume (compression) and a change in its shape (shear). When the induced shear stresses become large enough, dislocations glide so that the shear strains are relieved, and the stress state becomes more isotropic. In order

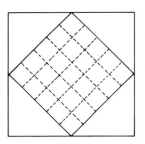
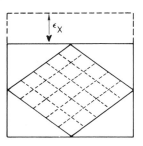
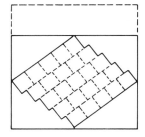

(a) UNDEFORMED STATE   (b) PLANE WAVE PRODUCES UNIAXIAL STRAIN ϵ_x (c) DISLOCATION MOTION RELAXES SHEAR STRAINS LEAVING ONLY VOLUME CHANGE

Fig. 4 Schematic representation (exaggerated) of dislocation-induced plastic flow resulting from uniaxial-strain plane-wave compression.

to translate this idea into mathematical terms three Cartesian coordinate systems identified by prime superscripts are introduced.[27] The crystallographic, loading, and glide coordinates correspond, respectively, to none, one, and two prime superscripts. In general, glide can occur on several different slip planes and in different directions. For example, slip in a face-centered-cubic metal such as copper may occur on any of the twelve $\{111\}\langle110\rangle$ slip systems.

The plastic strain rate $^{\alpha}\dot{\gamma}$ for dislocation motion on a particular slip system α is given by the well-known formula,[21]

$$^{\alpha}\dot{\gamma} = b\, {}^{\alpha}N_m\, {}^{\alpha}\bar{v}({}^{\alpha}\tau) \quad , \tag{10}$$

where b is the Burgers vector, $^{\alpha}N_m$ is the density of mobile dislocations, and $^{\alpha}\bar{v}$ is the average dislocation velocity. The resolved shear stress $^{\alpha}\tau$ acting to move dislocations on the α slip system is

$$^{\alpha}\tau \equiv {}^{\alpha}S_{13}'' = {}^{\alpha}a_{1r}'''\, {}^{\alpha}a_{3s}'''\, S_{rs}' \quad , \tag{11}$$

where S_{ij} is the stress tensor, $^{\alpha}a_{ij}'''$ is the transformation matrix between the (prime) loading and (double prime) glide coordinate systems, $^{\alpha}x_1''$ is in the direction of the Burgers vector, and $^{\alpha}x_3''$ is normal to the glide plane. In the loading coordinate system, the total symmetric plastic strain-rate tensor $\dot{P}_{k\ell}'$ resulting from dislocation motion on all slip systems is

$$\dot{P}_{k\ell}' = \frac{1}{2}\sum_{\alpha} {}^{\alpha}a_{mk}'''\, {}^{\alpha}a_{n\ell}'''\, {}^{\alpha}\dot{\gamma}\, [\delta_{1m}\delta_{3n} + \delta_{3m}\delta_{1n}] \quad , \tag{12}$$

where δ_{ij} is the Kronecker delta. If it is assumed (i) that the total strain-rate tensor $\dot{T}_{k\ell}'$ in the loading coordinate system is the sum of the elastic strain-rate tensor $\dot{E}_{k\ell}'$ and the total plastic strain-rate tensor $\dot{P}_{k\ell}'$, and (ii) that only the elastic strain-rate tensor $\dot{E}_{k\ell}'$ gives rise to a stress-rate tensor $\dot{S}_{ij}'$ through Hooke's law, then

$$\dot{S}_{ij}' = C_{ijk\ell}'\, \dot{E}_{k\ell}' = C_{ijk\ell}'\, (\dot{T}_{k\ell}' - \dot{P}_{k\ell}') \quad , \tag{13}$$

where $C_{ijk\ell}'$ are the elastic moduli referred to the loading coordinate system. Combining Eqs. (12) and (13) yields the general infinitesimal stress-strain constitutive relation for dislocation-induced plastic flow in single crystals,

$$\dot{S}_{ij}' - a_{ip}'\, a_{jq}'\, a_{kr}'\, a_{\ell s}'\, C_{pqrs}\, \dot{T}_{k\ell}'$$

$$= -\sum_{\alpha} a_{ip}'\, a_{jq}'\, C_{pqrs}\, {}^{\alpha}a_{1r}''\, {}^{\alpha}a_{3s}''\, {}^{\alpha}\dot{\gamma} \tag{14}$$

where a_{ij}' is the transformation matrix relating the (unprime) crystallographic and (prime) loading coordinate systems, $^{\alpha}a_{ij}''$ relates the (unprime) crystallographic and (double prime) glide coordinate systems, and $C_{ijk\ell}$ are the elastic moduli in the crystallographic coordinate system.

As an example, Eqs. (10), (11), and (14) will be applied to the case of uniaxial-strain dynamic yielding in a face-centered-cubic metal, such as copper.[22,27] It is reasonable to assume that the dislocation density is the same on each of the twelve different $\{111\}\langle 110\rangle$ slip systems; therefore, $^{\alpha}N_m = N_m/12$, where N_m is the total mobile dislocation density. Also, the dislocation mobility relation $^{\alpha}\bar{v}(\tau)$ is assumed to be the same for each slip system. Further, only uniaxial-strain, longitudinal plane-wave propagation in the [100] crystallographic direction will be considered. By letting $T_{11}' = \epsilon_x$ (all other T_{ij}' are zero), $S_{11}' = \sigma_x$, and $S_{22}' = S_{33}' = \sigma_r \equiv \sigma_\theta$, Eq. (14) becomes

$$\dot{\sigma}_x - C_{11}\dot{\epsilon}_x = -2(C_{11} - C_{12})b\, N_m \bar{v}(\tau)/(3\sqrt{6}) \quad , \qquad (15)$$

$$\dot{\sigma}_r - C_{12}\dot{\epsilon}_x = (C_{11} - C_{12})b\, N_m \bar{v}(\tau)/(3\sqrt{6}) \quad , \qquad (16)$$

where $\tau = (\sigma_x - \sigma_r)/\sqrt{6}$. Similar equations may be derived for propagation in the [110] and [111] crystallographic directions, which are the only other "specific" directions in which a pure longitudinal wave may be propagated in a cubic crystal. Equations (15) and (16) are elastic-viscoplastic constitutive equations of the Malvern-type; when they are combined with the one-dimensional Lagrangian equation of motion, $-\rho_0 \partial u/\partial t = \partial \sigma_x/\partial x$, it is possible to calculate the flow field.

If the elastic precursor is assumed to rise instantaneously to its peak value $\hat{\sigma}_x$ (cf. Fig. 3), the jump is elastic since no time is available for plastic flow to occur. In this case, the total strain ϵ_x is elastic and the decay of $\hat{\sigma}_x$ with propagation distance x in the [100] direction is[13]

$$\frac{d\hat{\sigma}_x}{dx} = -\hat{F}/2c \equiv -(C_{11}-C_{12})b\, N_{mo}\bar{v}(\hat{\tau})\sqrt{\rho_0}/(3\sqrt{6C_{11}}) \quad , \qquad (17)$$

where N_{mo} is the initial mobile dislocation density, and $c\,(=\sqrt{C_{11}/\rho_0})$ is the elastic wave velocity in the [100] direction. By specifying the dislocation mobility relation $\bar{v}(\tau)$, precursor decay curves may be calculated from Eq. (17) with N_{mo} as a parameter. The dashed curves in Fig. 5 are the result of such calculations based on a modified phonon-viscosity mobility relation[27] of the form $\bar{v}(\tau) = (v_s^2 B/2b\tau)\{[1 + (2b\tau/v_s B)^2]^{1/2} - 1\}$, where v_s, the limiting shear velocity in the slip plane is the limiting dislocation velocity, and B is the dislocation damping constant. Also shown are experimental values of $\hat{\sigma}_x$.[22] An initial mobile dislocation density N_{mo} of between 10^8 and 10^9cm^{-2} would be required to fit the experimental results. This is at least an order-of-magnitude greater than the initial total dislocation density N_0 of the crystals which was estimated to be between 10^6 and 10^7cm^{-2}.

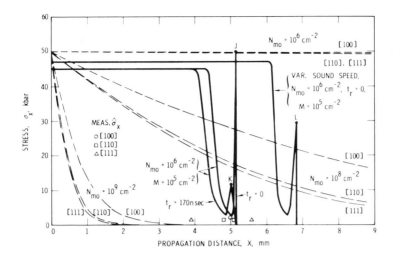

Fig. 5 Predicted [100], [110], and [110] precursor decay curves[22,27] (dashed lines) and [100] stress-position wave profiles[28] (solid lines) for shock-induced, longitudinal plane-wave propagation in copper single crystals. The initial mobile dislocation density N_{mo}, the dislocation multiplication constant M, and the load application risetime t_r at x = 0 are parameters. The decay curves correspond to t_r = 0. Experimentally-determined precursor amplitudes[22] are identified by symbols. For t_r = 0 and constant sound speed, the velocity jump used for the [100] stress-position profiles produces an initial elastic impact stress of 50 kbar (point J) which subsequently relaxes to about 45 kbar due to dislocation motion.

Similar comparisons have been made[27] between values of $\hat{\sigma}_x$ predicted in the manner just described and those measured experimentally for tungsten,[23] sodium chloride,[24] and lithium fluoride.[25] In all cases the elastic precursor theory properly predicts the relative precursor amplitudes for different wave propagation directions. However, in all cases, the initial mobile dislocation densities required to predict the observed precursor decays are two to three orders-of-magnitude greater than the measured <u>total</u> dislocation densities. Explanations for this systematic discrepancy have been discussed,[22,27] but the difficulty remains unresolved. Herrmann, Hicks, and Young[28] have investigated some of these difficulties for [100] propagation in copper single crystals by numerically calculating the entire wave profile in order to determine effects of dislocation multiplication, load application risetime, and variable sound speed on the precursor decay. In their calculations, the particle velocity u of the surface x = 0 was parabolically increased from zero to u_o in a time t_r. Experimentally it is expected that the loading risetime t_r will be nonzero because of surface roughness, nonplanarity between the plates in an impact experiment, or detonation wave nonplanarity in an explosive experiment. Three calculated wave profiles for initial mobile dislocation densities of $N_{mo} = 10^6/cm^2$ are shown as solid curves in Fig. 5.

For $t_r = 0$ and constant sound speed, the numerically calculated precursor amplitude $\hat{\sigma}_x$ at point J corresponds to the value predicted analytically from Eq. (17). Linear dislocation multiplication was included in Eqs. (15) and (16) by taking $N_m = N_{mo}(1 + M\gamma)$, where M is a dislocation multiplication constant and $\dot{\gamma} = b\,N_m\,\bar{v}(\tau)$. For M = 0, corresponding to no dislocation multiplication with increasing strain, stress relaxation was not predicted behind the elastic precursor wave. Since stress relaxation was observed experimentally,[22] multiplication must be included, and the profiles in Fig. 5 were calculated with $M = 10^5 cm^{-2}$. As shown, extreme stress relaxation occurs behind the precursor peak at J, and the resulting precursor spike is so narrow that, if actually present, it might not be resolved experimentally. In view of this, the value of $M = 10^5 cm^{-2}$ was selected to fit the experimentally observed stress minimum for a physically reasonable value of $N_{mo} = 10^6 cm^{-2}$.

If the loading risetime t_r is nonzero, then dislocation multiplication can occur during the precursor rise and thus serve to reduce the value of $\hat{\sigma}_x$. For $t_r = 170$ nsec, which is far too large from experimental considerations, the precursor peak corresponds to point K in Fig. 5. Although its amplitude is dramatically reduced from point J, the remainder of the profile is little changed.

Values of $\hat{\sigma}_x$ can be large enough to warrant incorporating third-order elastic constants into Eqs. (15) and (16), and this greatly enhances the decay rate of the precursor spike.[28] For $t_r = 0$, Fig. 5 shows that $\hat{\sigma}_x$ is then reduced from point J to point L by the overtaking stress release wave.

Recent comprehensive experimental studies by Asay and coworkers[25] on [100] shock propagation in LiF single crystals of several purities (hardnesses) showed no evidence for a sharp precursor spike. Also, dislocation multiplication in the wavefront did not seem to be responsible for the large values of N_{mo} (above the initial total dislocation density) required to fit the experimental decay curves. Instead, the experimental results suggested that heterogeneous dislocation nucleation at impurity clusters might be the source of the excess dislocations.

In general, the dislocation dynamics theory of shock-induced dynamic yielding in single crystals has provided a rational basis for qualitatively understanding experimental results and for identifying important effects and variables. Quantitative predictions are presently not possible primarily because of inadequate knowledge of the behavior of the dislocation density N_m and mobility $\bar{v}$ in Eq. (10) at very high stresses and strain rates. The numerical results of Herrmann et al[28] suggest that empirical determination of the details of Eq. (10) from experimental data will be difficult, if not ambiguous, because the functional dependencies of N_m and $\bar{v}$ are interwoven. Thus, it may be desirable to pursue independent means of deducing information about dislocation mobility and multiplication under shock-loading conditions. One possibility for gaining such additional information is to recover shock-loaded specimens and to measure the distance dislocations have moved and their final density; however, this depends on being able to attribute the results to shock compression and unloading, rather than to the subsequent radial release process.

In all of the above discussion it has been tacitly assumed that wave propagation is in a specific crystalline direction for which a pure longitudinal wave may be propagated.[29] For example, in a cubic crystal this occurs only for propagation in [100], [110], or [111] directions. Recently, Johnson[30] has employed the single-crystal constitutive relation of Eq. (14) to analyze both rate-independent and rate-dependent plastic wave propagation in nonspecific crystalline directions. Rate-independent plastic flow corresponds to the shear stress α_T not exceeding a fixed value when Eqs. (11) and (14) are combined, while for rate-dependent flow $\alpha \dot{\gamma}$ is given by Eq. (10). For the case of an elastically isotropic solid that is plastically anisotropic with a single slip plane and direction, both quasilongitudinal and quasitransverse plastic waves may be propagated for loading in a nonspecific direction. This situation approximates the behavior of a hexagonal single crystal, such as beryllium, for which the effects of elastic anisotropy are small[30] and slip occurs predominantly on the basal plane. Fig. 6 shows the effects of nonspecific orientations ($\theta \neq 0°$, $90°$) on rate-independent quasilongitudinal and quasitransverse plastic wave propagation in beryllium. Such effects have been observed experimentally[26] and are discussed elsewhere in this proceedings.

Thus, in spite of quantitative difficulties, real progress has been made in the past few years in understanding shock-induced dynamic yielding. There is little question that further advances will come from single crystal studies.

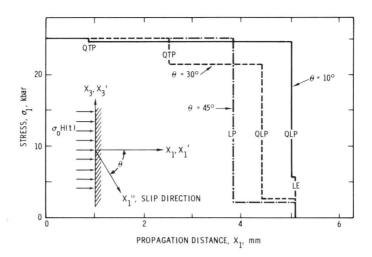

Fig. 6 Stress-position wave profiles for plane-wave propagation in an elastically isotropic single crystal that is plastically anisotropic with a single slip direction which makes an angle θ with the propagation directions. Depending on θ, longitudinal (LP), quasilongitudinal (QLP), or quasitransverse (QTP) plastic waves may be propagated in addition to the longitudinal elastic precursor (LE). (After Johnson.[30])

IV. RESIDUAL EFFECTS: SHOCK HARDENING AND PULSE DURATION

With care it is possible to recover a specimen which has been shock-loaded and to examine its residual properties corresponding to state E in Fig. 2(a). These properties are of interest in the processing and fabrication of materials because the passage of a shock pulse through a metal can result in substantial increases in the hardness and strength of the material.[3-5] The unique feature of this hardening phenomenon is that it is associated with only slight permanent dimensional changes of the workpiece, whereas to achieve similar hardness levels by conventional processes such as rolling or forging would require substantial deformation. This occurs because the shock front that separates compressed material from that which is uncompressed is very abrupt; hence, the movement and generation of large numbers of dislocations, as discussed in Section III, is strongly enhanced. Such effects can find direct commercial application in explosive hardening of a surface by detonation of a thin sheet of explosive in contact with it, usually at grazing incidence.[4]

These are some of the reasons for active metallurgical interest in the effects of shock-loading on the mechanical properties of metals and alloys.[3,5] The relationship of hardness to the occurrence of dislocations, stacking faults, deformation twins, and phase transformations has been extensively investigated. In most of these studies the observed terminal phenomena have been related only to the peak shock stress experienced by the metal. However, if plastic flow is rate-dependent, as discussed in Section III, then shock-pulse duration might also be expected to be an important variable. For example, if the metal initially responds elastically until dislocation motion and generation can produce plastic strain, then the shear stresses acting on dislocations will be much greater than predicted by the rate-independent description summarized in Fig. 2(a). The initial shear stresses will correspond to the offset between the extension of the elastic unloading line AH and the pressure curve P, and these shear stresses obviously exceed their rate-independent counterparts which correspond to the offset between the Rayleigh line HB and the curve P. Subsequently, they relax (cf. Fig. 4), allowing the impact stress σ_x to relax to state B in Fig. 2(a). This behavior is apparent in the wave profiles of Fig. 5. Since a relaxation time is involved, the degree of completion may be expected to be determined by the shock pulse duration.

Appleton and Waddington[31] attributed terminal hardness effects in shock-loaded copper to wave profile shape and shock-pulse duration. Champion and Rohde[32] systematically investigated the effect of shock pulse amplitude and duration on the hardening of Hadfield steel, an alloy which has been explosively-hardened on a commercial scale. Typically, the alloy contains about 13 percent manganese and one percent carbon. It exhibits an extraordinary work-hardening capacity due in part to its low stacking fault energy which inhibits cross-slip of dislocations, and is characteristically used for parts subjected to severe impact and abrasion such as railroad switch frogs, rock crusher jaws, and power shovel buckets. Figure 7 shows the residual hardness of the shock-loaded surface x = 0 as a function of peak shock stress for pulse durations of 65 nsec (curve A), 230 nsec (curve B), and 2200 nsec (curve C). It is obvious that pulse duration is important: for a given shock stress, a shorter duration shock pulse produces less hardening. With increasing shock stress, the terminal hardness reaches a saturation value (curve C).

Representative electron micrographs of the material are shown in Fig. 8. The microstructure of the metal in its initial, unshocked state appears in Fig. 8(a). The initial total dislocation density is about $1 \times 10^{10}/cm^2$. The micrographs in Figs. 8(b), (c) and (d) represent specimens of nearly the same terminal hardness, about Rockwell C20, but with different shock loading histories. The dislocation microstructure is quite similar for these three specimens and corresponds to a density of about $2.5 \times 10^{10}/cm^2$.

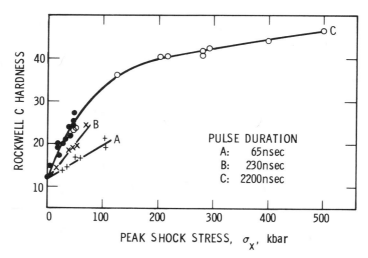

Fig. 7 Effect of peak shock amplitude and duration on the residual hardness of the impacted surface of Hadfield steel. The symbols denote results obtained by different experimental methods. (After Champion and Rohde.[32])

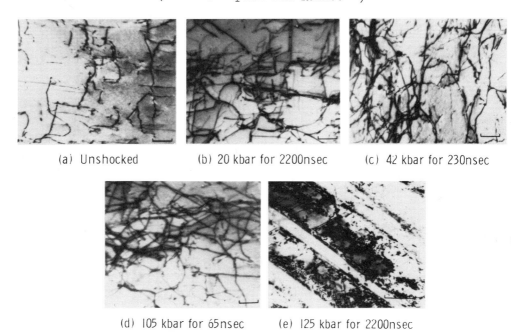

(a) Unshocked   (b) 20 kbar for 2200nsec   (c) 42 kbar for 230nsec

(d) 105 kbar for 65nsec   (e) 125 kbar for 2200nsec

Fig. 8 Electron micrographs of shock-loaded Hadfield steel. Markers indicate 1 μ. (After Champion and Rohde.[32])

Planar arrays of dislocations have developed and no cellular dislocation structure is observed. The lack of cellular structure is indicative of a low stacking-fault energy material. A dramatic effect of shock-pulse duration is apparent from Figs. 8(d) and (e). In both cases the peak shock stress was about 100 kbar; however, the shock pulse duration for Fig. 8(d) was 65 nsec as compared to 2200 nsec for Fig. 8(e). For the longer pulse duration, numerous deformation twins have developed and twinning, rather than slip, has become the dominant deformation feature. Thus, there appears to be a threshold time for the initiation of twinning. Johnson and Rohde[33] have subsequently made a detailed investigation of dynamic deformation twinning in commercially pure iron, and have shown it to be competitive with the dislocation dynamics description of Eq. (10). The analogous expression for the plastic shear strain γ due to twin formation is $\gamma = k(f - f_o)$, where k is the twinning shear and f, the volume fraction of twinned material, depends on the growth dynamics of twin platelets.

Summarizing, time-dependent phenomena occurring during transient dynamic yielding can profoundly influence residual postmortem effects produced by shock-loading. Studies of shock-induced hardening of metals require attention to both shock pulse profile and duration as important experimental variables in addition to the peak shock amplitude.

V. RADIAL RELEASE EFFECTS ON RESIDUAL PROPERTIES

As implied in Section IV, residual properties observed following uniaxial-strain shock loading are tacitly assumed to result from the shock compression and unloading cycle AHBCD in Fig. 2(a), even though the final state really corresponds to E. Residual effects produced by the radial release process in going from state D to E are usually ignored. However, experimental evidence exists that convergent radial release waves can cause significant effects. Figure 9, for example, shows a concave dimple on the face of an aluminum single crystal disk which was impacted and recovered. The concavity, unexplainable in terms of one-dimensional uniaxial-strain phenomena, suggests that tensile stresses due to converging radial release waves were large enough to induce observable plastic flow.

Recently, Stevens and Jones[34] have studied this dynamic unloading process by using the TOODY II-A two-dimensional Lagrangian finite-difference computer program[35] to model sixteen physical experiments representative of common practice. They investigated the generic case, shown in Fig. 10, of a thick circular flyer disk of thickness ℓ_F and radius r_o impacting a thin circular target disk of thickness ℓ_T and identical radius. The thickness ratio ℓ_F/ℓ_T is greater than one ($\ell_F/\ell_T > 1$), and both disks have large diameter-to-thickness ratios ($d_o/\ell_F \gg 1$ and $d_o/\ell_T \gg 1$). A guard

Fig. 9 Concave dimple on face of single-crystal aluminum disk recovered after 12 kbar impact. Initial disk diameter and thickness were 38.1 and 6.35 mm, respectively. (After Stevens and Jones.[34])

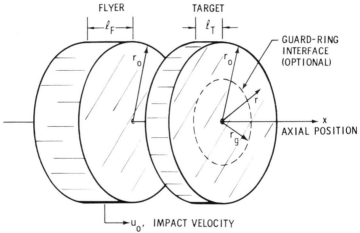

Fig. 10 Impact-release experiment of a thick flyer disk impacting a thin flyer disk (axial symmetry).

ring interface may be introduced at $r = r_g$ as a cylindrical zone of zero tensile strength. The disks are assumed to be made of the same homogeneous, isotropic material: 6061-T6 aluminum represents materials which exhibit significant yield strength and little work-hardening, whereas 1100-0 aluminum models materials which show a small yield stress and pronounced work-hardening. Isotropic work-hardening was assumed, with the deviatoric component of the stress tensor being limited by the von Mises yield criterion of Eq. (3).

The plastic flow rule was chosen such that plastic stretching occurred orthogonally to the yield surface.

The initial compressive longitudinal plane waves generated at impact reflect, respectively, from the free surfaces of the target and flyer disks as unloading, or rarefaction, waves. Subsequently they interact in the flyer disk to produce tensile waves. Because the impact interface between the two disks cannot support tension, the disks separate when the tensile wave arrives at the interface. Thus, before arrival of the lateral radial release wave, the material in the center of the target disk has been axially shock compressed and then unloaded while in a state of uniaxial strain.[36] For a rate-independent material, this corresponds to the cycle AHBCD in Fig. 2(a). However, even though the axial stress σ_x is unloaded to zero at state D, the radial stress σ_r is nonzero because of plastic flow and continues to drive the radial release wave. This is immediately evident from Eq. (3): when σ_x is zero, the in-plane radial stress σ_r in the yielded state may be as large as the yield stress Y. This is true regardless of whether a guard ring is placed around the target disk.

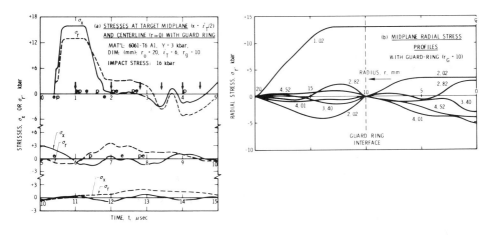

Fig. 11 Axial stress σ_x and radial stress σ_r histories for typical impact-release experiment. (Bold arrows in (a) refer to plot times in (b)).

Typical results are shown in Fig. 11. The axial and radial stress histories at the target disk midplane ($x = \ell_T/2$) and axis ($r \approx 0$) are shown in Fig. 11(a) for the case of a 6-mm thick, guard-ringed, 6061-T6 target disk impacted at 16 kbar. The "e" and "p" subscripts along the abscissa indicate the onset of elastic or plastic response, respectively. The bold arrows along the abscissa indicate times at which plots of the midplane radial stress distribution are shown in Fig. 11(b). Referring to Fig. 11(b), the radial release wave propagates inward and causes separation of the guard-ring and the inner target disk. At 4.01 μsec it arrives at the target disk axis and results in a radial tension stress of about 5 kbar which initiates plastic yielding, (see Fig. 11(a)). It then propagates outward, reflects as a compressive wave from the target disk perimeter (at $r = 10$ mm), and again converges on the axis at about 7.8 μsec and produces a small amount of additional plastic flow. Thereafter the response is elastic and the plate simply "rings" elastically. By using special techniques and precautions, the target disk can be gradually decelerated and recovered without incurring further plastic deformation.[37] Thus, the plastic flow accumulated at about 7.9 μsec is representative of the residual properties of the disk after recovery.

Table I summarizes the sixteen numerical experiments. The plastic work per unit volume produced by the uniaxial-strain compression and unloading cycle (corresponding to AHBCD in Fig. 2(a)) at the target axis ($r \cong 0$) and midplane ($x = \ell_T/2$) is denoted as W_p^*. If W_p is the total plastic work at time t, then the additional plastic work $(W_p - W_p^*)/W_p^*$ produced following the uniaxial-strain cycle is a simple figure of merit for judging whether permanent property changes should be attributed to the uniaxial-strain cycle or the radial release process. For example, since plastic work is associated with dislocation generation and motion, $(W_p - W_p^*)/W_p^*$ should be small if residual effects are to be attributed primarily to uniaxial-strain shock compression. Values of $(W_p - W_p^*)/W_p^*$ for maximum problem running times t_m are given in Table I.

Although Table I represents only a small portion of the total results, it is apparent that the use of a guard ring (or spall ring) reduces the additional plastic work due to the radial release process, although an assignment of residual effects to only the uniaxial-strain cycle may still be questionable. Increasing the diameter-to-thickness ratio d_o/ℓ_T reduces the additional plastic work due to the radial release process, while, for a given yield stress, the additional plastic work decreases with increasing impact stress. For a large diameter-to-thickness ratio target and a large impact-to-yield stress ratio, the radial release wave is "trapped," or dissipated, in the region of large target radii. This phenomenon is similar in effect to a guard ring, but is not so efficient.

In general, Stevens and Jones[34] found that the assumption that the radial release process can be ignored is highly doubtful in many cases. For example, in unfavorable circumstances the additional plastic work can be 1000 percent of the plastic work produced by uniaxial-strain shock loading and unloading. However, as may be seen from Table I, their studies also suggest that it is possible, with care, to design recovery experiments in which the uniaxial-strain shock effects clearly dominate radial release phenomena. This means that new methods of testing the dislocation dynamics models of Section III may be possible, such as testing multiplication models on the basis of initial and terminal dislocation densities.

TABLE I. Impact-Release Problems Solved and Results

Configuration[a]	A. No guard ring B. With guard ring				A. No guard ring B. With guard ring			
Target and flyer disk dimensions[a] (mm)	A. $\ell_T = 6$, $r_o = 20$, $\ell_F = 12$ B. $\ell_T = 6$, $r_o = 20$, $r_g = 10$, $\ell_F = 12$				A. $\ell_T = 3$, $r_o = 20$, $\ell_F = 6$ B. $\ell_T = 3$, $r_o = 20$, $r_g = 10$, $\ell_F = 6$			
Material - Aluminum[b]	6061-T6		1100-0		6061-T6		1100-0	
Impact stress (kbar)	16.0	101.0	15.0	100.3	16.0	101.0	15.0	100.3
Midplane ($x = \ell_T/2$) plastic work due to uniaxial strain cycle, W_p^* (joules/cm^3)	3.31	36.8	0.863	11.2	3.39	36.9	0.844	11.0
Running time, t_m (μsec) after impact	15.0	5.0	14.0	5.0	15.0	5.0	11.0	5.0
Additional midplane plastic work at disk axis, at time t_m, due to radial release process, $(W_p - W_p^*)/W_p^*$ (percent)	A. 490 B. 170	A. 28 B. 15	A. 800 B. 44	A. 38 B. 11	A. 310 B. 71	A. 26 B. 8	A. 200 B. 29	A. 9 B. 3

[a] Refer to Fig. 10
[b] Uniaxial-stress engineering stress-strain curves were used to represent the deviatoric yield behavior of 6061-T6 and 1100-0 aluminums. Both materials were assumed to be rate-independent. The 6061-T6 aluminum was approximated as an elastic-perfectly plastic material with $Y_1 = 3$ kbar. For 1100-0 aluminum, parabolic strain hardening was assumed with $Y = 0.0316 + 3.861\ e^{\frac{1}{2}}$ for $Y \geq Y_o$, where $Y_o = 0.069$ kbar.

REFERENCES

*Work supported by the U. S. Atomic Energy Commission. This tutorial is derived from an earlier paper, O. E. Jones, in Engineering Solids Under Pressure, H. L. D. Pugh, Ed. (Institution of Mechanical Engineers, London 1971) p. 75.

1. W. Herrmann, in Wave Propagation in Solids, J. Miklowitz, Ed. (The American Society of Mechanical Engineers, N.Y. 1969), p. 129.

2. J. J. Burke and V. Weiss, Eds., Shock Waves and the Mechanical Properties of Solids (Syracuse U. Press, Syracuse, N.Y., 1971).

3. P. G. Shewmon and V. F. Zackay, Eds., Response of Metals to High Velocity Deformation (Interscience Publishers, N. Y., 1961).

4. J. S. Rinehart and J. Pearson, Explosive Working of Metals (The Macmillan Co., N. Y. 1963).

5. A. H. Jones, C. J. Maiden, and W. M. Isbell, in Mechanical Behavior of Materials Under Pressure, H. L. D. Pugh, Ed. (Elsevier Publ. Co. Limited, Essex, England, 1970), p. 680.

6. G. E. Duvall and G. R. Fowles, in High Pressure Physics and Chemistry, R. S. Bradley, Ed. (Academic Press, N. Y., 1963), Vol. 2, p. 271.

7. F. S. Marshall, in Response of Metals to High Velocity Deformation, P. G. Shewmon and V. F. Zackay, Eds. (Interscience Publ., N. Y., 1961), p. 266.

8. O. E. Jones, F. W. Neilson and W. B. Benedick, J. Appl. Phys. $\underline{33}$, 3224 (1962).

9. J. W. Taylor and M. H. Rice, J. Appl. Phys. $\underline{34}$, 364 (1963).

10. R. W. Rohde, Acta Met. $\underline{17}$, 353 (1969).

11. B. M. Butcher and D. E. Munson, in Dislocation Dynamics, A. R. Rosenfield, G. T. Hahn, A. L. Bement, Jr., and R. I. Jaffee, Eds. (McGraw-Hill Book Co., N. Y. 1968), p. 591.

12. P. P. Gillis, K. G. Hoge, and R. J. Wasley, J. Appl. Phys. $\underline{42}$, 2145 (1971).

13. J. W. Taylor, J. Appl. Phys. $\underline{36}$, 3146 (1965).

14. L. M. Barker, B. M. Butcher, and C. H. Karnes, J. Appl. Phys. $\underline{37}$, 1989 (1966).

15. J. N. Johnson and W. Band, J. Appl. Phys. $\underline{38}$, 1578 (1967).

16. M. L. Wilkins, in Behavior of Dense Media Under High Dynamic Pressures (Gordon and Breach, N. Y., 1968), p. 267.

17. J. M. Kelly and P. P. Gillis, J. Appl. Phys. $\underline{38}$, 4044 (1967).

18. O. E. Jones and J. R. Holland, Acta Met. $\underline{16}$, 1037 (1968).

19. J. N. Johnson, J. Appl. Phys. $\underline{40}$, 2287 (1969).

20. J. N. Johnson and L. M. Barker, J. Appl. Phys. $\underline{40}$, 4321 (1969).

21. J. J. Gilman, Appl. Mech. Rev. $\underline{21}$, 767 (1968).

22. O. E. Jones and J. D. Mote, J. Appl. Phys. $\underline{40}$, 4920 (1969).

23. T. E. Michaels, PhD. Thesis, Washington State University, 1972.

24. W. J. Murri and G. D. Anderson, J. Appl. Phys. 41, 3521 (1970).

25. J. R. Asay, G. R. Fowles, G. E. Duvall, M. H. Miles, and R. F. Tinder, J. Appl. Phys. 43, 2132 (1972).

26. L. E. Pope and A. L. Stevens, Bull. Amer. Phys. Soc. 107, 1083 (1972). Also, see this Proceedings.

27. J. N. Johnson, O. E. Jones, and T. E. Michaels, J. Appl. Phys. 41, 2330 (1970).

28. W. Herrmann, D. L. Hicks, and E. G. Young, in Shock Waves and the Mechanical Properties of Solids, J. J. Burke and V. Weiss, Eds. (Syracuse Univ. Press, Syracuse, N. Y., 1971), p. 23.

29. J. N. Johnson, J. Appl. Physics 42, 5522 (1971).

30. J. N. Johnson, J. Appl. Physics 43, 2074 (1972).

31. A. S. Appleton and J. S. Waddington, Acta Met. 12, 956 (1964).

32. A. R. Champion and R. W. Rohde, J. Appl. Phys. 41, 2213 (1970).

33. J. N. Johnson and R. W. Rohde, J. Appl. Phys. 42, 4171 (1971).

34. A. L. Stevens and O. E. Jones, J. Appl. Mech. 39, 321 (1972).

35. S. E. Benzley, L. D. Bertholf, and G. E. Clark, Sandia Laboratories Development Report No. SC-DR-69-516, Albuquerque, New Mexico, 1969.

36. The converse configuration of a thin flyer disk impacting a thick target disk produces tension in the target disk, and is used in spallation studies of dynamic fracture.

37. W. F. Hartman, J. Appl. Phys. 35, 2090 (1964).

NUMERICAL ANALYSIS METHODS

W. Herrmann and D. L. Hicks

Sandia Laboratories, Albuquerque, New Mexico

INTRODUCTION

Concentrated research over the last decade has led to rapid development of an understanding of the response of solid materials to high amplitude dynamic loads. Constitutive equations have been developed for such real materials as engineering alloys, fiber composites, porous earth materials, and polymers. Together with the conservation laws, these equations have been incorporated into numerical solution methods which have allowed analysis of such problems as ballistic penetration, high velocity impacts, explosive devices and components, and many others which were intractable only a few years ago. Stress wave codes have also become an indispensable tool in further research on the dynamic behavior of materials.

The ready accessibility of computers and stress wave codes has made it possible for many engineers and scientists to perform their own computational analyses. Often the stress wave code user is unaware of the power of the mathematical techniques implicit in the computer code, or of their limitations. In this tutorial paper, we will attempt to illustrate a few of the elementary concepts in numerical analysis of stress wave propagation, in the hope of providing some insight for the casual stress wave code user. Most of the concepts can be illustrated in one dimension, but a connection to multi-dimensional codes will be made at the end of the paper.

PROBLEM SPECIFICATION

Before proceeding to a discussion of solution methods, it is, of course, necessary to give a concise mathematical statement of the problem. Choosing the Lagrangian coordinate X as appropriate to the description of the deformation of solid materials, the laws of conservation of mass, momentum and energy governing smooth motions are

$$\frac{\partial}{\partial t}\left(\rho_0 e\right) + \frac{\partial}{\partial X}\left(\rho_0 u\right) = 0$$

$$\frac{\partial}{\partial t}\left(\rho_0 u\right) + \frac{\partial}{\partial X}\left(\sigma\right) = 0 \quad (1)$$

$$\frac{\partial}{\partial t}\left(\tfrac{1}{2}\rho_0 u^2 + \rho_0 \mathcal{E}\right) + \frac{\partial}{\partial X}\left(\sigma u\right) = 0$$

where ρ_0 is the density in some reference configuration, e is the engineering strain ($e = 1 - \rho_0/\rho$), σ is the stress, (e and σ taken positive in compression), u is the material particle velocity, and $\mathcal{E}$ is the specific internal energy.

Each of the above equations is in so-called conservation law form, relating the time rates of change of a quantity (mass, momentum or total energy) to its spatial flux. In fact, integrals of Eq. 1 over a finite amount of material are valid even if the motion is discontinuous.

Equations 1 may be simplified by elementary manipulations to

$$\frac{\partial e}{\partial t} + \frac{\partial u}{\partial X} = 0$$

$$\rho_0 \frac{\partial u}{\partial t} + \frac{\partial \sigma}{\partial X} = 0 \quad (2)$$

$$\rho_0 \frac{\partial \mathcal{E}}{\partial t} - \sigma \frac{\partial e}{\partial t} = 0$$

Of course, Eqs. 2 are applicable only if all variables are continuous.

The problem specification must include a description of the behavior of the particular class of materials which we wish to study. As an initial illustration, we begin with the class of simple thermoelastic materials, i.e., materials whose current response does not depend on their previous history. The constitutive equation for such materials takes the general form

$$\sigma = f(e, \mathcal{E}, X) \tag{3}$$

that is, the stress is an ordinary (perhaps nonlinear) function of the current values of strain, internal energy and material particle, the latter allowing treatment of non-homogeneous materials or layers of different materials.

Finally, we require a specification of initial and boundary conditions. As an illustration we consider a homogeneous slab of material with a specified velocity history applied to its left face and a specified stress history applied to its right face. The initial/boundary data are then specified by

$$u(X, t_1) = u_o(X) \qquad u(X_1, t) = \hat{u}(t)$$
$$\sigma(X, t_1) = \sigma_o(X) \qquad \sigma(X_2, t) = \hat{\sigma}(t) \tag{4}$$
$$\mathcal{E}(X, t_1) = \mathcal{E}_o(X)$$

for $X_1 \leq X \leq X_2$ and $t_1 \leq t \leq t_2$. Note that the initial distribution of strain $e_o(X)$ follows implicitly from $\sigma_o(X)$ and $\mathcal{E}_o(X)$ and the constitutive equation, Eq. 3.

The above three elements; conservation laws, constitutive equations, and initial/boundary conditions are all required for a complete specification of the problem and must be provided at the outset. The solution method must be tailored to the entire problem specification. Of course, we also require some assurance that the problem we have posed has a solution, otherwise we might embark on a fruitless exercise in constructing a solution method. It is also important for us to know if there exists just one solution, or if we are faced with the task of finding the solution we want from among several.

Mathematicians have given much thought to the question of existence and uniqueness of solutions for various classes of problems. In order to cast our problem into recognizable form, we note that for a thermoelastic material, the following thermodynamic relation holds

$$\rho_o d\mathcal{E} = \rho_o T dS + \sigma de \tag{5}$$

where S is the specific entropy and T the absolute temperature. The energy equation in Eq. 2 therefore implies that, at constant X, $\partial S/\partial t = 0$, i.e., each material particle carries unchanged entropy throughout the motion. If we rewrite the constitutive equation in terms of the entropy, i.e., $\sigma = h(e, S, X)$ and define the intrinsic or Lagrangian sound speed C by

$$C^2 = \frac{1}{\rho_o} \frac{\partial h}{\partial e} \tag{6}$$

then we may use the chain rule of differentiation to eliminate derivatives of e with respect to t at constant X from Eq. 2 to obtain

$$\frac{\partial \sigma}{\partial t} + \rho_0 C^2 \frac{\partial u}{\partial X} = 0$$

$$\rho_0 \frac{\partial u}{\partial t} + \frac{\partial \sigma}{\partial X} = 0 \tag{7}$$

$$\rho_0^2 C^2 \frac{\partial \mathcal{E}}{\partial t} - \sigma \frac{\partial \sigma}{\partial t} = 0$$

Our equations now correspond to the general form of quasi-linear partial differential equations with variable coefficients, which may be written in matrix notation as

$$\underset{\sim}{A} \frac{\partial \underset{\sim}{\psi}}{\partial t} + \underset{\sim}{B} \frac{\partial \underset{\sim}{\psi}}{\partial X} + \underset{\sim}{d} = \underset{\sim}{0} \tag{8}$$

where $\underset{\sim}{A}$ and $\underset{\sim}{B}$ are coefficient matrices and $\underset{\sim}{d}$ a coefficient vector, all of which may be functions of the vector of dependent variables $\underset{\sim}{\psi}(X, t)$. In our particular case, $\underset{\sim}{\psi} = \{\sigma, u, \mathcal{E}\}$.

Quasi-linear equations have been studied in great detail (for example, see Ref. 1 and 2). In our particular case we find that the characteristic equation has three real roots, and Eqs. 7 are hyperbolic. There exist three characteristic directions in X,t space along which partial derivatives combine to form total directional derivatives. If the characteristic directions are defined by

$$\alpha = \alpha(X, t) \qquad \beta = \beta(X, t) \qquad \gamma = \gamma(X, t)$$

then Eqs. 7 can be replaced by ordinary differential equations with respect to α, β and γ

$$\frac{d\sigma}{d\alpha} + \rho_0 C \frac{du}{d\alpha} = 0 \qquad \frac{dX}{d\alpha} - C \frac{dt}{d\alpha} = 0$$

$$\frac{d\sigma}{d\beta} - \rho_0 C \frac{du}{d\beta} = 0 \qquad \frac{dX}{d\beta} + C \frac{dt}{d\beta} = 0 \tag{9}$$

$$\frac{d\mathcal{E}}{d\gamma} - \frac{\sigma}{\rho_0^2 C^2} \frac{d\sigma}{d\gamma} = 0 \qquad \frac{dX}{d\gamma} = 0$$

Solutions of Eqs. 9 can be shown, under proper conditions, to correspond to solutions to our original equations. In fact, in some cases, exact solutions can be found utilizing Eqs. 9, and

NUMERICAL ANALYSIS METHODS

such solutions will be used later to investigate the performance of numerical solution methods.

Consideration of the properties of characteristics leads to the fact that if the initial/boundary data are sufficiently smooth, then at least for a short time there exists a unique solution which is smooth and depends continuously on the initial/boundary data. With the assurance that the problem is properly posed in the above classical sense, we can proceed to find numerical solution methods.

NUMERICAL ALGORITHMS

One way of proceeding to find discrete approximations to the partial differential equations is as follows. If ψ represents any one of the dependent variables, we will replace the continuous function $\psi(X,t)$ by a finite set of values at discrete points $\psi_j^n = \psi(j\Delta X, n\Delta t)$, $1 \leq j \leq J$, $1 \leq n \leq N$. If $\psi(X,t)$ is continuous and has continuous derivatives up to second order in X, then we may expand ψ in a Taylor series with remainder

$$\psi_{j+1}^n = \psi_j^n + \left(\frac{\partial \psi}{\partial X}\right)_j^n \Delta X + \frac{1}{2}\left(\frac{\partial^2 \psi}{\partial X^2}\right)_k^n \Delta X^2$$

where the derivative in the remainder term is evaluated at k with $j < k < j+1$. Rearranging

$$\frac{\psi_{j+1}^n - \psi_j^n}{\Delta X} = \left(\frac{\partial \psi}{\partial X}\right)_j^n + \frac{1}{2}\left(\frac{\partial^2 \psi}{\partial X^2}\right)_k^n \Delta X \tag{10}$$

The expression on the left represents the first derivative at n,j to within an error term of $O(\Delta X)$. It is called a forward difference algorithm of first order accuracy. A backward difference algorithm of the same order can be constructed by starting with a Taylor series expansion for ψ_{j-1}^n.

If $\psi(X,t)$ has continuous derivatives through third-order, we may write Taylor series expansions for $\psi_{j+½}^n$ and $\psi_{j-½}^n$ as

$$\psi_{j+½}^n = \psi_j^n + \left(\frac{\partial \psi}{\partial X}\right)_j^n \frac{\Delta X}{2} + \frac{1}{2}\left(\frac{\partial^2 \psi}{\partial X^2}\right)_j^n \left(\frac{\Delta X}{2}\right)^2 + \frac{1}{6}\left(\frac{\partial^3 \psi}{\partial X^3}\right)_k^n \left(\frac{\Delta X}{2}\right)^3$$

where $j < k < j+½$, and

$$\psi_{j-½}^n = \psi_j^n - \left(\frac{\partial \psi}{\partial X}\right)_j^n \frac{\Delta X}{2} + \frac{1}{2}\left(\frac{\partial^2 \psi}{\partial X^2}\right)_j^n \left(\frac{\Delta X}{2}\right)^2 - \frac{1}{6}\left(\frac{\partial^3 \psi}{\partial X^3}\right)_\ell^n \left(\frac{\Delta X}{2}\right)^3$$

where $j-\frac{1}{2} < \ell < j$. Subtracting these two expansions provides

$$\frac{\psi^n_{j+1/2} - \psi^n_{j-1/2}}{\Delta X} = \left(\frac{\partial \psi}{\partial X}\right)^n_j + \frac{1}{24}\left(\frac{\partial^3 \psi}{\partial X^3}\right)^n_m \Delta X^2 \qquad (11)$$

where $j-\frac{1}{2} < m < j+\frac{1}{2}$.

The expression on the left of Eq. 11 is a centered difference algorithm of second-order accuracy. The process may be continued to higher order, but if the result is to be expressed in a form analogous to Eq. 11, the highest order derivative of $\psi(X,t)$ in the error term must be continuous. Similar expressions may be developed for derivatives with respect to t.

A set of difference equations results when all partial derivatives in the governing equations are replaced by suitable difference algorithms. We could choose to start with the equations in the forms Eq. 1, 2, 7 or 9 since these are all equivalent. Obviously, there are a great many difference equations which can be developed for our sample problem, depending on the starting point and the order and centering chosen for each derivative.

For example, if we use second-order centered difference algorithms in each derivative in Eqs. 2, we might obtain

$$\rho_0 \left(\frac{u^{n+1/2}_j - u^{n-1/2}_j}{\Delta t}\right) + \left(\frac{\sigma^n_{j+1/2} - \sigma^n_{j-1/2}}{\Delta X}\right) = 0$$

$$\left(\frac{e^{n+1}_{j-1/2} - e^n_{j-1/2}}{\Delta t}\right) + \left(\frac{u^{n+1/2}_j - u^{n+1/2}_{j-1}}{\Delta X}\right) = 0 \qquad (12)$$

$$\rho_0 \left(\frac{\varepsilon^{n+1}_{j-1/2} - \varepsilon^n_{j-1/2}}{\Delta t}\right) - \sigma^{n+1/2}_{j-1/2} \left(\frac{e^{n+1}_{j-1/2} - e^n_{j-1/2}}{\Delta t}\right) = 0$$

The constitutive equation Eq. 3 can be written in discrete form

$$\sigma^{n+1/2}_{j-1/2} = f\left(e^{n+1/2}_{j-1/2}, \varepsilon^{n+1/2}_{j-1/2}\right) \qquad (13)$$

In Eqs. 12 and 13, the momentum equation has been centered about n,j, but the mass and energy equations have been centered about $n+\frac{1}{2}$, $j-\frac{1}{2}$ in order to achieve equations which can be solved explicitly for quantities at the next time step in terms of

quantities at the previous time step.* If initial values are
provided at each meshpoint, then the algebraic difference equations
may be solved with appropriate boundary conditions in order to
produce the solution for all meshpoints at successive time steps.

Care is necessary with the initial/boundary conditions since
u is centered at j, $n+\frac{1}{2}$ while other quantities are centered at
$j-\frac{1}{2}$, n. One specification is

$$e^{0}_{j+1/2} = e_0\left((j+\tfrac{1}{2})\Delta X, 0\right) \qquad u^{n+1/2}_1 = \hat{u}\left((n+\tfrac{1}{2})\Delta t\right)$$

$$\mathcal{E}^{0}_{j+1/2} = \mathcal{E}_0\left((j+\tfrac{1}{2})\Delta X, 0\right) \qquad \frac{\sigma^{n}_{J+1/2} + \sigma^{n}_{J-1/2}}{2} = \hat{\sigma}(n\Delta t) \qquad (14)$$

$$u^{-1/2}_j = u_0(j\Delta X, -\tfrac{1}{2}\Delta t)$$

for $1 \le j \le J$ and $1 \le n \le N$.

If the error terms are carried along in the development of
the difference representations of the conservation equations
and initial/boundary conditions, then we find that the error terms
are all of $O(\Delta X^2, \Delta t^2)$, provided that ΔX and Δt are constants.
Of course if ΔX or Δt vary, or if liberties are taken with the
boundary conditions, then first-order error terms might appear.

The difference equations, Eqs. 12, 13 and 14, are essentially
equivalent to those of von Neumann and Richtmyer.[3,8] They remain
the basis for many current production codes for the solution of
one-dimensional stress wave propagation problems, for example
PUFF[4] and WONDY.[5]

Evidently, the difference equations, Eqs. 12, 13 and 14, will
approach the original differential equations as ΔX and Δt jointly
go to zero, provided that the solution is sufficiently smooth so
that the derivatives in the error terms are bounded. A set of
difference equations with this property is termed consistent. The
order of accuracy† is defined to be the same as that of the
lowest order error terms in ΔX and Δt. Thus the above set of
difference equations is of second-order accuracy in both ΔX and
Δt.

---

* Normally, values of e and $\mathcal{E}$ at $t^{n+1/2}$ in the constitutive equation
are obtained by linear interpolation between their values at t^n
and t^{n+1}. Consequently, the energy and constitutive equations
must be solved simultaneously for $\mathcal{E}^{n+1}_{j-1/2}$ and $\sigma^{n+1}_{j-1/2}$. This can be
done explicitly if the constitutive equation is linear in $\mathcal{E}$, as
it appears to be for many real materials.

† Definitions of order of accuracy vary in the literature, and may
not correspond precisely to that given here.

CONVERGENCE AND STABILITY

While we have discussed consistent difference equations which converge to the differential equations we are trying to solve, it remains to be seen if the solutions of the difference equations converge to the solution of the differential equations as ΔX and Δt are reduced. In order to discuss this matter, we need some measure of the difference between two solutions. We shall define a norm in order to determine the distance between two solutions in the solution space. Many suitable norms may be defined, each has somewhat different implications for the definition of convergence.

Suppose that we have two solutions $\underline{F}$ and $\underline{G}$ which are vector valued functions of X and t whose components are some or all of the dependent variables. If $\underline{F}$ and $\underline{G}$ are defined over some domain $\Re$ of X,t space, we could define the distance between $\underline{F}$ and $\underline{G}$ over $\Re$ to be the L_p norm

$$\|\underline{F} - \underline{G}\|_p = \left\{ \iint_\Re |\underline{F} - \underline{G}|^p \, dX \, dt \right\}^{1/p} \tag{15}$$

for $p \geq 1$. Here $|\underline{F} - \underline{G}|$ is a suitable vector distance between the two solutions at a given X and t. It is common to normalize the result by dividing by $\|F\|_p$.

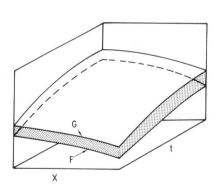

FIG. 1. SCHEMATIC OF SOLUTION SURFACES

For $p = 1$, the above norm may be given the simple geometric interpretation of the volume contained between the two solution surfaces F and G over the region $\Re$. This is illustrated schematically in Fig. 1, for the case when $\underline{F}$ and $\underline{G}$ each have a single scalar component. For $p = \infty$, the norm reduces to the essential supremum of $|\underline{F} - \underline{G}|$. This may be interpreted geometrically as the greatest vertical distance between the two solution surfaces $\underline{F}$ and $\underline{G}$ over the region $\Re$.

Evidently there are many choices of norm, corresponding to different choices of the value of p, the choice of components of $\underline{F}$ and $\underline{G}$, the choice of the region $\Re$, and the means by which the result is non-dimensionalized and normalized. The norm in Eq. 15 was defined under the assumption that $\underline{F}$ and $\underline{G}$ are defined for all X and t. We wish to compare a continuum solution $\underline{F}$ with a discrete numerical solution G. In order to do this, the numerical solution could be extended to cover all X and t in the domain $\mathcal{R}$ by adducing an interpolation scheme. Alternatively, we could consider the exact solution only at discrete meshpoints, and replace the integral in Eq. 15 by a finite summation.

We will illustrate concepts of convergence and stability by considering a specific problem below. For purposes of illustration, we will find it sufficient to arbitrarily restrict attention to the L_1 and L_∞ norms defined over the discrete set of X,t points $j\Delta X$, $1 \leq j \leq J$ at a single fixed time $n\Delta t = t_2$ and furthermore limit consideration to a single component of $\underset{\sim}{F}$ and $\underset{\sim}{G}$, namely the material particle velocity u.

We will consider the specific problem of a homogeneous slab of material, initially at rest in the reference configuration, subjected to a velocity history on its left hand face

$$\hat{u}(t) = \frac{a}{2}\left\{1 - \cos(2\pi t/\tau_0)\right\} \tag{16}$$

where a is the amplitude and τ_0 the period of the forcing function. The constitutive equation is taken in the form of an "ideal solid" equation of state*

$$\sigma = \frac{\rho_0 C_0^2 e + (\xi-1)\rho_0 \mathcal{E}}{1 - e} \tag{17}$$

where ρ_0, C_0 and ξ are material constants.

In order to simplify the problem, the following non-dimensional variables are introduced: $\overline{X} = X/C_0\tau_0$, $\overline{t} = t/\tau_0$, $\overline{u} = u/C_0$, $\overline{\sigma} = \sigma/\rho_0 C_0^2$, $\overline{\mathcal{E}} = \mathcal{E}/C_0^2$. It is then unnecessary to specify values of ρ_0, C_0, and τ_0. The values of the remaining constants were chosen to be $a = 0.01$, $\xi = 4$. The exact solution, at least for a short time, is easily found[6] by solving the characteristic equations, Eqs. 9.

---

* This constitutive equation corresponds to a Mie-Grueneisen equation of state with constant Grueneisen ratio and isentropes in the Murnaghan form.

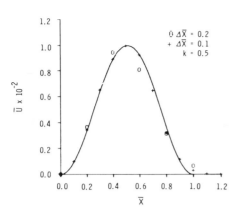

FIG. 2. SOLID LINE IS THE EXACT SOLUTION
PLOTTED POINTS ARE NUMERICAL SOLUTIONS

Two discrete numerical solutions with different values of $\Delta \overline{X}$ are compared with the exact solution in Fig. 2. In each case, $\Delta \overline{t}$ was chosen such that the ratio

$$k = \frac{C_o \Delta t}{\Delta X} \qquad (18)$$

had the value $k = 0.5$. The L_1 and L_∞ error norms for the two numerical solutions are shown in Table I.

It may be seen that when ΔX and Δt are reduced by a factor of two, the L_1 error norm is reduced by very nearly a factor of four. Intuitively, we feel that the error will continue to diminish as ΔX and Δt are reduced further, and that the numerical solution will converge to the exact solution as ΔX and Δt jointly go to zero. In fact, we might define convergence as follows: having chosen a particular norm, we consider a sequence of calculations with mesh sizes $\epsilon \Delta X$ and $\epsilon \Delta t$. Convergence in the particular normed space means that $\| \underset{\sim}{F} - \underset{\sim}{G} \| \to 0$ as $\epsilon \to 0$, where $\underset{\sim}{F}$ is the exact solution and $\underset{\sim}{G}$ the numerical solution with mesh sizes $\epsilon \Delta X$ and $\epsilon \Delta t$.

Table I

Error Norms for $k = 0.5$

$\Delta \overline{X}$	L_1	L_∞
0.2	0.145	0.150
0.1	0.038	0.061

The above example suggests that L_1 convergence of the numerical solution to the exact solution may be of second-order in ΔX and Δt, i.e., that the order of convergence and order of accuracy are the same. However, in Fig. 3 are shown two numerical solutions in which a larger value of k was used. The error norms for these solutions, listed in Table II, show that smaller mesh sizes lead to increased error norms. Growth of errors is suggestive of an instability. Intuitively, we feel that in this case, errors might continue to grow as the mesh sizes are reduced, and perhaps become unbounded as the mesh sizes go to zero. This concept leads to one definition of stability: boundedness of the discrete solution up to a given finite time for all mesh sizes $\epsilon \Delta X$ and $\epsilon \Delta t$ as $\epsilon \to 0$.

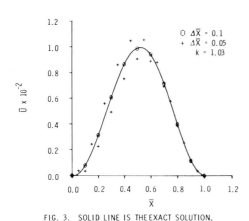

FIG. 3. SOLID LINE IS THE EXACT SOLUTION.
PLOTTED POINTS ARE NUMERICAL SOLUTIONS

Table II

Error Norms for $k = 1.03$

$\Delta \overline{X}$	L_1	L_∞
0.1	0.005	0.006
0.05	0.104	0.116

Stability is a property of the difference equations. The fact that instability is usually connected with the growth of spurious oscillations in the numerical solution leads to another concept of stability. If the governing equations are linear with constant coefficients, i.e., if in our case $C = C_o$, then the numerical solution may be subjected to a Fourier analysis. The requirement that no component of the solution be allowed to grow in one time cycle by a factor whose magnitude is greater than $1+O(\Delta t)$ leads to a stability condition. This condition, first used by von Neumann[8] demands in the present case that k defined in Eq. 18 be less than or equal to unity.

Of course, the equations we are considering have variable coefficients, since C is not, in general, equal to C_o. Von Neumann's definition of stability is sometimes heuristically extended to the nonlinear case by considering the growth of high frequency perturbations superimposed on a solution which is considered to be approximately uniform in space and non-varying in time. Needless to say, such a linearized stability analysis is not rigorous. In the present case, the result is

$$k_c = \frac{C \Delta t}{\Delta X} \leq 1 \qquad (19)$$

An intuitive explanation of the growth of errors is provided by the properties of hyperbolic partial differential equations. As shown schematically in Fig. 4 the solution at a point A in X,t space depends on information in the domain of dependence APQ enclosed by characteristic curves through A. In the momentum equation in Eq. 12, the velocity at A, $u_j^{n+\frac{1}{2}}$ is calculated from $\sigma_{j+\frac{1}{2}}^n$ and $\sigma_{j-\frac{1}{2}}^n$. As long as the time step size Δt is sufficiently small so that the domain of dependence falls within the available data, all is well. If the time step is too large, however, the domain

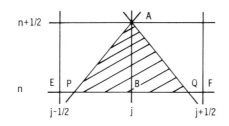

FIG. 4. SCHEMATIC OF THE VON NEUMANN RICHTMYER DIFFERENCE MESH

of dependence extends beyond the available data, and spurious results may be obtained. Since characteristics have a slope C from Eq. 9, we see that the time step size must be restricted precisely by Eq. 19. This argument was first provided by Courant, Friedrichs and Lewy[7] and k_c is frequently termed the CFL number.

Numerical examples, such as those above, while suggestive, cannot establish convergence of numerical methods nor stability, since these are properties of the difference equations in the limit as mesh sizes go to zero. They do suggest a connection between convergence and stability, and such a connection was long surmised. An important theorem due to Lax and Richtmyer[8] finally provided a rigorous connection between stability and convergence, but only for linear equations and properly posed problems: consistency and stability together are necessary and sufficient for convergence. This result has been extended by Strang[9] to nonlinear equations but only under the assumption of the existence of sufficiently smooth solutions. For this case, Strang also showed that the order of accuracy and the order of convergence are the same if the method is stable.* While many subsequent contributions have been made to questions of stability and convergence, few have simultaneously removed the restrictions to linear equations and smooth solutions, and most of these have been limited to specially simple forms of the constitutive equations.

Important as stability and convergence proofs are, the stress wave code user would like more than an assurance that the numerical solution is bounded and approaches the exact solution as the mesh sizes go to zero (and hence the computer times goes to infinity). Of more utility would be the establishment of reasonably tight error bounds on the numerical solution obtained with given finite mesh sizes. If such error bounds were available, then the analyst would have a means of choosing mesh sizes (and hence computer time) appropriate to the accuracy desired in the solution. No such

---

* Strang's definition of stability, however, differs slightly from that of Lax.

results have been given to date, but work towards this end has been started. Preliminary results are encouraging.

All that exist at present are heuristic error estimates based on particular sample calculations, such as those above, for which exact solutions can be found. While these cases are necessarily very restricted and special (else we would not need numerical solution methods) they do illustrate some rough rules of thumb for choosing mesh sizes appropriate to various problems. One of these is apparent from Fig. 2: for the present second-order accurate method, about ten spatial meshes per wavelength are required for the highest frequency component in the solution which is to be reproduced with any fidelity. Another is suggested by Tables I and II: the accuracy of the solution is dependent on CFL number, the highest possible value for stability providing the most accurate solution.

DISCONTINUITIES

The discussion so far has been largely limited to problems with smooth solutions, in particular, to properly posed problems in the classical sense. Such problems are extremely rare in practice. Even the rather contrived illustrative solution of Figs. 2 and 3 exhibits a discontinuity in second derivatives at the front of the wave, i.e., on the characteristic curve through $X = X_1$, $t = t_1$. On this curve, third derivatives are unbounded. Consequently, the arguments concerning second-order accuracy are not applicable in the vicinity of the front of the wave. That is, in fact, suggested by Table I. The L_∞ error norm provides the maximum difference between the numerical solution and exact solution, and this maximum can be seen, in Fig. 2, to occur at meshes near the front of the wave. Table I shows that the L_∞ error norm is roughly halved when the mesh sizes are halved. This suggests that the method is only first-order accurate in the vicinity of the discontinuity.

To see why this should be so, we must return to the derivation of difference algorithms by Taylor series in Eq. 11. There, we assumed that derivatives of quantities were continuous through third-order. In our case, only first derivations are continuous, second derivatives are discontinuous but bounded, while third derivatives are unbounded. Consequently, Taylor series for $\psi^n_{j+1/2}$ and $\psi^n_{j-1/2}$ can only be carried out to

$$\psi^n_{j+1/2} = \psi^n_j + \left(\frac{\partial \psi}{\partial X}\right)^n_j \frac{\Delta X}{2} + R^+$$

$$\psi^n_{j-1/2} = \psi^n_j - \left(\frac{\partial \psi}{\partial X}\right)^n_j \frac{\Delta X}{2} + R^-$$

where remainders are given by

$$R^+ = \int_0^{\frac{1}{2}\Delta X} \frac{\partial^2}{\partial X^2} \psi\left((j+\tfrac{1}{2})\Delta X - \xi\right) \xi \, d\xi$$

$$R^- = \int_0^{-\frac{1}{2}\Delta X} \frac{\partial^2}{\partial X^2} \psi\left((j-\tfrac{1}{2})\Delta X - \xi\right) \xi \, d\xi$$

It may be noted that since second derivatives are bounded, the remainder R^+ is

$$R^+ = O(\Delta X^2) \leq \left\| \frac{\partial^2 \psi}{\partial X^2} \right\|_\infty \frac{\Delta X^2}{8}$$

where $\|\cdot\|_\infty$ is the maximum value of the second derivative over the interval $j\Delta X$ to $(j+\tfrac{1}{2})\Delta X$. A similar expression holds for R^-, except that $\|\cdot\|_\infty$ is the maximum over the interval $(j-\tfrac{1}{2})\Delta X$ to $j\Delta X$.

Subtracting the above two Taylor series provides the difference algorithm

$$\frac{\psi_{j+1/2}^n - \psi_{j-1/2}^n}{\Delta X} = \left(\frac{\partial \psi}{\partial X}\right)_j^n + R \tag{20}$$

where

$$R = O(\Delta X) \leq \left\| \frac{\partial^2 \psi}{\partial X^2} \right\|_\infty \frac{\Delta X}{4}$$

and $\|\cdot\|_\infty$ is now the maximum over the interval $(j-\tfrac{1}{2})\Delta X$ to $(j+\tfrac{1}{2})\Delta X$. Consequently, in the vicinity of a second-order discontinuity, the difference algorithms in Eqs. 12, 13 and 14 can indeed be expected to be only first-order accurate.

Most real problems in stress wave propagation involve even stronger discontinuities. Impacts, explosions, etc., produce shock waves in which quantities themselves are discontinuous. Reflection of shock waves at free boundaries produce rarefactions which contain discontinuities in first derivatives of quantities, called acceleration waves. Even for the sample problem above, stronger discontinuities will eventually form as the compressive portion of the smooth wave steepens due to material nonlinearities. On such stronger discontinuities, arguments based on Taylor expansions fail entirely.

If attempts are made to use the numerical method of Eqs. 12, 13 and 14 without modification on rarefaction waves and shock waves, severe oscillations appear in the numerical solution.

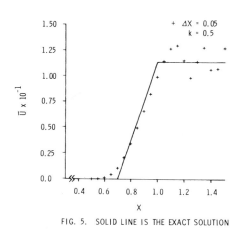

FIG. 5. SOLID LINE IS THE EXACT SOLUTION

Fig. 5 shows a comparison of a numerical solution with an exact solution for a homogeneous slab of material initially subjected to a strain $e_o = 0.047$ and internal energy $\bar{\mathcal{E}}_o = 0.00125$ but with a free boundary on the right hand side. A rarefaction propagating to the left imparts a velocity to the material which was initially at rest. Fig. 6 shows a numerical solution for the problem of a step change in velocity to $\hat{u} = 0.05$ on the left hand boundary of a slab which is initially at rest in the reference configuration. A shock wave propagates to the right into the material. The numerical oscillations do not indicate an instability. In both cases calculations with reduced mesh sizes yield smaller error norms, and the oscillations do not grow as the calculations are continued to longer times.

There are several ways in which this problem may be overcome. Since the basic assumptions of continuum mechanics are such that only a finite number of discrete discontinuities may appear in the solution, one obvious device is to introduce internal floating boundaries at discontinuities. Shock jump relations or acceleration wave equations[10] are introduced to join the smooth solutions on either side of the discontinuities. While this has not been attempted for acceleration waves, a number of codes have been written in which a technique of this type, called shock fitting, is used, for example that of Hicks and Holdridge.[11] When numerous shocks appear in the solution, the logic required in the computer code is complicated. However, this is the only method capable of very fine resolution in the vicinity of discontinuities.

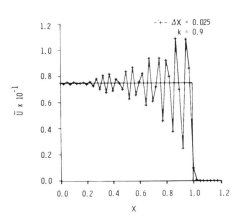

FIG. 6. SOLID LINE IS THE EXACT SOLUTION

ARTIFICIAL VISCOSITY

One means of overcoming problems associated with strong discontinuities is to replace the physical problem for which exact solutions admit discontinuities by an analogous physical problem for which exact solutions are smooth. Von Neumann and Richtmyer[3] proposed to do this by adding a viscous term, dependent on the strain rate, to the constitutive equation. Several forms of such artificial viscosity have been proposed. In general, these are all special cases of the constitutive equation for a thermoelastic material with viscosity of the rate type

$$\sigma = g(e, \dot{e}, \mathcal{E}, X) \tag{21}$$

where $\dot{e} = \partial e/\partial t$ is the strain rate.

The general behavior of materials governed by constitutive equations of this type has been explored in some detail.[10] It can be shown that discontinuities in first derivatives (i.e., acceleration waves) cannot propagate in such materials. It is surmised, although general proofs are lacking, that discontinuous shocks also cannot propagate. However, higher order discontinuities may be formed and propagated in these materials, in general. Consequently, it is expected that exact solutions will be continuous, have continuous first derivatives and bounded second derivatives when Eq. 21 is used.

While shocks are expected to be absent, solutions may, under certain circumstances, exhibit steady compressive waves,[10] that is, waves which do not alter their shape as they propagate. The steady wave thickness depends on the magnitude of the viscous coefficients, and exact steady wave solutions have been shown to converge to discontinuous shock solutions in the limit as the viscous coefficients go to zero. In the case of linear equations, it is easily shown that all viscous solutions converge to inviscid solutions when the viscosity coefficients are reduced to zero. On these bases it is surmised that the same is true in the general nonlinear case.

When a constitutive equation of the type of Eq. 21 is used with the conservation laws, Eqs. 2, the equations are no longer hyperbolic. The viscous terms introduce a diffusive or parabolic behavior. Existence and uniqueness proofs can still be given for the mixed initial/boundary value problem, and finite difference equations may again be developed in much the same way as before, except that some care is necessary in centering the strain rate term in the constitutive equation.

Von Neumann and Richtmyer[3] in effect introduced the specific constitutive equation

$$\sigma = f(e, \mathcal{E}, X) + \Lambda_1^2 \frac{\rho_0}{1-e} \frac{\partial e}{\partial t} \left| \frac{\partial e}{\partial t} \right| \tag{22}$$

NUMERICAL ANALYSIS METHODS

where f is the same function as in Eq. 3 and where Λ_1 is a viscous coefficient* with dimensions of length. By making the viscous term quadratic in the strain rate, the viscous stress is expected to be non-negligible only in regions of high strain rate, where discontinuities might be expected to form in the inviscid solution. Using the mass equation in Eqs. 2 this may be set into finite difference form as

$$\sigma_{j-1/2}^{n+1/2} = f\left(e_{j-1/2}^{n+1/2}, \mathcal{E}_{j-1/2}^{n+1/2}\right) - \left(\frac{\Lambda_1}{\Delta X}\right)^2 \frac{\rho_0}{1-e_{j-1/2}^{n+1/2}} \Delta u \, |\Delta u| \tag{23}$$

where $\Delta u = \left(u_j^{n+1/2} - u_{j-1}^{n+1/2}\right)$. When used with Eqs. 12 and 14, this again requires interpolation for e and $\mathcal{E}$ at $t^{n+1/2}$ but explicit equations result as before if f is linear in $\mathcal{E}$.

The stability arguments of Courant, Friedrichs and Lewy fails in this case, but a linearized stability analysis of the von Neumann type suggests that

$$k_c = \frac{C \Delta t}{\Delta X} \leq \frac{1}{\sqrt{1+\eta^2} + \eta} \tag{24}$$

where

$$\eta = \left(\frac{\Lambda_1}{\Delta X}\right)^2 \frac{|\Delta u|}{(1-e)\,C}$$

Since $\eta \geq 0$, the introduction of artificial viscosity reduces the allowable time step for stability.

An exact closed form solution for steady waves is easily obtained for the von Neumann-Richtmyer viscosity[3] when Eq. 17 is used for f. The steady wave is found to have a finite thickness given by $\ell = \Lambda_1 \pi \sqrt{2/(\xi+1)}$. For $-\ell/2 \leq X \leq \ell/2$ the solution is

$$u = \frac{u_f + u_i}{2} + \frac{u_f - u_i}{2} \sin\left(\pi \frac{X}{\ell}\right) \tag{25}$$

while for $X > \ell/2$, $u = u_i$ and for $X < -\ell/2$, $u = u_f$, u_i and u_f being the material particle velocities in front of and behind the wave respectively. Note that the steady wave solution contains discontinuities in second derivatives at $X = \pm \ell/2$.

---

* Note that the definition of Λ_1 differs slightly from that of von Neumann and Richtmyer.

A numerical solution of a steady wave is compared with the exact solution in Fig. 7. The solution has been non-dimensionalized, in this case, by defining $\overline{X} = X/\ell$ and $\overline{t} = t\, C_o/\ell$. It is seen that as few as five spatial meshes in the thickness of the wave provide good resolution.

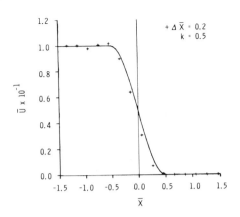

FIG. 7. SOLID LINE IS THE EXACT SOLUTION

Landshoff[12] proposed adding a viscous term which is linear in $\dot{e}$ in order to provide more smoothing of small numerical oscillations

$$\sigma = f(e, \mathcal{E}, X) + \Lambda_2\, \rho_o C\, \frac{\partial e}{\partial t} \tag{26}$$

where Λ_2 is a viscous coefficient with dimensions of length. When this viscosity is introduced, either alone or in combination with that of von Neumann and Richtmyer, steady wave solutions may still be obtained in closed form, but the forms are complicated. The steady wave solutions in these cases are analytic (and therefore have continuous derivatives to all orders) and approach u_i and u_f asymptotically.

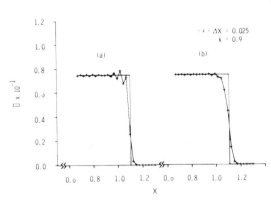

FIG. 8. SOLID LINE IS THE EXACT SOLUTION
(a) $\Lambda_1 = 0.7 \Delta X$ $\Lambda_2 = 0.05 \Delta X$
(b) $\Lambda_1 = 1.0 \Delta X$ $\Lambda_2 = 0.1 \Delta X$

Fig. 8 shows steady wave solutions, using both viscosities, compared to the inviscid shock solution (c.f. Fig. 6). In this case X has not been non-dimensionalized. These solutions clearly show the dependence of wave thickness on viscous coefficients. If Λ_1 and Λ_2 are small, then the wave is steep, but spurious

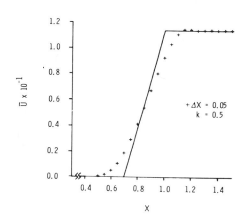

FIG. 9. SOLID LINE IS THE EXACT SOLUTION
$\Lambda_1 = 1.0\ \Delta X \qquad \Lambda_2 = 0.1\ \Delta X$

oscillations appear in the solution behind the wave.

In Fig. 9 is shown a numerical rarefaction wave solution* with both linear and quadratic viscosity compared to an exact solution (c.f. Fig.5). Further calculations show that if Λ_1 and Λ_2 are reduced, then the numerical solution is steepened, but spurious oscillations behind the wave are increased in this case also.

Clearly, the wave code user must choose the magnitude of the viscous coefficients in such a way as to compromise between errors introduced by the viscosity, and errors introduced by spurious oscillations. While the choice may vary with the application, the above examples provide a general rule of thumb: viscous coefficients should be chosen to spread shocks over about five spatial meshes to minimize errors.

It is useful to consider the numerical solution in the nonlinear case by analogy with a linearized Fourier analysis. Viscous dissipation essentially leads to attenuation and dispersion of high frequency components in the solution. We have seen from the numerical examples of Figs. 2 and 8 that modes with a wavelength less than about ten mesh widths are seriously falsified in the solution by the numerical method. Artificial viscosity therefore serves to limit spurious numerical errors by eliminating higher frequency modes from the solution.

The question of convergence of the numerical solution to the exact inviscid solution may be considered in two parts when artificial viscosity is introduced. One may first inquire into the convergence of the numerical solution to the exact viscous solution, with the viscous coefficients Λ_1 and Λ_2 held constant. Assuming that the exact solution is sufficiently smooth, the previous discussion on convergence applies. However, since jump discontinuities

---

* In both Fig. 8 and Fig. 9 large mesh sizes have been used to illustrate the effects of viscosity. In practice, very much smaller mesh sizes would be used to reduce errors.

in second derivatives may occur in the exact solution, convergence may only be first-order near these discontinuities. Secondly, one may inquire into the convergence of the exact viscous solution to the exact inviscid solution as the viscosity coefficients Λ_1 and Λ_2 are reduced. Such convergence is conjectured, in general. If this conjecture is true, then by making Λ_1 and Λ_2 proportional to ΔX, the numerical solution will converge to the inviscid exact solution as ΔX and Δt jointly go to zero. If the linear viscosity, Eq. 26, is included, this convergence can be only first-order, while if only quadratic viscosity, Eq. 22, is used, convergence can be at most second-order. Since second-order discontinuities are admissible even when viscosity is used, we are led to question the utility of employing difference algorithms of order higher than the second for stress wave problems.

OTHER DIFFERENCE METHODS

So far, all of the numerical examples have involved the von Neumann-Richtmyer difference equations. Many other effective numerical techniques have been developed; a few of the more important will be discussed briefly here.

Lax and Wendroff[13] proposed a method based on the differential equations in conservation law form, Eq. 1. Richtmyer[8] subsequently introduced modifications to render the difference equations simpler to apply, and arrived at the so-called Lax-Wendroff two-step method. It may be noted that each of the conservation laws in Eqs. 1 relates the time rate of change of a quantity $\underline{U}$ to the gradient of its flux $\underline{F}$

$$\frac{\partial}{\partial t}\left(\underline{U}\right) + \frac{\partial}{\partial X}\left(\underline{F}(\underline{U})\right) = 0 \qquad (27)$$

where in our case $\underline{U}$ is the vector $\{\rho_0 e,\ \rho_0 u,\ (\tfrac{1}{2}\rho_0 u^2 + \rho_0 \mathcal{E})\}$ and $\underline{F}$ is the vector $\{\rho_0 u,\ \sigma,\ \sigma u\}$. Noting that σ is a function of e and $\mathcal{E}$ through the constitutive equation, it is obvious that $\underline{F} = \underline{F}(\underline{U})$. Eqs. 27 are set into finite difference form as

$$\begin{aligned}
\underline{U}_{j-1/2}^{n+1/2} &= \tfrac{1}{2}\left(\underline{U}_{j+1}^{n} + \underline{U}_{j}^{n}\right) - \frac{\Delta t}{2\Delta X}\left(\underline{F}_{j}^{n} - \underline{F}_{j-1}^{n}\right) \\
\underline{U}_{j}^{n+1} &= \underline{U}_{j}^{n} - \frac{\Delta t}{\Delta X}\left(\underline{F}_{j+1/2}^{n+1/2} - \underline{F}_{j-1/2}^{n+1/2}\right)
\end{aligned} \qquad (28)$$

where $\underline{U}_{j-1/2}^{n+1}$ is a provisional value used to evaluate $\underline{F}_{j-1/2}^{n+1/2}$.

The time integration essentially employs a two-step Runge-Kutta method, and an error analysis shows that the scheme is second-order accurate. Boundary conditions pose somewhat of a problem, and first-order accurate boundary schemes are sometimes used for convenience in computation.

One advantage of beginning with the partial differential equations in conservation law form is that the difference equations exactly conserve total mass, momentum and energy in a problem. By contrast, the von Neumann-Richtmyer scheme is not exactly conservative, and errors in total energy may accumulate as the calculation proceeds.

Another advantage is that consistent finite difference forms of the equations in conservation law form reduce exactly to the shock jump relations. Consequently, it is considered that schemes of the Lax-Wendroff type should handle discontinuous solutions without the need of an artificial viscosity. Calculations at shocks in fact produce results closely resembling those of Fig. 8a. An examination of numerical solutions shows that Lax-Wendroff methods lead to attenuation of high frequency components in solutions in much the same way as methods employing artificial viscosity. Richtmyer[8] has shown that the Lax-Wendroff difference equations, in effect, contain dissipative terms of fourth-order. Lax and Wendroff[13] introduced added dissipation in the form of an artificial viscosity in order to reduce spurious oscillations, and it is found, in fact, that such added dissipation is needed to prevent numerical oscillations in certain cases. When such viscous terms are added, methods of the Lax-Wendroff type behave very much like those of the von Neumann-Richtmyer type, and the previous rough rules of thumb appear to apply.

Higher order methods are easily generated using multi-step Runge-Kutta time integrations and higher order spatial gradient algorithms. For the reasons stated previously, however, it seems unlikely that these offer advantages for stress wave problems, and they have not been used extensively.

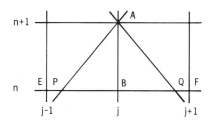

FIG. 10. SCHEMATIC OF CIR DIFFERENCE MESH

Courant, Isaacson and Rees[14] proposed a quite different approach based on the differential equations in characteristic form, Eq. 9. A typical first-order method of this type may be described qualitatively with reference to Fig. 10. First-order difference algorithms are used for the ordinary differential equations along characteristics PA, BA and QA, with the sound speed C evaluated at point B. Positions P and Q are found from the sound speed evaluated at B, and values of σ and u at P and Q are found by linear interpolation between their values at E,B and B,F respectively. If the resulting difference equations are combined with a first-order difference representation of Eq. 9, then they may be put into the form

$$\rho_o \left(\frac{u_j^{n+1} - u_j^n}{\Delta t} \right) + \left(\frac{\sigma_{j+1}^n - \sigma_{j-1}^n}{2\Delta X} \right) = \rho_o C_j^n \left(\frac{u_{j+1}^n - 2u_j^n + u_{j-1}^n}{2\Delta X} \right)$$

$$\left(\frac{e_j^{n+1} - e_j^n}{\Delta t} \right) + \left(\frac{u_{j+1}^n - u_{j-1}^n}{2\Delta X} \right) = \frac{1}{\rho_o C_j^n} \left(\frac{\sigma_{j+1}^n - 2\sigma_j^n + \sigma_{j-1}^n}{2\Delta X} \right) \qquad (29)$$

$$\rho_o \left(\frac{\mathcal{E}_j^{n+1} - \mathcal{E}_j^n}{\Delta t} \right) - \sigma_j^n \left(\frac{e_j^{n+1} - e_j^n}{\Delta t} \right) = 0$$

It is easily seen that these equations involve terms which are dissipative. The smoothing action of the internal dissipation depends on CFL number, and it is found that the dissipation increases rapidly as the CFL number is reduced.

One advantage of the method is that difference equations at boundaries and interfaces are generated in a natural way with accuracy comparable to those at an interior mesh point. The method lends itself to shock fitting, since the equations to be solved at a discontinuity can be obtained directly from the characteristic equations and the shock jump relations. The method is, perhaps, most useful in this connection. If shock fitting is not used, an artificial viscosity must be introduced to smooth strong discontinuities, and methods of this type again behave like those of the von Neumann-Richtmyer type.

Second-order accurate methods can easily be generated, using the same approach, for example that of Hartree.[15] The difference equations along characteristics can be centered by using interpolated values of sound speed between A,P and A,Q respectively, and quadratic interpolation can be used for quantities at P and Q. This renders the equations implicit, and it is usually necessary to resort to numerical iteration at each meshpoint. The increased computer time required for each meshpoint calculation largely offsets the gain in accuracy in many cases.

Another class of methods uses implicit time difference algorithms in order to ease restrictions on the time step size imposed by stability requirements. One such scheme[16] uses second-order central difference algorithms in Eqs. 2

$$\left(\frac{e_j^{n+1} - e_j^n}{\Delta t}\right) + \left(\frac{u_{j+1}^{n+1} + u_{j+1}^n - u_{j-1}^{n+1} - u_{j-1}^n}{4\Delta X}\right) = 0$$

$$\rho_0 \left(\frac{u_j^{n+1} - u_j^n}{\Delta t}\right) + \left(\frac{\sigma_{j+1}^{n+1} + \sigma_{j+1}^n - \sigma_{j-1}^{n+1} - \sigma_{j-1}^n}{4\Delta X}\right) = 0 \quad (30)$$

$$\rho_0 \left(\frac{\mathcal{E}_j^{n+1} - \mathcal{E}_j^n}{\Delta t}\right) - \left(\frac{\sigma_j^{n+1} + \sigma_j^n}{2}\right) \left(\frac{e_j^{n+1} - e_j^n}{\Delta t}\right) = 0$$

together with a constitutive equation containing an artificial viscosity of the von Neumann-Richtmyer type. This set of equations is implicit, since quantities at X_{j+1}, X_j and X_{j-1} appear simultaneously at t^{n+1}, and iterative sweeps over all spatial meshes are required in order to obtain a solution. While it is not obvious from Eqs. 30, these particular equations have the advantage of being exactly conservative.

A linearized stability analysis of the von Neumann type suggests that there is no restriction on the time step size, and that the method is unconditionally stable in the sense of von Neumann. However, an investigation of the conditions under which numerical solutions remain bounded leads to restrictions on the time step size which essentially come from a necessary limitation on the growth of quantities from one cycle to another. In fact, it is found that growth restrictions required to preserve accuracy in the solution lead to a time step restriction comparable to that of Courant, Friedrichs and Lewy. Consequently, implicit unconditionally stable schemes of this sort offer little, if any, advantage for one-dimensional problems in which the solution is changing rapidly with time. However, they offer an advantage for one-dimensional problems with solutions which are changing slowly with time. The use of such schemes in multi-dimensional problems, where they promise distinct advantages, will be mentioned later.

There are, of course, a great many numerical methods which have been developed besides the few examples which have been given. The absence of error bounds for any of these methods has fostered lengthy rhetoric concerning the relative merits of various methods. From the viewpoint of the wave code user, the elegance of the derivations of various methods is not a real issue. The wave code user is primarily concerned with the computer time required to produce an approximate numerical solution with acceptable accuracy.

Extensive testing has been carried out in an effort to identify the most efficient numerical methods for particular types of problems.

There are many factors which affect the relative efficiency of numerical methods. First of all, it is the total computer time required to produce a solution which is of interest to the wave code user. Therefore, a given numerical method cannot be considered separately from the computer code written to implement it. The skill of the programmer creating the computer software can have a marked effect on computational efficiency. For very large problems, the specific computer hardware configuration and systems software may also have an important influence. Secondly, the particular type of problem of interest may affect the results. Some methods are more efficient when solutions are relatively smooth. Shock fitting techniques have a great advantage when details are required near a discontinuity. Finally, such factors as the choice of viscosity coefficients and CFL number obviously affect the accuracy of the solution, and hence computational efficiency.

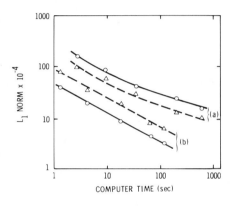

FIG. 11. COMPUTATIONAL EFFICIENCY COMPARISON
- ─○─ VON NEUMANN RICHTMYER
- ─△─ LAX WENDROFF
- (a) SHOCK WAVE PROBLEM
- (b) RAREFRACTION WAVE PROBLEM

Fig. 11 shows some sample results[17] which are typical of many tests of various computer codes. The L_1 error norm relative to a known exact solution is shown as a function of computer time for two codes and two simple problems. Fig. 11 indicates that the code based on the Lax-Wendroff method is more efficient for a shock wave problem, but the von Neumann-Richtmyer code is more efficient for a problem involving a rarefaction wave. However, the Lax-Wendroff code was run without added viscosity, while the viscous coefficients used in the von Neumann-Richtmyer code were quite large. The results for the two codes can easily be reversed by different choices of viscous coefficients.

The above results are typical. It is found that codes based on some methods can be made to approximate certain exact solutions with high efficiency if viscous coefficients, CFL number, and other factors are "tuned" appropriately. However, if "tuning" is disallowed, then no one method appears to have any significant advantage for a broad variety of problems.

SPECIAL TECHNIQUES

There are a number of things which can be done to realize very large savings in computer time for certain types of problems. One of these, involving shock fitting, has already been mentioned. In problems where fine resolution is required near discontinuities, dissipative methods which smear the discontinuity may require prohibitively small mesh sizes. Shock fitting may allow very accurate solutions to be obtained at the discontinuity with mesh sizes appropriate to the accuracy required in the smooth portions of the motion.

One method, which may lead to very large increases in computational efficiency for certain problems, dispenses with a regular difference mesh on lines of constant X and t in favor of a computational mesh on characteristic lines. The characteristic equations, Eqs. 9, are set into finite difference form, but interpolation to regular mesh points in X,t space is not used. Rather, positions where characteristic lines intersect each other, or reflect at shocks, interfaces and boundaries are calculated explicitly, and used as a basis for the computational mesh. Such a characteristic mesh is diagrammed in Fig. 12. In this problem a jump increase in velocity at $t = 0$ followed by a jump decrease in velocity at $t = t^*$ is applied to the left-hand boundary of a homogeneous slab of material. Computations are only performed at points where the initial shock reaches the right-hand stress-free boundary, and where the discretized characteristic lines in the ensuing rarefaction waves cross each other or reflect at boundaries. Barker[18] has developed a particularly versatile computer code based on this method. He noted that finite difference forms of Eqs. 9 can be centered in such a way that the equations coincide with the shock jump relations. Characteristics can then be treated as weak shocks, greatly reducing the number of types of wave interactions which must be accommodated. Special steps are required to place a lower limit on the characteristic mesh size in the computation. When compared with conventional finite difference codes, this method is capable of computer time reductions of two orders of magnitude or more for comparable accuracies when the solution contains large regions in uniform equilibrium, as in Fig. 12.

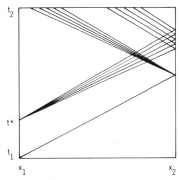

FIG. 12. CHARACTERISTIC DIFFERENCE MESH

Returning to difference methods based on a regular computational mesh in X,t space, computer time can often be saved without compromising accuracy by using non-uniform mesh sizes. Small meshes may be introduced in regions of X,t space where fine resolution is needed in the solution, larger meshes being used elsewhere. It must be remembered that the use of non-uniform mesh sizes introduces error terms of first-order. Consequently, minimum mesh sizes may have to be reduced to maintain accuracy if methods based on Eq. 11 are used, somewhat offsetting the gains to be made in computational efficiency by the use of this technique.

A limitation to the use of non-uniform mesh sizes is that regions of the solution requiring fine resolution usually propagate as time proceeds. What is needed is a technique which dynamically reconfigures the finite difference mesh as the computation proceeds, so that small mesh sizes are always provided where needed. Such a feature, generally termed dynamic rezoning, has been implemented in several computer codes. In one method, large spatial meshes are subdivided and small meshes are combined as needed to provide appropriate resolution. Mason and Thorne[19] developed a very simple idea which makes the method feasible. In their method, the ratio of ΔX in adjacent meshes can never exceed a factor of two. Imposing this "two-to-one" rule forces subdivision of meshes well ahead of propagating waves containing large gradients, and delays recombination of meshes behind such waves without the necessity of providing logic to determine in which direction the waves are propagating. Consequently, the method can be applied to virtually any problem amenable to solution by numerical means. A sample solution, showing two steep rarefaction waves approaching each other, is shown in Fig. 13, in which the variations in spatial mesh size are obvious. For a wide variety of problems, computer time savings approaching an order of magnitude may be realized, compared to methods using a fixed spatial mesh.

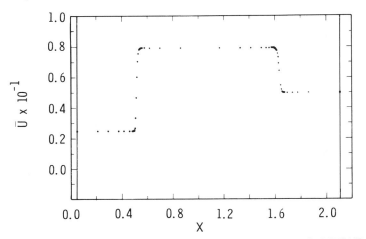

FIG. 13. NUMERICAL SOLUTION WITH DYNAMIC REZONE

OTHER CONSTITUTIVE RELATIONS

All of the previous discussion has centered on the solution of problems involving simple thermoelastic materials. This description includes gas dynamics as well as the historically important one of "hydrodynamics" in which solid materials are treated as if they were compressible liquids. We have already noted how viscosity of the rate type may be handled, whether artificial or physical. However, our knowledge of material behavior has long since progressed to the point where other types of constitutive laws must be considered.

An important class of materials can be described by elastic-plastic constitutive relations. This includes structural metals[20] and ductile porous materials.[21] Provided that the constitutive equation is rate independent, and that the materials are elastically compressible, the character of the governing equations is not changed. In the simplest case of perfect plasticity (constant yield stress) the solution is divided into elastic and plastic regions, each of which is described by equations of form identical to those in Eqs. 2, 3 and 4 above. Some added logic is required to determine whether the material in a given mesh is elastic or plastic in order to provide, in effect, the correct secant modulus. If strain hardening is included, internal state variables representing strain hardening parameters must be provided. The material in each mesh may have a different effective secant modulus, depending on their values. However, the forms of the partial differential equations are not effectively altered.

When time derivatives representing relaxation processes are introduced into the constitutive equations, then the equations are altered, and some care is required in developing numerical solution methods. We will illustrate this point with one example. Many materials are described by constitutive relations of the general form

$$\sigma = f(e, \dot{e}, \underset{\sim}{\alpha}) \qquad (31)$$

where α is the vector whose components α_i, $i = 1, 2 \cdots n$ are internal state variables whose evolution is governed by differential equations of the form

$$\frac{\partial \alpha_i}{\partial t} = g_i(e, \dot{e}, \underset{\sim}{\alpha}) \qquad (32)$$

Typical internal state variables of this type might be the plastic strain in viscoplasticity, history functions in viscoelasticity, extents of reaction or mass fractions in theories involving phase changes, chemical reactions or relaxation of excited states.

In order to provide a specific example, we will omit dependence on thermodynamic quantities (in this case on $\mathcal{E}$), and restrict attention to a single internal state variable α. If Eq. 31 is invertible in α, then differentiating Eq. 31 with respect to t and setting

$$F(\sigma,e) = \tilde{F}\bigl(e, \alpha(\sigma,e)\bigr) = \partial f(e,\alpha)/\partial e$$
$$G(\sigma,e) = \tilde{G}\bigl(e, \alpha(\sigma,e)\bigr) = g(e,\alpha)\, \partial f(e,\alpha)/\partial \alpha$$

we obtain the constitutive equation

$$\frac{\partial \sigma}{\partial t} = F(\sigma,e) \frac{\partial e}{\partial t} + G(\sigma,e) \tag{33}$$

This is the general form of the Maxwell equation, which has been used successfully to describe rate-dependent metals[20] and porous materials.[21]

With this constitutive relation, the governing equations, Eq. 2 and 33, are found to be hyperbolic, and the usual existence and uniqueness proofs can be given for solutions to the initial/boundary value problem defined by Eq. 4. Perhaps the most convenient method of numerical solution for this case is one of the type of Courant, Isaacson and Rees,[14] using first-order forward difference algorithms and shock fitting. Such a method is capable of excellent results.[22] Fig. 14 shows a sample solution for the case when F is constant and $G = \sigma/\tau$ where τ is a relaxation time. In this case the equations take the form of the telegraph equation for which an exact solution can be found. The numerical solution using the Courant-Isaacson-Rees method cannot be distinguished from the exact solution in Fig. 14.

The von Neumann-Richtmyer method can also be used in the present case, but care is necessary in order to construct a difference equation for the constitutive relation, Eq. 33. Strictly, one should use second-order centered difference algorithms.

$$\left(\frac{\sigma_{j-1/2}^{n+1} - \sigma_{j-1/2}^{n}}{\Delta t}\right) = F \left(\frac{e_{j-1/2}^{n+1} - e_{j-1/2}^{n}}{\Delta t}\right) + G \tag{34}$$

where F, G are calculated from σ and e interpolated between their values at t^n and t^{n+1}. Together with Eqs. 12, this choice leads to a set of implicit equations which require time consuming iterative solution if F and G are complicated functions.

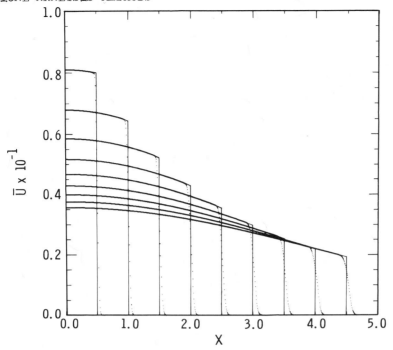

FIG. 14. THE SOLID LINES SHOW THE EXACT SOLUTION AND COURANT ISAACSON REES METHOD WITH SHOCK FITTING. THE DOTS REPRESENT THE VON NEUMANN RICHTMYER METHOD RESULTS.

An alternative is to use first-order forward differences in Eq. 34 by evaluating F and G from $\sigma^n_{j-1/2}$ and $e^n_{j-1/2}$. The difference equations are then explicit, but contain error terms of $O(\Delta t)$. Second-order accuracy in ΔX is retained. Consequently, the spatial mesh size may not need to be reduced, but the time step size may require reduction, in order to retain accuracy comparable to a second-order accurate method. If G is a strongly nonlinear function of its arguments, such as an exponential function, then the time step size may need to be reduced drastically.

One way of retaining the advantage of explicit forward differences in Eq. 34, but ameliorating the need for small time steps is to use subcycling. The entire calculation is advanced by the largest time step allowed by stability. Meshes in which changes occurred in F and G which are too large to retain accuracy are then subcycled, i.e., calculations are repeated in these meshes using smaller time steps. If error terms are retained in Eq. 34, then a criterion for the maximum subcycle time step required to retain accuracy is easily found.[23] In most problems, subcycling is found to be necessary at only a few meshes at each time step, and this method is found to require much less computer time, for given accuracy, than a second-order method using iterative solution, or a first-order method using a uniformly small time step.

Since solutions to the equations based on Eq. 33 admit strong discontinuities, artificial viscosity must be added if shock fitting is not used. A solution, using the von Neumann-Richtmyer method with subcycling and artificial viscosity is also shown in Fig. 14. The usual effect of artificial viscosity at the discontinuity is evident, but the numerical solution is indistinguishable from the exact solution elsewhere. For many problems, fine resolution is not needed at shocks, and in this case this method is suitable.

MULTI-DIMENSIONAL METHODS

Finally, we will connect the remarks we have made with regard to one-dimensional methods to methods designed for more than one space dimension. All of the methods we have mentioned can be extended directly to two and three dimensions. For example, Wilkins has extended the method of von Neumann and Richtmyer to the two-dimensional difference equations which form the basis of such widely used codes as HEMP[24] and TOODY.[25] Further obvious extensions have been made to three dimensions. Richtmyer[8] has provided a straightforward extension of the Lax-Wendroff two-step method to two dimensions. Characteristic methods may also be extended to higher dimensions.

One direct means of extending one-dimensional difference methods to higher dimensions involves operator splitting. We may express a set of first-order difference equations in the general operator form

$$\underline{U}^{n+1} = \underline{U}^n + \Delta t \, D[\underline{U}^n] \tag{35}$$

where $\underline{U}$ is the vector of dependent variables, and $D[\cdot]$ is a difference operator. Godunov and Bagrinovskii[26] noted that the three-dimensional difference operator can be decomposed additively into three terms each involving differences in only one coordinate direction.

$$D[\cdot] = D_x[\cdot] + D_y[\cdot] + D_z[\cdot]$$

They therefore approximated Eq. 35 by the three separate equations

$$\begin{align}
\underset{\sim}{U}^{(1)} &= (I + \Delta t\, D_x)\, [\underset{\sim}{U}^n] \\
\underset{\sim}{U}^{(2)} &= (I + \Delta t\, D_y)\, [\underset{\sim}{U}^{(1)}] \\
\underset{\sim}{U}^{n+1} &= (I + \Delta t\, D_z)\, [\underset{\sim}{U}^{(2)}]
\end{align} \qquad (36)$$

where I is the identity operator and where $\underset{\sim}{U}^{(1)}$ and $\underset{\sim}{U}^{(2)}$ are intermediate values. These three equations are equivalent to

$$\underset{\sim}{U}^{n+1} = \underset{\sim}{U}^n + \Delta t(D_x + D_y + D_z)\,[\underset{\sim}{U}^n] + O(\Delta t^2) \qquad (37)$$

Each of the operations in Eqs. 36 is equivalent to a one-dimensional calculation in the X, Y and Z directions respectively. The method therefore may extend any one-dimensional method to higher dimensions, to first-order accuracy. Gottlieb[27] has shown how operator splitting may be made second-order accurate by appropriately alternating the directions of the one-dimensional calculations, provided the one-dimensional algorithms are at least second-order accurate.

One very promising use of operator splitting involves the use of the unconditionally stable implicit method of Eqs. 30. Since many multi-dimensional problems involve large regions which are relatively quiescent, it is feasible to use a large time step with subcycling in regions where changes are occurring. Tests have indicated reductions in computer time of one or two orders of magnitude may be realized, compared to calculations using a uniform time step.

The above remarks should make it clear that all of the previous discussion of one-dimensional methods carries over directly to higher dimensions. The qualitative rules of thumb regarding resolution of the difference methods apply. The chief difficulty concerns the provision of sufficiently many meshes to provide desired accuracy, and particular care is usually necessary in the construction of computer codes to optimize data flow, handling and storage to minimize computer time.

The urgency of reducing computer time in higher dimensional methods places special emphasis on such features as dynamic rezoning. However, there is another factor which makes such features necessary. The increased freedom from geometric restraint may allow large distortions to develop. In the Lagrangian methods which we have been considering, the computational mesh distorts with the material, and may become so disordered as to make the computation meaningless. One means of alleviating this problem is the provision of sliding interfaces,[28] i.e., mesh lines which allow the material (and computational mesh) on one side to shear or flow over that on the other side. However, more effective means are usually necessary to prevent the computational mesh from becoming disordered. Dynamic rezoning,[29] the use of arbitrarily convected computational meshes[30] and the use of Eulerian computational meshes [31,32] have all been implemented with varying degrees of success. It is beyond the scope of this paper to consider these techniques here.

CLOSURE

We have discussed a few of the considerations in the numerical solution of stress wave problems. Numerous numerical schemes have been proposed for one-dimensional problems. Among those which give good results, it is found that particular methods are best suited to problems of a particular kind. No wave code has been found which is a panacea. That is, on the basis of extensive testing, no one of these methods has emerged as significantly superior in performance, as measured in terms of computer time to achieve a given accuracy, when a broad spectrum of problem types is considered. Rather, computational time can be considerably reduced in many problems by special techniques such as shock fitting, characteristic differencing and dynamic rezoning.

One-dimensional methods may be extended to multi-dimensional techniques. Most of the features and limitations of the one-dimensional techniques are carried over intact. That this is so is especially evident when operator splitting is considered. Substantial gains appear possible in efficiency of multidimensional methods, and further work in their development is promising. Many unsolved problems and open questions still remain, however.

REFERENCES

1. R. Courant and D. Hilbert
   Methods of Mathematical Physics, Vol. 2
   Interscience Publishers, New York (1962)

2. P. R. Garabedian
   Partial Differential Equations
   John Wiley & Sons, Inc., New York (1964)

3. J. von Neumann and R. D. Richtmyer
   A Method for the Numerical Calculation of Hydrodynamic Shocks
   J. Appl. Phys., 21; 232 (1950)

4. R. N. Brodie and J. E. Hormuth
   The PUFF 66 and P PUFF 66 Computer Programs
   Air Force Weapons Laboratory, AFWL-TR-66-48, May 1966

5. W. Herrmann, P. Holzhauser and R. J. Thompson
   WONDY A Computer Program for Calculating Problems of Motion in One Dimension
   Sandia Laboratories, SC-RR-66-601, Feb. 1967

6. R. Courant and K. O. Friedrichs
   Supersonic Flow and Shock Waves
   Interscience Publishers, New York (1948)

7. R. Courant, K. O. Friedrichs and H. Lewy
   Über die Partiellen Differenzengeichungen der Mathematischen Physik
   Math. Ann, 100; 13 (1928)

8. R. D. Richtmyer and K. W. Morton
   Difference Methods for Initial-Value Problems
   Interscience Publishers, New York (1967)

9. W. G. Strang
   Accurate Partial Difference Methods II: Non-linear Problems
   Numer. Math, 6; 37 (1964)

10. W. Herrmann and J. W. Nunziato
    Nonlinear Constitutive Equations
    in "Dynamic Behavior of Materials under Impulsive Loading" ed. P. C. Chou (in press)

11. D. L. Hicks and D. B. Holdridge
    The CONCHAS Wavecode
    Sandia Laboratories, SC-RR-72-0451, Sept. 1972

12. R. Landshoff
    A Numerical Method for Treating Fluid Flow in the Presence of Shocks
    Los Alamos Scientific Laboratory, LA-1930, Jan. (1955)

13. P. D. Lax and B. Wendroff
    Systems of Conservation Laws
    Comm. Pure Appl. Math, 13; 217 (1960)

14. R. Courant. M. Isaacson and M. Rees
    On the Solution of Nonlinear Hyperbolic Differential
    Equations by Finite Differences
    Comm. Pure Appl. Math, 5; 243 (1952)

15. D. R. Hartree
    Some Practical Methods of Using Characteristics in
    the Calculation of Non-Steady Compressible Flow
    Atomic Energy Commission, AECU-2713, Sept. 1953

16. D. L. Hicks
    Analysis of CAVIL, A Conservative Artificial
    Viscosity Implicit and Lagrangean Wavecode
    Sandia Laboratories, SC-RR-71 0863, July 1972

17. B. J. Thorne and D. A. Dahlgren
    A Comparison of Numerical Techniques for Wave
    Propagation Problems in Solids
    Sandia Laboratories, SC-RR-70-571, Nov. 1971

18. L. M. Barker
    SWAP-9: An Improved Stress Wave Analyzing Program
    Sandia Laboratories, SC-RR-69-233, Aug. 1969

19. D. S. Mason and B. J. Thorne
    A Preliminary Report Describing the Rezoning Features
    of the WONDY IV Program
    Sandia Laboratories, SC-DR-70-146, March 1970

20. W. Herrmann
    Nonlinear Stress Waves in Metals
    in "Wave Propagation in Solids" ed. J. Miklowitz
    The American Society of Mechanical Engineers (1969)

21. W. Herrmann
    Constitutive Equations for Compaction of Porous
    Materials
    in "Applied Mechanics Aspects of Nuclear Effects in
    Materials" ed. C. C. Wan
    The American Society of Mechanical Engineers (1971)

22. W. Herrmann, D. L. Hicks and E. G. Young
    Attenuation of Elastic-Plastic Stress Waves
    in "Shock Waves and the Mechanical Properties of
    Solids" ed. J. J. Burke and V. Weiss
    Syracuse University Press (1971)

23. W. Herrmann, R. J. Lawrence and D. S. Mason
    Strain Hardening and Strain Rate in One-Dimensional
    Wave Propagation Calculations
    Sandia Laboratories, SC-RR-70-471, Nov. 1970

24. M. L. Wilkins
    Calculation of Elastic Plastic Flow
    in "Methods in Computational Physics, Vol. 3
    Fundamental Methods in Hydrodynamics"
    ed. B. Alder, S. Fernbach and M. Rotenberg
    Academic Press, New York (1964)

25. B. J. Thorne and W. Herrmann
    TOODY, A Computer Program for Calculating Problems
    of Motion in Two Dimensions
    Sandia Laboratories, SC-RR-66-602, July 1967

26. S. K. Gudunov and K. A. Bagrinovskii
    Difference Schemes for Multidimensional Problems
    Doklady Akad Nauk, 115; 431 (1957)

27. D. Gottlieg
    Strang-type Difference Schemes for Multidimensional
    Problems
    Siam J. Numer. Anal., 9; 650 (1972)

28. L. D. Bertholf and S. E. Benzley
    TOODY II--A Computer Program for Two-Dimensional
    Wave Propagation
    Sandia Laboratories, SC-RR-68-41, Nov. 1968

29. B. J. Thorne and D. Holdridge
    The TOOREZ Lagrangian Rezone Code
    Sandia Laboratories, SC-RR, in press

30. R. T. Walsh
    Finite Difference Methods
    in "Dynamic Response of Materials under Impulsive
    Loading" ed. P. C. Chou, in press

31. F. H. Harlow
    The Particle-in-cell Computing Method for Fluid
    Dynamics
    in "Methods in Computational Physics, Vol. 3,
    Fundamental Methods in Hydrodynamics" ed. B. Alder,
    S. Fernbach and M. Rotenberg
    Academic Press, New York (1964) p. 319

32. W. E. Johnson
    OIL - A Continuous Two-Dimensional Hydrodynamic Code
    General Atomic/General Dynamics, GAMD-5580, Oct. (1964)

CONTINUUM PLASTICITY IN RELATION TO MICROSCALE DEFORMATION MECHANISMS

James R. Rice

Division of Engineering, Brown University

Providence, R. I. 02912

In principle, constitutive laws for continuum plasticity are derivable from quantitative descriptions of deformation mechanisms, such as dislocation motion, operative on the microscale. The present paper reviews progress on rigorous approaches to this micro-macro transition problem. First, a general thermodynamic framework is established by which the macroscopic plastic strain tensor is related to corresponding structural rearrangements, such as slip, taking place locally on the microscale of a heterogeneous sample of material (e.g., a polycrystal).

Next, kinetic relations governing the rates of these rearrangements are considered, together with second law restrictions upon them. Typical kinetic relations for slip entail that at a given plastically deformed state of a crystalline element, the instantaneous velocity of dislocation motion on any particular slip system is stress-state dependent only through the shear stress component exerted on that system. This feature (or, more generally, the feature that the rate of any particular structural rearrangement is governed by its conjugate thermodynamic 'force') leads to a remarkable 'normality' structure for macroscopic constitutive laws: a scalar 'flow potential' exists, dependent on the macroscopic stress state and on the prior plastic deformation, so that components of the macroscopic plastic strain rate are given by its derivatives on corresponding stress components.

In the final section, some attempts at specific predictions of the macroscopic behavior of polycrystals are reviewed. These range from detailed 'self-consistent model' calculations to simpler approaches which are merely motivated by microscale considerations.

A GENERAL FRAMEWORK FOR ELASTIC-PLASTIC BEHAVIOR

The aim in this part of the paper is to present a general framework by which microstructural deformation mechanisms are related to macroscopic plastic straining. The formalism is wide enough to include crystalline slip, which will be the mechanism of primary concern here, and also phase transformations, twinning, diffusional transport, etc. For maximum generality, a collection of "internal variables" are introduced to describe the local, microstructural rearrangements of a material sample by such mechanisms. The approach followed is due to Rice[1]; it is related to, and provides a unified setting for similar general studies by Havner[2], Hill[3], Kestin and Rice[4], Lin[5], Mandel[6], Rice[7], and Zarka[8].

Consider a representative macroscopic sample of material, having volume V is an unloaded reference state. This is subjected to boundary loadings causing macroscopically homogeneous deformation; σ_{ij} and ε_{ij} are the macroscopic stresses and strains thus induced, and these are supposed to satisfy

$$V\sigma_{ij}d\varepsilon_{ij} = \text{work increment of boundary loadings} \qquad (1)$$

Here we consider infinitesimal strain and isothermal behavior; neither simplification is essential and results for the general case[1] will be noted when fruitful.

The material sample may deform by: (i) elastic stretching of lattice bonds, and (ii) local microstructural rearrangements of its constituent elements by slip, etc. Let H denote symbolically the current pattern of microstructural rearrangement; this pattern is due to the prior history of inelastic deformation experienced by the sample, and H may equally be thought of as representing this history. The free energy Φ of the sample depends on ε and H. When processes at fixed H are considered, we have

$$V\sigma_{ij}d\varepsilon_{ij} = [d\Phi]_{H \text{ constant}}, \text{ and } \sigma_{ij} = \frac{1}{V}\frac{\partial\Phi(\varepsilon,H)}{\partial\varepsilon_{ij}}. \qquad (2)$$

Alternatively, if we introduce a dual potential

$$\Psi = V\sigma_{ij}\varepsilon_{ij} - \Phi, \text{ then } \varepsilon_{ij} = \frac{1}{V}\frac{\partial\Psi(\sigma,H)}{\partial\sigma_{ij}} \qquad (3)$$

Consider two neighboring patterns of microstructural rearrangement denoted by H, H+dH. We suppose that a set of incremental internal variables $d\xi_1, d\xi_2, \ldots, d\xi_n$ characterize the specific local rearrangements, which are represented collectively by dH, at sites throughout the sample. The required number of such variables increases in proportion to the size of the sample. Indeed, we shall see that an essentially infinite number of continuous variables defined piecewise throughout the sample is required for, say, a description of crystalline slip[1], but the structure of the theory is made evident more simply in terms of the discrete $d\xi$'s and there is no loss of generality. In analogy with the definition of the thermodynamic "force" on a dislocation line or crack front, we define a set of forces $f_\alpha = f_\alpha(\varepsilon,H)$ conjugate to the variables by

$$\sum f_\alpha(\varepsilon,H)d\xi_\alpha = -[\Phi(\varepsilon,H+dH) - \Phi(\varepsilon,H)] . \qquad (4)$$

The analogous expression in terms of the dual potential is

$$\sum f_\alpha(\sigma,H)d\xi_\alpha = [\Psi(\sigma,H+dH) - \Psi(\sigma,H)] . \qquad (5)$$

We shall want to define the elastic part of any strain increment as that due to changing σ at fixed H, and the plastic part as that due to changing H at fixed σ. Hence, by definition

$$d^P\varepsilon_{ij} = \varepsilon_{ij}(\sigma,H+dH) - \varepsilon_{ij}(\sigma,H) . \qquad (6)$$

But from (3), this strain difference is given by differentiating the corresponding Ψ difference, which is expressed in terms of the conjugate forces by (5). Thus we obtain the following fundamental relation between a macroscopic plastic strain increment and the corresponding set of microstructural rearrangements:

$$d^P\varepsilon_{ij} = \frac{1}{V} \sum \frac{\partial f_\alpha(\sigma,H)}{\partial \sigma_{ij}} d\xi_\alpha \qquad (7)$$

This plays a key role in establishing a 'normality' structure to constitutive laws.

Evidently, a general strain increment may be written as

$$d\varepsilon_{ij} = \frac{\partial \varepsilon_{ij}(\sigma,H)}{\partial \sigma_{k\ell}} d\sigma_{k\ell} + d^P\varepsilon_{ij} = M_{ijk\ell}d\sigma_{k\ell} + d^P\varepsilon_{ij} \qquad (8)$$

where M is the symmetric tensor of incremental elastic stiffness. Within the small strain formulation, M may be assumed independent of σ. It is frequently independent of H as well (e.g., slip leaves lattice elastic moduli virtually unaltered). In such cases

$$d^P\varepsilon_{ij} = d\varepsilon^P_{ij} , \text{ where } \varepsilon^P_{ij} = \varepsilon_{ij} - M_{ijk\ell}\sigma_{k\ell} \qquad (9)$$

is the strain that would remain upon unloading at fixed H. Evidently, ε^P can only depend on H and therefore the $d^P\varepsilon$ corresponding to a given dH cannot depend on σ. From (7) this means that each f is then linearly dependent on σ.

To account for temperature, which is an additional canonical variable in Φ and Ψ, we need only consider it as another variable held constant in the differentiations of (1,2), on the right in (4,5,6), and in the differentiation of (7)[1]. An additional term representing thermal straining at fixed σ,H is then to be added to the right of (8). The entire development applies for finite deformation if ε is a materially objective strain measure and σ its conjugate stress measure[1,9]. Of course, M will then not be independent of H or, if finite lattice stretching is involved, of σ either.

Crystalline slip. Suppose that the transition between H and $H+dH$ can be described as due to incremental glide motions of the dislocations in a metal, where the dislocations are regarded as line defects and dn is a continuous variable along each dislocation loop, denoting the local advance of the line normal to itself in its slip plane. Then in (4)

$$\sum f_\alpha d\xi_\alpha \rightarrow \int_L [q\, dn] dL , \tag{10}$$

where q is the force per unit length of dislocation line and L denotes an integration along all lines in the material sample. With this representation (7) becomes[1]

$$d^p\varepsilon_{ij} = \frac{1}{V} \int_L \left[\frac{\partial q(\sigma,H)}{\partial \sigma_{ij}} dn\right] dL . \tag{11}$$

Here the notation means that q is a function of the macroscopic stress and of the entire current pattern of dislocations within the sample. However, within the linear elastic treatment

$$q = q_o(h) + \bar{\tau} b \tag{12}$$

where $\bar{\tau}$ is that contribution local shear stress, acting on the slip plane in the direction of the Burgers vector b, which is induced elastically by the macroscopic applied stress, and where q_o represents the effect of the self stressing. Thus[7]

$$d^p\varepsilon_{ij} = \frac{1}{V} \int_L \left[\frac{\partial \bar{\tau}}{\partial \sigma_{ij}} b\, dn\right] dL \tag{13}$$

More generally, we shall average out the individual dislocations, specifying instead the local amounts of shear $d\gamma^{(1)}, d\gamma^{(2)}, \ldots$ on the operative slip systems of the crystalline sub-element encompassing any considered point of the material sample. If (13) were applied to a single crystal under macroscopically homogeneous deformation we would have

$$d\gamma^{(k)} = \frac{1}{V} \int_{L^{(k)}} [b^{(k)} dn] dL = \rho^{(k)} b^{(k)} <dn>^{(k)} \tag{14}$$

where now the integral extends only over the dislocations on system (k), and we express the result in terms of dislocation density $\rho = L/V$, b, and average advance $<dn>$ on that system. The same interpretation is adopted locally within the heterogeneous material sample in that, e.g., we consider the ρ's and γ's to be defined locally throughout each grain of a polycrystal. Thus in (4)

$$\sum f_\alpha d\xi_\alpha \rightarrow \int_V [\sum \tau^{(k)} d\gamma^{(k)}] dV , \qquad (15)$$

where this defines thermodynamic "stresses" conjugate to the $d\gamma$'s, and from (7)

$$d^p\varepsilon_{ij} = \frac{1}{V} \int_V [\sum \frac{\partial \tau^{(k)}(\sigma,H)}{\partial \sigma_{ij}} d\gamma^{(k)}] dV . \qquad (16)$$

Here the τ's may be shown to take the form
$$\tau^{(k)} = \tau_o^{(k)}(H) + \bar{\tau}^{(k)}$$
where again when the elasticity is linear, $\bar{\tau}$ represents the contribution to the local shear stress, on the slip plane in the slip direction, induced elastically by σ, and where τ_o includes a part which is the residual stress contribution to the local shear stress, from misfits of adjacent grains, etc. It, like q_o, is the source of the 'stored energy of plastic deformation'.

Exceptionally, (15) will be insufficient to represent the H change in Φ; e.g. during annealing, dislocations may annihilate one another without creating $d\gamma$'s. In this case the local $d\rho$'s are appropriately added as internal variables. At least within the linearly elastic formulation, their conjugate forces do not depend explicitly on σ and hence they do not appear in (16), but instead affect only the stored energy.

Now, the results of this section are strictly valid during deformation processes only when these may be viewed as 'sequences of constrained equilibrium states'[1,4,7]. Some degree of approximation is entailed in adopting them, as we shall, for actual processes and this may become important at very high strain rates. The same approximation is tacitly made, e.g., in dislocation theory when static elasticity calculations are adopted to describe the interactions and strain fields of moving dislocations.

KINETIC RELATIONS AND PLASTIC NORMALITY

The framework is completed in principle by a specification of kinetic relations for the rates $d\xi/dt$ of local rearrangement. The thermodynamic requirement of positive entropy production restricts these only by[1,4]

$$\sum f_\alpha d\xi_\alpha/dt \geq 0 , \qquad (17)$$

and this is re-expressed through (10,15) for the two slip models. When each f is linear in σ, it may be shown from (7,5,3) that an equivalent inequality is

$$V \sigma_{ij} d\varepsilon^p_{ij}/dt \geq d\Phi_o/dt \qquad (18)$$

for isothermal deformation, where Φ_o is the free energy that would remain if σ were reduced to zero with H held at its instantaneous state (i.e., Φ_o = 'stored energy'). This shows that the macroscopic plastic work rate can conceivably be negative, as for a strong Bauschinger effect, provided that stored energy is being given off.

Now consider the class of kinetic relations which is conventionally adopted for crystalline slip: The instantaneous dislocation velocities on a given slip system are considered to depend on the prevailing state of stress only through the local shear stress component which acts on that system in the slip direction. Of course, the functional relationship depends additionally on the temperature θ and current dislocated state as parameters. For example, the average dislocation velocity $<dn/dt>$ on a given system may be thought of as depending on the shear stress on that system, on θ, and on parameters such as the ρ's and γ's accumulated on that and other systems. Hence with θ and H as parameters, such relations are equivalent to assuming that the local dislocation velocity dn/dt is dependent on the applied stress σ only through the conjugate force q, or that $d\gamma^{(k)}/dt$ at a point depends on σ only through the conjugate stress $\tau^{(k)}$, or, in terms of the general framework, that kinetic relations are of a kind

$$d\xi_\alpha/dt = r_\alpha(f_\alpha, \theta, H) ,\quad (19)$$

where the rate of each specific structural rearrangement depends on σ only through the conjugate thermodynamic force f.

This class of kinetic relations has the remarkable property that a macroscopic 'flow potential' Ω exists such that the instantaneous plastic strain rate is given by[1,4,7]

$$d^p\varepsilon_{ij}/dt = \frac{\partial \Omega(\sigma,\theta,H)}{\partial \sigma_{ij}} . \quad (20)$$

Indeed, the microscopic representation of Ω is

$$\Omega(\sigma,\theta,H) = \frac{1}{V} \sum \int_0^{f_\alpha(\sigma,\theta,H)} r_\alpha(f_\alpha,\theta,H)df_\alpha , \quad (21)$$

where the integration is done at fixed θ and H, and with this (20) may be proven as a direct application of (7):

$$\frac{\partial \Omega}{\partial \sigma_{ij}} = \frac{1}{V}\sum \frac{\partial f_\alpha}{\partial \sigma_{ij}} r_\alpha = \frac{1}{V}\sum \frac{\partial f_\alpha}{\partial \sigma_{ij}} d\xi_\alpha/dt = d^p\varepsilon_{ij}/dt .$$

If we specialize these results to the slip model which averages out the individual dislocations, the corresponding equations to (19,21) are

$$d\gamma^{(k)}/dt = \Gamma^{(k)}(\tau^{(k)},\theta,H) , \quad (22)$$

$$\Omega(\sigma,\theta,H) = \frac{1}{V}\int_V \sum \left\{ \int_o^{\tau^{(k)}(\sigma,\theta,H)} \Gamma^{(k)}(\tau^{(k)},\theta,H)d\tau^{(k)} \right\} dV \quad , \quad (23)$$

and now (20) may be proven directly from the corresponding special version (16) of (7). This shows also that the macroscopic flow potential is just the volume average of local flow potentials for each slip system of each individual crystallite of the material sample.

Consider the significance of (20) from a purely macroscopic standpoint: Now instead of requiring six separate constitutive relations for the stress and history dependence of the instantaneous plastic strain rate components, we require only one for the scalar Ω, from which the others are generated. Further, (20) has a geometric interpretation in a stress space having coordinate axes which are the components of σ. At each epoch in the history of deformation, a family of surfaces of the form Ω = constant exist in this space, and have the property that the instantaneous plastic strain rate has a direction 'normal' to the Ω surface through the current stress point, and a magnitude equal to the gradient between neighboring Ω surfaces. Provided that the local rates are steadily increasing functions of the conjugate forces for any given θ and H, and that conditions for the forces to be linear in σ are met, each Ω surface may be shown[7] to be convex, in that a plane which is tangent to the surface at any point will never cross it.

Often the kinetic relation (22) is strongly nonlinear: At any given H and θ, an essentially zero $d\gamma/dt$ results for a certain range of τ values, whereas $d\gamma/dt$ takes on very large magnitudes for values of τ only slightly beyond the limits of this range. Of course, these limits change with accumulating H. Evidently, if we consider a restricted range of deformation rates, then the resulting behavior is well described by a time-independent idealization in which the limits represent critical shear stresses for yielding a slip system, and in which the changes in the limits with H represent strain hardening.

By directly adopting this idealization in connection with a somewhat more restricted slip model than necessitated in the present framework, Hill[3], Mandel[6], and Rice[10] independently derived a corresponding normality rule. This states that the plastic strain increment $d^P\varepsilon$ has the direction of the outward normal to the current macroscopic yield surface in σ space when that surface is smooth, and has a direction within the cone of limiting normals at a vertex. In fact, as Hill has emphasized, a vertex is to be expected on subsequent yield surfaces at points of sustained deformation. This is because the macroscopic yield surface is the envelope of an infinite family of yield planes in σ space, each corresponding to a critical shear stress on a local slip system. These plane may translate as residual stress contributions to τ build up and as direct or latent hardening occurs, but their normals remain of fixed orientation. Every individual plane corresponding to a slip system active in the sustained deformation must pass through the current stress state, and this creates the vertex. Hill[3] has also proposed that vertex-free large offset "yield surfaces" can be interpreted as families of plastic limit states as defined in terms of the local distribution of hardness in a polycrystalline aggregate. This is tantamount to treating the material as rigid-plastic, as in a study by Bishop and Hill[11].

Rice[7] has discussed the time-independent idealization in terms of a clustering of Ω surfaces outside the non-yielding domain of σ space. Also, he has shown that if the relation between $d\gamma/dt$ and τ on each slip system is continuous, then the Ω surfaces do not contain vertices except possibly when a surface corresponds to zero strain rate. This latter case arises when a non-yielding domain, in which Ω = constant , exists.

The dual potential to Ω has not been studied previously. Let us suppose that the relation of $d^P\varepsilon/dt$ to σ is invertible to the extent that a function

$$\Lambda = \Lambda(d^P\varepsilon/dt,\theta,H) = \sigma_{ij} d^P\varepsilon_{ij}/dt - \Omega \qquad (24)$$

may be defined. Then by (20) when θ and H are considered fixed,

$$d\Lambda = \sigma_{ij} d(d^P\varepsilon_{ij}/dt) \quad , \quad \text{or} \quad \sigma_{ij} = \frac{\partial \Lambda(d^P\varepsilon/dt,\theta,H)}{\partial (d^P\varepsilon_{ij}/dt)} \qquad (25)$$

if components of $d^P\varepsilon/dt$ can be varied independently. Usually they cannot be, because plastic straining is incompressible. In this case it is easy to see that the differential form of (25) allows solution for the deviatoric part of σ .

By using (7,21) and by writing r_α for $d\xi_\alpha/dt$ and integrating by parts, the microstructural interpretation of Λ is

$$\Lambda = \frac{1}{V} \sum \left[\int_0^{r_\alpha} f_\alpha(r_\alpha,\theta,H) dr_\alpha - (f_\alpha - \sigma_{ij} \frac{\partial f_\alpha}{\partial \sigma_{ij}}) r_\alpha \right] \, , \qquad (26)$$

where for purposes of the integration at fixed θ,H , the kinetic law (19) is supposed to have been inverted to obtain f in terms of r . Of course, when f is linear in σ this becomes

$$\Lambda = \frac{1}{V} \sum \left[\int_0^{r_\alpha} f_\alpha(r_\alpha,\theta,H) dr_\alpha - f_{\alpha o}(\theta,H) r_\alpha \right] \, , \qquad (27)$$

where $f_{\alpha o}$ is the value of f_α when $\sigma = 0$. This would, for example, take the form

$$\Lambda = \frac{1}{V} \int_V \sum \left[\int_o^{\eta^{(k)}} \tau^{(k)}(\eta^{(k)}, \theta, H) d\eta^{(k)} - \tau_o^{(k)}(\theta, H) \eta^{(k)} \right] dV \qquad (28)$$

for the slip model which averages out the individual dislocations, where $\eta^{(k)}$ is written for $d\gamma^{(k)}/dt$, and where the rate law of (22) is supposed to have been inverted in the integrand.

For the time-independent idealization, f_α will have a definite value (namely, the current yield value) if the associated r_α or $d\xi_\alpha/dt$ is non-zero. Hence each of the integrals in (26) amounts in a term $f_\alpha r_\alpha$, and so

$$\Lambda = \frac{1}{V} \sum \sigma_{ij} \frac{\partial f_\alpha}{\partial \sigma_{ij}} r_\alpha = \sigma_{ij} d^p\varepsilon_{ij}/dt \qquad (29)$$

by (7). Thus (25) reproduces a known result for time-independent materials satisfying the normality rule: that components of σ are derivatives of the rate of plastic working with respect to corresponding components of $d^p\varepsilon/dt$. This identification of Λ could also be developed directly from (24) in the rate-insensitive limit.

POLYCRYSTAL MODELS

Here we shall review some attempts at predicting the plastic behavior of polycrystals, on the assumption that kinetic relations of the kind (22) are given, a priori, from dislocation dynamics considerations and/or experiment for the operative slip systems within individual crystals of the aggregate. Of course, these relations are not known precisely in general and very simple forms have been employed in the studies under review. Nevertheless, they do presumably show the way that factors such as the constraints and residual stresses induced by neighboring grains affect the macroscopic constitutive relations of polycrystals.

Local stress and strain fields within individual crystalline elements of the aggregate are denoted by s and e. The plastic strain is given by

$$e^p_{ij} = \sum \mu_{ij}^{(k)} \gamma^{(k)} , \qquad (30)$$

where the summation extends over all operative slip systems of the element and where

$$\mu_{ij} = \frac{1}{2}(n_i m_j + n_j m_i) , \qquad (31)$$

with n and m being unit vectors describing the slip plane normal and slip direction for a given system. The mechanical shearing stress acting on a given system is

$$\tau^{(k)} = s_{ij} \mu_{ij}^{(k)} ; \qquad (32)$$

this differs from the thermodynamic shear stress of (15) only in that the latter contains an additional part accounting for energy which would remain stored in the dislocation substructure even if the local stress were reduced to zero $(s \to 0)$[1]. Hence τ as we now use it includes the long range residual shear stress as well as that induced elastically by σ, and called $\bar{\tau}$ earlier.

Taylor[12] and Bishop and Hill[11] considered a single phase polycrystal and neglected elastic strains (rigid-plastic model), further supposing that each individual grain sustains the macroscopic strain ε^p of the aggregate. While their considerations were for time-independent behavior only, we can in fact consider the general time-dependent case, presuming that by inversion of (22), rate laws are given in the form

$$\tau^{(k)} = \tau^{(k)}(\eta^{(k)}, H) \quad, \quad \text{where} \quad \eta^{(k)} = d\gamma^{(k)}/dt \qquad (33)$$

The procedure is to directly calculate the potential Λ from (28) as

$$\Lambda = \frac{1}{V} \int_V \sum \left[\int_o^{\eta^{(k)}} \tau^{(k)}(\eta^{(k)}, H) d\eta^{(k)} \right] dV \quad . \qquad (34)$$

Now the term with $\tau_o^{(k)}$ of (28) has seemingly disappeared. This is because its part representing stored energy in the dislocation substructure has already been incorporated, due to the discussion following (32), and because the remaining long-range residual stress part does no net work on the $d\gamma$'s by the principle of virtual work, which can here be applied because elastic strains are neglected and hence the γ's give the total strain.

To calculate each $\eta^{(k)}$ of an individual grain so that Λ may be computed, one recognizes that these are to be constrained by the approximation that each grain sustains the same strain. Hence

$$\sum \mu_{ij}^{(k)} \eta^{(k)} = d\varepsilon_{ij}^p/dt \qquad (35)$$

for each. We must further choose the η's so that the associated set of τ's as computed from (33) are, in fact, derivable from a local stress field s by (32). The correct η's are given by minimizing the bracketed terms in (34) subject to the constraint (35), for by the method of Lagrange multipliers, this is equivalent to

$$\delta \left\{ \left[\sum \int_o^{\eta^{(k)}} \tau^{(k)}(\eta^{(k)}, H) d\eta^{(k)} \right] - \lambda_{ij} \left[\sum \mu_{ij}^{(k)} \eta^{(k)} \right] \right\} = 0 \quad ,$$

or

$$\left\{ \tau^{(k)}(\eta^{(k)}, H) - \lambda_{ij} \mu_{ij}^{(k)} \right\} \delta \eta^{(k)} = 0 \quad ,$$

where λ_{ij} are the multipliers. Evidently, the equation is solved when

$$\tau^{(k)} = \lambda_{ij} \mu_{ij}^{(k)} \quad , \qquad (36)$$

which is the same as saying that the τ's are derivable from a stress field. In fact, $\lambda = s$.

By performing this constrained minimization, the bracketed term of (34) is determined as a function of $d\epsilon^p/dt$ for each grain orientation. The remaining volume integral means that Λ is given by the average of this function over all grain orientations, and σ is computed from (25). The time-independent version of this general approach is exactly that employed by Bishop and Hill[11]. The net result is that Taylor orientation factors have been determined showing, for example, that the flow stress of an fcc polycrystal loaded in simple tension is approximately 3 times the corresponding shear strength on its (1,1,1)(1,-1,0) slip systems (assumed equal for all). Lin[13] has further extended this approach to the elastic-plastic case by assuming that the total strain is constant in each grain; this allows an estimate of the entire stress-strain curve. Of course, the constraint that each grain deforms the same makes it impossible for stress equilibrium to hold and also causes an overestimate of the resistance to flow.

Batdorf and Budiansky[14] have proposed a slip theory of plasticity which, if reinterpreted in the present context, can be seen as complementing the above approach by assuming that each individual grain carries the same stress, equal to σ. Hence (32) becomes

$$\tau^{(k)} = \sigma_{ij}\mu_{ij}^{(k)} \quad (37)$$

and with this together with rate laws of the type (22), phrased in terms of the mechanical shear stress, one may directly calculate Ω from (23) as the average over all orientations of the flow potential which an individual grain would have if subjected to the stress σ. The corresponding plastic strain rate is then given by (20). Of course, this approach does not satisfy displacement continuity between adjacent grains, nor can it account for development of residual stresses which tend to build up preferentially on the systems of greatest slipping[7]; hence it underestimates the resistance to flow.

Recently Clough and Simmons[15] have proposed an approach to rate-dependent flow which, on examination may be seen as amounting to the formalism outlined above with a rate law for which $d\gamma/dt$ varies as a hyperbolic sine of τ, with no effect of H on the relation. The original Batdorf-Budiansky application was to rate-independent slip, with hardening of active systems, but no latent hardening or reverse hardening. This led to a pronounced vertex at the current load point. Also, for any stress path which continuously activated every slip system, once initially activated, the total strain was seen to depend only on the stress -- i.e., 'deformation theory' applied[16].

Lin and Ito[17,18] analyzed by methods of three dimensional

elasticity the behavior of a polycrystalline model of 4 x 4 x 4 square blocks, each containing one permissible set of slip planes with three equally spaced slip directions. Orientations were chosen to simulate a macroscopically isotropic polycrystal. They showed that a vertex formed at the current load point when a zero offset strain definition of yield was adopted, but they also showed that this vertex became a rounded bulge when a small but finite offset definition was used.

The bulk of work on predicting elastic-plastic behavior of polycrystals has been based on the self-consistent model of Kroner[19] and Budiansky and Wu[20]. It considers s and e to take on constant values within each grain. Apart from any constitutive connection between the two, these are related to σ and ε by the same formulae that would apply if the grain were a homogeneous spherical inclusion imbedded in an infinite homogeneous matrix, having the overall elastic properties of the aggregate, and carrying the remotely uniform fields σ and ε. Thus, if these overall properties are isotropic with shear modulus G and Poisson ratio ν,

$$s_{ij} = \sigma_{ij} - \frac{3-5\nu}{4-5\nu} G \delta_{ij}(e_{kk} - \varepsilon_{kk}) - \frac{7-5\nu}{4-5\nu} G (e_{ij} - \varepsilon_{ij}) . \tag{38}$$

Here, ε, σ are volume averages of e, s,

$$\varepsilon_{ij} = \frac{1}{V} \int_V e_{ij} \, dV \quad , \quad \sigma_{ij} = \frac{1}{V} \int_V s_{ij} \, dV \tag{39}$$

and the first of these will imply the second by (38).

Now, in the special case when each grain is idealized as being elastically isotropic with the same constants ν and G, (38) may be re-written solely in terms of plastic strain as

$$s_{ij} = \sigma_{ij} - \frac{2}{15} \frac{7-5\nu}{1-\nu} G (e_{ij}^p - \varepsilon_{ij}^p) , \tag{40}$$

and in this case, although not generally [7,21], ε^p is the volume average of e^p. Hence, using (30,32) the shear stress associated with a slip system in a given grain, having the orientation parameter $\mu^{(k)}$, is

$$\tau^{(k)} = \sigma_{ij}\mu_{ij}^{(k)} - \frac{2}{15} \frac{7-5\nu}{1-\nu} G \left[\sum \mu_{ij}^{(\ell)} \mu_{ij}^{(k)} \gamma^{(\ell)} \right.$$
$$\left. - \frac{1}{V} \int_V \sum \mu_{ij}^{(\ell)'} \mu_{ij}^{(k)} \gamma^{(\ell)'} dV' \right] , \tag{41}$$

where the first sum, on (ℓ), extends over all slip systems of the same grain and the second, on $(\ell)'$, extends over all systems of every grain as it is encountered in the volume integral (or orientation average); the primes distinguish those variable quantities in the integration. This gives an explicit representation for the long range residual stress, as a linear function

of all the γ's in all the grains. The procedure is then to
solve the kinetic relations for the γ's, given a history of σ
variation, and to thereby compute ε^p.

Hutchinson[22] has applied this procedure to time-independent
calculations, both without hardening and with Taylor hardening,
for fcc and bcc polycrystals. His results include the calculation of Bauschinger effects and of the response to proportional loading under combined stress. Bui[23] has adopted the
model to compute subsequent yield surfaces and shows a clear vertex formation. Brown[24] and Zarka[8] have considered time dependent behavior; the former has adopted a power law relation between $d\gamma/dt$ and τ for fcc polycrystals, and has computed
the surfaces of constant flow potential in tension-torsion stress
space for various deformation histories. These seem, to a fair
approximation, to show kinematic translation without much shape
change. Brown[25] has also attempted direct experimental measurement of Ω surfaces.

Hill[26] has suggested a more elaborate self-consistent model
which is intended to take account of directional weaknesses developing with continuing deformation in a time-independent plastic
framework; the corresponding generalization for time-dependence
is, however, unclear. Recently Hutchinson[27] has given an extensive review, contrasting the Hill model with that of Kroner-Budiansky-Wu. The latter gives limit states which agree with
the Taylor model, and are thus overestimates, whereas the Hill
model seems to give lower values. Hutchinson also calculates the
plastic moduli governing increments of shear after tensile loading, and shows that these are considerably nearer to the predictions of 'deformation' theory than to those of 'flow' theory
with a smooth yield surface.

These calculations reveal the central features of combined
stress behavior but they are, in general, too complex to consider as a basis for constitutive laws in structural analysis.
What seems to be needed is a phenomenological formulation which
aims to incorporate the most important features of such detailed
models, such as vertex formation on yield surfaces, Bauschinger
effects, kinematic-like translations of Ω surfaces, etc., but
which involves far simpler means of calculating the effects of
deformation history. Some possible candidates have been reviewed by Rice[7], but in general little work has been directed
to this aim.

Acknowledgement. I wish to acknowledge the support of the
Atomic Energy Commission under contract AT(11-1)-3084 with
Brown University.

REFERENCES

1. J. R. Rice, J. Mech. Phys. Solids, v. 19, p. 433, 1971.
2. K. S. Havner, Int. J. Solids and Structures, v. 5, p. 215, 1969.
3. R. Hill, J. Mech. Phys. Solids, v. 15, p. 79, 1967.
4. J. Kestin and J. R. Rice, in A Critical Review of Thermodynamics, ed. E. B. Stuart et al., Mono Book Corp., Baltimore, p. 275, 1970.
5. T. H. Lin, Theory of Inelastic Structures, Chp. 4, Wiley, 1968.
6. J. Mandel, in Proc. 11th Int. Cong. Appl. Mech. (Munich,1964), ed. H. Görtler, Springer-Verlag, Berlin, p. 502, 1966.
7. J. R. Rice, J. Appl. Mech., v. 37, p. 728, 1970.
8. J. Zarka, J. Mech. Phys. Solids, v. 20, p. 179, 1972.
9. R. Hill, Proc. R. Soc. Lond. A., v. 326, p. 131, 1972.
10. J. R. Rice, Tech. Rept. ARPA SD-86 E-31, Brown Univ., 1966.
11. J. F. W. Bishop and R. Hill, Phil. Mag., v. 42, p. 414 and 1298, 1951.
12. G. I. Taylor, J. Inst. Metals, v. 62, p. 307, 1938.
13. T. H. Lin, J. Mech. Phys. Solids, v. 5, p. 143, 1957.
14. S. B. Batdorf and B. Budiansky, NACA TN 1871, 1949.
15. R. B. Clough and J. A. Simmons, in Rate Processes in Plastic Deformation, ed. J. C. M. Li and A. K. Mukherjee, ASM, in press.
16. B. Budiansky, J. Appl. Mech., v. 26, p. 259, 1959.
17. T. H. Lin and M. Ito, J. Mech. Phys. Solids, v. 13, p. 103, 1965.
18. T. H. Lin and M. Ito, Int. J. Engng. Sci., v. 4, p. 543, 1966.
19. E. Kröner, Acta Met., v. 9, p. 155, 1961.
20. B. Budiansky and T. T. Wu, in Proc. 4th U. S. Nat. Cong. Appl. Mech., ASME, N. Y., p. 1175, 1962.
21. R. Hill, Prik. Mat. Mekh., v. 35, p. 31, 1971.
22. J. W. Hutchinson, J. Mech. Phys. Solids, v. 12, p. 11 and 25, 1964.
23. H. D. Bui, Sc.D. thesis, Paris, 1970.
24. G. M. Brown, J. Mech. Phys. Solids, v. 18, p. 367, 1970.
25. G. M. Brown, J. Mech. Phys. Solids, v. 18, p. 383, 1970.
26. R. Hill, J. Mech. Phys. Solids, v. 13, p. 89, 1965.
27. J. W. Hutchinson, Proc. R. Soc. Lond. A., v. 319, p. 247, 1970.

EXPERIMENTAL METHODS IN SHOCK WAVE PHYSICS

J. W. Taylor

University of California

Los Alamos Scientific Laboratory

I. Introduction

The study of the response of solid materials to high velocity deformation, also called shock wave physics, stems from several objectives and draws on a wide spectrum of technologies. As so frequently happens in science, the whole subject derived considerable impetus from the relatively sudden availability of new tools in the form of high quality solid explosives and plane wave explosive lenses shortly after World War II. Further developments in tools to generate analyzable shock waves and the means of analysis have resulted from the space program, which led to modern gas gun technology, and space, military and nuclear weapons programs which have speeded the development of modern submicrosecond resolution electronics, optics and radiography.

The methods which may be used for production and analysis of large amplitude stress waves in solids depend on several factors, not the least of which may be availability and cost. Frequently, however, there are sound technical reasons for preferring some techniques over others in a particular experiment. Such tradeoffs should become apparent in the following pages, in which we shall outline the present situation. In general, the reader is referred to the literature for details.

The earliest high precision work in which strong shock waves in solids were studied was directed toward studies of the thermodynamic equations of state, first of metals[1],[2] and later of nonmetals, compounds and even mixtures.[3] The technique employed was to place large (frequently 30 cm x 30 cm x 20 cm) accurately machined slabs of high explosive in direct contact with the material of interest, detonate the explosive by means of a "plane-wave explosive lens" which converts a point detonation into a moderately flat detonation front*, and record the arrival of the shock at various levels in the inert material by means of electrical contactor pins or high speed rotating mirror cameras. In addition, the velocity of the free surface of the inert specimen was monitored by the same techniques and was theoretically related to the material particle velocity behind the shock front.

The theoretical basis for such a technique lies in the fact that for any material whose elastic modulus is a monotonically increasing function of pressure; the front of the compressive disturbance must steepen until dissipative mechanisms such as viscosity and heat conduction force it to remain of finite thickness. Therefore, if the pressures generated sufficiently exceed the yield strength of the material, and if sufficient explosive is used to maintain an essentially "flat topped" wave, one visualizes a shock wave in which a discontinuity in pressure (actually normal stress) and particle velocity separates two zones of constant state. Under these conditions, it is readily shown that the shock velocity U_s and particle velocity U_p are related to the initial pressure and density (P_o, ρ_o) and the final pressure and density (P, ρ) by the equations

$$P - P_o = \rho_o U_s U_p \tag{1}$$

$$\rho_o U_s = \rho (U_s - U_p) \tag{2}$$

which follow from the conservation laws for mass and momentum. If we label the specific internal energy E, the third Rankine-Hugoniot relation

$$E - E_o = \frac{P + P_o}{2} (V_o - V) \tag{3}$$

in which $(V = 1/\rho)$ is specific volume follows from the conservation of energy. It is Eq. (3) which really makes such experiments interesting, because one can hope to define the complete thermodynamic state of a material at each point reached in a shock experiment. This statement is subject in part to the assumption that knowledge of one principal stress implies knowledge of the other two. Notice also that from these experiments one cannot define the temperature directly.

In the mid and late 1950's significant attention began to be turned toward the study of shock waves of intermediate strength

---

*Modern plane wave explosive lenses, which may be as large as 30 cm in diameter, can produce detonation fronts which are simultaneous within about 0.1 μsec on a slowly varying profile.

in solids. By intermediate, we mean considerably greater than the dynamic elastic limit but not strong enough to result in essentially discontinuous shock fronts, so that the bulk compressibility and the flow stress are of comparable importance. It was quickly realized that such shocks are not generally steady in time, even if driven by constant driving parameters. More recently, however, it has been found that they may be "quasi steady". In addition, the past ten years have shown an increasing importance of the details of the relief wave which propagates into preshocked solids. These considerations have stimulated the development of a variety of techniques for observing, to the extent possible, time resolved "in material" stresses and particle velocities, and free surface velocities.

In order to understand the basis for this plethora of technology it is necessary to understand what must happen as a plane stress wave passes through a solid material. Macroscopically, the increases in particle velocity and compression are strictly one dimensional, but because of the fact that microscopically one dimensional elastic strains would result in an ever increasing difference between the longitudinal stress (parallel to the direction of shock propagation) and the transverse stress (parallel to the shock front) the shear stress on planes inclined at any angle to the direction of shock propagation will exceed the material failure stress and mass readjustments will take place in such directions that the local compressions tend toward a more isotropic configuration. The stress level at which this begins to happen is called the Hugoniot elastic limit, and the rate at which it happens typically results in shock wave structure which extends over times of the order of 0.01 to 1.0 μsec. The detailed mechanism of this mass readjustment in particular cases is the subject of much current research.

A moment's reflection shows that Eqs. (1) and (2) can be generalized to any differential disturbance, so that a continuous record of the particle velocity of a material point can be transformed with good approximation to a continuous measure of the stress (in the shock propagation direction) at that point. Furthermore, it can be shown that in the absence of viscosity and other dissipative mechanisms, the particle velocity change in a rarefaction (relief wave) is related to the stress change through the Riemann integral

$$U - U_o = \int_{\sigma_o}^{\sigma} \frac{d\sigma}{\rho C} \qquad (4)$$

where

U = particle velocity
σ = stress
ρ = density
$C = (\partial \sigma / \partial \rho)_s^{1/2}$ is the sound speed.

As a consequence, it turns out that if the initial shock is not sufficiently strong that the thermal energy makes gross changes in the material elastic modulus, the rarefaction particle velocity defined by Eq. (4) is very nearly equal to the shock particle velocity and hence equal to one-half the free surface velocity.

It follows that, to a good degree of approximation, a time resolved measurement of free surface velocity can at least in principle be used to deduce a stress-time history. In particular, such measurements can be used to deduce spall tensile strengths with a precision which is quite likely to be as good as such quantities are reproducible from sample to sample.[3]

II. Methods of Producing Strong Plane Shock Waves in Solids-- Some Advantages and Disadvantages

It has already been mentioned that one convenient method for producing strong plane shock waves involves solid high explosives and "plane wave lenses". Cast or pressed explosives must be of high quality with minimal (<1%) density variations and opposing flat surfaces must be machined to tolerances for flatness and parallelness of the order of 0.025 mm or better. Four fairly standard explosives which produce a reasonable range of pressures in inert materials are cyclotol (75% RDX, 25% TNT), Composition B (60% RDX, 40% TNT), TNT and baratol, which is 24% TNT and 76% barium nitrate. Another material which has been used extensively is nitroguanadine because its detonation pressure can be conveniently varied from approximately 20 kbars to 100 kbars by varying the pressing density. The quality of the detonation front in such a material is, however, subject to question.

In using such explosives it is commonly assumed that their properties are adequately described by assuming a steady state detonation resulting in the Chapmen-Jouguet (C-J) pressure just behind the detonation front in a material which subsequently obeys the caloric equation of state[4]

$$E = \frac{PV}{\gamma - 1} \qquad (5)$$

where

E = specific internal energy
P = pressure
V = specific volume
γ = adiabatic exponent (assume nearly constant).

The C-J pressure is obtained by applying Eqs. (1) and (2) with the result that

$$P_{C-J} = \frac{\rho_0 D^2}{\gamma - 1} \tag{6}$$

where

ρ_0 = initial explosive density
D = (constant) detonation velocity.

Using these relations, the thermodynamic relation

$$dE_s = -p\,dV_s \tag{7}$$

and the characteristic solutions of the Euler-Langrange equations, one can calculate the locus of all possible isentropic release states from the C-J state. Further use of Eq. (5) and the Rankine-Hugoniot equations enables one to calculate the locus of all possible reflected shocks from the C-J state. One can then design explosive experiments by making use of the impedance matching concept, which we briefly describe.

The Rankine-Hugoniot equations (1, 2 and 3) together with a material equation of state in the form

$$E = E(P, V) \tag{8}$$

defines for any material a Hugoniot locus which can be represented in terms of any two variables and in particular in terms of pressure and particle velocity. If a shock in any material crosses into a second material and the result is steady state at the boundary, the pressure and particle velocity must be continuous across the boundary. Hence, as illustrated in Fig. 1, one can use the pressure-particle velocity plane to define the interaction between an explosive and an inert material, or two inert materials.

It is quite important to note that in actual practice one cannot assume that the result of any particular experiment with explosive will agree precisely with the predictions of this impedance match solution, even if the explosive reflected Hugoniot and release isentrope are measured rather than calculated. First, the pressure behind the detonation front in any explosive decays more or less rapidly depending on the length of the charge, so the average pressure in an inert sample of finite thickness is certainly not expected to be constant. Second, the actual pressure at the detonation front in an explosive is known to be a function of the distance the detonation has propagated and the details of the initiation method. Hence, one must calibrate specific systems.

The pressure range available with explosives can be materially extended by using the explosive to accelerate thin metal plates. In this manner a significant fraction of the explosive energy is

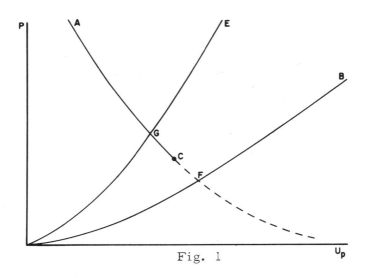

Fig. 1

The impedance matching method applied to a detonated explosive (line ACF) and two hypothetical inert materials (OFB and OGE). The hugoniots of the inert materials are curves starting at the origin in the pressure-particle velocity (p-U_p) plane. The explosive C-J state is point C. The locus of reflected shocks is AC and the locus of release states is CB. The intersection points F and G define the pressures and particle velocities at which the explosive will equilibrate with the two inert materials.

---

concentrated in the metal plate, which results in plate velocities as high as 5 or 6 km/sec. The theoretical limit is of course the escape velocity of the explosive reaction products into a vacuum, approximately 9 km/sec. Extensive experiments at Los Alamos have demonstrated that for best results and minimal breakup of the accelerated plates, a small gap of the order of 0.5 cm should be used between the explosive charge and the plate. The choice of plate material is dictated by the required collision pressure in the sample, the necessary time duration before rarefaction signals overtake the shock, the resistance of the plate to breakup, and the cost of materials. Generally, it is easy to reduce plate velocities and one looks for ways to get maximum pressure. This implies a relatively inexpensive material with the maximum shock impedance per unit density and good toughness. These requirements are best met by 300 series stainless steel sheet stock.

The use of smooth bored guns for producing shock waves in the amplitude range from one to several hundred kbars is becoming increasingly common. The projectile velocities required for these pressures range up to 2 km/sec. Projectile velocities below 1 km/sec are quite conveniently achieved using helium at initial pressures up to about 350 bars as the driving medium.[5] Higher velocities (up to 2 km/sec) can be achieved with propellant fired guns. Projectile velocities up to 8 km/sec are currently being routinely achieved in several laboratories by means of two-stage light gas guns which were originally developed to their present performance levels by the General Motors group at Santa Barbara.[6]

Compared with explosives, guns suffer from two disadvantages: The working area is generally more restricted; the worst case being a working diameter of approximately 2 cm in a practical two-stage light gas gun fired at maximum velocities. A second disadvantage, which is becoming increasingly less important, is the fact that guns cannot conveniently be synchronized to high speed cameras.

On the other hand, guns offer several distinct advantages over explosives. First, projectile velocities are continuously and easily variable over the entire range, and the projectile velocity in a particular shot can typically be predicted with a precision of a few percent at worst. Second, projectile velocities are easily measured with a precision of 0.1%.[7] Third, it is relatively easy to attach almost any material to the projectile so that one can study collisions of like and unlike materials with varying thickness ratios. Hence, on the one hand, it is possible by using identical materials for the driver and the target to know the initial shock particle velocity as precisely as the projectile velocity. On the other hand, by using dissimilar materials, one can shape the stress pulse in the target and study signal propagation in precompressed material. It is also possible to use x-cut quartz crystals on the projectile or piezoresistive pressure transducers at the projectile-target interface and thus measure stress directly at the collision interface.

III. Free Surface and Shock Particle Velocity Measurements

Several basic techniques are available for making free surface velocity measurements.

The oldest, historically, is to set a very large number of electrical contactor pins at carefully measured distances from the surface and record the shorting times on one or more oscilloscopes.[8] While this method is excellent for projectile velocity measurements, its application to time-resolved free-surface velocity data requires such extreme care that it is rarely used.

A second relatively straightforward but still fairly crude technique is the slant-wire resistor.[9] One simply records the time-resolved resistance as the free surface contacts the wire. The problem is that these two techniques both record distance vs. time, which data must then be differentiated to get velocity vs. time.

Sweeping image cameras may be used to record either the closure of a gap between a surface and, for example, a razor blade, or to follow the change in position of a carefully collimated light beam reflected from a tilted surface.[2] The latter has, for example, enabled McQueen, et al, to observe grain-sized variations in the elastic precursor in iron.

A technique which has proven most valuable is the D. C. capacitor, first developed by Rice.[10] A parallel plate capacitor is made with a grounded shield and accurately spaced at distances of the order of one or two millimeters from the surface to be studied. It has been found that if the capacitor diameter is at least ten times the spacing, the capacitance is very accurately represented by

$$C = \frac{C_o D_o}{D_o - X} \qquad (9)$$

where D_o is an initial spacing and X is the amount of motion. The author has found that such a device can quite conveniently be calibrated by means of the arrangement shown in Fig. 2. The appropriate circuitry for use with the capacitor is represented schematically in Fig. 3. The differential equation[7] which describes the capacitor-response free-surface velocity is readily integrated by computer. The precision of the technique is in general a few percent. The time resolution is 10 to 20 nsec.

Figure 4 shows a typical oscilloscope record from such a device. This is in fact comparable to the record from any good analog device. The electronically generated grid which shows on the record provides a very useful method for eliminating all distortions, including nonlinearities in the electronic components, oscilloscope nonlinearities, and camera optics. This grid is derived from a very accurate voltage standard into which time markers are mixed. The film is translated between calibration and shot. The calibration is dynamic and is done through all the circuitry. The computer program then does a two-dimensional quartic fit in which the X-Y film space is mapped into voltage time space and the results can be used to analyze any arbitrary tranducer record.[7]

Figure 5 shows the result of analysis of the record from Fig. 4.

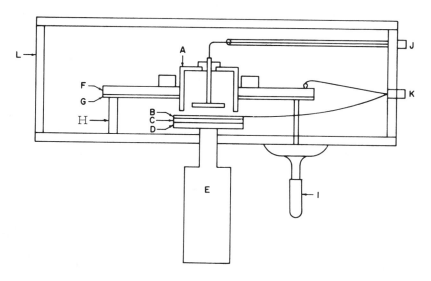

Fig. 2

This device is conveniently used for calibration of D. C. capacitors whose magnitude ranges between roughly 0.5 pf and 10 pf. The capacitor A is held in the mounting and centering fixture F which is insulated by a plastic spacer G from the support points H and the leveling micrometer I. The second plate of the capacitor B is insulated from "case ground" L by insulator C which is supported on plate D, attached to the high precision micrometer E. In the arrangement used, connectors J and K are used with coaxial cables to connect to the high impedance and low impedance terminals respectively of a General Radio model 1615 capacitance bridge. The bridge is used in the three-terminal mode, which automatically zeros out capacitance to case ground. One then measures capacitance as a function of the position of plate B.

The capacitor technique is applicable to virtually all experiments involving shocks in metals as long as the main-shock velocity does not exceed the material longitudinal sound speed. It has been found that stronger shocks in metals produce an internally-generated electrical signal (whose origin and mechanism remains a complete mystery) which mixes with the capacitor record and cannot readily be unfolded. The technique can also be used with insulators whenever it is possible to evaporate or otherwise deposit a thin metal film on the surface of the material. If such materials are severly inhomogenous and some penalty in time resolution is acceptable, one may attach a foil of aluminum

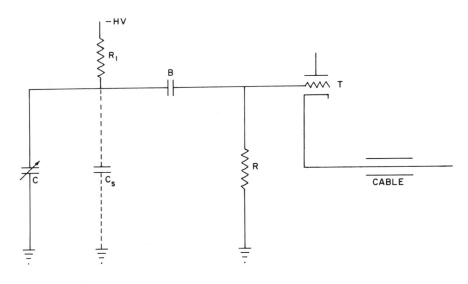

Fig. 3

Basic circuitry used with the d. c. capacitor technique. The variable capacitance C (and the stray shunt capacitance C_s) are polarized through the isolation resistor R_1. During data recording, the fixed capacitor B serves as a battery and a voltage on the load resistor R is impedance transformed through T for transmission on the cable to an oscilloscope. Typical values are: C = 2 pF, C_s = 12 pF, R_1 = 2MΩ, B = 0.01 μF, R = 11kΩ.

---

or copper to the surface by epoxy or other cement.

An electromagnetic velocity transducer which was first used extensively by Dremin[11] is useful for in material time-resolved particle-velocity measurements in insulators. A short length of conductor embedded in the material produces a voltage proportional to velocity and to the magnitude of an externally applied transverse magnetic field. Dremin used large magnets and went to considerable pains to keep the magnetic field homogeneous; McQueen[12] has found that entirely satisfactory results are obtained with inexpensive expendable magnets. The magnetic field is then carefully measured over the region of interest by means of a Hall probe. This technique is valuable but has the disadvantage that it is frequently difficult to retain electrical contact to the embedded foil for as long as one would wish.

It has recently been found possible to develop a highly precise "zero-contact" electromagnetic velocity transducer. A descrip-

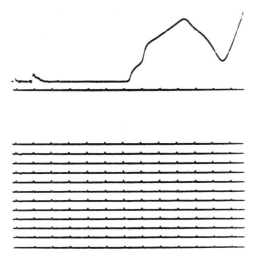

Fig. 4

Original oscillograph of a d.c. capacitor record from an impacted Vascomax 250 plate. The elastic and plastic waves and the tension wave terminating due to spall are evident. The early pulse is an impact fiducial. The calibration grid, applied through the cathode follower, serves dynamically to account for all electronic and optical nonlinearities in the system. (Courtesy of John W. Hopson).

tive paper by J. N. Fritz and J. A. Morgan[13] is in press. The basic principle is illustrated in Fig. 6. This technique, which is applicable to any nonferromagnetic conductor, has the advantage that it is intrinsically such a low impedance device that it is apparently unaffected by the internally generated signals mentioned above and is hence useful for measurements at very high pressures. In addition, since no electrical contact to the moving surface is required, the technique can be used to measure the particle velocity in an insulator by insertion of a thin metal foil. This technique has been demonstrated by Fritz and Morgan in a measurement of complex wave profiles in B_4C. The original transducer record is shown in Fig. 7. The small "blips" are caused by an evaporated film on the free surface. The calibration of this type of transducer requires only an accurate mapping of the magnetic field of the small magnet over the range of interest and an accurate knowledge of the pertinent physical dimensions. The data must be analyzed by means of a rather elaborate computer code in which the complete electrodynamic problem is solved, including the effect of finite conductivity of the metal surface.

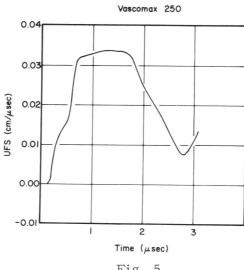

Fig. 5

Free surface velocity vs. time obtained by computerized data reduction from the oscillograph record shown in Fig. 4. (Courtesy of John W. Hopson). The large decrease in the velocity following the main shock indicates a spall tensile strength of the order of 45 kbars.

This technique appears to be capable of 1% precision or better. The time resolution is better than 10 nsec.

An extremely versatile tool which has been developed over the last several years is laser interferometry. This technique is used in two ways. For low velocities (10^{-3} to 10^{-2} km/sec) the interferometer simply measures spatial position.[14] As the interference fringes go through maxima and minima, the oscilloscope records free-surface motion measured in wave lengths of light.

The second modification, the velocity interferometer, makes use of the Doppler shift in the laser light which is reflected from the moving surface.[15] The experimental configuration is illustrated in Fig. 8. The laser light is focused on the moving surface and the return paths to the photomultiplier consists of one straight path through the beam splitters and a second longer path or "delay leg". If the wave length of the light is λ, complete constructive interference occurs at the photomultiplier whenever the delay leg path is $N\lambda$, where N is any integer. If the surface velocity is $U(t)$, the Doppler shift in the wave length is

$$\Delta\lambda = \frac{-2\lambda}{C} U(t) \tag{10}$$

where C is the velocity of light. If T is the transit time of the light around the delay leg, we can write

$$N\lambda = CT \tag{11}$$

so that

$$\Delta N(t) = \frac{-CT}{\lambda^2} \Delta N(t) \tag{12}$$

and

$$U(t) = \frac{\lambda}{2T} \Delta N(t) \tag{13}$$

is the relation between the surface velocity and the number of fringes which have changed in time t. Notice that because the photomultiplier actually records light intensity changes continuously, one can in fact discern changes of fractions of fringes. The precision of such "analog interpolation" is obviously less than of "quantized counting".

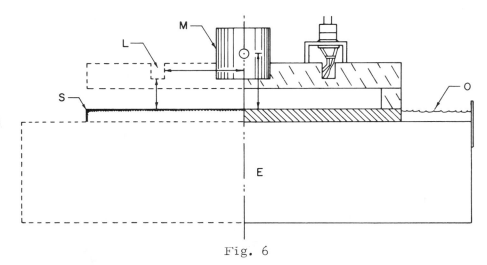

Fig. 6

Schematic of a typical assembly exploying the axisymmetric magnetic probe. The small ceramic magnet M is surrounded by a single turn of wire (L) which connects to a cable. All are mounted on a premachined lucite plate. The double arrows indicate critical dimensions. The sample S rests on explosive E and is surrounded by mineral oil O which helps reduce electrical noise from the explosive. The sample-explosive joint is sealed with grease. (Courtesy of Fritz and Morgan).

In order to employ the laser velocity interferometer effectively, one must pay close attention to the length of the delay leg because the rate of formation of fringes cannot be allowed to exceed the time resolution of the oscilloscope, yet must be sufficiently great to yield the desired detail. If, for example, a He-Ne laser is used, $\lambda = 6.328 \times 10^{-7}$ m. The delay leg transit time is 1/C where L is the delay leg length. Hence, for a one-meter delay leg, the system sensitivity is

$$\frac{\lambda C}{2L} = 94.9 \text{ m/sec per fringe.}$$

If, therefore, the free surface velocity were to change from zero to 10^3 m/sec in 10 nsec, the required oscilloscope writing speed would be 1000 MHz. On the other hand, any velocity less than 94.9 m/sec does not produce a complete fringe shift and is correspondingly imprecisely measured. Properly used, the technique is capable of 2% precision.

Fig. 7

Original oscillograph record of the motion of an aluminum foil used as a particle velocity analyzer between two B_4C plates. The axisymmetric magnetic probe technique was used. (Courtesy of Fritz and Morgan)

A second point to notice is that the technique is sensitive only to changes in velocity and does not intrinsically distinguish the sense of the change. As a consequence, decelerations caused by overtaking tension signals must be recognized by independent anticipation of their time of arrival. This is not ordinarily a serious problem and does not detract from the usefulness of the technique.

One very great advantage of the laser technique, which it holds in common with the electromagnetic induction probe and the capacitor, is that nothing need be actually attached to the surface. Further, in common with the electromagnetic technique, the laser may be used to view a moving surface through an impedance matching insulator. In the case of the laser, however, the insulator must be transparent and the data must be corrected for refractive index changes.

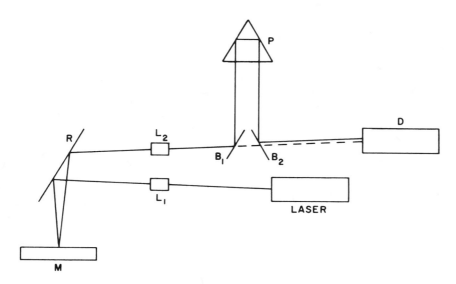

Fig. 8

Schematic layout of the laser velocity interferometer. The laser beam, which must have good spatial and temporal coherence, is focused by lens L_1 through mirror R on the mirrorized moving surface M. The beam reflected by M and R is recollimated by lens L_2 and split by beam splitter B_1, one part goes directly to the detector D through B_2. The other part goes through prism P on the delay leg. The length of this delay leg is adjusted for optimum precision in any shot.

The original velocity interferometer technique is limited to use with specularly reflecting surfaces. This follows from the fact that in a long path length interferometer spatial coherence of the light must be maintained. It has recently [16] been demonstrated, however, that if the moving surface is used as the light source for a wide angle Michelson interferometer (WAMI) which incorporates an etalon to produce time delay, a diffusely reflecting surface will do. This has resulted in the VISAR (velocity interferometer system for any reflector). The price, however, is considerable complication in the required optics and painstaking care in alignment of the system. Nevertheless, this technique holds forth great promise. The use of polarized light, for example, enables the sign of a velocity change to be unambiguously determined. For experimental details, the reader is referred to the literature.

It appears to the author that two very great advantages of the laser technique are the relative insensitivity to almost arbitrarily large ambient electromagnetic noise and the fact that the target sample of interest may be initially at almost any temperature consistent with the preservation of a good optical alignment. In addition, the VISAR may be used in conjunction with the axisymmetric magnetic probe to obtain simultaneous particle velocity and free surface velocity data.

IV. Analog Stress Measurements

One of the most widely used and historically the earliest developed stress gauge is the x-cut quartz crystal transducer.[17] The material is a linear piezoelectric at stresses up to at least its Hugoniot elastic limit of approximately 25 kbar and is useful, at somewhat higher stresses, in the hands of a very careful and knowledgeable experimenter. The acoustic impedance is very near to that of aluminum. Hence, when it is used in contact with low impedance materials it provides a very good time-resolved stress history with a minimum of data analysis. The very low effective electrical impedance makes it a voltage source which is not seriously affected by external electrical noise. The material has been calibrated over a considerable range of temperatures. A disadvantage lies in the fact that very large crystals must be used if data are to be recorded over times much in excess of one or two μsec.

In recent years considerable effort has been expended in the development of manganin, ytterbium and carbon pierzoresistive gauges. These materials, used in thicknesses of the order of 0.002 cm and protected by epoxy, polyimide, mica and even flame-sprayed alumina insulation, have been successfully used over the stress range from roughly 1 kbar to 400 kbars. It has been found that such gauges are reliable even when pulsed with several amperes of current for several hundred microseconds. There are two basic electrical and mechanical configurations in which such a gauge is used. In both configurations it is common practice to pulse the voltage by using silicon-controlled rectifiers (SCR's) to switch precharged capacitors across the active circuit a few microseconds before data are to be recorded.

The most basic configuration for piezoelectric gauges may be termed the constant current mode.[18] Several variations have been used successfully but the common feature is that the active element consists of a relatively low (one to five ohm) resistance which is driven by a pulsed "constant current" source. The voltage across the gauge is then simply proportional to its resistance, which changes with the applied pressure. This mode of use has the considerable advantage of intrinsic simplicity but the disadvantage that a very small initial resistance must be known with high precision and that the signal of interest is a differential on an initially large signal. The latter disadvantage is slight if the stresses of interest are of the order of 100 kbar or greater, but becomes serious if one is interested in the range below a few tens of kbars. This has led to various methods for "automatic zeroing" of the signal.

A very common configuration is the wheatstone bridge circuit,[19] one basic design of which is shown in Fig. 9. In this particular configuration the input impedance of the bridge circuitry is designed to match, a 50-ohm cable to avoid reflections. Since the majority of the data are collected when the bridge is seriously unbalanced, data analysis virtually requires a small computer program. The equation which must be unfolded for this particular circuit is:

$$\frac{V_s}{E_o} = \frac{AX}{1 + BX} \qquad (14)$$

The parameters are:

$$A = \frac{R_g \, R \, R_{10}}{RR_B + R_o \, (2R+R_B) \quad R+R_o+2R_g} \, ; \qquad (15)$$

$$B = \frac{R_{10}(2R+R_B)(R_o+R_g) + R_o R(R \pm R_B)}{(R+R_o+2R_g) \, R_o(2R+R_B) + RR_B} \, ; \qquad (16)$$

$C = \Delta R/R_{10}$ is the fractional change in gauge resistance;

V_s = signal voltage;

E_o = supply voltage.

Note that the initial resistance R_o of the variable portion is not R_{10} because the signal cable and gauge leads may introduce as much as a few ohms of constant series resistance.

The magnitudes of the constant resistors are so chosen to make the input impedance of the network 50 ohms and to isolate the bridge from the large capacitance to ground of the floating power supply.

If a 50-ohm manganin gauge is used in the circuit shown and the supply voltage is 300 volts, the output sensitivity is approximately 0.03 V/kbar. The recently developed ytterbium and carbon gauges increase this number by about a factor of ten.

The physical dimensions of active gauge elements used in the

50-ohm bridge are of the order of 1/2 cm x 1/2 cm x 0.01 cm or less. Ordinarily, the connecting leads are copper plated to maximize their conductivity. The most difficult aspects of their use derive from their extreme delicacy and susceptibility to short circuiting. Their use demands considerable care during installation, but if such care is exercised, they are remarkably durable. When these gauges are used in rigid materials such as metals, mica or alumina insulation is recommended for best response to details of wave profiles.

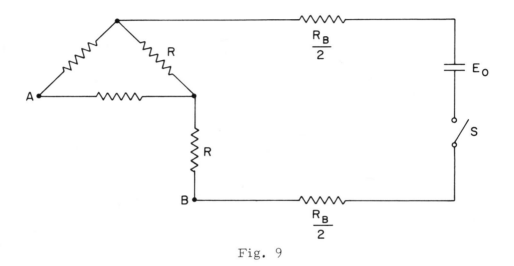

Fig. 9

Basic schematic of a wheatstone bridge and power supply for use with a 50-ohm piezoresistive gauge. The resistance R_g is actually a terminated 50-ohm cable to the oscilloscope. The resistors $R_B/2$ serve to isolate the power supply capacitance to ground from the circuit.

TEMPERATURES IN SHOCKED SOLIDS

As mentioned previously, the usual mechanical measurements on shocked solids tell us nothing directly about temperatures. Very little hope is held for measuring the internal temperature of shocked opaque solids because thermoelectric elements cannot be trusted in such an enviornment. The internal temperature of shocked transparent solids can be measured by optical means. Kormer, et al[20] reported measurements of the temperatures of shocked alkali halides. The radiated light was passed through a spectroscope and the relative intensity at several wavelengths was measured with photomultipliers. The results were fitted to a blackbody formula and the temperatures calculated. Davis[21] has successfully used photographic film blackening to measure the temperature of shocked liquids.

In the case of opaque solids, only the free-surface temperature after passage of the shock wave is available at all. McQueen, et al, made such measurements on shock-loaded iron by observing the temper colors.[22] Taylor [23] obtained what appears to be reasonable data from the free-surface of copper in the temperature range above 700°C by measuring total radiation with precalibrated photomultipliers. King, et al [24], on the other hand, using very similar techniques and infrared detectors, found that in the low temperature range (100 to 400°C) the experimental data indicated temperatures which exceeded theoretical predictions by as much as 180°C. An experiment performed with lead at 440°C, however, resulted in measurements which agreed well with calculations. The authors noted that the copper emissivity may have changed sufficiently to influence their results. The conclusion seems to be that such measurements are possible, but that generally it is very desirable to compare optical emission at several wavelengths to eliminate the possibility of emissivity changes.

MAGNETIC MEASUREMENTS

Some very interesting measurements of shock induced magnetic effects have been reported by Royce.[25] The experimental arrangement consists simply of two ferrite cores which complete the magnetic circuit between the sample and a bar magnet. A few turns of wire wrapped around one of the ferrite cores senses any flux change. The method enabled Royce to show definitely that the iron phase transition at 130 kbars results in nonferromagnetic iron. Interesting results were also obtained with several ferrites. It is obviously applicable for use both with explosives and guns.

RECOVERY TECHNIQUES

There are frequent requirements for recovering shock-loaded materials in such a way that relatively little damage beyond the effects of the first shock occur. Such experiments necessitate ensuring first that lateral relief waves do not enter the sample before the plane unloading waves have reduced the stresses to very low levels. This is accomplished by embedding the sample (pressed, if possible) in a closely fitting "spall ring" which may itself be further inserted into a larger plate. The spall ring and "carrier plate" and the drive system, whether it be explosive or a projectile, must all have sufficient lateral extent to keep the loading and unloading in the sample "one dimensional". In addition, it is frequently desirable to avoid tension stresses in the sample. In this case, either the driver system must be thick enough to prevent rarefaction signals from reaching the sample before it has been relieved from its free surface, or a "spall plate" must be added to the free surface to carry off the momentum in the shock pulse. Note that the hydrodynamic impedance of the spall plate should closely match that of the sample.

In explosive experiments, the sample may conveniently be relatively slowly decelerated by interposing several layers of increasing densities of foamed plastic between the sample and a pile of wet sand. In experiments with guns, similar arrangements, supplemented in some cases with aluminum honeycomb, will serve to stop the sample. Recovery experiments with guns involve the additional requirement to stop the projectile. The author has found that the obvious method is the best. A large and heavy billet of steel, with a hole which the sample will enter and the projectile will not, is placed immediately behind the target assembly and used as the recovery vessel. The utility of this method decreases rapidly as the projectile velocity increases beyond a few tenths millimeter per microsecond because of projectile breakup and for higher velocities the experimenter is on his own.

V. Acknowledgements

The author wishes to express his gratitude to J. N. Fritz, J. A. Morgan, J. W. Hopson, R. G. McQueen, and L. M. Barker for their assistance and cooperation during the preparation of this review. The work was performed under the auspices of the U. S. Atomic Energy Commission.

REFERENCES

1. M. H. Rice, R. G. McQueen, and J. M. Walsh, Solid State Physics, $\underline{6}$, 1-63 (1958).

2. R. G. McQueen, "Laboratory Techniques for Very High Pressures and the Behavior of Metals under Dynamic Loading", (K. A. Gschneider, Jr., M. T. Hepworth and NAD Parlee, eds.), Metall. Soc. Conf., Vol. 22, Gordon and Bready, New York, 1964.

3. R. G. McQueen, S. P. Marsh, J. W. Taylor, J. N. Fritz and W. J. Carter, "The Equation of State of Solids from Shock Wave Studies in High Velocity Impact Phenomena", Ray Kinslow, ed., Academic Press, New York and London, 1970, pp. 293-417.

4. J. Taylor, "Detonation in Condensed Explosives", Oxford, 1952.

5. G. R. Fowles, et al, Rev. Sci. Instr., $\underline{41}$, 784 (1970).

6. A. H. Jones, W. M. Isbell and C. J. Maiden, J. Appl. Phys., $\underline{37}$, 3493 (1966).

7. J. W. Taylor and M. H. Rice, J. Appl. Phys., $\underline{34}$, 364 (1963).

8. D. Bancroft, E. L. Peterson and F. S. Minshall, J. Appl. Phys., $\underline{27}$, 291 (1956).

9. L. M. Barker, "Measurement of Free Surface Motion by the Slanted Resistor Technique", SCDR-78-61, Sandia Corporation, Albuquerque, N. Mex. (1961).

10. M. H. Rice, Rev. Sci. Instr., $\underline{32}$, 449 (1961).

11. A. N. Dremin, S. V. Penshin and V. F. Pogarelov, Combust. Explos. Shock Waves, $\underline{1}$, No. 4, 1 (1965).

12. R. G. McQueen, private communication.

13. J. N. Fritz and J. A. Morgan, Rev. Sci. Instr., in press.

14. L. M. Barker and R. E. Hollenbach, Rev. Sci. Instr., $\underline{36}$, 1617 (1965).

15. L. M. Barker, "Fine Structure of Compressive and Release Wave Shapes in Aluminum Measured by the Velocity Interferometer Technique in Behavior of Dense Media Under High Dynamic Pressures", IUTAM symposium, Paris, 1967,

pp. 483-503.

16. L. M. Barker and R. E. Hallenbach, J. Appl. Phys., 43, 4669 (1972).

17. W. J. Halper, O. E. Jones and R. A. Graham in "Symposium Dynamic Behavior of Materials", ASTM Special Technical Publication No. 336 (American Society for Testing Materials, Philadelphia, 1963).

18. D. D. Keogh and J. Y. Wong, J. Appl. Phys., 41, 3508 (1970).

19. M. H. Rice, "Calibration of the Power Supply for Manganin Gauges", AFWL-TR-70-120.

20. S. B. Kormer, M. V. Sinitsyn, G. A. Kirillov and V. D. Urlin, J. Exp. Theoret. Phys., 48, 1033 (1965); Soviet Phys. - JETP (English Transl.), 21, 689 (1965).

21. W. C. Davis, Private communication.

22. R. G. McQueen, E. G. Zukas and S. P. Marsh, "Residual Temperatures of Shock Loaded Iron", Special Tech. Publ. No. 336, pp. 306-316 (Am. Soc. for Testing and Materials, Philadelphia, 1962).

23. J. W. Taylor, J. Appl. Phys., 34, 2727 (1963).

24. P. J. King, D. F. Cotgrove and P. M. B. Slate, "Infra-Red Method of Estimating the Residual Temperature of Shocked Metal Plates", ref. 13, pp. 513-520.

25. E. B. Royce, "Shock-Induced Demagnetization of Nickel Ferrite, Yttrium Iron Garnet, and Iron", ref. 13, pp. 419-429.

METALLURGICAL EFFECTS OF HIGH ENERGY RATE FORMING

R. N. Orava

Metallurgy and Materials Science Division, Denver Research Institute, University of Denver, Denver, Colorado 80210

INTRODUCTION

The authors of preceding tutorial papers in this volume have discussed the physical nature of shock-wave propagation in solids, the experimental techniques involved, and some aspects of transient mechanical behavior. The purpose of the present paper is to consider, firstly, the application of stress and shock waves to the technology of material fabrication and processing, and secondly, to assess the ensuing effects on material properties.

The working of metals by high-velocity techniques has achieved considerable interest, prominence, and success in recent years. High-velocity forming processes can be classified according to the means by which energy is generated: (1) chemically (explosive forming); electrically by high-voltage discharge (electrohydraulic forming); magnetically via high-voltage discharge also (electromagnetic forming); and fast-moving masses (impact extrusion, dynamic blanking, etc.). All of these techniques produce high rates of strain characteristic of stress-wave propagation; under certain circumstances, such as when explosives are placed in contact with a workpiece, or when the workpiece impacts a die at a high velocity, then the Hugoniot elastic limit can be exceeded and a plastic shock wave introduced. In most forming-type processes, however, one is not dealing with shock-wave phenomena but rather with plastic stress waves. Moreover, this treatment will be restricted primarily to the first three of the above group, i.e. explosive forming (EX) using high-velocity explosives, electrohydraulic forming (EH), and electromagnetic forming (EM).

This author and many others in the field have chosen to identify EX, EH, and EM metalworking operations as "high energy rate forming" (HERF) in spite of the opposition which periodically is voiced to the appropriateness of this terminology (e.g. Ref. 1, 2). The argument is based on the fact that the energy rate must be taken to mean the rate at which work is done. This defines power, and a high power can be generated by a large mass or force in conjunction with a low velocity although the characterizing deformation feature of HERF processes is a high workpiece velocity. However, it seems certain that the term HERF originated from a consideration of the rate at which energy is released from the primary energy source such as an explosive. Thus, while HERF is more correctly written as "high energy release rate forming," the meaning of high energy rate in this context should be quite clear without getting unduly pedantic about the matter.

Following a brief outline of the mechanics of EX, EH, and EM forming, their effects on metallurgical behavior will be reviewed. It was felt that the treatment of terminal characteristics would be more meaningful if some basis of comparison other than the undeformed state were selected. Accordingly, the discussion centers on the assessment of the metallurgical effects of HERF relative to conventional deformation processing. It is reasonable to presume that a component would normally have been produced in this manner if HERF had not been undertaken. Whereas material subjected to HERF will have experienced strain rates in the range 10^2 to 10^4 sec^{-1} at workpiece speeds of 50 to 600 ft/sec, conventional deformation processing involves deformation rates at least an order of magnitude or more lower[1-3]. In addition, stress waves are not involved in the latter and the high rates which can be developed are the result of a localization of deformation.

HIGH ENERGY RATE FORMING

The following advantages are offered by HERF over more conventional metalworking processes:

(1) large integral parts can be produced with little tooling and capital equipment expense when only one or few are required;

(2) difficult-to-form metals can be formed more readily;

(3) instability strains and accordingly, draw depths, tend to be higher;

(4) spring-back is minimized;

(5) components can be fabricated which could otherwise not be formed by any other process.

The first of the preceding items characterizes mainly EX forming. The forming of large parts utilizing electrical or magnetic phenomena is limited by the power requirements since the capacitor banks necessary would be prohibitive. On the other hand, for the fabrication of many, but smaller, components, the establishment of an EM or EH forming facility has obvious merit.

Explosive Forming

The fabrication of metals by explosives can be divided into two main groups, depending on the position of the explosive charge relative to the workpiece. In standoff operations, the charge is displaced from the workpiece surface and energy is transferred through a suitable medium, usually water. In contact operations, the explosive is detonated while in contact with the workpiece. The orders of magnitude of the pressure generated, metal velocity, and time period over which events occur in the workpiece are different, as typified in Figure 1[4]. Since EX forming is almost invariably accomplished by charge standoff, and since contact operations, as related to explosion welding and cladding, are discussed elsewhere in this volume, the present treatment is directed toward the former.

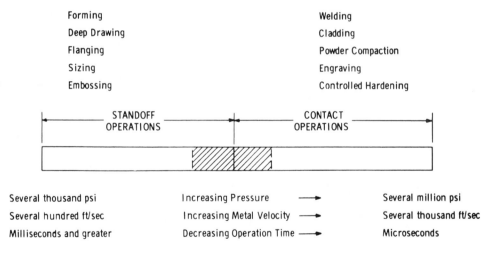

Figure 1. Range of Explosive Fabrication Processes (Ref. 4).

In standoff operations, the forming die, workpiece, and explosive, are immersed in water to a specified head depth, as shown in Figure 2[5]. Two general types of die configurations are the most common although mandrels are required for certain applications. Free forming (Figure 2a) is accomplished by bulging into an open die with the depth of draw controlled by the type and weight of explosive, standoff distance, and edge clamping pressure which governs the allowable edge pull-in. The alternative is to control the part shape by means of a female die (Figure 2b); die forming is understandably more popular in industrial applications. The effects of die impact, however, may be a significant factor in terminal material behavior.

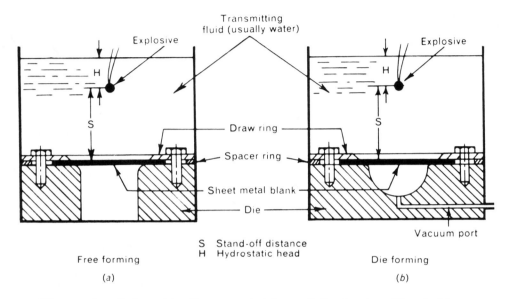

Figure 2. Schematic Representation of Explosive Standoff Free-Forming (a) and Die-Forming (b) Systems (Ref. 5).

The choice of explosive is dictated by availability, handling safety, cost, physical characteristics, and inactivity in the energy transfer medium. The detonation velocity of explosives ranges from about 2000 to 8500 m/sec and EX forming is commonly done with explosives in the upper range of this spectrum, typically TNT (6900 m/sec), RDX (8350 m/sec), PETN (7000-8300 m/sec), Composition C-4 (8040 m/sec), and Composition B (7800 m/sec). The amount of energy which can be produced by 1 gram of high explosive is approximately 1 kcal or 4000 joules, released in times of 1 to 2 μsec. While combustible fuels, say, have a somewhat higher specific energy output the release rate is about 7 orders of magnitude lower. It is noteworthy that an indoor EX forming facility is reasonably limited to 50 g of high explosive without confinement. If the energy utilization efficiency is 50%, then the maximum energy available for workpiece deformation is roughly 100,000 joules -- a figure to bear in mind relative to current EM and EH forming capacity. The reader is referred to Cook[6] and others[7,8] for additional information on explosive characteristics.

Methods have been developed for calculating forming conditions in the case of simple parts such as hemispherical domes and cylindrical tubes. The equations, derived from energy balance considerations, include as parameters standoff distance, charge weight and heat of combustion, pressure head of the energy transfer medium which controls efficiency, dimensions of the workpiece and final part, and the mechanical properties of the workpiece material. When more complex part configurations are involved, estimates become less accurate. Trial and error, and scaling, then play a much larger role in achieving the final product since calculational models have not reached the stage of sophistication required to accurately predict full-scale loading conditions. Details may be found in a number of treatises on the subject[1,5,7-11] and specific citations given therein.

Two factors are primarily responsible, to various degrees, for the ultimate displacement of the workpiece -- a pressure pulse or shock wave generated by the detonation of the explosive, and a pressurized gas bubble formed from the gaseous detonation products. The pressure pulse, containing about 60% of the explosive energy is reflected as a compression wave at the front face of the higher impedance metal workpiece. This corresponds to the principal forming force of the primary pressure pulse. When the gas bubble is permitted by design (short standoff distance) to break over the workpiece, then some of its pressure can be released by displacement of the workpiece. If the interaction between bubble and workpiece occurs during the expansion in the first of several oscillations of the bubble, then the deformation possible may be several times that due to the pressure pulse alone[4]. Some forming force also can be generated if the rarefaction wave, reflected originally from the lower-impedance air at the back surface of the workpiece causes cavitation of the water which has little tolerance for a tensile stress[4]. Collapse of the cavity can then produce a water hammer effect.

In order to enhance the utilization and potential of EX forming, a great deal of work is still required in the areas of scaling laws, energy transfer, die configurations and materials, appropriate explosives, preforming material conditions, and transient material behavior. At the present time, for example, it is often necessary, because of the lack of data, to use quasi-static strength and work-hardening properties for loading estimates, although dynamic rates of strain are involved. It would seem that the talents and experience of many individuals presently engaged in the mechanics and physics of wave propagation in solids could be gainfully applied to the solution of many of the problems still being encountered in the field of HERF.

Electromagnetic Forming

Both EM and EH forming have the advantage over high-explosive forming of more economical multi-part capability, substantially lower noise levels, and less hazardous operation. Indeed, many industrial plants now use EM or EH techniques to produce parts for automobiles, electrical equipment, office machinery, ordnance, home appliances, and aerospace assemblies. Equipment for EM[12] and EH[13,14] forming is now available commercially, ranging up to 150 kilojoule capacitor bank energy storage capacity. It must be kept in mind, however, that even with a 25% efficiency of energy utilization, the working potential is unlikely to exceed that of 15 grams of, say, Composition B.

The basis of the associated EM and EH forming processes is the rapid release, within a few microseconds, of stored electrical energy. In the former, the electrical pulse suddenly generates a magnetic field, up to the order of 100 kilogauss in one or more work coils located adjacent to the workpiece surface (Figure 3). The interaction of the magnetic field with the workpiece induces in it a current which in turn interacts with the compressed magnetic field to generate a force normal to the workpiece surface. The kinetic energy created in the workpiece by this impulse is then dissipated in a matter of a few milliseconds by deformation when appropriate constraints are present.

Part-size limitations aside, EM forming does offer the following advantages.

(1) There is no physical contact between the pressure generator (work coils) and the workpiece thereby precluding friction contact and lubrication.

(2) Since the magnetic field can pass through nonconductive coatings, protective surfaces present no restriction to the forming of the composite material.

(3) The pressure is uniform since the force is effectively normal to all surfaces except in the case of relatively sharp surface discontinuities.

(4) No torques or bending moments are produced.

(5) The forming pressure is relatively insensitive to variations in the separation between the work coil and workpiece.

(6) Since the workpiece is the only moving part, the repetition rate, which can reach 1000 per minute under certain conditions, is not limited by the inertia of any extraneous solid or liquid; the transfer medium is air.

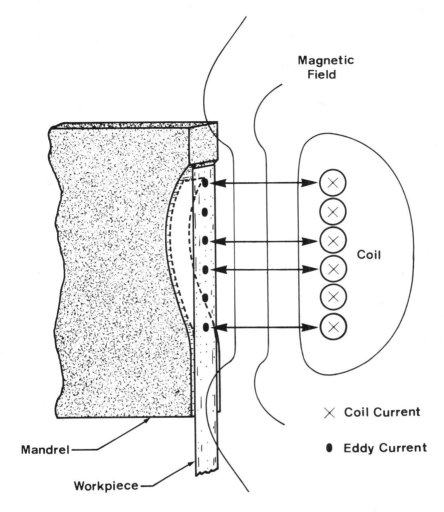

Figure 3. Underlying Principles of Electromagnetic Forming.

(7) The timing and the strength of the magnetic pulse is precisely controllable.

(8) The potential is great for elevated temperature and controlled environment operation since there is no thermal and intimate contact with the workpiece.

There are, as would be expected, disadvantages also. The physical strength of the work coil provides a restriction to peak pressures and the repetition rate. It must be capable of withstanding the same forces as are exerted on the workpiece. Configurations with slotted parts are more difficult to form although field shapers can be incorporated into the system to counteract this problem to some degree. Direct forming is most efficient for materials with a high conductivity, in excess of about 0.17 µmho/in. These include copper and aluminum base alloys, gold, silver, iron and mild steels. Austenitic stainless steels and titanium alloys are generally not readily formable electromagnetically. The method commonly adopted for the EM working of low-conductivity materials involves the utilization of a high-conductivity metal driver sheet. However, this does increase material costs, introduces the possibility of arcing between driver and workpiece which can cause bonding, and requires precautionary measures to avoid the entrapment of air between the driver and workpiece.

Additional information on EM or magnetic-pulse forming, including facility design, component configurations which can be produced, and general principles of the process, may be found in References 1, 5, 9, 15, 16, among many others.

Electrohydraulic Forming

In EH metalworking processes, the energy stored in large capacitor banks is rapidly released either to a discharge in the gap between electrodes, or to vaporize and explode a wire bridging an electrode gap. The spark discharge rapidly ionizes the liquid of the transfer medium, usually water, to produce a shock wave and a pressurized gas bubble; the effect is the same when the bridge wire explodes[1]. Thereafter, the events leading to workpiece displacement are similar to those in EX forming. However, EH forming lends itself rather better to control since reflectors can be used to direct and concentrate the pressure pulse toward a selected region of the workpiece. Overall control is better with bridge wires than spark gaps but replacement of the wire will raise unit costs.

Systems for forming flat sheets and tubes is depicted schematically in Figure 4[5]. Many of the advantages peculiar to EM forming pertain to EH fabrication except for the repetition rate, which will be lower, and those features relating to environment and temperature control. The storage capacity does not currently exceed 150,000 joules in commercially available equipment. A wide variety of small parts have been formed by EH techniques which are described more fully in a number of surveys[1,5,9]. The greatest advantage appears to be in the capability for localizing the energy to create, effectively, a small-scale liquid punch. This provides then a means for shaping complicated configurations.

METALLURGICAL EFFECTS OF HIGH ENERGY RATE FORMING 137

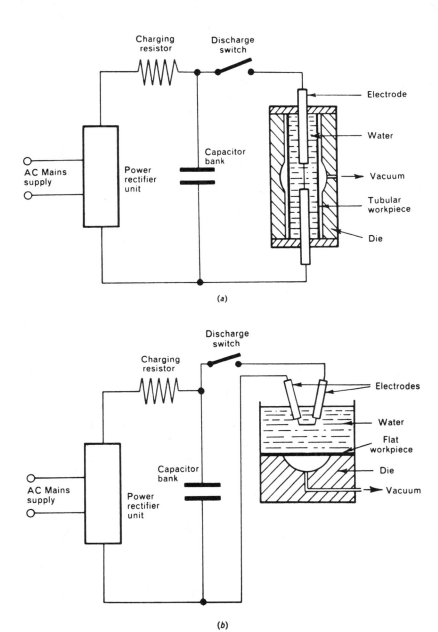

Figure 4. Electrohydraulic Forming Systems for (a) Tubular Parts and (b) Parts Produced from Flat Sheet (Ref. 5).

METALLURGICAL EFFECTS

General Overview

A factor which has proved to be one of the major blockages to the wider application of HERF, particularly for relatively large, critical components requiring explosives, is the reservation that the mechanical properties could be affected adversely. As stated previously, any deterioration of residual behavior is sensibly viewed in relation to some more conventional means of fabricating a component, for which normal design data accounting for work hardening, texturing, etc. should be plentiful. Accordingly, only differential metallurgical effects of HERF relative to lower velocity operations are examined. Although a uniaxial tensile test is not forming in the strictest sense, information pertinent to the problem can be derived by the examination of the influence of quasi-static and dynamic tensile prestraining on terminal material behavior and substructures.

The residual characteristics after forming are conveniently studied according to the grouping of the post-forming material history into four categories. Firstly, the component may be used in the as-formed state. Alternatively, a terminal thermal treatment may be applied which is intended to: relieve large-scale residual stresses, or provide some control of the deformation substructure for property optimization, a procedure often termed thermal-mechanical treatment (T_{MT}); remove all effects of prior work, i.e. full heat treatment or recrystallization anneal; complete a heat treatment sequence, usually one or more aging or tempering steps, a procedure commonly termed thermomechanical processing (TMP) or treatment (TMT). At the present time, relative data are most numerous for the as-formed condition which is subject to all of the constraints, limitations, and advantages introduced by strain hardening.

Prior to 1970, the only paper devoted exclusively to the detailed examination of the metallurgical effects of HERF, specifically EX forming, was that of Rowden[17]. Based on the data available to him, Rowden reasoned that the influence of explosive forming on mechanical properties is largely dependent on the degree of strain hardening, whence he concluded that explosive forming would be expected to affect subsequent behavior in a way similar to other deformation processes. No apparent consideration was given to the possibility of any strain-rate sensitivity of work hardening, i.e. a failure of the mechanical equation of state.

Various reports[18,19] and papers[20-23] contain brief summaries of previous findings. All cover much of the same earlier work. Strohecker et al.[19] emphasized the fact that terminal property data were so few in 1964 that overall conclusions were impossible.

Others[20,22,23] merely reported results with little definitive comment and some[24] have stated conclusions without reporting data.

On occasion, broad generalizations have been implied, with little apparent reservation, from investigators' own studies of specific alloys[18,21,24]. All favor the view that explosive forming is not detrimental to post-forming properties, and in some instances, can lead to improvements[24]. No support seemed to exist at the time for the contention[17] that properties can be restored generally to an acceptable level by a simple heat treatment, but this statement relies heavily on one's criterion of acceptability. A survey of the terminal behavior of aluminum alloys did disclose an absence of forming-rate sensitivity and moreover, a restoration of unformed properties by appropriate heat treatment.

One of the earlier problems encountered was the prevalent misunderstanding that EX forming standoff operations could be classified right along with contact or shock loading operations in the equivalence of their effect on residual behavior. This seemed to reflect a failure to recognize the differences between the basic nature of the wave motion responsible for deformation in the two cases, and that the material rate of strain could be at variance by as much as four or five orders of magnitude.

The above inconsistencies and uncertainties led to the publication in 1970 of a review paper by Orava and Otto[25] in which an attempt was made to place into proper perspective the significance of the data to which the authors had access on the metallurgical effects of EX, EH, and EM forming relative to more conventional fabrication techniques. While there was little disagreement with some of the observations made by Rowden[17], and his overall impressions concerning the hopeless inadequacy of vital information, the authors were able to be somewhat more detailed and specific in their final assessment of the current status of terminal behavior. They reached the general conclusion that components formed by high energy release rate techniques could be used with little reservation in most cases, relative to conventionally processed counterparts, if design requirements were limited by hardness, strength, ductility, fatigue, and/or stress corrosion characteristics. However, adverse differences in the following properties of dynamically, as opposed to quasi-statically, deformed materials could represent the source of some concern: strength of 17-7 PH semi-austenitic stainless steel and some mild steel; ductility of 5456-0, 2219-T87, and 7039 (after STA heat treatment) aluminum alloys; ductile-brittle transition temperature of 0.18% C mild steel (even after stress relief); fatigue life of unalloyed aluminum and 5052-0 aluminum alloy; stress corrosion life of Cr-Ni 300-series stainless steel, although a full anneal may restore resistance to undeformed levels. The question of whether many other alloys in the families to which the preceding belong are similarly affected could not be answered. It was also emphasized that overaging could result unless standard aging practice is modified in precipitation-hardenable alloys sensitive to strain-accelerated aging effects. However, whether the rate itself plays a significant role in this respect could not be incontrovertibly established. It was found that the dependence of microstructural changes such as dislocation substructure, mechanical twins, and strain-induced transition phases, on forming rate could be substantial. On the other hand, these had neither been documented well enough, nor their effects sufficiently well understood to permit any reasonable correlation between terminal structure and properties.

Since the above survey article[25] was written, the results of a number of investigations have been reported which have a direct or indirect bearing on the state of our knowledge and understanding of the relative metallurgical effects of HERF and conventional forming. These will be incorporated into a brief summary of the state of knowledge of each important property, as extracted from the previous paper. To conserve space, and with apologies to the authors, references which have been cited previously[25] will not be repeated. Again, unless otherwise specified, it is to be understood that all comments refer to the effects of HERF or dynamic deformation relative to conventional processing or quasi-static deformation.

Microstructure

The single most important factor which will govern mechanical behavior after forming is the terminal microstructure. This, in turn, depends on the transient response of the material to deformation at high rates of strain. Microstructural differences reside primarily in variations in the density and distribution of lattice defects such as dislocations, vacancies and interstitials, stacking faults, and mechanical twins, and also in the amount of strain-induced transformation products in alloys susceptible to such transitions. In all of these respects, the existing state of knowledge and understanding must be considered inadequate.

Dislocation Substructure.
A feature often associated with dynamically loaded materials is a refinement of the dislocation substructure (Table 1). This can take the form of a decreased slip-band spacing and width as determined by the observation of polished surfaces. Alternatively, there may be a decrease in the cell diameter and/or decrease in the length of dislocation segments which are more numerous and more uniformly distributed. An example of the latter effect of forming rate on the substructure of unalloyed titanium[31] is shown in Figure 5. However, as Table 1 indicates, this refinement is not universally observed. Concurrently, it has been found that multiple slip tends to be less prevalent. It is interesting to note that the dispersal of slip is accompanied in some cases[28,32] by an increase in the total dislocation density and in

Table 1. Incidence of Slip Refinement at High Strain Rates.

Stress State	Method of Observation	Refinement Observed*	
		Yes	No
Uniaxial	Surface	Fe, Fe-Si 0.24C Steel 0.32C Steel Cu, Al(25,26,27)	Al 301 S.S. Zn
	TEM	Mo, Nb Al(26,28,29)	Cu Ni(30)
Biaxial	Surface	Al 0.025C Steel Nimonic 90	
	TEM	347 S.S. Al Ti(31)	

* Work not cited is referenced in Reference 25.

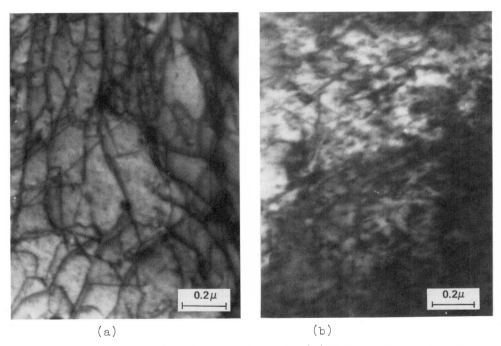

(a)   (b)

Figure 5. Dislocation Substructure in (a) Rubber Pressed and (b) Explosively Free-Formed Ti-50A Titanium. Effective Strain: 4.0%.

others[26,33] by no significant change, for comparable low and high rate prestrains.

The strain-rate dependence of the dislocation distribution has been attributed to the activation of more sources at the higher strain rates, with less activity per source, than at lower rates. This can be understood when the standard expression for shear strain rate,

$$\dot{\gamma} = \rho_m b \bar{v}$$

where ρ_m is the mobile dislocation density, b the Burgers vector and $\bar{v}$ the average dislocation velocity, is rewritten as

$$\dot{\gamma} = N\bar{L}b\bar{v}$$

Here, N is now the number of mobile dislocation segments with average length $\bar{L}$. If ρ_m is unchanged by increasing the rate[32], then a decrease in L, related to source length, requires that N, related to the number of sources, increase. Since the flow stress has a positive rate sensitivity, the effective stress for source operation is greater and the source length need not be as large to be activated.

<u>Twinning</u>. The incidence of mechanical twinning exhibits a positive dependence on the rate of strain, as demonstrated in Table 2 for a number of metals. The author is unaware of any data which indicate that twin densities are lowered by dynamic straining. Three points are noteworthy. The second column reveals that a number of materials which might be expected to twin at dynamic rates (exclusive of shock loading rates) apparently do not. The results also show that the limiting carbon content above which mild steels will be unlikely to twin at rates characteristic of room-temperature HERF processes is approximately 0.2%. Lastly, a discrepancy concerning the nature of observed deformation products in austenitic 300-series stainless steels is evident from the table. These have been attributed variously to HCP ε-phase, stacking faults, and twins. However, after this question was raised[25], the results of an investigation of the terminal substructure of explosively expanded thin-walled cylinders of metastable 304 stainless steel have been published by Murr et al.[36]. Their incontrovertible identification of the deformation products as thin mechanical twins appears to resolve the controversy. Only traces of HCP ε-phase were found, while α' martensite could not be detected at all. Stacking faults at such were clearly discernible only in material which had received the lowest strain (0.5%).

The distinction between the contributions of dislocation substructure and twinning to strength and hardness still remains somewhat of a problem since the two deformation modes coexist. However, the close correlation of strength properties with twin density in

304 stainless steel[36,37] and manganese steel[38], does suggest indirectly that twin boundaries do act as significant strengthening barriers. However, since twins have a simple orientation relationship relative to the matrix, twin boundaries are not impenetrable obstacles to all slip dislocations in BCC[39], FCC[40], and HCP[41] structures. Strengthening contributions derive from the blockage of some dislocations, and the energy required for dissociation of those which are faborably oriented to traverse the twin boundaries. One can regard the effect of the presence of twins as a grain-size refinement but with a Hall-Petch slope less than that if the twin boundaries were actually grain boundaries. Accordingly, if HERF generates twins then the terminal strength could exceed that of conventionally formed components which contain a dislocation substructure only.

It has been established that twins can be responsible for the nucleation of cleavage cracks during deformation at low temperature[42] or high strain rates[43]. If the twins are generated by prior work, it is not entirely clear how they will influence subsequent fracture properties. If the deformation has been imparted at a high rate, it is entirely possible that some microscopic flaws will have been developed. One would then have to utilize fracture toughness concepts to allow for such defects in design. If such microcracks cannot be detected and, for the sake of argument, do not exist, then the pre-existing twins are unlikely to lead to crack nucleation unless subsequent service conditions are conducive either to their growth or to the nucleation of new twins, and where conditions favor fracture over slip and twinning, i.e. low temperatures and high rates.

Small prestrains prior to HERF can serve to inhibit or eliminate twinning if it is felt to be a problem[25].

Point Defects. Several aspects of terminal behavior, such as otherwise inexplicable strengths[36,37] and apparent enhanced aging effects[25] due to HERF, and higher density of dislocation loops, jogs, and dipoles after dynamic deformation, all tend to suggest indirectly that the production of point defects is greater during high strain rate loading. However, there is, as yet, no direct evidence to confirm this implication. Kressel and Brown's resistivity study has demonstrated such an effect in the shock regime[4], but whether it is true for rates four to five orders of magnitude lower is open to speculation. Nevertheless, a contribution to differential strength from point defects can be expected if an excess exists.

Phase Transitions. The revelation by Murr et al.[36] that mechanical twins rather than HCP ε-phase are most likely responsible for the controversial strain markings in explosively formed 304 stainless steel reverses the prediction advanced previously[25] that the high strain rates characteristics of HERF will result in more HCP ε-phase than lower rates and no mechanical twins in austenitic stainless steel. However, lower forming rates are more conducive to the strain-induced transition to BCC martensite than HERF rates, as demonstrated in 300-series[25,36] and 17-7PH stainless steels[45], accounting for the higher strengths after conventional working at appropriate strain levels[25,37,45].

Adiabatic heating effects during high strain rate deformation can play a role in the introduction of undesirable localized microstructural phases. For example, narrow untempered martensitic zones were identified in explosively-loaded thick-walled cylinders of 4337 HSLA steel[46] and in a low-alloy, projectile-impacted armor plate[47]. Since these exist in a tougher tempered martensitic matrix, they can represent potential paths for brittle fracture. An excellent treatment of thermoplastic effects during dynamic deformation has been presented by Culver[48]. He disclosed that thermal instability is unlikely to be encountered in other than titanium alloys at HERF rates.

Mechanical Properties

Attention is directed in this section to as-formed properties and the effects of subsequent heat treatments. Thermal treatments which could reasonably be considered as part of a thermomechanical processing schedule are dealt with in the subsequent section.

<u>Hardness and Tensile Behavior</u>. A great deal of information is available on dynamic stress-strain behavior[2]. Unfortunately, differences between static and dynamic curves traditionally reflect differences both in the work-hardening component of the flow stress, if the mechanical equation of state is not obeyed, and a thermal stress component since the time available for thermal activation is changed. Whereas the former will contribute to relative strength levels during subsequent static testing, the latter will not. Therefore, a simple comparison of static and dynamic stress-strain curves is not sufficient for present purposes.

A qualitative picture is presented in Table 3 of the influence of dynamic uniaxial loading or HERF on terminal strength levels, grouped according to whether the property is unchanged ($\leq \pm 5\%$), decreased, or increased, relative to the corresponding value after straining at a lower rate.

Attention should be focussed on the last column. In addition to the aforementioned concern about the strengths of the higher carbon mild steels and 17-7PH stainless steel formed in the solution treated condition, the yield strength of explosively formed Beta III is nearly 20% lower than that of cold-rolled material.

Table 3. Qualitative Influence of Dynamic Uniaxial or Multiaxial (HERF) Deformation On Terminal Hardness and Strength Levels in the As-Deformed Condition, Relative to Low Rate Straining.

Property	Stress State	No Change	Increase	Decrease
		(a) Fe, Fe Alloys		
Hardness	Uniaxial	0.2C Steel	0.05C Steel (?)	Armco Fe 0.24C Steel
	Multiaxial		17-7PH: T, R-100 (45) 316, 321, 347 304 (25, 37)	0.025C Steel Mild Steel 17-7PH: A, A1750(45)
Yield	Uniaxial		301, 304, 310	Armco Fe Steel: 0.025, 0.22, 0.24, 0.32C 17-7PH: A
	Multiaxial	HY-80: contact, air (49)	304, 316, 321 4130 (49), 4340 (49)	HY-80: contact, H$_2$O (49)
Ultimate	Uniaxial	Armco Fe 0.025C, Fg, Steel 310, 17-7PH: A	301, 304	0.025C, cg, Steel
	Multiaxial	321: EH 4130 (49), 4340 (49) HY-80 (49)	304, 316 321: EX	

Table 3. (Continued)

Property	Stress State	No Change	Increase	Decrease
		(b) Al, Al Alloys		
Hardness	Uniaxial	Al		
	Multiaxial	2014-0, T6 (25, 50) H.S. 15 SA, WP DTD 687A		
Yield	Uniaxial	1100	Al	
	Multiaxial	2014-0 (25, 50) 2014-T6 (50)		5052-0, 5456-0
Ultimate	Uniaxial	Al, 1100, 5456-0		5052-0
	Multiaxial	2014-0 (25, 50) 2014-T6 (50)		
		(c) Ti, Ti Alloys		
Hardness	Multiaxial			Ti
Yield	Uniaxial	6Al-4V: ST 13V-11Cr-3Al: ST		
Yield and Ultimate	Multiaxial	Ti (35) 6Al-4V: ann (35)		Beta III: (51)
		(d) Ni, Ni Alloys, Copper		
Hardness	Multiaxial		Cu, Nimonic 75 Nimonic 90	

There is a general tendency for the terminal strengths of FCC metals to be higher after dynamic loading than static deformation, while the reverse trend prevails for basically BCC and HCP structures. One might regard the behavior of 17-7PH steel, and 4130 and 4340 HSLA steels, as evidence to the contrary. However, it should be noted that 17-7PH is more susceptible to a strain-induced martensitic transformation at low than at high strain rates[35] which would necessarily introduce a negative prestrain rate sensitivity of terminal strength. Also, the 4130 and 4340 experienced die impact during explosive forming, a factor which could have provided an additional work-hardening contribution.

The difference in the effect of strain rate on the dislocation distribution in FCC and BCC could account for the difference in the sign of the prestrain rate dependence of terminal strength. In aluminum, for example, a cell structure presists during high rate deformation but the cell size is decreased. Since the effect on strength is similar or more pronounced than refining the grain size, in keeping with a Hall-Petch type of relation[52], the strength after dynamic prestraining can be expected to be higher than after static prestraining. Conversely, it has been shown that the tendency for cell formation is curtailed during the high-velocity deformation of BCC metals, e.g. molybdenum[32] and columbium[33]. If, as pointed out by Edington[33], a cell or tangled structure is a more effective strengthener than a uniform distribution of dislocations, then one can expect a lower strength after high rate deformation than after low.

Evidence generated since 1970[25] on explosively formed 2014-0 and T6 alluminum alloy[50], unalloyed titanium[35], Ti-6Al-4V[35] and Beta III[51] titanium alloys, and HY-80, 4130 and 4340 steels[49], indicates that tensile ductility is either the same (within ± 10%), or better than after conventional processing and does not alter the conclusion regarding any need for undue concern about this parameter.

A full terminal hardening heat treatment does not lead to any significant differences in tensile properties between conventionally and EX formed 2014[50] and 2219[25] aluminum alloys; 4130 and 4340 steels[49] are also essentially unaffected except for an 18% relative reduction in the elongation of 4340[49]. However, a quenching and tempering treatment resulted in 10 to 15% lower strengths in EX formed HY-80 steel, compared with the cold-rolled and similarly heat treated counterpart; ductility was also lowered. Consequently, it would be unwise to accept, without experimental confirmation, the intuitive feeling that a thermal treatment involving high temperature recrystallization or solutionizing will automatically alleviate the effects of prior work irrespective of its rate of application.

Fracture Characteristics. Fracture properties presented previously[25] were all evaluated relative to the parent metal values. Information on forming rate dependence has been generated only recently. Otto and Mikesell[49] documented significant differences between EX formed and cold rolled Charpy impact strengths and transition temperatures in three HSLA steels. The results were complicated by initial plate rolling orientation effects, the resolution of which are presently under study. Nevertheless, it was found that the impact resistance (strength and DBTT) of EX formed 4130 (L) was better, of 4340 (L and T) equivalent or better, and of 4130 (T) and HY-80 (T) worse than that of cross-rolled material. Where comparisons were possible, the above trends persist after a full heat treatment, being reversed only in the case of 4340 (T). An appreciation for the magnitude of the differences can be gained from the data in Figure 6 for HY-80 (L) which was quenched and tempered after working.

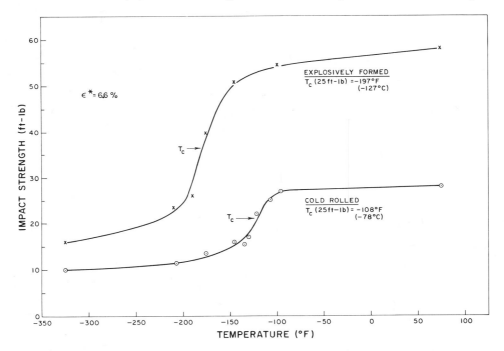

Figure 6. Impact Strength of HY-80 Steel Austenized, Quenched, and Tempered (1000°F) After Forming; Longitudinal Orientation (Ref. 49).

In a forthcoming publication on pressure vessel steels, it will be reported that the impact resistance of EX formed and stress-relieved A285 is better than its tensile prestrained and stress-relieved counterpart, while the opposite trend prevails for A515.

The observation by Watters et al.[53] that pre-existing twins generated during the shock loading of low carbon steel benefit both crack initiation and propagation may prove to have a bearing on the future understanding of the behavior of dynamically processed materials with high twin densities.

It is unfortunate that the increasing indispensibility of fracture toughness as a measure of unstable crack extension has failed to stimulate any publicized interest in its application to the evaluation of terminal fracture behavior.

Fatigue Resistance. With the exception of the wider dissemination[50] of some information cited previously[25] relating the fatigue life of EX formed 2014-T6, and 2014-T6 EX formed in the 0 condition, to rubber pressed counterparts, no new information on the forming rate dependence of fatigue resistance has appeared. Reliable and carefully documented evidence of terminal fatigue behavior is still very limited, existing only for some austenitic stainless steels and 2014 aluminum. Nonetheless, these data, and claims for several other materials, suggest that explosive forming produces changes in fatigue properties which either differ little from changes produced by more conventional processing, or represent a substantial improvement.

Creep and Stress Rupture Behavior. There is no available information on the long-term elevated temperature behavior of HERF materials compared either with undeformed stock or with conventionally formed counterparts. This deplorable situation is eased somewhat by the knowledge that static prestraining can lead to an improvement in creep resistance under certain circumstances[52]. Moreover, it was disclosed that the stress-rupture life of shock-hardened 304 stainless steel was raised by over an order of magnitude, and the minimum creep rate reduced by three orders of magnitude[54]. Clearly some studies are warranted in this area.

Stress Corrosion Behavior. While it has been shown that the smooth-specimen stress corrosion resistance of several alloys[25], including the methanol cracking of unalloyed titanium and Ti-6Al-4V[35], is virtually unaffected by forming rate, the relative detrimental effect of HERF on the cracking behavior of Cr-Ni austenitic stainless steels in hot chlorine solutions remains a problem. It has been claimed[25] that this deterioration can be completely alleviated by a full annealed. Conflicting and comforting evidence will be presented later this year by Van Wely[55] to illustrate that cold-pressed and EX formed 304 L exhibit the same reduction in cracking resistance from the annealed state.

The comment concerning the evaluation of fracture toughness pertains also to environmental subcritical crack growth. Studies are presently in progress at this laboratory to assess the resistance of EX formed 304 and 310 stainless steel to crack extension in NaCl solutions and to establish thermal treatments intended to restore resistance to acceptable levels.

Explosive Thermomechanical Processing

When an alloy is amenable to synergistic strengthening effects due to the combined selected application of mechanical and thermal treatments (TMP), it seems sensible, rather than perform a straightforward HERF operation, to utilize this as the mechanical working stage in a cold (ambient temperature) TMP schedule. Once again, the rationale for so doing should be based on gaining improvements in terminal properties, over and above those which would normally derive from conventional TMP.

There have been implications that dynamic TMP by HERF might be potentially advantageous to the strength of aluminum alloys[25]. However, a critical examination[25,56] of the evidence reveals that a forming rate dependence of aging response and substructure strengthening, and therefore, TMP strengthening, of precipitation-hardenable aluminum alloys has not been established. Moreover, the early results of Stein and Johnson[57] showed that the effectiveness of the ausworking of D6-AC steel was independent of whether the deformation was imparted conventionally or by EX forming. This investigation has understandably served for a number of years as one of the deterrents to further studies in this area.

More recent attempts at dynamic TMP by the EX forming of 17-7PH stainless steel and Beta III titanium alloy have been equally disappointing[45]. Explosive TMP was responsible for a small improvement in the tensile properties of 17-7PH steel for some of several TMP schedules examined, compared with conventional TMP by cold rolling. However, it was concluded that none of the benefits were sufficient to warrant the purposeful application of explosive TMP specifically for strengthening. Conversely, the tensile properties of solution treated, EX formed, and aged (900°F) Beta III titanium were poorer than after cold rolling and aging. Heat treatment schedules other than standard have not been examined. Before TMP by HERF is discarded as a viable method of strengthening, thermal treatment modifications, and a number of other alloys should be investigated.

In contrast to the above findings, the few results which have been generated thus far have indicated that TMP by shock loading may prove to be an attractive strengthening technique. Its physical advantages include small dimensional changes, and uniformity and isotropy of hardening.

Relative beneficial effects of shock TMP have been observed for Fe-3Cr-0.4C steel (strength)[57], A-286 stainless steel (strength)[58], 17-7PH stainless steel (strength)[45], AISI 1008 carbon steel (strength)[59], 6061 aluminum alloy (strength and microstructural stability)[60], 7075 aluminum alloy (stress corrosion resistance[61], strength[62], and fracture toughness[62]), Beta III titanium (strength)[51] and Udimet 700 nickel-base alloy (strength, ductility, stress-rupture life, creep rate)[63,64]. Typical engineering creep curves for Udimet 700 are presented in Figure 7, comparing the influence of shock TMP, conventional TMP, and thermal treatment only[63]. Another advantage of the shock-aging of Udimet 700 is the inhibition of cellular recrystallization which can result from normal cold work[64]. The property improvements were believed to result from a refinement of slip by shock loading. Slip is still largely planar but the separation between bands is appreciably reduced[64].

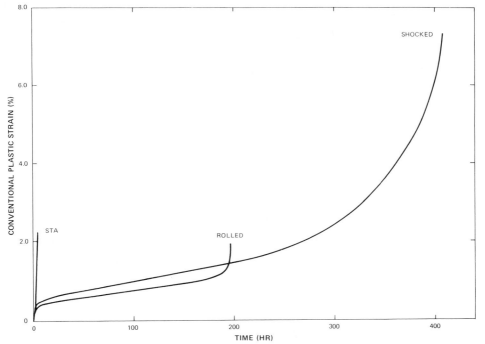

Figure 7. Constant-Load Creep Curves for Udimet 700 After Thermal Processing (STA), Conventional TMP, or Shock TMP (Ref. 63)/

The future of shock TMP appears to be a bright one but considerable ingenuity and imagination is going to be necessary to transfer the concept into technological applications. Moreover, the nature of shock deformation substructure points to the possibility of success in the area of MTT, or substructure control[52].

SUMMARY AND CONCLUSIONS

In this paper, the basic principles of three high energy (release) rate forming techniques -- explosive, electromagnetic, and electrohydraulic -- have been outlined. Subsequently, their effect on terminal material behavior was reviewed, with emphasis on studies reported since late 1969.

All in all, it can be concluded that the microstructural changes which are introduced by HERF have much the same influence on terminal properties and thermal response as conventional metalworking, with some notable exceptions. In addition to the reservations itemized in the preceding general overview, it should be emphasized that if an austenitic steel is sufficiently unstable to be highly susceptible to the formation of BCC martensite during deformation, then its occurrence is less pronounced at high rates of strain. This will be reflected in lower terminal strengths after HERF unless thermal steps are taken to complete the transformation. In addition, while fracture resistance of HSLA steels can often be better after HERF than after conventional forming, care should be taken to ensure that the transverse orientation, relative to the original plate rolling direction, does not exhibit adverse effects. Recent evidence indicates that HERF may not be as damaging, comparatively, to the stress corrosion resistance of Cr-Ni austenitic stainless steels as has hitherto been feared. Nevertheless, the possibility of a detrimental influence cannot be ignored.

Although thermomechanically processing by utilizing HERF instead of conventional forming has not been shown as yet to hold any significant advantages, shock TMP on the other hand has been demonstrated to be an attractive alloy strengthening technique. Mechanical-thermal, and multimechanical-thermal treatments involving shock loading also warrent further investigation.

REFERENCES

1. M. C. Noland, et al., High-Velocity Metalworking, NASA Technology Utilization Report No. NASA SP-5062.

2. K. Bitans and P. W. Whitton, Intl. Met. Review 17, 66 (1972).

3. G. E. Dieter, Fundamentals of Deformation Processing, Proc. 9th Sagamore Army Materials Research Conf., ed. W. A. Backofen, et al., Syracuse University Press, 145 (1964).

4. J. Pearson, Proc. Symp. on Behavior and Utilization of Explosives in Engineering Design, L. Davison et al., eds. New Mexico Section, American Society of Mechanical Engineers, 69 (1972).

5. R. Davies and E. R. Austin, Developments in High Speed Metal Forming, Industrial Press Inc., New York (1970).

6. M. A. Cook, *The Science of High Explosives*, Reinhold Publishing Corp., New York, (1958).

7. J. S. Rinehart and J. Pearson, *Explosive Working of Metals*, Pergamon, New York (1963).

8. A. Ezra, ed., *Principles and Practice of Explosive Metalworking*, Fuel and Metallurgical Journals, Ltd, London, England (1973).

9. E. J. Bruno, ed., *High Velocity Forming of Metals*, American Society of Tool and Manufacturing Engineers, Dearborn, Michigan (1968).

10. A. A. Ezra, J. Materials $\underline{4}$, 338 (1969).

11. W. Johnson, et al., *Advances in Machine Tool Design and Research Conference*, Proc. 4th Intl. M.T.D.R. Conf., Pergamon Press, New York, 257 (1964).

12. "Operation and Application of MAGNEFORM Machines," Bulletin No. 100, Magneform, Maxwell Laboratories, Inc., San Diego, Calif.

13. "Electrohydraulic Forming," brochure, SoniForm, Inc., La Mesa, California.

14. "The Cincinnat Electroshape," Manual ES-3, The Cincinnati Shaper Co., Cincinnati, Ohio (1969).

15. D. F. Brower, Metals Progress, $\underline{93}$ (4), 95 (1968).

16. W. N. Lawrence, Proc. 2nd Intl. Conf. of the Center for High Energy Forming, ed. A. A. Ezra, Denver Research Institute, University of Denver, p. 3.5.1 (1969).

17. G. Rowden, Metallurgia $\underline{75}$, 199 (1967).

18. F. C. Pipher, G. N. Rardin and W. L. Richter, "High Energy Rate Metal Forming," Lockheed Aircraft Corporation, Report No. AMC TR 60-7-588, October (1960); AD 254 776.

19. D. E. Strohecker, et al., "Explosive Forming of Metals", DMIC Report No. 203, Metals and Ceramics Information Center, Battelle-Columbus, Ohio, May (1964).

20. H. G. Baron and E. de L. Costello, Met. Reviews $\underline{8}$, 369 (1963).

21. E. K. Hendricksen, et al., Proc. Symp. on *Dynamic Behavior of Materials*, ASTM STP 336, 104 (1964).

22. H. Ll. D. Pugh, Bulleid Memorial Lectures, $\underline{IIIA}$ (2), 50 (1965).

23. M. T. Watkins, J. West. Scotl. Iron Steel Inst. $\underline{74}$, 148 (1966-1967); Contemp. Physics $\underline{9}$, 447 (1968).

24. R. Gorcey, J. Glyman and E. Green, "Advanced Fabrication Techniques," Aviation Conference, Los Angeles, California, March (1961), ASME Paper No. 61-AV-13.

25. R. N. Orava and H. E. Otto, J. Metals $\underline{22}$ (2), 17 (1970).

26. L. Taborsky, phys. stat. sol. 35, K5 (1969).
27. T. Kovacs, Met. Trans. 2, 961 (1971).
28. A. Korbel and K. Swiatkowski, Met. Science J. 6, 60 (1972).
29. H. J. McQueen and J. E. Hockett, Met. Trans. 1, 2997 (1970).
30. I. A. Grindin, et al., Soviet Phys. - Solid State 10, 1738 (1969).
31. R. N. Orava, University of Denver, Denver, Colorado, unpublished results (1971).
32. A. Gilbert, B. A. Wilcox, and G. T. Hahn, Phil. Mag. 12, 649 (1965).
33. J. W. Edington, Behavior of Metals Under Dynamic Loads, ed. U.S. Lindholm, Springer-Verlag, New York, 191 (1968); Phil. Mag. 20, 531 (1969).
34. T. Muller, J. Mech. Eng. Science 14, 161 (1972).
35. R. N. Orava and P. C. Khuntia, Proc. 3rd Intl. Conf. of the Center for High Energy Forming, ed. A. A. Ezra, Denver Research Institute, University of Denver, Colorado, p. 4.1.1 (1971).
36. L. E. Murr, J. V. Foltz, and F. D. Altman, Phil. Mag. 23, 1011 (1971).
37. J. A. Korbonski and L. E. Murr, Metals Eng. Quart. 11(3), 47(1971).
38. K. S. Ragavan, A. S. Sastri, and M. J. Marcinkowski, Trans. TMS-AIME, 245, 1569 (1969).
39. S. Mahajan, Phil. Mag. 23, 781 (1971).
40. S. Mahajan and G. Y. Chin, Acta Met., to be published (1973).
41. S. Mahajan and D.F. Williams, Intl. Met. Rev., to be published (1973).
42. D. F. Williams and C. N. Reid, Acta Met. 19, 931 (1971).
43. J. Harding, Proc. Roy. Soc. (London), A299, 464 (1967).
44. H. Kressel and N. Brown, J. Appl. Phys. 38, 1618 (1967).
45. R. N. Orava, "Explosive Thermomechanical Processing," Chapter 12, Final Report No. AMMRC-CR-66-05/31(f), Center for High Energy Forming, ed. I. R. Kramer et al. Martin Marietta Corp., Denver, Colorado, June (1972).
46. P. A. Thornton and F. A. Heiser, Met. Trans. 2, 1496 (1971).
47. R. C. Glenn and W. C. Leslie, Met. Trans. 2, 2945 (1971).
48. R. S. Culver, source cited in Ref. 35, p. 4.3.1.
49. H. E. Otto and R. Mikesell, source cited in Ref. 35, p. 4.2.1.
50. R. N. Orava, H. E. Otto, and R. Mikesell, Met. Trans. 2, 1675 (1971).

51. M. B. de Carvalho and R. N. Orava, University of Denver, Colorado, unpublished results (1972).
52. R. J. McElroy and Z. C. Szkopiak, Intl. Met. Rev. 17, 175 (1972).
53. J. L. Watters, T. R. Wilshaw, and A. S. Tetelman, Met. Trans. 1, 2664 (1970).
54. M. Kangilaski, et al., Met. Trans. 2, 2324 (1971).
55. F. E. Van Wely, 4th Intl. Conf. of the Center for High Energy Forming, Vail, Colorado, July (1973).
56. R. N. Orava and H. E. Otto, source cited in Ref. 16, p. 1.2.1.
57. B. A. Stein and P. C. Johnson, Trans. TMS-AIME 227, 1188 (1963).
58. B. G. Koepke, R. P. Jewett, and W. T. Chandler, "Strengthening Iron-Base Alloys by Shock Waves," Report No. ML TDR 64-282, A.F. Materials Laboratory, WPAFB, OH, October (1964).
59. R. H. Wittman and R. N. Orava, University of Denver, Colorado, unpublished results (1972).
60. R. H. Wittman, this volume.
61. A. J. Jacobs, Proc. Conf. on Fundamental Aspects of Stress Corrosion Cracking, ed. R. W. Staehle et al., NACA, Houston, Texas (1969).
62. D. Voss and T. M. F. Ronald, A. F. Materials Laboratory, WPAFB, OH, unpublished results (1972).
63. R. N. Orava, Mat. Sci. Eng., to be published (1973).
64. R. N. Orava, "Thermomechanical Processing of Nickel-Base Superalloys by Shock Deformation," Final Report, U.S. Naval Air Systems Command Contract No. N00019-72-C-0138, University of Denver, Colorado (1973).

A PRACTICAL GUIDE TO ACCURATE GRÜNEISEN EQUATIONS OF STATE

David J. O'Keeffe and D. John Pastine*

Naval Ordnance Laboratory
White Oak
Silver Spring, Maryland 20910

The equation of state of a solid is a formula relating some of the thermodynamic variables of that solid (usually the pressure, volume, temperature, and energy) over some range of interest. This paper will focus on the Grüneisen Equation of State, the most widely used theoretical equation for a solid. However, before proceeding with the body of the paper we consider it advisable to demonstrate the usefulness of such a theoretical equation of state with a few examples.

Many problems require a knowledge of the behavior of metals when stressed beyond their normal elastic limit. In this region conventional elasticity theory fails, but the Grüneisen theory can be utilized to give an accurate pressure-volume relationship at various temperatures.

In general, the equation of state can be exploited in the prediction of phase changes through the Gibbs free energy which must be the same for both phases along the phase line. In particular, the Grüneisen Equation has been successfully used in conjunction with the Lindemann law of melting to compute melting curves.

Other applications can be found in the area of high pressure shock wave research where frequently more information is required than is provided by the shock measurements. For instance, a theoretical equation of state is useful for shock wave impedance calculations where reflected shocks and rarefaction release waves are determined.

Furthermore, with a theoretical equation of state it is possible to make a reasonably accurate calculation of quantities not amenable to measurement, such as the pressure dependence of the coefficient of thermal expansion.

Three distinct advantages of the theoretical approach over static experimental measurements are: (1) the expense of measuring physical properties in the kilobar region of pressure is far greater than the cost of a theoretical calculation; (2) the range of the theoretical data is far more extensive in terms of pressure; and (3) more accurate information is often generated by the theory. As a matter of fact, theoretical equations of state actually function as standards for static high pressure experiments.

In recent years a technique has been developed for the calculation of the Grüneisen Equation of State of a solid which offers certain definite advantages to the experimentalist: (1) it is relatively straightforward; (2) it is based on experimental data readily available in the literature; (3) the calculation is accurate; and (4) it has wide applications.

The method is based on the reaction of a solid to shock compression, and in this paper we examine the two most common cases: (a) the solid described by a linear $U_s - u_p$ (i.e., shock velocity-particle velocity relationship); and (b) the solid whose $U_s - u_p$ curve is quadratic. Both treatments require a knowledge of the volume dependence of γ the temperature independent Grüneisen parameter. For the linear situation we employ an expression for γ derived by Pastine and Forbes[1] which minimizes the number of approximations. The quadratic case is handled by a Taylor expansion for γ accurate to at least second order in the compression. The equation of state is obtained from the shock Hugoniot and the solution of a first order differential equation in the compression.

Numerical results are given for copper and silver, which compare favorably with experimental data. Copper and silver were chosen since their $U_s - u_p$ curves are most accurately represented by linear and quadratic fits, respectively[2]. The theoretical 300° K isotherm for copper yields pressures accurate to within ± 2% at a given volume for pressures up to 550 kbar; and for silver the 300° K isotherm has an accuracy of ± 2.5% for pressures up to 290 kbar.

GUIDE TO ACCURATE GRÜNEISEN EQUATIONS OF STATE

Two calculations were carried out on silver, the second assuming a linear $U_s - u_p$. A comparison of the two sets of results indicates that the linear relation leads to discrepancies in the two theoretical isotherms for silver in the low pressure range. This conclusion is in accord with earlier work on the relation between shock and particle velocities in metals.[3]

BASIC THEORY

Since many of the details of the calculation can be found in reference 4, we will only give a brief outline of the work. First of all there is no serious limitation in restricting the discussion to a quasi-harmonic solid in the classical temperature region. The Grüneisen equation of state for such a solid is given exactly by[5]

$$P = P_o(x) + [\rho_o \gamma(x) E_T(x,T)]/x \tag{1}$$

where the Grüneisen parameter γ is independent of temperature. Equation (1) relates the pressure P specific volume V and temperature T; $E_T(x,T)$ denotes the specific thermal energy of the solid and $P_o(x)$ the pressure along the zero degree isotherm. The compression ratio $x = V/V_o$ where V_o is the specific volume at $T = 0$ and $P = 0$, and ρ_o is $1/V_o$. We can express the total specific energy E as

$$E = E_T(x,T) + \phi(x) \tag{2}$$

with $E_T(x,T)$ as the specific thermal energy and $\phi(x)$ the specific energy at 0° Kelvin. The Rankine Hugoniot energy conservation relation states that

$$E_h - E_i = P_h(x_i - x)/2\,\rho_i x_i \tag{3}$$

where the subscripts h and i refer respectively to the compressed and uncompressed shock states. From Equations (1), (2), and (3) we obtain the following relation for the Hugoniot pressure,[6]

$$P_h = \frac{P_o(x) + \rho_o \gamma(x)[\phi(x_i) - \phi(x) + E_T(x_i,T_i)]/x}{1 - \gamma(x_i - x)/2x} \tag{4}$$

It is possible to derive a differential equation in terms of P_h and x through the zero degree isotherm $P_o(x)$ which is given by

$$P_o(x) = -\rho_o \, d\phi/dx \tag{5}$$

For convenience we define the quantity $\psi(x)$ as

$$\psi(x) = \rho_o [\phi(x_i) - \phi(x) + E_T(x_i, T_i)] \qquad (6)$$

which implies that

$$P_o(x) = d\psi/dx \equiv \psi' \qquad (7)$$

Making use of Equations (6) and (7), we see that (4) reduces to

$$P_h = \frac{\psi' + \gamma(x)\psi(x)/x}{1 - \gamma(x_i - x)/2x} \qquad (8)$$

The solution of (8) is contingent on the explicit form of $P_h(x)$ which is a different function for the linear and quadratic $U_s - u_p$, as we shall see.

CONSEQUENCES OF A LINEAR $U_s - u_p$ RELATION

Many materials exhibit a linear $U_s - u_p$ relation when subjected to shock compression. In other words,

$$U_s = C_i + b u_p \qquad (9)$$

where C_i is the bulk sound speed in the uncompressed state and b is a constant that depends on the initial conditions.

We can express P_h as a function of x by means of Equation (9) and the remaining two Rankine Hugoniot equations:

$$P_h = \rho_i U_s u_p \qquad (10)$$

and

$$x_i (U_s - u_p) = x U_s \qquad (11)$$

that is,

$$P_h(x) = \frac{\rho_i C_i^2 (x_i - x) x_i}{[x_i - b(x_i - x)]^2} \qquad (12)$$

Finally, the differential equation (8) takes the form

$$\psi' + \gamma(x)\psi(x)/x = \frac{\rho_i C_i^2 x_i (x_i - x)[2x - \gamma(x_i - x)]}{2x [x_i - b(x_i - x)]^2} \qquad (13)$$

where $\psi' \equiv d\psi/dx$.

This is a non-homogeneous first order ordinary differential equation in x with the boundary condition,

$$\psi'(x_i) = -\gamma(x_i) \rho_o E_T(x_i, T_i)/x_i \tag{14}$$

following from the fact that

$$P_h(x_i) = 0$$

Once $\gamma(x)$ is known (13) and (14) can be solved for $\psi(x)$, and then Equation (7) exploited to furnish the zero degree isotherm $P_o(x)$.

Pastine and Forbes[1] have derived the following expression for $\gamma(x)$:

$$\gamma(x) = \frac{2x\eta[2bx - \delta(x_i - b\eta)] + 2x^2(x_i - b\eta) + 4x\eta^2 \nu}{2\eta^3 \nu + \eta^2[2bx - \delta(x_i - b\eta)] + \eta(x + x_i)(x_i - b\eta) + 2(x_i - b\eta)^3 C_p/\alpha C_i^2} \tag{15}$$

which is exact provided (9) holds exactly for all u_p, and that P_h is a true shock hydrostat (see reference 6).

In Equation (15) $\eta = x_i - x$, C_p is the specific heat at constant pressure, and α is the coefficient of thermal expansion. The quantities ν and δ (the second Grüneisen parameter) are defined by

$$\nu = [(\partial b/\partial T_i)|_{P_i = 0}]/\alpha \tag{16}$$

and

$$\delta = 1 - \frac{2}{\alpha C_i}(\partial C_i/\partial T_i)|_{P_i = 0} \tag{17}$$

Reference 4 describes the procedure for the evaluation of the parameters C_p, α, C_i, ν, and δ under ambient conditions from experimental data available in the literature.

QUADRATIC $U_s - u_p$

In the case of a quadratic $U_s(u_p)$ we know that

$$U_s = C_i + b_1 u_p + b_2 u_p^2 \tag{18}$$

and with (10) this allows us to write for the shock pressure

$$P_h = \rho_i (C_i u_p + b_1 u_p^2 + b_2 u_p^3) \tag{19}$$

Substituting (18) into (11) we find

$$u_p = \left[\frac{1 - b_1 (x_i - x)}{2b_2 (x_i - x)}\right] \left\{1 - \left[1 - \frac{4b_2 C_i (x_i - x)^2}{[1 - b_1 (x_i - x)]^2}\right]^{1/2}\right\} \tag{20}$$

A power series in $b_1(x_i - x)$ is used to eliminate the radical in (20) and the simplified expression for u_p is substituted into Equation (19) to yield

$$P_h = \frac{\rho_i C_i^2 (x_i - x)[1 - b_1 (x_i - x)]^2 + 2\rho_i b_2 C_i^3 (x_i - x)^3}{[1 - b_1 (x_i - x)]^4} \tag{21}$$

The differential equation (8) then assumes the following form for the quadratic $U_s - u_p$,

$$\psi' + \gamma(x) \psi(x) =$$

$$\frac{\rho_i C_i^2 (x_i - x)\left\{[1 - b_1 (x_i - x)]^2 + 2b_2 C_i (x_i - x)^2\right\}[2x - \gamma(x_i - x)]}{2x[1 - b_1 (x_i - x)]^4} \tag{22}$$

with (14) again serving as the boundary condition.

One needs to know $\gamma(x)$ in order to solve (22). Since the analytic form analogous to Equation (15) proved intractable, we found it expedient to resort to a series expansion of the type

$$\gamma(x) = \gamma_i - \gamma_i' (x_i - x) + \frac{\gamma_i''}{2} (x_i - x)^2 - \frac{\gamma_i'''}{6} (x_i - x)^3 + \cdots \tag{23}$$

where γ_i', γ_i'', γ_i''' are derivatives evaluated at $x = x_i$, and $\gamma_i = \gamma(x_i)$. More explicitly,

$$\gamma_i = \alpha C_i^2 / C_p$$

$$\gamma_i' = \gamma_i (\gamma_i + 2 - \delta - 4b_1)$$

$$\gamma_i'' = 2\gamma_i \Big\{ 2\nu + 1 - b_1 [8 + 2\delta + 7\gamma_i - 9b_1]$$

$$+ \delta (1.5 \gamma_i + 1) + 2.5 \gamma_i + \gamma_i^2 + 6 c_i b_2 \Big\} \qquad (24)$$

Equation (22) is then solved in the same manner as in the preceding section.

RESULTS

Copper

Since the $U_s - u_p$ curve for copper is linear[7], for $U_s \leq$ 7 km/sec, Equation (15), the exact expression for $\gamma(x)$, was utilized in the calculation. The sources for the experimental data are listed in reference 4. We used the Debye expression for the thermal energy, $E_T(x,T)$ with a Debye temperature of 297° K.

Equation (13) was solved by the Runge-Kutta technique to give $P_o(x)$, the zero degree isotherm. In the process we effectively separate the thermal and volume contributions to the pressure in view of the form of the Grüneisen equation (1) (i.e., the thermal pressure is given by $\rho_o \gamma(x) E_T(x,T)/x$). Table 1 lists the values of $\gamma(x)$ for copper along with $P_o(x)$, and the 300° K isotherm. A thorough error analysis indicated that the error in γ should be less than $\pm 7\%$ for $0.8 \leq x \leq 1.01$, while the pressures along the theoretical 300° K isotherm should be accurate to within $\pm 2\%$ for pressures up to 550 kbar. Figure 1 shows a comparison of this isotherm with two sets of experimental data[8,9], and the agreement is satisfactory.

A list of the parameters employed in this calculation reads as follows: $x_i = 1.01$; $\gamma_i = 2.02$; $\delta = 3.57$ and $\nu = 0.727$. All of these parameters were evaluated under ambient conditions. It should be noted that two of the above parameters δ and ν differ from those quoted in reference 4. There are two reasons for this: (1) the computation of the temperature and pressure derivatives of the elastic constants (necessary for the evaluation of ν) were improved; (2) an algebraic error was found in reference 4 and corrected.

The resultant change in $\gamma(x)$ is less than 3% at $x = 0.9$, and less than 10% at $x = 0.8$, whereas the pressures are changed by less than 1% at $x = 0.8$.

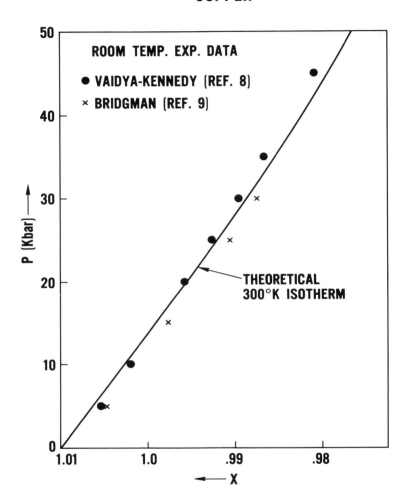

Figure 1. A Comparison of Room Temperature Experimental Data and the Theoretical 300° Kelvin Isotherm for Copper. The compression ratio, $x = V/V_o$ where V is the specific volume of the solid, and V_o is the specific volume at P = 0 and T = 0° Kelvin.

x	$\gamma(x)$	$P_o(x)$ (kbar)	$P_{300}(x)$ (kbar)
1.01	2.02	-13.8	0.0
1.00	1.99	0.0	13.6
0.98	1.93	30.3	43.6
0.96	1.87	64.4	77.3
0.94	1.82	102.8	115.3
0.92	1.77	146.0	158.2
0.90	1.72	194.7	206.6
0.88	1.68	249.5	261.2
0.86	1.65	311.5	322.8
0.84	1.62	381.4	392.6
0.82	1.59	460.5	471.5
0.80	1.57	550.0	560.8
0.78	1.56	651.4	662.1
0.76	1.55	766.4	777.0
0.74	1.55	896.7	907.2
0.72	1.54	1044.6	1055.0
0.70	1.55	1212.2	1222.5

Table 1. Grüneisen Parameter $\gamma(x)$; Zero Degree Isotherm $P_o(x)$; & 300° Isotherm $P_{300}(x)$ for Copper

Silver

Evidence for the curvature of the $U_s(u_p)$ function for silver was presented in reference 3. Consequently, we resorted to the second order expansion for $\gamma(x)$ given by Equation (23). A least squares fit to experimental data of the form of (18), (i.e., $C_i = 3.141$ km/sec, $b_1 = 1.757$, and $b_2 = -.0597$ sec/km) was published in reference 1 along with other experimental data suitable for our purposes. In order to compute a value for the specific thermal energy $E_T(x_i, T_i)$ with the boundary condition expressed in (14) we chose a Debye temperature of 227.2° K.

From this point on the calculation was executed exactly as for copper, and the results: $\gamma(x)$, $P_h(x)$, $P_o(x)$, and $P_{300}(x)$ can be found in Table 2. If experimental errors are neglected, the second order expansion in $\gamma(x)$ should be accurate to within ±10.5% for $0.85 \leq x \leq 1.0124$ inasmuch as the term in γ_i''' contributes approximately 0.10 to γ at 15% compression. However, the effect of experimental uncertainty on the parameters in the expansion cannot be ignored. A study of the ensuing errors indicates that at worst the total uncertainty in γ is ± 11.5% at

$x = 0.85$. The pressures along the theoretical 300° K isotherm should be accurate to within ±2.5% for pressures up to 290 kbar. In Figure 2 we have plotted this isotherm as well as two sets of experimental data[8,9] and the theoretical isotherm compares favorably with the experimental results.

We should point out that the value of ν cited in reference 1 is in error. Rather than being equal to -1.35, our computation found $\nu = 1.67$. Other parameters of interest are: $x_i = 1.0124$; $\gamma_i = 2.38$; and $\delta = 3.83$.

It has been demonstrated in reference 3 that for some cubic metals the assumption of a linear $U_s(u_p)$ function leads to discrepancies in values of the initial slope (i.e., $(dU_s/du_p)|_{u_p = 0})$ calculated from shock data as opposed to values predicted from ultrasonic measurements.

Since silver was one of those metals, we thought it should be examined to see if the linear assumption would lead to similar discrepancies in its equation of state. A linear equation of the form (9) (i.e., $C_i = 3.265$ km/sec and $b = 1.572$) was therefore determined by a least squares fit to the same experimental data which gave rise to the quadratic $U_s - u_p$ of the previous section. With this linear relation established it was possible to calculate $\gamma(x)$ from (15). A Debye temperature of 190° K was chosen to meet the boundary condition (14), and the calculation of the equation of state was implemented just as for copper. We have tabulated the resulting $\gamma(x)$, $P_h(x)$, $P_o(x)$, and $P_{300}(x)$ in Table 3. A comparison of Tables 2 and 3 seems to substantiate one of the conclusions reached in reference 3; namely, that the linear $U_s(u_p)$ assumption is not completely satisfactory in the range $\frac{u_p}{U_s} < 0.1$. This is evident since the pressures given by both 300° K isotherms only begin to coincide within a few percent at roughly $x = 0.88$ (or $U_s \approx 8 u_p$). The fact that this value of x approximates the lowest datum point for which the $U_s - u_p$ fits were made is no coincidence. As stated in reference 3: for many cubic metals the linear assumption is only tenable in ranges of U_s and u_p for which experimental shock data exist. The work on silver described in this paper lends support to that conclusion.

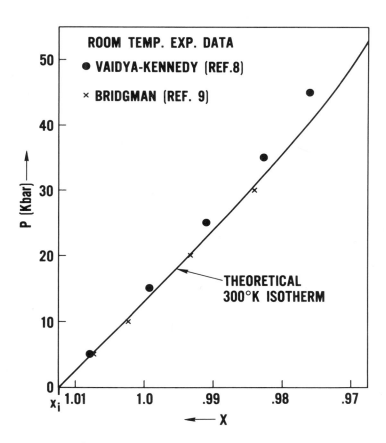

Figure 2. A Comparison of Room Temperature Experimental Data and the Theoretical 300° Kelvin Isotherm for Silver. The compression ratio, $x = V/V_o$ where V is the specific volume of the solid and V_o the specific volume at $P = 0$ and $T = 0°$ Kelvin.

x	γ(x)	$P_h(x)$ (kbar)	$P_o(x)$ (kbar)	$P_{300}(x)$ (kbar)
1.0124	2.38	0	-12.9	0
1.0	2.34	13.3	0	12.7
0.98	2.29	37.2	23.3	35.8
0.96	2.25	64.8	50.0	62.4
0.94	2.22	96.7	80.5	92.7
0.92	2.19	133.5	115.6	127.7
0.90	2.16	176.2	155.6	167.6
0.88	2.14	225.9	201.3	213.2
0.86	2.13	283.7	253.4	265.3
0.85	2.13	316.2	282.0	293.9
0.84	2.13	351.2	312.6	324.5
0.82	2.13	430.2	379.6	391.5
0.80	2.14	522.6	455.2	467.1

Table 2. Grüneisen Parameter γ(x); Shock Pressure $P_h(x)$; Zero Degree Isotherm $P_o(x)$ and 300° K Isotherm $P_{300}(x)$ for Silver with a Quadratic $U_s - u_p$ Curve.

x	γ(x)	$P_h(x)$ (kbar)	$P_o(x)$ (kbar)	$P_{300}(x)$ (kbar)
1.0124	2.38	0.0	-13.9	-0.3
1.0	2.33	14.3	0.0	13.3
0.98	2.26	39.7	24.9	38.0
0.96	2.20	68.7	53.0	65.8
0.94	2.15	101.6	84.9	97.5
0.92	2.11	139.2	120.8	133.3
0.90	2.08	182.4	161.5	173.9
0.88	2.06	232.0	207.5	219.8
0.86	2.04	289.2	259.5	271.8
0.85	2.04	321.1	288.0	300.3
0.84	2.03	355.4	318.3	330.6
0.82	2.03	432.6	384.7	397.1
0.80	2.04	522.8	459.5	471.9

Table 3. Grüneisen Parameter γ(x); Shock Pressure $P_h(x)$; Zero Degree Isotherm $P_o(x)$ and 300° K Isotherm $P_{300}(x)$ for Silver with a Linear $U_s - u_p$ Curve.

REFERENCES

* Supported by the Naval Ordnance Laboratory Independent Research Fund.

1. D. J. Pastine and J. W. Forbes, Phys. Rev. Lett. **21**, 1582 (1968).

2. In the case of silver it is possible to represent the experimental data by a linear fit, a quadratic fit, or even a cubic fit, but only the quadratic fit can be forced to meet the condition that the intercept of the $U_s - u_p$ curve be equal to C_i, the bulk sound speed.

3. D. J. Pastine and D. Piacesi, J. Phys. Chem. Solids **27**, 1783 (1966).

4. D. J. O'Keeffe, J. Geophys. Res. **75**, 1947 (1970).

5. G. Liebfried and W. Ludwig, Solid State Phys. **12**, 275 (1961).

6. This implies that the pressures are far enough above the yield point to allow the shock pressure P_h to be closely identified with a pure hydrostatic pressure. Recent experimental evidence (see reference 10) seems to indicate this is the case.

7. W. J. Carter, S. P. Marsh, J. N. Fritz, and R. G. McQueen, Accurate Characterization of the High-Pressure Environment, edited by E. C. Lloyd, p. 219, National Bureau of Standards Special Publication 326, Washington, D. C. (1971)

8. S. N. Vaidya and G. C. Kennedy, J. Phys. Chem. Solids **31**, 2329 (1970).

9. P. W. Bridgman, Proc. Am. Acad. Arts Sci. **77**, 189 (1949).

10. Q. Johnson, A. Mitchell, R. N. Keeler, and L. Evans, Phys. Rev. Lett. **25**, 1099 (1970).

RELATION BETWEEN DYNAMIC AND STATIC PHASE TRANSFORMATION STUDIES

William J. Carter

Los Alamos Scientific Laboratory

Los Alamos, New Mexico

Introduction

Phase transformations have always held a special fascination for high-pressure physicists. The past decade especially has seen a tremendous expansion in the amount of work performed under both static and dynamic conditions. Most of this has concentrated on simple elements or compounds which have geophysical significance, leaving those materials of most interest to the working metallurgist (e. g., binary and ternary alloys) virtually untouched. Equally unfortunate is the fact that the two sources of data (static and dynamic) often seem to be treated as separate entities with no real effort made to correlate results from the two, even though the accessible pressure and temperature ranges now overlap in many cases. There has perhaps been some justification for this because of the understandable reluctance of many workers to believe that thermodynamic equilibrium can be attained under shock-wave conditions. The good agreement between the Hugoniot intercept and zero-pressure sound speeds for many materials was an early indication that, despite strain rate differences of many orders of magnitude, static and dynamic measurements are compatible. The recent work of Vaidya and Kennedy[1] among others has shown that properly reduced shock wave data and static compressibility data agree to well within experimental error up to 50 kb or more. Furthermore, the recent dynamic x-ray diffraction work of Johnson and Mitchell[2] shows that a crystal structure differing from that of the undisturbed material can exist immediately behind a shock front and is created on a very short time scale. On the basis of these results it is probably safe to assume that shock-wave data can normally be successfully treated on the basis of equilibrium thermodynamics. On this very basic assumption we shall attempt to relate some recent static and dynamic experiments. Transformation kinetics are of course important in any study of phase boundaries; for the cases we have studied however, the excellent agreement between shock work and static work are further indications that shock wave data can be used to define equilibrium phase boundaries. This is probably because the exceedingly large number of dislocations introduced by even modest shocks provide enough nucleation centers, and the very large deviatoric shear stress along the slip planes provides a driving force sufficiently in excess of any activation barriers, that the transformation proceeds quite rapidly. This effect has been seen before; in some cases, transformations which ordinarily require hours or days to proceed to completion occur in microseconds or less under shock wave conditions. For example, the well known 130 kb $\gamma \rightarrow \epsilon$ transition in iron is observed readily using shock waves, but is extremely sluggish under static loading.

In order to present the calculations more clearly and to avoid the difficulties of large mixed phase regions, we shall restrict ourselves to elements, and for the most part to their melting phase lines. The reason for this latter restriction is simply that many of the closed low-pressure solid phases are not accessible to the Hugoniot centered at standard conditions, and special experimental techniques which hardly seem justified would be necessary to explore these phases dynamically. Ordinarily, melting is difficult to detect dynamically, since the energy-volume relation which is measured in shock-wave work is only slightly affected by the melting transformation. However there are several cases where the onset of melting is clearly detectable. Methods of analysis exist for the cases where the actual beginning of the phase change is not detectable.

The Method

The goal of this study is to construct a phase diagram for a material which is thermodynamically self-consistent, using all the data available. The method is extremely straightforward and is certainly not new,[3,4] but perhaps it is not sufficiently well known to those outside the shock-wave field nor is its power fully appreciated. Basically, the idea is to establish experimentally the Hugoniot equation of state of the two phases separately and then extend these equations of state to regions off the Hugoniot by use of the Mie-Grüneisen relation. This allows calculation of complete thermodynamic information, including the relative Gibbs free energy G = H-TS, throughout the P-V field for each phase separately. The relative scales of the entropy and energy for the two phases are fixed through the Clapeyron equation by specifying a P, T point on the phase line and the slope at that point, and the phase line is then uniquely determined by the requirement $\Delta G = 0$.

Figure 1 is a ficticious but by no means unrepresentative phase diagram for a monocomponent system. The Hugoniot is indicated by the heavy line and intersects the α-, β-, and liquid phases, each characterized by their own thermodynamic parameters. Whether or not these phase boundaries are detectable on the Hugoniot of course depends on the relative energies, volumes, and compressibilities of the phases. The γ-, δ-, and ϵ- phases are inaccessible to this Hugoniot but could in principle be reached by Hugoniots centered at different states.

In brief, the calculations proceed as follows. We start with the thermodynamic relations

$$dE = TdS - PdV \tag{1}$$

$$TdS = (\partial E/\partial T)_V dT + T(\partial P/\partial T)_V dV \tag{2}$$

and the Rankine-Hugoniot equations valid within any phase

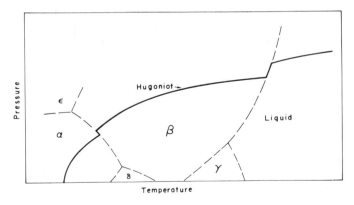

Fig. 1. Behavior of Hugoniot through a phase change.

$$V_h/V_o = (u_s - u_p)/u_s \tag{3}$$

$$P_h - P_o = u_s u_p / V_o \tag{4}$$

$$E_h - E_o = (P_h + P_o)(V_o - V_h)/2 \tag{5}$$

where the subscript h refers to the value on the Hugoniot and the other symbols have their usual meanings. By introduction of the Grüneisen parameter γ, defined as

$$\gamma = V(\partial P/\partial E)_V \tag{6}$$

equation (2) becomes

$$dS = C_V \left[\frac{dT}{T} + (\gamma/V)\, dV \right] \tag{7}$$

Since our knowledge of both γ and C_V over the field of interest is extremely limited, further progress requires some basic assumptions about their high-pressure behavior. It is known from theoretical considerations that γ is temperature-independent until anharmonic effects become significant, so that γ will be taken to be a function of volume alone. Experimental work using porous materials has shown that a γ linear in volume is adequate over a large range of compression, and we will make this assumption for the rest of this work. That is,

$$\gamma = \left(\frac{V}{V_o}\right) \gamma_o \tag{8}$$

where γ_o is the thermodynamic value at standard conditions given by $\gamma_o = 3\alpha c_o^2 / C_p$. The assumption that γ is a function only of volume implies that the specific heat C_V is a function only of entropy. We shall further require that C_V be constant, certainly a valid assumption for those materials with a low characteristic temperature. Equations (1), (5) and (7) can then be combined to yield

$$dT_h = \frac{V_o - V_h}{2 C_V} dP_h + \left[\frac{P_h - P_o}{2 C_V} - T_h \left(\frac{\gamma_o}{V_o}\right) \right] dV_h \tag{9}$$

which is used in turn to determine the entropy through equation (7). Assuming a linear $u_s - u_p$ relation of the form $u_s = c_o + s\, u_p$, the temperature and entropy on the Hugoniot are written in closed form as

$$T_h(V) = T_o \left\{ 1 + \frac{c_o^2}{C_V T_o} \int_1^\xi d\xi\, \frac{s(\xi - 1)^2}{[s - \xi(s-1)]^3} e^{-\gamma\left(\frac{\xi-1}{\xi}\right)} \right\} e^{\gamma\left(\frac{\xi-1}{\xi}\right)} \tag{10}$$

and

$$S_h(V) = S_o + C_V \ln \left\{ 1 + \frac{C_o^2}{C_V T_o} \int_1^\xi d\xi \frac{s(\xi-1)^2}{[s-\xi(s-1)]^3} e^{-\gamma(\frac{\xi-1}{\xi})} \right\} \quad (11)$$

where $\xi = V_o/V$. Equations (10) and (11) can be integrated numerically to obtain a complete thermodynamic description along the Hugoniot for each of the two phases.

Under the assumption of constant $(\rho\gamma)$ and C_V, thermodynamic quantities off the Hugoniot can be computed through the relation

$$(\partial P/\partial T)_V = \rho\gamma C_V = \text{constant} \quad (12)$$

or

$$P - P_h(V) = \rho\gamma C_V (T - T_h(V)) \quad (13)$$

The quantity $T_h(V)$ determined from eq. (10), coupled with the known $P_h(V)$ and eq. (13), serves to determine $V(P,T)$ implicitly. Eq. (11) can then be used to determine the entropy, and the Mie-Grüneisen equation

$$E(P,V) = E_h(V) + [P-P_h(V)]/(\rho\gamma) \quad (14)$$

serves to determine the internal energy at every point off the Hugoniot.

To connect the two phases, the differences in energy and entropy across the phase boundary at some point are needed. These differences may be computed from the Clapeyron equation if the slope of the phase line $(dP/dT)^*$ is known at some point (P^*, T^*). This serves to fix the energy and entropy scales of the two phases. The rest of the phase line is then determined. Other quantities on the phase line are calculated in a straightforward manner by assuming ideal mixing and specifying the mole fraction of the two components.

The principal difficulty in this approach, assuming that static phase line data and sufficient thermodynamic data throughout the P-T field are available, is to obtain a good Hugoniot equation of state for the two phases. If the thermodynamics of the transition is of the correct type, a break in the u_s-u_p curve is observed which identifies the transition point precisely. This is often not the case, in particular for the solid-liquid or liquid-vapor transitions. Even if a break is observed, it is still convenient to recenter the two Hugoniots to a common (P_o, T_o)

centering point, which requires a calculation, described elsewhere,[5] somewhat similar to that above. For this, a knowledge of the slope of the phase line at its intersection with the Hugoniot is necessary, which implies that an iterative procedure is called for. Other methods have been used as well to extract the upper phase Hugoniot from sometimes ambiguous data; these will be discussed as they arise.

Experimental Results

1. <u>Europium</u>. Perhaps the simplest and most believable application of the method is to the melting phase line of europium. Europium has no known solid phase other than bcc, although a pronounced resistance discontinuity has been detected at room temperature and about 160 kb. Jayaraman[6] has measured the melting phase line statically to 65 kb, whereas our Hugoniot data shown in Fig. 2 indicates a transition at 106 kb and 950 °K. The location of the transition is well established. The parameters for the Hugoniots of the two phases, as given in Table 1, are relatively simple to extract from the data. When these are used in the calculations described earlier, together with a T^* and dP/dT^* at $P^* = 0$ from the static data, the phase line shown in Fig. 2 is produced. The intersection of the calculated phase line and the Hugoniot is indeed at the observed transition point. It is significant that the calculated phase line is relatively insensitive to where the static data are centered; in Fig. 2 the values of P^*, T^*, and dP/dT^* were taken at the zero pressure melting point, and the resulting phase line matches Jayaraman's data reasonably well. It should be noted

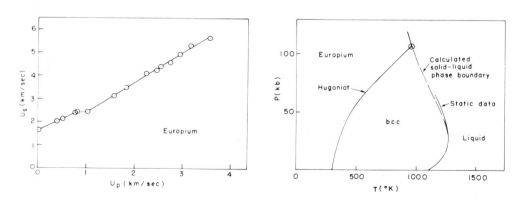

Fig. 2. Hugoniot data and P-T diagram for europium.

that the small flat section at $u_p = .9$ on the Hugoniot is an indication of a small negative volume change which is compatible with the negative slope of the phase line at the Hugoniot intersection.

2. Erbium. Erbium also has only one solid phase, the hcp phase, but undergoes a transition to a less compressible phase on the Hugoniot at ~ 440 kb and 2070 °K as shown in Fig. 3. The melting phase line of Jayaraman[7] extends to only 8 kb, as the melting point at zero pressure is quite high (1770 °K) and static high pressure work is difficult at these temperatures. The parameters of the two phases are given in Table 1, and the resulting calculated phase line is shown in Fig. 3. The agreement of the calculated transition point with the experimental data is quite satisfying.

A check on these calculations can be made by using porous samples in a manner somewhat similar to that previously done for the melting phase line of copper.[4] Here, use is made of the large $P\Delta V$ heating obtained by shocking porous materials. We have used samples with 79% of theoretical crystal density which were obtained by cold pressing a fine-mesh powder. The experimental data are shown in Fig. 3. Calculations from eq. (10) indicate that temperatures attained on the Hugoniot at about 200 kb are sufficient to melt this porous material. Hence, if the parameters listed in Table 1 and used for the phase line calculation are truly representative of the liquid phase, then a simple recentering of the liquid Hugoniot to the initial porous density using Eq. (6) should reproduce the porous Hugoniot data (it should be noted that ΔH of melting is very small compared to the total energy on the Hugoniot

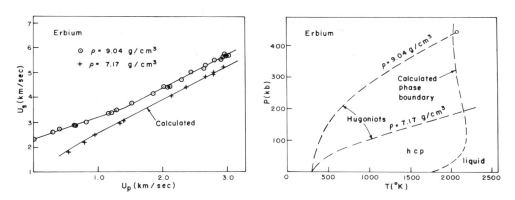

Fig. 3. Hugoniot data and P-T diagram for erbium.

even at the lowest experimental pressures). The results of this calculation are shown in Fig. 3; the liquid Hugoniot does indeed reproduce the porous data quite well.

It is interesting to speculate on the nature of phase line maxima observed in virtually all the rare earths. It is becoming increasingly obvious that anamolous melting, that is, melting characterized by a negative slope of the fusion curve, is as normal as that characterized by a positive slope when one carries the data to sufficiently high pressures. Simple relations such as the Simon equation or that of Kennedy-Kraut are hopelessly inadequate to handle these negative slopes or to predict maxima or minima in the fusion curve. The explanation of the phenomena undoubtedly lies in understanding the properties of the liquid near the phase boundaries rather than the behavior only of the solid, which is the basis of most of the present semi-empirical descriptions. Since the entropy change upon melting probably does not vary appreciably with pressure, a negative phase line slope means that the density of the liquid continuously increases relative to that of the contiguous solid as one proceeds up the phase line. This can come about through one of two mechanisms: either the coordination number of the liquid becomes greater than that of the solid, or else the ionic radius characteristic of the liquid is less than that of the solid. High coordination numbers, at least on a local scale, are not unknown in liquids under pressure. However, in view of the close spacing between 4f, 5d and 6s energy bands and the sensitivity of the effective ionic radius to the occupation of these bands for the lanthanides, the second possibility seems more likely. Any change in ionic radius upon melting then implies a corresponding change in the electronic structure of the liquid metallic ion. Since the ionic radii of the lanthanides are known to decrease in the order divalent > trivalent > tetravalent, one conclusion is that an additional f-electron is forced into the valence band from the 4f shell by the perturbation introduced by compression. The curvature of the phase line toward the pressure axis is then accounted for by an extension of the two-fluid model proposed by Klement,[8] in which an increasingly greater proportion of atoms in the liquid undergo a 4f-5d electron promotion as one proceeds up the phase boundary. It is likely that there is a continuous change in the amount of s-d bonding with pressure in the solid as well, as evidenced by the very small initial slopes of the Hugoniots. The anomalous melting, however, is associated with the more rapid completion of this electronic transition in the liquid state adjacent to the solid material.

3. Cerium. Cerium has been the subject of a great deal of high pressure work because it is one of the few elements in which an electronic transition has been unambiguously identified. This transition, which is fcc → fcc and involves a very large volume change, is identified with a decrease in the d-band population as determined by neutron scattering experiments. The effect of this transition on the Hugoniot of cerium is unmistakable, as seen in Fig. 4, since the lower-phase Hugoniot intercept misses zero pressure sound speed by more than a factor of two. Cerium is unique among the rare earths in that its Hugoniot indicates a transition to a more compressible phase at a pressure of about 480 kb. Hence, we must deal with three phases. This is particularly simple for cerium, however, since both the zero pressure sound speed and the details of the low pressure electronic transition are already known. It is then rather simple to recenter the second phase to zero pressure. The resulting parameters are shown in Table 1. This recentered metastable Hugoniot can then in turn be used in the manner described before to calculate the solid-liquid phase line. The result, shown in Fig. 4, is quite satisfying. It will be noted that, at the intersection of the phase line and the Hugoniot, dP/dT is positive.

4. Gadolinium. As a further illustration of the technique of treating consecutive phase changes, consider the phase line of gadolinium. The usual calculation, assuming the transition observed on the Hugoniot of gadolinium is indeed of the solid → liquid type, fails in this case. An inspection of the statically-determined phase diagram[9] gives a hint why it fails: the bcc phase, which for

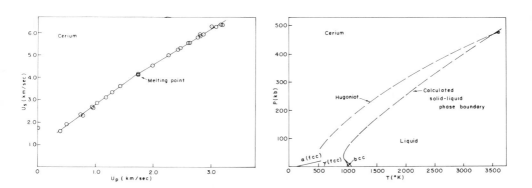

Fig. 4. Hugoniot data and P-T diagram for cerium.

rare earths usually occurs slightly before melting and is ordinarily closed at relatively low pressures, is not closed for gadolinium. A calculation using the static data for the rhomb → bcc transition reproduces the Hugoniot transition reasonably well. Close inspection of the shock data reveals the possibility of another discontinuity in slope at considerably higher pressure, which could be fit using the melting phase line static data and the metastable bcc Hugoniot. The resulting phase diagram in shown in Fig. 5. We have ignored the hcp → rhomb transition at much lower pressure, which has a very small volume change and which presumably does not greatly affect the thermodynamic parameters C_v and γ. This phase change is not detected on the Hugoniot. Hence, for the case of gadolinium, it would appear that the electronic rearrangement responsible for anamolous phase trajectories in the other rare earths either occurs in conjunction with the rhomb → bcc transition or gradually throughout the solid regions. It is significant that the gadolinium electronic core, having a half-filled shell ($4f^7$), is particularly tightly bound. In view of the extremely restrictive assumptions on the behavior of the thermodynamic parameters of these three phases and the uncertainties inherent in both the shock and static data, it is remarkable that such a self-consistant picture can be obtained.

5. __Lead.__ Because of its low melting point, lead is a particularly appealing candidate for a phase line calculation. It should be possible to obtain a good liquid Hugoniot directly by preheating the sample to the liquid phase, but this has not been done to our knowledge. Experiments using preshocked Pb to obtain the Hugoniot of the liquid phase lacked sufficient precision to do this analysis. However, good measurements of both the liquid

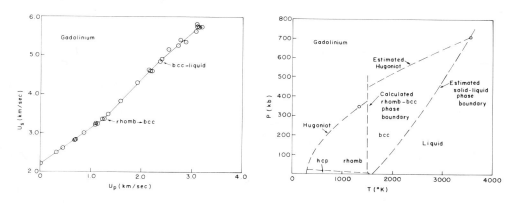

Fig. 5. Hugoniot data and P-T diagram for gadolinium.

sound speed and density are available. Russian work[10] has indicated that lead melts under shock loading at about 230 kb; inspection of our data shown in Fig. 6 does indeed reveal a slight change in slope in this general region. By using only the Hugoniot data well above this region, which can safely be assumed to be liquid, with an initial slope of the u_s-u_p determined from ultrasonic measurements, the Hugoniot centered at its melting temperature was calculated. This was used together with the initial slope of the phase line measured by Kennedy and Newton[11] and the low pressure Hugoniot data to calculate the phase line shown in Fig. 6. We have ignored the 140 kb fcc-hcp transition,[12] which has a very small associated ΔV and is also undetectable on the Hugoniot. These calculations indicate that lead begins to melt at about 275 kb and 1210 °K when subjected to shock. This point is indicated on the Hugoniot of Fig. 6, as well as the point at which melting is completed and the Hugoniot emerges into the liquid region. Extrapolation of the two Hugoniots coupled with the Grüneisen functions and specific heat specify the volume vs. temperature relation along the zero pressure isobar for both the liquid and the solid phase. This agreement is shown in Fig. 7 and reflects the adequacy of the assumption $\rho\gamma$ = constant. An independent check on the Hugoniots and Grüneisen parameters of the two phases is the calculated sound velocity-temperature relation along the zero pressure isobar which can be compared with Russian[13] data. The agreement is good to about 2%, which probably is well within the accuracy of the data. Although evidence from the Hugoniot data alone is not convincing, the agreement with the Russian work as well as the success the method has found with the rare earths strengthens the argument. It should be noted also that for our

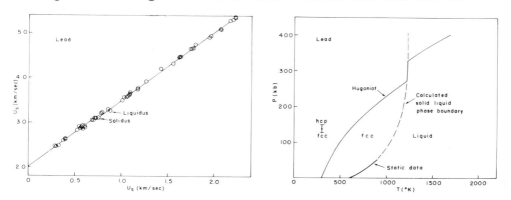

Fig. 6. Hugoniot data and P-T diagram for lead.

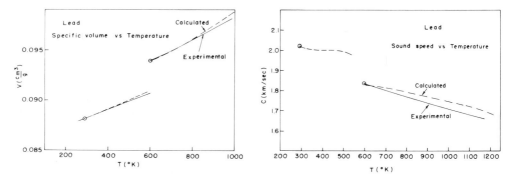

Fig. 7. Volume and sound speed vs. temperature for lead.

calculations only a single set of (P^*, T^*, dP/dT^*), in this case chosen at zero pressure, reproduces Kennedy's experimental data with great accuracy up to 45 kb.

6. Other Materials. No attempt has been made here to correlate all the available static and dynamic data. The examples shown illustrate the success of this method. Similar calculations have been made for a number of other materials, including most of the rare earths and the alkaline earth metals. We have applied these calculations to the melting phase line of iron, which is of great interest in geophysical applications. However, because of complications caused by several phase changes and inadequate data we can put little credance in the results. The group IV-A metals titanium, zirconium and hafnium have proven intractable but particularly interesting. The Hugoniot u_s-u_p data and the available static phase line data[14] for zirconium are shown in Fig. 8; the data for titanium are quite similar. A well-defined transition is found on the Hugoniots of each of these metals which appears to be unrelated to any phases yet found by static methods. There are good reasons for believing these transitions to be electronic in nature, and it is possible that the phase lines end in critical points like that of cerium, but without any manifestations at low pressures. A more likely possibility, however, is that the static phase line data shown in Fig. 8 is either incomplete or inaccurate; these phase boundaries have by no means been fully investigated, and some investigators think the triple point lies much too high in pressure.[15] A calculation based on the static data of Fig. 8 and taken at the hcp-bcc-ω triple point with the slope of the bcc-ω boundary is shown. Although the calculation results in a value for

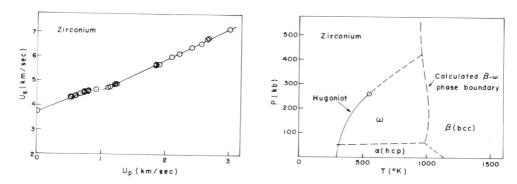

Fig. 8. Hugoniot data and P-T diagram for zirconium.

the transition pressure which is much too high, the phase line does indeed bend back toward the pressure axis in a manner generally required to explain the shock wave data. Believable calculations will require more extensive and accurate static data.

It would be of great practical interest to perform these calculations for alloys if the required experimental data existed. We have performed these calculations for some minerals of geophysical interest. Of particular interest would be to compare this straightforward technique with the phase line calculations of Kaufmann[16] and others. The kinetics of shock-induced phase transformations are of great interest, and the detailed study of transformation mechanisms will probably play a major role in the shock-wave work of the next several years. However, it is apparent that these studies are not crucial to the determination of equilibrium phase boundaries.

TABLE I - Parameters used in phase line calculations.

Material	Lower Phase Parameters					Upper Phase Parameters					Static Phase Line Parameters		
	ρ_0 (gm/cm³)	c_0 (km/sec)	s	γ_0	C_v (erg/gm × 10⁻⁶)	ρ_0 (gm/cm³)	c_0 (km/sec)	s	γ_0	C_v (erg/gm × 10⁻⁶)	P^* (kb)	T^* (°K)	$(dP/dT)^*$ (kb/°K)
Europium	5.27	1.62	0.99	1.02	1.65	4.82	1.03	1.29	1.0	1.65	0	1099	.067
Erbium	9.039	2.31	0.90	0.92	1.61	7.68	1.36	1.31	1.0	1.61	0	1770	.067
Cerium													
(fcc-fcc)	6.729	1.73	1.50	0.37	2.06	7.90	1.66	1.75	1.0	2.06	7	293	.038
(fcc-liq.)	7.90	1.66	1.75	1.0	2.06	7.80	1.50	1.50	1.0	2.06	70	1020	.28
Gadolinium													
(rhomb-bcc)	7.877	2.20	.92	0.31	2.99	8.24	1.86	1.33	1.0	2.99	8	1518	10
(bcc-liq.)	8.24	1.86	1.33	1.0	2.99	9.00	2.10	1.25	1.0	2.99	0	1585	.15
Lead	11.34	2.02	1.53	2.72	1.298	10.65	1.83	1.53	2.48	1.298	0	600.5	.129
Zirconium	6.505	3.76	1.02	1.09	2.72	6.00	2.71	1.35	1.09	2.72	62	983	1.25

Acknowledgments

The author would like to thank his colleagues at Group M-6, LASL, in particular Joseph Fritz, Robert McQueen and Stanley Marsh, for much encouragement and many helpful discussions.

References

1. S. N. Vaidya and G. C. Kennedy, J. Phys. Chem. Solids **31**, 2329 (1970).
2. Q. Johnson and A. C. Mitchell, Phys. Rev. Lett. **29**, 1369 (1972).
3. J. N. Johnson, Sandia Laboratories Report SC-RR-72 0626 (1972).
4. R. G. McQueen, W. J. Carter, J. N. Fritz and S. P. Marsh in "Accurate Characterization of the High-Pressure Environment," E. C. Lloyd, ed., N.B.S. Special Publication 326 (1971).
5. R. G. McQueen, S. P. Marsh, J. W. Taylor, J. N. Fritz, and W. J. Carter in "High-Velocity Impact Phenomena," R. Kinslow, ed. (Academic Press, New York, 1970).
6. A. Jayaraman, Phys. Rev. **135**, A1056 (1964).
7. A. Jayaraman, Phys. Rev. **139**, A690 (1965).
8. E. Rappaport, J. Chem. Phys. **48**, 1433 (1968).
9. A. Jayaraman, Phys. Rev. 137, A179 (1965).
10. L. V. Belyakov, V. P. Valitskii, N. A. Zlatin, and S. M. Mochelov, Soviet Physics - Doklady **11**, 808 (1967).
11. G. C. Kennedy and R. C. Newton in "Solids Under Pressure," W. Paul and D. Warschauer, ed. (McGraw-Hill, New York, 1963).
12. T. Takahashi, H. K. Mao, and W. A. Bassett, Science **165**, 1352 (1969).
13. M. B. Gitis and I. G. Mikhailov, Soviet Physics - Acoustics **11**, 372 (1966).
14. V. V. Evdokimova, Societ Physics - Uspekhi **9**, 54 (1966).
15. B. Olinger (private communication).
16. L. Kaufman and H. Bernstein in "Phase Diagrams: Materials Science and Technology," Vol. 1, A. M. Alper, ed. (Academic Press, New York, 1970).

CHARACTERISTICS OF THE SHOCK INDUCED TRANSFORMATION IN BaF_2

D. P. Dandekar and G. E. Duvall

Shock Dynamics Laboratory, Physics Department

Washington State University, Pullman, Washington

I. INTRODUCTION

Two of the several structure types commonly displayed by the solids with the formula AB_2, in which B is fluorine at room temperature and one atmospheric pressure, are quasi-six-coordinated rutile and eight-coordinated fluorite. The rutile structure is tetragonal and the fluorite structure is cubic. In situ high pressure x-ray diffraction studies and related post mortem analyses of quenched materials of the AB_2 compounds have revealed a large number of polymorphic transitions in these compounds at high pressures.[1] Most of these studies have been carried out at room temperature and the transition pressures have been only roughly established. The situation is further complicated by varying rates of transformation in the different AB_2 compounds, which has made the study of these compounds by static high pressure methods difficult. Since the shock compression method had been successful in determining the pressure of transition in various materials,[2] it prompted us to use this technique to establish transition pressures in these compounds.

The starting material chosen was barium fluoride because the pressure of transition for it is one of the lowest among the fluorides studied so far.[3] The shock compression experiments reported in this work were performed on the gas gun facility at Washington State University.

In order to make this paper self-contained, we first summarize the information available on barium fluoride from static compression studies and then present the results of our shock compression work.

II. RESULTS OF STATIC COMPRESSION IN BARIUM FLUORIDE

Phase transformations in barium fluoride (BaF_2) at elevated pressure under static conditions have been studied by several investigators.[1,4,5,6,7,8,9,10] Of these, the investigations of Seifert,[5,6] and of Dandekar and Jamieson[9] consisted in analyzing the quenched specimens of BaF_2 subjected to high pressure and temperature by the x-ray diffraction technique. The rest pertain to in situ high pressure investigations of BaF_2. Both types of investigations established that (1) BaF_2, which crystallizes in the fluorite structure (from now on identified as β-BaF_2) at ambient condition, transforms to the orthorhombic α-$PbCl_2$ structure (from now on also identified as α-BaF_2) at elevated pressures.[11] (2) The transition from β- to α-BaF_2 is extremely sluggish at room temperature, i.e. the transition is characterized by a large hysteresis. (3) α-BaF_2 is 10% to 11% denser than the fluorite phase at the ambient conditions.

The results of these investigations vary in three respects: (1) with regard to recovery of the high pressure polymorph of barium fluoride on release of pressure; (2) with regard to the magnitude and direction of change in pressure of transition with increasing temperature; and, (3) with regard to the pressure of transition at room temperature. These differences are described below in that order.

Under hydrostatic pressure, α-BaF_2 transforms back to β-BaF_2 on release of pressure.[8] In anvil type pressure generating systems, α-BaF_2, once formed, is metastably recovered.[1,5,9]

Samara[8] reports that the pressure of transition decreases with increasing temperature. For example, whereas the transition pressure at 22°C is 26.8 kbars, the transition pressure at 200°C is around 20 kbars. The values of dP/dT are $\sim -2.6 \times 10^{-2}$ kbars/°C below 100°C and $\sim -2.2 \times 10^{-2}$ kbars/°C between 100° and 300°C. The results of ref. 9 indicate that the transition pressure tends to decrease with increasing temperature. However, Seifert's work[5] indicates that the pressure of transition tends to increase with increasing temperature. He reports an increase of 22 kbars in transition pressure when temperature is increased from room temperature to 200°C. It is not clear if the increased pressure is an equilibrium pressure.

Values of transition pressure at room temperature reported by various investigators are listed in Table 1. It is seen that variations in transition pressure are quite large.

Table 1 -- Transformation Pressure of BaF_2 from Fluorite to Orthorhombic Structure at Ambient Temperature (β to α phase)

Investigators		Transformation Pressure (kbars)
Minomura and Drickamer	(1961)	30.8
Chen and Smith	(1966)	36.0
Seifert	(1968)	25.0
Dandekar and Jamieson	(1969)	<20.0
Samara	(1970)	26.8
Kessler and Nicol	(1972)	23.0

If temperature of the press containing α-BaF_2 is quickly reduced from T to room temperature, and pressure is subsequently released, α-BaF_2 is found in the sample if T < 470°C but not for greater T.[5,9] This temperature dependence of the effects of quenching is inexplicable on the basis of information given above.

It is very difficult to provide a clear explanation for disagreements among the investigators with regard to the pressure of transition at room temperature and its variation with temperature, and the variation in quenchability of α-BaF_2. However, one can easily think of a few of the variables which may have brought about those disagreements. For example the type of high pressure apparatus used in an investigation, the magnitude of hysteresis, impurities in the BaF_2 used could easily differ from one experiment to another and the experimental procedure to establish transition pressure, i.e. whether the pressure was observed by in situ examination of the sample or in a quenched sample. Inability to quench the orthorhombic phase at any pressure beyond 470°C tends to indicate that there may be a third phase of BaF_2 present at high temperature but below the melting temperature of BaF_2. That such a phase may exist even at one atmosphere is suggested by an observed transformation in the isostructural compound CaF_2 between 1047°C and 1100°C by Naylor.[12] The structure of this high temperature phase of CaF_2 remains undetermined.[13]

Finally the equation of state of the orthorhombic phase of barium fluoride is as yet undetermined.

III. SHOCK COMPRESSION EXPERIMENTS

Shock compression experiments were performed with the Washington State University gas gun. A detailed description of the gun and its operating characteristics are documented in ref. 14. Briefly, the gun is 44' long and four inches in diameter. It can propel a four inch diameter, 1 kg projectile with any velocity in

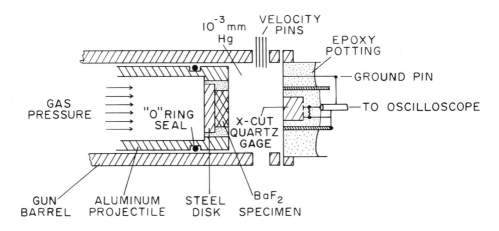

Fig. 1 -- An experimental assembly for measuring impact stress-time profiles. For measuring transmitted stress-time profiles a specimen is bonded to a quartz gage and mounted in the target. In recent experiments on BaF_2, the back-up steel disk was found not to be necessary for the impact stress-time profile measurements.

the range of 0.1 mm/μsec to approximately 1.4 mm/μsec. Angular misalignment between the impacting surfaces of the projectile and target is of the order of 5×10^{-4} radians or less. The entire impact chamber is evacuated to 10^{-3} torr to prevent an air cushion forming between the two impacting surfaces. The electronic recording instruments consist of Tektronix 454, 585 and 519 oscilloscopes. Parameters measured in the experiments are projectile velocity, misalignment of the projectile and target, and stress history either at the impact interface or at the specimen surface opposite the impact surface. A schematic of a finished experimental assembly for measuring stress history at the impact face is shown in Fig. 1. The stress gages used were x-cut quartz disks with 1/2" diameter and 3.2 mm thickness. These gages were used in the shorted configuration.

Shock compression experiments were performed on single crystals of BaF_2 oriented along <111> and <100> directions.[15] These specimens were in the shape of circular disks of diameter 22-25 mm and thickness ranging between 1.1 and 4 mm. Basically three types of experiments were performed. Experiments of the first type yielded stress history in the quartz gage after the shock wave was transmitted through the specimen. Experiments of the second type measured the stress history at the impact surface of a BaF$_2$ specimen before reflections returned. Experiments of the third type were designed to recover a specimen of BaF$_2$ subjected to shock compression stresses well above the pressure of transformation at room temperature. The results of these experiments are described in the following section.

IV. EXPERIMENTAL RESULTS

Front Surface Impact

In these experiments a stress gage records the jump in stress (σ_i) impressed upon a specimen at the impact surface because the quartz gage impactor is also the transducer.[16,17,18,19] The magnitude of this jump is calculated directly from the piezoelectric response of the gage.[20] For the configuration shown in Fig. 1, particle velocity in the quartz gage at the interface, is determined by the relation

$$u_q = \sigma_i / \rho_q D_q \tag{1}$$

where ρ_q and D_q are density and elastic longitudinal velocity in quartz, respectively. Since both stress and particle velocity are continuous at the impact interface, the pair of values (σ_i, u_q) also apply to the BaF$_2$. However, since BaF$_2$ has the initial velocity, u_0, the physically significant velocity is

$$u_s = u_0 - u_q. \tag{2}$$

Therefore the values tabulated as being characteristic of BaF$_2$ under the impact conditions are (σ_i, u_s). The characterization of a material in the above manner is valid whether the profile of the wave is steady or not, or whether there are any rate effects present or not.

As shown by Hayes,[19] in front surface impact the rate of stress relaxation at the impact surface immediately after impact is directly related to initial transformation rate at the impact surface. If a relaxation is present, the post-relaxation stress at the impact surface is propagated into the specimen with a steady profile.

In the present set of experiments, steady stress profiles were obtained only below ∼24 kbars, i.e. probably below the stress at which β-BaF$_2$ transforms to α-BaF$_2$. At higher stresses, stress profiles obtained at the impact surface for specimens oriented along <111> show a unique type of profile never before reported. These show that the stress initially rises, then goes through a minimum, and finally attains a steady state. A profile so obtained, with 40 kbar peak stress, is shown in Fig. 2, as recorded on two different oscilloscopes. Figure 2 also shows that the initial stress rise is very sensitive to the response of the electronic instrument used in the experiment. Figure 2(a) was recorded on a Tektronix oscilloscope Type 454 with 150 MHz bandwidth, and Fig. 2(b) was recorded on a Tektronix oscilloscope Type 519 with 1000 MHz bandwidth. This is important to remember because it implies that the initial record of a stress history does not necessarily reflect the mechanical property of the material. It may reflect the speed of the oscilloscope response to a signal. However, the magnitude of minima and subsequent steady state stress levels are insensitive to the type of oscilloscope used. Moreover, these stress profiles have been reproduced in identical experiments. Calculation of the impact stress for elastic compression of the β-phase indicates that the magnitude of initial stress is equal to the steady state value. However, the data present a difficult problem of interpretation because, on the basis of a standard stress-particle velocity (σ-u) diagram for a material which exhibits a phase transition under high pressure, shocking up of the stress profile after an initial relaxation is impossible. For example, consider a typical σ-u diagram for a front surface impact experiment in which the specimen is stationary, (Fig. 3). Let OA and ABC, parts of the curve OABC, represent the stable and metastable part of the σ-u profile for phase I of a material. Let AB'C' represent a similar profile for stable phase II of the material. Let the transition stress be σ_A. If a quartz gage with a velocity u_0 is impacted on the specimen in phase I so as to generate a stress σ_B, then if the transformation to phase II is instantaneous or faster than the rise time of the recording instrument, the impact stress profile would be of magnitude $\sigma_{B'}$. However, if the rate of transformation is slower than

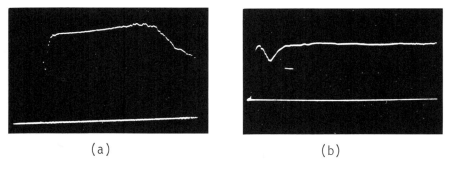

(a) (b)

Fig. 2 -- Stress-time profile for shock compression of BaF_2 in <111> direction. The impact stress in both cases is 40 kbars.
(a) The profile recorded on Tektronix Type 454 oscilloscope.
(b) The profile recorded on Tektronix Type 519 oscilloscope.

SHOCK INDUCED TRANSFORMATION IN BaF$_2$

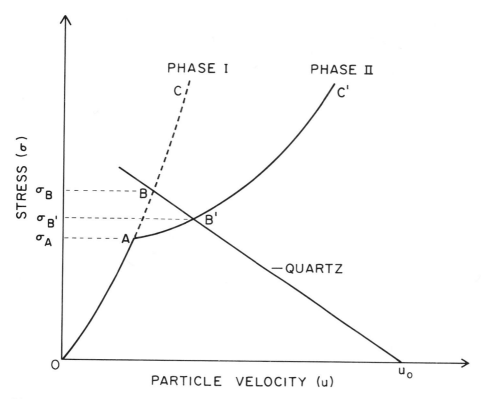

Fig. 3 -- Stress-particle velocity diagram for a material which transforms to a new phase at stress σ_A, being shocked to a stress $\sigma_B > \sigma_A$ by an impactor, moving with a velocity u_0. In the diagram impactor is a quartz gage.

the rise time of the recording instrument the impact stress would relax from σ_B to $\sigma_{B'}$ in some finite time. In no case would it shock up. Thus for a phase-transforming material impact stress cannot shock up as it does in our profile. Hence, the first question that arises is whether the profile represents a material property of BaF$_2$ or whether the quartz gage is behaving in a peculiar manner, since these shots develop a stress of 40 kbars in the gage, a value which is near or above the upper limit for reliable performance. Three experiments were performed to rule out one of the above mentioned alternatives. In the first of these experiments an aluminum 6061-T6 projectile was impacted on a quartz gage with a sufficient velocity to guarantee an impact stress of 40 kbars. The stress profile obtained was steady and showed no peculiarities whatsoever. In the second experiment, a projectile generated 55 kbars in the gage. At this stress the quartz gage was found to break down. The profile obtained is shown in Fig. 4(a). The third experiment consisted of impacting a specimen of BaF_2 on a quartz gage to generate a stress of 55 kbars once more. The stress profile obtained in this experiment is shown in Fig. 4(b). It may be seen from Figs. 4(a) and 4(b) and the result of the first experiment that, except for distortions due to response time and nonlinearity of the quartz gage response, the total stress profile shown in Fig. 2 does represent the mechanical behavior of BaF_2. The type of profile shown in Fig. 2 continues to be present at impact stress down to ∼25 kbars. Below 25 kbars the impact stress profiles are steady, as shown in Fig. 5.

A plot of stress and particle velocity for BaF_2 specimens with <111> orientation (Fig. 6) shows that (1) the values of steady state stress and the corresponding values of particle velocity lie along the stress vs particle velocity curve of BaF_2 <111> in the β-phase,

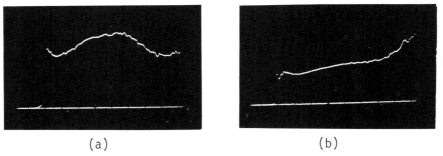

(a)   (b)

Fig. 4 -- Stress-time profile at 55 kbars. (a) By impacting a quartz gage on aluminum. (b) By impacting a quartz gage on a BaF_2 specimen.

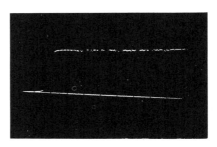

Fig. 5 -- Stress-time profile obtained for shock compression of BaF_2 in <111> direction for an impact stress below 25 kbars.

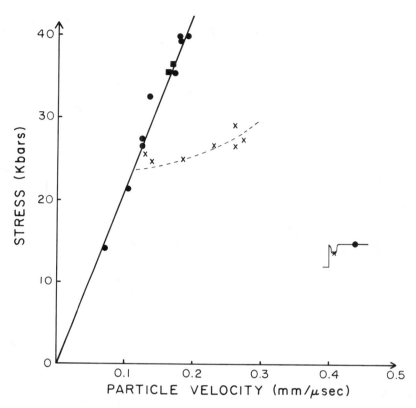

Fig. 6 -- Stress-particle velocity (σ-u) plot of BaF$_2$ as obtained from the front surface impact experiments. The line marked <111> denotes the estimated σ-u dependence of β-BaF$_2$ in <111> direction. It is identical with σ-u plot for <100> direction. ● steady state stress realized in the experiments. x the pair of σ, u values corresponding to minima in the experiments. ■ the steady state (σ,u) values for <100> shots.

as obtained from the elastic constant data of BaF$_2$.[21] In the same vein, if we plot the magnitude of the stress corresponding to minima and related particle velocity, these two sets of data form a cusp at 22 to 24 kbars.

The stress profiles obtained for BaF$_2$ specimens oriented along the <100> direction show that a steady state stress is reached immediately. A plot of stress and particle velocity indicates that these pairs of points lie along the curve of elastic compression of the β-phase of BaF$_2$. A stress profile obtained around 40 kbars for the <100> orientation is shown in Fig. 7.

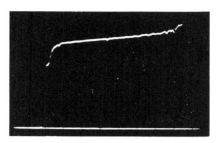

Fig. 7 -- Stress-time profile obtained for shock compression of BaF_2 in <100> direction with an impact stress of 40 kbars.

Transmission Experiments

Only three transmission experiments have been performed. A representative stress profile recorded by a quartz gage at 38 kbars is shown in Fig. 8. Stress profiles obtained around 20 kbar show the same character as that shown in Fig. 5. The nature of the profile obtained at higher stresses makes it difficult to analyze in a straightforward manner, and no further transmission experiments have been performed as of this date.

Recovery Experiments

Two recovery shots were performed. The details of design and construction, being of only ancillary interest, are not described here. Both experiments were performed at 35 to 40 kbars, well above the stress at which the minimum occurs in the stress profile.

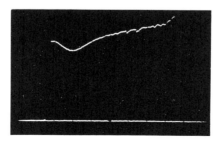

Fig. 8 -- Transmitted stress-time profile for BaF_2 in <111> direction with an impact stress of 40 kbars.

Post mortem examination of the shocked BaF_2 specimens by x-ray diffraction techniques showed that the materials were in the β-BaF_2 phase. There was no trace of the α-BaF_2 in either specimen.

V. DISCUSSION OF EXPERIMENTAL RESULTS

In the previous section we established that (1) stress profiles observed in BaF_2 reflect its mechanical properties within the limitation of the recording instruments, (2) shock recovered specimens of BaF_2 were found to be in the β- phase and (3) information from the transmission shots is too meagre to yield any significant insight in interpreting the data.[22] In light of the above, we shall try to elucidate the behavior of BaF_2 on the basis of experimental observations resulting from front surface impact and recovery experiments.

It is useful to make some remarks about the time scale of the events recorded in these experiments. The minima occur between 5 and 12 nsec after impact, and their times of occurrence decrease with increasing impact pressure. These minima are sharp and have no temporal spread. The times taken to reach steady state bear no relation to the value of the steady state impact stress. The steady state stress is reached within a time interval of 23-62 nanoseconds for impact stresses above 25 kbars. These time intervals which add to less than 80 nsec, are well within the time at which the disturbing effects of using shorted quartz gages would seriously affect the recorded stress-time profile in these experiments.

If the presence of a cusp is taken to indicate the onset of a transition in BaF_2, one may conclude that the observed transition in BaF_2 is shear induced. This is because in a fluorite structure shear stresses developed in specimens oriented along the <111> and the <100> directions are equal to 26.1 and 0.0 per cent of the applied normal stress, respectively.[23,24,25] Since recovered specimens of BaF_2 were found to be in β-phase, one may conclude that the transition is a reversible one. Thus, in this regard, the response of BaF_2 to shock compression is like the response of BaF_2 to a hydrostatic stress environment.

Information available from experiments reported here is not enough to enable us to characterize the BaF_2 under these conditions. We do note, however, that the upturn of stress following its initial decay, shown in Fig. 2, is consistent with a simple relaxation model of the material. To see this, consider the configuration in which a stationary quartz gage is struck by a moving BaF_2 sample. Let the constitutive relation for BaF_2 be of the form[26]

$$\frac{d\sigma_x}{dt} - a^2 \left(\frac{d\rho}{dt}\right) = -F \qquad (3)$$

where a is the speed of elastic compression in material of density ρ, σ_x is the longitudinal stress, and F is a function characterizing the rate at which an equilibrium stress is reached in the shocked material. The specific form of F depends upon the type of process or processes through which the material relaxes. In general F can be a function of static as well as rate dependent variables. Let h be the Lagrangian space coordinate for the problem. Then for arbitrary h, equation (3) becomes

$$(\partial \sigma_x / \partial t)_h - a^2 (\partial \rho / \partial t)_h = -F. \tag{4}$$

The equations expressing mass and momentum conservation, in Lagrangian coordinates are, respectively

$$(\rho_0/\rho^2) \cdot (\partial \rho / \partial t)_h + (\partial u / \partial h)_t = 0, \tag{5}$$

$$(\rho_0/\rho) \cdot (\partial u / \partial t)_h + (1/\rho) \cdot (\partial \sigma_x / \partial h)_t = 0. \tag{6}$$

Combining Eqs. (5) and (4) we obtain

$$(\partial \sigma_x / \partial t)_h = -(a\rho)^2 \cdot (1/\rho_0) \cdot (\partial u / \partial h)_t - F. \tag{7}$$

At the interface, we have, referring to Fig. 1,

$$(\partial \sigma_x / \partial t)_{h=0} = Z_q (\partial u / \partial t)_{h=0} \tag{8}$$

where $Z_q = \rho_q D_q$. Substituting the value of $(\partial \sigma_x / \partial t)_{h=0}$ from Eq. (8) into Eq. (7) we obtain

$$Z_q (\partial u / \partial t)_h = -(a\rho)^2 \cdot (1/\rho_0) \cdot (\partial u / \partial h)_t - F. \tag{9}$$

The stress profile obtained for BaF_2 indicates that after the initial rise, say at time $t = 0$, stress and particle velocity decrease up to a time τ_1 where they both attain an extremum value. From time τ_1 to some time τ_2, stress and particle velocity increase and beyond τ_2 the stress and particle velocity are constant. For these time intervals, we have

$$(\partial u / \partial t)_{h=0} < 0 \quad 0 \leq t < \tau_1 \tag{10a}$$

$$(\partial u / \partial t)_{h=0} = 0 \quad t = \tau_1 \text{ and } \tau_2, \text{ and} \tag{10b}$$

$$(\partial u / \partial t)_{h=0} > 0 \quad \tau_1 < t < \tau_2 \tag{10c}$$

Since F is always positive, it implies that in order to obtain the observed profile one must have for

$$0 \leq t < \tau_1, \quad (a\rho)^2 \cdot (1/\rho_0) \cdot (\partial u / \partial h)_{t, h=0} < F \tag{11a}$$

$$t = \tau_1 \text{ or } \tau_2 \quad (a\rho)^2 \cdot (1/\rho_0) \cdot (\partial u/\partial h)_{t,h=0} = F \quad (11b)$$

$$\tau_1 < t < \tau_2 \quad (a\rho)^2 \cdot (1/\rho_0) \cdot (\partial u/\partial h)_{t,h=0} > F \quad (11c)$$

In the derivation of the conditions expressed by Eq. (11), it has been assumed that particle velocity decreases with increasing propagation distance. These conditions do not seem to violate any fundamental relations. In terms of the rate of change of density, the above sets of conditions imply that during the time interval 0 and up to τ_1, the rate of change of density at the interface may be positive or negative. In the case it is positive, its magnitude must be less than the value of F/a^2. During the time interval τ_1 and up to τ_2, the rate of change of density must be positive and its magnitude must be larger than the value of F/a^2. At the time points τ_1 and τ_2 either both the rate of change of density and F are identically zero or the rate of change of density must be positive and balance with the value of F/a^2. None of the situations described above violates any physical conditions, especially since a negative rate of change of density is observed in an elastic-plastic relaxing material as a shock compression wave propagates in it.

In BaF_2, the situation may be more complicated if above 25 kbars it behaves both plastically and starts to transform to its α-phase. We plan to pursue the present study of polymorphic behavior of BaF_2 further by performing the following experiments. (1) Experiments where stresses generated would be such that β-BaF_2 would almost instantaneously transform to α-BaF_2, i.e., probably at stresses in excess of 70 kbars; (2) experiments on the samples at high temperatures; (3) experiments measuring stress profile at various depths in the specimens of BaF_2 at varying initial temperatures; and (4) the above type of experiments on annealed specimens of BaF_2. The data from these experiments will permit us to build a model for the behavior of BaF_2 under shock compression.

ACKNOWLEDGMENT

This work was supported by the United States Air Force Office of Scientific Research Contract Number 71-2037A.

REFERENCES

1. D. P. Dandekar and J. C. Jamieson, Proceedings of Symposium on Crystal Structures at High Pressures, American Crystallographic Association (1969) pp. 19, and the references in the article.

2. O. E. Jones and R. A. Graham, <u>Accurate Characterization of the High-Pressure Environment</u>, National Bureau of Standards Special Publication 325, U.S. Government Printing Office, Washington, D.C., 1971, pp. 229 and references therein.

3. Lead fluoride which is isostructural to BaF_2 transforms at 4.8 kbars at 23°C. Hence, lead fluoride would have been the ideal material to start with, but for the unavailability of a good large single crystal of the same. On the other hand BaF_2 crystals are readily available and economical to perform experiments with.

4. S. Minomura and H. G. Drickamer, J. Chem. Phys. <u>34</u>, 670 (1960).

5. K. F. Seifert, Ber. Bunsenger, Physik. Chem. <u>70</u>, 1041 (1966).

6. K. F. Seifert, Fortschrift Miner. <u>45</u>, 214 (1968).

7. J. H. Chen and H. I. Smith (Report AFCRL-66-601), (1966) and Bull. APS <u>11</u>, 414 (1966).

8. G. A. Samara, Phys. Rev. <u>B1</u>, 4194 (1970).

9. D. P. Dandekar and J. C. Jamieson (Unpublished).

10. J. R. Kessler and M. Nicol, Bull. APS II, <u>17</u>, 123 (1972).

11. Both structures contain 4 molecules per unit cell. In the fluorite structure (space group $Fm3m-O_h^5$) barium atoms are in a cubic close-packed type of arrangement with fluorine atoms occupying all the tetrahedral sites. Each fluorine atom is surrounded by 4 barium atoms disposed tetrahedrally, and each barium atom by 8 fluorine atoms disposed toward the corners of a cube. In the α-$PbCl_2$ structure (space group $Pbnm-v^{16}$) each barium atom is surrounded by 9 fluorine atoms, but all of these atoms are not equidistant from the cation they surround. This structure may be considered as a distorted close-packing of fluorine atoms with the barium atoms accommodated in the same plane with them. For details see: R. W. Wyckoff, <u>Crystal Structures</u>, Vol. 1, (Interscience, New York, 1963). Chapter IV.

12. B. F. Naylor, J. Am. Chem. Soc. <u>67</u>, 150 (1945).

13. A similar transition has been observed in $SrCl_2$, $BaCl_2$, $SrBr_2$ below their respective melting points. These compounds are isostructural to CaF_2 and BaF_2. The possible nature of this transition may be inferred from the x-ray diffraction work of Croatto and Bruno on $SrCl_2$. They suggested that the resulting high temperature structure shows a disordering in the anion

semilattice. For further information see the following references. (1) U. Croatto and M. Bruno, Gass. Chem. Ital. 76, 246 (1946), and (2) A. S. Dworkin and M. A. Bredig, J. Physical Chem. 67, 697 (1963).

14. G. R. Fowles, G. E. Duvall, J. Asay, P. Bellamy, F. Feistman, D. Grady, T. Michaels and R. Mitchell, Rev. Sci. Instr. 41, 984 (1970).

15. BaF_2 crystals were bought from Harshaw Chemical Co., Solon, Ohio and Optovac Inc., North Brookfield, Mass. These crystals were of optical quality and free from any visual defects. Back reflection x-ray pattern indicated that misorientation of specific specimen could be accurately represented by $\theta = \pm 0.5°$ and $\phi = \pm 0.5°$.

16. A. R. Champion, J. Appl. Phys. 42, 5546 (1971).

17. W. J. Halpin and R. A. Graham, "Shock Wave Compression of Plexiglass from 3 to 20 kbars," in Fourth Symposium on Detonation, Office of Naval Research Report #ACR-126 (1965).

18. W. J. Halpin, O. E. Jones and R. A. Graham, "A Submicrosecond Technique for Simultaneous Observation of Input and Propagated Impact Stresses," in Dynamic Behavior of Materials, ASTM Special Technical Publication #336 (1963).

19. D. B. Hayes, Ph.D. Thesis, Washington State University (1972).

20. R. A. Graham, F. W. Nielson and W. B. Benedick, J. Appl. Phys. 36, 1775 (1965).

21. C. E. Wong and D. E. Schuele, J. Phys. Chem. Solids 29, 1309 (1968).

22. This conclusion is not based on the three shots reported in here alone but includes the transmitted stress profiles obtained previously in this laboratory by a graduate student in 1969-70.

23. J. N. Johnson, O. E. Jones and T. E. Michaels, J. Appl. Phys. 41, 2330 (1970).

24. An independent way of confirming the conclusion that the transformation in BaF_2 from fluorite to orthorhombic structure is shear induced would be to subject a small sample of either BaF_2 or preferably lead fluoride to a prolonged grinding in a mechanical mortar.

25. J. R. Kessler (private communication) finds that β-BaF$_2$ is relatively easily and reasonably completely transformed to α-BaF$_2$ phase at 23 kbars where the pressure is applied normal to <1$\bar{1}$1> direction in β-BaF$_2$.

26. G. E. Duvall, in <u>Stress Waves in Anelastic Solids</u>, H. Kolsky and W. Prager, Eds., Springer-Verlag, Berlin (1964), pp. 20-32.

A THEORY OF THE α→ε TRANSITION IN Fe AND OF POSSIBLE HIGHER PRESSURE TRANSITIONS IN Fe AND IN THE LIGHTER ELEMENTS OF THE FIRST TRANSITION SERIES

D. John Pastine

Naval Ordnance Laboratory
White Oak, Silver Spring, Maryland 20910

ABSTRACT

Estimates have been made of the volume variation of the total energy of α-Fe as a function of magnetic moment, μ, and electronic configuration. The calculations were done using Wood's theoretical state densities scaled in such a way that the product of the d-band width and the density of states is held constant. This scaling method is consistent with the results of earlier work by Stern. The overall results indicate that, with increasing pressure, μ→o slowly in α-Fe and that the electronic structure moves toward d^8. The energy line of a paramagnetic phase of Fe is shown to intersect that of α-Fe and become the phase of lowest free energy. This occurs in the region of the α-ε transition and suggest that ε-Fe is paramagnetic. An analysis of the behavior of the d-energies suggests that the lighter transition or pre-transition elements may form stable transition phases under pressure which are also stable at zero pressure.

INTRODUCTION

In recent years a number of measurements have been made of the high pressure properties of iron (Fe) in the region of the α→ε (bcc to hcp) transition. Among the more recent are, static measurements made by Basset and Takahashi[1], Giles[2] et al and by Pipkorn[3] et al. These indicate that at room temperature (here 298°K) the transition occurs in the region of 130 kbars and that it is abaric, initiating at about 130 kbar and going to completion at pressures >170 kbars. The observed specific volume change associated with the transition is $\Delta v \approx .0066$ cc/gm and the compression ratio at the onset of the transition is $x \approx .943$. Here x is the specific volume, v, divided by the specific volume, v_o, of α-Fe at 0°K and zero pressure, P.

Mossbauer measurements[3] indicate that the magnetic moment per atom μ is practically constant for the α phase up to the pressure of its disappearance and that the ε phase is non-ferromagnetic. These measurements also imply that the number N_s of s-electrons per atom is invariant with pressure for the α phase whereas for the ε phase N_s decreases with increasing pressure.

Upon the reduction of pressure the ε phase persists down to a pressure of about 45 kbar causing hysteresis in the compressional behavior of iron in the vicinity of the transition. At pressures below 45 kbar the ε phase disappears completely[2].

Recent dynamic measurements, using shock waves, have been made by Keeler[4] et al. These confirm the non-ferromagnetic nature of the ε phase and in addition show some demagnetization (an apparent reduction of μ) at pressures as low as 50 kbar.

In this work there is an attempt to explain some of these effects from the viewpoint of theory. The overall results suggest the following:

(1) As pressure increases μ should approach zero for all metals in the first transition series. This may occur either gradually or suddenly.

(2) In Fe, μ ought to go to zero suddenly. The theoretical pressure at which this event occurs is in the region of the observed α→ε transition and implies the presence of a new phase.

(3) As pressure increases well beyond the transition point there is a tendency in Fe to approach the electronic structure d^8.

(4) This tendency toward a completely d-like structure (with increasing pressure) should also exist in the lighter transition metals and in some of these it may lead to the formation of stable transition phases.

METHOD

The approach here is to make computations of the volume variation of the total energy of Fe as a function of the electronic structure and of μ. The energy variation is assumed to arise primarily from the eight d and s-like valence electrons per atom and these are described in the itinerant electron picture. Complete band calculations are not done. Instead, use is made of Wood's[5] computed density of electron states, $n(\eta)$, for bcc Fe (η here is the energy taken as zero at the bottom of Wood's band). Wood's computations have recently been revised by Cornwell[6] et al and it is the latter data which are used in subsequent calculations. In order to find energies as a function of volume Wood's revised values of $n(\eta)$ are scaled with d-band width, Δ, in such a way that $\Delta n(\eta)$ is held constant. This assumption will be seen to provide a good description of α-Fe and is consistent with the work of Stern[7] (see Appendix). The volume variation of the bandwidth, Δ, is assessed by monitoring the volume variation of the energy separation, $\Delta_{2,3}$, of special symmetry points (in B.S.W.[8] notation) N_2 and N_3 in the Brillouin zone. The energies of the latter are computed in the Wigner-Seitz[9] (W-S) approximation and the total bandwidth, Δ, is assumed to be proportional to their energy separation. The appropriate bandwidth at $0°K$ and $P = 0$ is found by requiring α-Fe to have an energy minimum at the observed W-S radius[10], $r_s = 2.662$ Bohr units. When this is done the result is found to be close to Wood's theoretical value of Δ.

The s-electrons of Fe are treated by relatively standard techniques. The procedural details for the computation of s energies and of total energies are given below.

THE EXPRESSION FOR THE TOTAL ENERGY

For Fe that part of the total energy per atom (in Ry.) which varies with volume in the region of r_s considered here is given by,

$$E_{tot} = N_d\{\bar{\varepsilon}_d - (N_d-1)\bar{V}_{dd}/2 - [U(U-1)+D(D-1)]\bar{J}_{dd}/2 - N_s\bar{V}_{ds}/2\} + E_{tot}^s \quad (1)$$

where, $\bar{\varepsilon}_d$ is the mean Hartree-Fock (H-F) energy[11] of the d electrons,

$\bar{V}_{dd}$ is the mean Coulomb interaction energy between two d electrons both normalized within a sphere of radius r_s,

$\bar{V}_{ds}$ is the mean Coulomb and exchange interaction energy between a d- and an s-electron each normalized within r_s,

$\bar{J}_{dd}$ is the mean exchange interaction energy between two d electrons each normalized within r_s,

N_d is the total number of d-electrons per atom,

N_s is the total number of s-electrons per atom,

U is the total number per atom of majority (up) spins,

D is the total number per atom of minority (down) spins,

and E_{tot}^s is the total energy per atom of s-electrons.

A number of simplifications may be employed to facilitate the evaluation of Eq. (1). As a first step, it is convenient to eliminate the requirement of calculating the d-energies in the potential field of the s-electrons.

In the solid the d-electron is assumed to see a sphericalized potential $v(r)$ which may be broken up into $v'(r) + v_{ds}(r)$ with the latter term arising from the d-s Coulomb and exchange interactions. The Schrödinger equation for an electron in the d-band with energy ε_d and wave function, ψ_d, (normalized within r_s and subject to certain boundary conditions) is, in atomic units,

$$-\nabla^2 \psi_d + [v' + v_{ds}]\psi_d = \varepsilon_d \psi_d \tag{2}$$

Now v_{ds} can be written as $\bar{v}_{ds} + v_{ds} - \bar{v}_{ds}$ where $\bar{v}_{ds}$ is the volume average of the s-d interaction in a W-S cell, i.e. $\bar{v}_{ds} = \int_0^{r_s} \psi_d^* v_{ds} \psi_d \, d\tau$. From this it follows that,

$$\nabla^2 \psi_d = [v' + \bar{v}_{ds} + (v_{ds} - \bar{v}_{ds})] \psi_d = \varepsilon_d \psi_d \tag{3}$$

or

$$\nabla^2 \psi_d = [v' + (v_{ds} - \bar{v}_{ds})] \psi_d = \varepsilon_d' \psi_d \tag{4}$$

where $\varepsilon_d' = (\varepsilon_d - \bar{v}_{ds})$.

Now note that if $(v_{ds} - \bar{v}_{ds})$ is treated as a perturbation of the state (subject to the same boundary conditions) described by,

$$-\nabla^2 \psi_d' + v' \psi_d' = \varepsilon_d'' \psi_d' \tag{5}$$

then the s-d interaction is removed in the Shrödinger equation to be solved. When Eq. (5) is solved for ε_d'' one may use first order perturbation theory to obtain ε_d' since the perturbation $(v_{ds} - \bar{v}_{ds})$ is small compared to v' everywhere within the W-S sphere. This is true even near r_s, where it is largest, since, for typical solid state d and s functions, $\bar{v}_{ds}$ is about $2.8N_s/r_s$ whereas, at r_s, $v'(r_s) = 2(N_s+1)/r_s$ and $v_{ds} = 2N_s/r_s$. Therefore at r_s the ratio $(v_{ds}-\bar{v}_{ds})/v'$ is approximately $-.8N_s/(2N_s+1)$. One should consequently expect that,

$$\varepsilon'_d \simeq \varepsilon''_d + \int_o^{r_s} \psi'^*_d (v_{ds}-\bar{v}_{ds})\psi'_d \, d\tau \tag{6}$$

The integral on the right of Eq. (6) however is very small compared to ε''_d and may be neglected in the eventual computation of $\bar{\varepsilon}_d$. This is because ψ'_d is very little different from ψ_d for which the integral is identically zero. This is so because the two functions must obey the same boundary conditions and because the potential v_{ds} is slowly varying (v_{ds} is $\simeq 3N_s/r_s$ at r=0 and $2N_s/r_s$ at $r=r_s$) and consequently has very little effect on the shape of ψ_d.

Since $\varepsilon'_d \simeq \varepsilon''_d$, then, ε_d, from Eqs. (4) and (6) is given by

$$\varepsilon_d \simeq \varepsilon''_d + \bar{v}_{ds} \tag{7}$$

Equation (7) indicates that one need not compute $\bar{\varepsilon}_d$ directly to obtain total energies but instead one can compute $\bar{\varepsilon}''_d$ and add in $N_s\bar{V}_{ds}$, i.e. $\bar{\varepsilon}_d = \bar{\varepsilon}''_d + N_s\bar{V}_{ds}$. In practice, rather than do a separate computation of $\bar{v}_{ds}$ for every computed ε''_d, it is easier to assign the total d-s interaction to the s-electrons since $\bar{V}_{ds}$ must equal $\bar{V}_{sd}$. Equation (1) may therefore be rewritten in the form

$$E_{tot} = N_d\{\bar{\varepsilon}''_d - (N_d-1)\bar{V}_{dd}/2 - [U(U-1)+D(D-1)]\bar{J}_{dd}/2\} + E'^s_{tot} \tag{8}$$

where $E'^s_{tot} = E^s_{tot} + N_dN_s\bar{V}_{sd}/2$.

Now assume that a mean value of the energies, $\bar{\varepsilon}'''_d$, is calculated within a W-S cell in the potential field of the free atom d functions and without d-s interactions. To obtain $\bar{\varepsilon}''_d$ from this energy it would have to be corrected (since it is not self consistent) to account for the change in the Coulomb and exchange potential fields due to the compression. This correction (if one at first neglects the exchange) may be written in the form,

$$\bar{\varepsilon}_d'' = \bar{\varepsilon}_d''' + (N_d-1)\left\{ \int_o^{r_s}\int_o^{r_s} \rho_f''(r_1)\frac{2}{r_{1,2}} \rho''(r_2)d\tau_1 d\tau_2 \right.$$

$$\left. - \int_o^{r_s}\int_o^{r_s} \rho_o(r_1)\frac{2}{r_{1,2}} \rho'''(r_2)d\tau_1 d\tau_2 \right\} \quad (9)$$

Where ρ_f'' is the mean charge density of a d-electron in the exact case normalized to -1 in r_s,

ρ'' is the charge density if the d-electron of mean energy $\bar{\varepsilon}_d''$ normalized to -1 in r_s,

ρ''' is the charge density of the d-electron of mean energy $\bar{\varepsilon}_d'''$ normalized to -1 in r_s

and ρ_o is the charge density of a d-electron in the free atom.

Note that the total energy per d-electron per atom $\bar{E}_d''$ may be obtained from Eq. (9) by subtracting out from the right side the integral,

$$\tfrac{1}{2}(N_d-1)\int_o^{r_s}\int_o^{r_s} \rho_f''(r_1)\frac{2}{r_{1,2}} \rho''(r_2)d\tau_1 d\tau_2 \quad \text{so that,}$$

$$\bar{E}_d'' = \bar{\varepsilon}_d''' + (N_d-1)\left\{ \tfrac{1}{2}\int_o^{r_s}\int_o^{r_s} \rho_f''(r_1)\frac{2}{r_{1,2}} \rho''(r_1)d\tau_1 d\tau_2 \right.$$

$$\left. - \int_o^{r_s}\int_o^{r_s} \rho_o(r_1)\frac{2}{r_{1,2}} \rho'''(r_1)d\tau_1 d\tau_2 \right\} \quad (10)$$

Now if, in Eq. (10), one substitutes $\rho_o+\Delta\rho_f''$, $\rho_o+\Delta\rho''$, and $\rho_o+\Delta\rho'''$ for ρ_f'', ρ'', and ρ''' respectively and then rearranges terms one can obtain,

$$\bar{E}_d'' = \bar{\varepsilon}_d''' + (N_d-1)\left\{ \int_o^{r_s}\int_o^{r_s} \rho_o(r_1)\frac{2}{r_{1,2}}\left(\frac{\Delta\rho_f''(r_2)+\Delta\rho''(r_2)}{2} - \Delta\rho'''(r_2)\right) \right.$$

$$+ \tfrac{1}{2}\int_o^{r_s}\int_o^{r_s} \Delta\rho_f''(r_1)\frac{2}{r_{1,2}} \Delta\rho''(r_2)d\tau_1 d\tau_2 \quad (11)$$

$$\left. - \tfrac{1}{2}\int_o^{r_s}\int_o^{r_s} \rho_o(r_1)\frac{2}{r_{1,2}} \rho_o(r_2)d\tau_1 d\tau_2 \right\}$$

On examination of Eq. (11) one can, as a first approximation, eliminate the first two double integrals on the right as they are small relative to the last. This is because the application here is to relatively narrow bands and the $\Delta\rho$'s are expected to be nearly equal and small next to ρ_o. An estimated[12] of the magnitude of the second integral in Eq. (11) shows it to be less than 1% of the magnitude of the last integral. In addition, the volume variation of the second integral is relatively small and is, for the time being, neglected.

This leaves the result,

$$\overline{E_d''} \simeq \overline{\epsilon_d'''} - \frac{1}{2}(N_d-1)V_o(r_s) \quad (12)$$

where $\quad V_o(r_s) = \int_o^{r_s} \int_o^{r_s} \rho_o(r_2)\frac{2}{r_{1,2}}\rho_o(r_2)d\tau_1 d\tau_2$.

Essentially Eq. (12) says that a good total energy may be obtained by calculating too low an energy $\overline{\epsilon_d'''}$ and then compensating for this by subtracting out too small an amount for the d-d Coulomb interaction. The use of this relation makes the problem of calculating total energies much simpler.

In deriving Eq. (12) the effects of exchange where not included. When exchange is included the result is,

$$\overline{E_d''} \simeq \overline{\epsilon'''} - \frac{1}{2}(N_d-1)V_o(r_s) - \frac{1}{2}[U(U-1) + D(D-1)]J_o(r_s) \quad (13)$$

where $J_o(r_s)$ is the exchange energy shared by 2 d-electrons of like spin in the free atom within a sphere of radius r_s. Equation (13) may be used to alter Eq. (8) to the form,

$$E_{tot} = N_d\{\epsilon_d''' - (N_d-1)V_o(r_s)/2 - [U(U-1)+D(D-1)]J_o(r_s)/2\} + E'^s_{tot} \quad (14)$$

This last equation is much easier to evaluate than the original of Eq. (1) because, as is mentioned above, ϵ_d''' is to be calculated in the field of (N_d-1) free atomic d-functions and to the exclusion of the d-s interaction. The quantity, E'^s_{tot}, on the other hand, is to be computed by including the total s-d interaction in the s-energies. The evaluation of V_o and J_o requires the calculation of free atom d-functions. The methods used to obtain these functions along with tables of V_o and J_o are given in the Appendix. It should be mentioned that the potential of the Argon core of Fe was obtained from the analytic wave functions of Watson[13].

THE EVALUATION OF $E_{tot}'^s(r_s)$

The energies $E_{tot}'^s(r_s)$ have been calculated for the configurations d^6s^2 and d^7s^1. Results are presented only for the d^7s^1 configuration which is commonly thought[3,5,7] to be close to the correct configuration for α-Fe. The method used is the relatively standard W-S method for nearly free s-electrons. $E_{tot}'^s$ is obtained from the relation,

$$E_{tot}'^s(r_s) = N_s \left\{ \epsilon_o^s(r_s) - \frac{1.2(N_s-2)}{r_s} + \frac{\alpha(r_s) 2.21 N_s^{2/3}}{r_s^2} - \frac{.916 N_s^{1/3}}{r_s} - \frac{.88 N_s^{1/3}}{[r_s + 7.8 N_s^{1/3}]} \right\} \quad (15)$$

where ϵ_o^s is the H-F energy[11] of the lowest lying s-electron calculated in an s-electron field of constant charge density, $-3(N_s-1)/4\pi r_s^3$; $1.2/r_s$ is half the Coulomb interaction energy between two s-electrons both normalized within a W-S cell, $\alpha 2.21 N_s^{2/3}/r_s^2$ is the mean kinetic energy of s-electrons due to their translation through the lattice (α is the ratio of the true to the effective electron mass, m/m^*, at the bottom of the s-band); $-.916 N_s^{1/3}/r_s$ is the mean exchange energy per s-electron per atom arising from the other s-electrons and $-.88 N_s^{1/3}/(r_s + 7.8 N_s^{1/3})$ is the correlation energy per s-electron per atom arising from the Coulomb repulsion of other s-electrons.

All the terms to the right of ϵ_o^s in Eq. (15) are calculated in the free electron approximation. This was found to be justifiable in the range of r_s of interest here since there is very little difference in the potential arising from a calculated s-function and that arising from the approximation of constant charge density.

The calculated values of $\alpha(r_s)$, $E_{tot}'^s(r_s)$, and the pressure contribution of $E_{tot}'^s(r_s)$ are given in Table 1 for various values of of r_s and the compression x. The values in Table 1 are for the d^7s^1 configuration only.

A THEORY OF THE α→ε TRANSITION IN Fe

r_s (Bohr units)	x	α	E'^s_{tot} (Ry.)	P'^s (kbar)
1.92	.3752	1.343	-.0021	7705.1
2.00	.4240	1.317	-.1707	5329.9
2.10	.4909	1.283	-.3239	3392.4
2.20	.5644	1.248	-.4317	2177.8
2.30	.6449	1.214	-.5077	1404.5
2.40	.7328	1.181	-.5611	905.1
2.50	.8283	1.149	-.5985	579.6
2.56	.8893	1.131	-.6151	441.3
2.58	.9104	1.125	-.6199	402.5
2.60	.9317	1.119	-.6243	366.8
2.62	.9534	1.113	-.6283	334.1
2.64	.9754	1.107	-.6321	304.0
2.66	.9977	1.101	-.6356	276.3
2.68	1.0204	1.095	-.6388	250.8
2.70	1.04344	1.090	-.6417	227.3

Table 1. The energy E'^s_{tot} and associated quantities. The quantity P'^s is the pressure associated with E'^s_{tot}. Numbers apply to d^7s^1 configuration only.

THE COMPUTATION OF d-ENERGIES

Since, in the end, the scaled density of states of Wood will be used, to obtain the total d-energies it is only necessary to know the widths, Δ^U and Δ^D, of the up and down spin bands and the energy value of one level in each band as a function of r_s. If one knows (for say the up spin band) the energy value (as a function of r_s) of one level, the width (as a function of r_s), and the density of states (relative to the band bottom) at any single value of r_s that is all that is necessary to calculate the mean energy $\bar{\varepsilon}^U$ for the up spin d-band if scaling is assumed. To obtain this information one can make use of the fact that the average energy, $\frac{1}{2}(\varepsilon'''_2+\varepsilon'''_3)$, associated with the symmetry points N_2 and N_3 in the Brillouin zone tends to be constant in the range of r_s of interest here for both the up and the down spin bands. If the spin structure is characteristic of some state of free ion then in each band the average, $\frac{1}{2}(\varepsilon'''_2+\varepsilon'''_3)$, remains close to the free ion energy of the electrons of up or down spin. Thus the average, $\frac{1}{2}(\varepsilon'''_2+\varepsilon'''_3)$, can be regarded as a fixed energy in either the up or the down spin d-band. In addition to this, because scaling is assumed, the width Δ of each d-band must be proportional to $\Delta_{2,3}=\varepsilon'''_3-\varepsilon'''_2$. This turns out to be enough information for a reasonable computation of total energies.

In the W-S scheme ε_3''' is evaluated[9] by solving the radial Schrödinger equation,

$$\nabla^2 \psi_d + [\varepsilon_3'''(r_s) - v'(r)]\psi_d = 0$$

with the boundary condition $\psi_d(r_1) = 0$ where r_1 is the half-distance between nearest neighbors (in bcc Fe, $r_1 = .87943\, r_s$). The energy ε_2''' may be found[9] by solving the same equation with the boundary condition $\frac{d\psi(r)}{dr}\big|_{r=r_1} = 0$. This has been done for the structure d^6s^2 with U=5 and D=1, for the structure d^7s^1 with U=5 and D=2 and for the structure d^8 with U=5 and D=3. The average energies $\frac{1}{2}(\varepsilon_2'''+\varepsilon_3''')$, and the widths, $\Delta_{2,3}$, are given for the up and down spin bands in the d^7s^1 configuration in Table 2 for various r_s. It should be mentioned that the separations $\Delta_{2,3}$ vary almost precisely as r_s^{-4}.

r_s (Bohr units)	$\frac{1}{2}(\varepsilon_2'''+\varepsilon_3''')^U$ U=5 (Ry.)	$\frac{1}{2}(\varepsilon_2'''+\varepsilon_3''')^D$ D=2 (Ry.)	$\Delta_{2,3}^U$ U=5 (Ry.)	$\Delta_{2,3}^D$ D=2 (Ry.)
2.36	-1.398	-1.207	.815	.863
2.40	-1.399	-1.209	.766	.814
2.50	-1.400	-1.213	.656	.704
2.56	-1.400	-1.214	.599	.647
2.58	-1.400	-1.214	.581	.629
2.60	-1.400	-1.214	.564	.612
2.62	-1.399	-1.215	.547	.595
2.64	-1.399	-1.215	.531	.578
2.66	-1.399	-1.215	.515	.562
2.68	-1.398	-1.215	.499	.547
2.70	-1.398	-1.215	.485	.532

Table 2. The averages $\frac{1}{2}(\varepsilon_2'''+\varepsilon_3''')$ and the widths $\Delta_{2,3}$ for the up (U=5) and down (D=2) spin bands of the d^7s^1 configuration.

The difference in the widths, $\Delta_{2,3}$, for the up and the down spin bands in Table 2 is due to the difference in the exchange potentials. The up spin electrons, having a lower potential energy due to exchange, have the narrower band width. One can estimate the effects of a change in spin configuration on the widths, $\Delta_{2,3}$, by interpolation, that is (for the d^7s^1 configuration),

$$\Delta_{2,3}^U(r_s) = \Delta_{2,3}^{U=5}(r_s) + (5-U)(\Delta_{2,3}^{D=2}-\Delta_{2,3}^{U=5})/3 \qquad (16)$$

and
$$\Delta_{2,3}^D = \Delta_{2,3}^{D=2} + (5-U)(\Delta_{2,3}^{U=5} - \Delta_{2,3}^{D=2})/3 \tag{17}$$

Likewise the mean energies, $\varepsilon_{2,3}^U = \frac{1}{2}(\varepsilon_2''' + \varepsilon_3''')^U$ and $\varepsilon_{2,3}^D = \frac{1}{2}(\varepsilon_2''' + \varepsilon_3''')^D$, for the up and down spin bands are given by

$$\varepsilon_{2,3}^U = \varepsilon_{2,3}^{U=5} + (5-U)(\varepsilon_{2,3}^{D=2} - \varepsilon_{2,3}^{U=5})/3 \tag{18}$$

and
$$\varepsilon_{2,3}^D = \varepsilon_{2,3}^{D=2} + (5-U)(\varepsilon_{2,3}^{U=5} - \varepsilon_{2,3}^{D=2})/3 \tag{19}$$

This interpolation procedure has been tested by applying it to the prediction of free atom energy levels which differ in the spin configuration (see Appendix).

Equations (16) through (19) permit the eventual calculation of the total energies and the widths, Δ^U and Δ^D, of the up and down spin d-bands of α-Fe as a function of μ.

In Wood's band calculation, which was done for paramagnetic α-Fe (i.e. μ=0 and $\Delta_{2,3}^U = \Delta_{2,3}^D$), the lowest lying d-level is found at the point N_1 and the highest level at N_3. The separation (total d-band width), $\varepsilon_{N_3} - \varepsilon_{N_1} = \Delta_{1,3}$, exceeds $\Delta_{2,3}$ by the ratio $\Delta_{1,3}/\Delta_{2,3} = 1.4397$. Therefore according to Wood's values of $n(\eta)$ the total up and down spin band widths are given by

$$\Delta^U = 1.4397\, \Delta_{2,3}^U \tag{20}$$

and
$$\Delta^D = 1.4397\, \Delta_{2,3}^D \tag{21}$$

In addition, since the mean energies, $\varepsilon_{2,3}^U$ and $\varepsilon_{2,3}^D$, are fixed for a given μ and N_d the lowest lying energies, ε_0^U, and ε_0^D, of the up and down spin bands are given by

$$\varepsilon_0^U = \varepsilon_{2,3}^U - \Delta_{2,3}^U/2 - .4397\, \Delta_{2,3}^U \tag{22}$$

and
$$\varepsilon_0^D = \varepsilon_{2,3}^D - \Delta_{2,3}^D/2 - .4397\, \Delta_{2,3}^D \tag{23}$$

With Wood's density of states and the bottoms and the widths of the up and down spin bands given as a function of μ by Eqs. (16) through (23) it would be possible to calculate the d part of the total energy in Eq. (14) if the actual width were known for either the up or down spin band at any value of r_s in the range of interest here. If this information were available the widths $\Delta_{2,3}^U$ and $\Delta_{2,3}^D$

could be increased or decreased by some common factor to provide the correct overall bandwidths for some r_s and then everything else would fall into place. Unfortunately calculated bandwidths vary considerably[14] and the information is not available. An appropriate bandwidth may however be found indirectly by finding that factor, F, which upon multiplying $\Delta_{2,3}^U$ and $\Delta_{2,3}^D$ produces a minimum in the total energy of α-Fe at the observed value of r_s at 0°K (r_s = 2.662). This is easily done and for the d^7s^1 configuration requires that $\Delta_{2,3}^U$ and $\Delta_{2,3}^D$ be multiplied by the factor F = .64288. The calculated values of E_{tot} and associated quantities corresponding to this value of F are shown, as case 1, in Table 3. These energies are minimized with respect to μ and show the variation of μ with r_s and x. Inspection of the results will show that the described behavior is close to that observed for α-Fe. For the magnetic moment per atom, μ(x), the value μ(1) = 1.6$μ_B$ is found at the minimum. This is to be compared with the observed value of 2.2$μ_B$ for α-Fe. This demonstrates the breakdown of Hund's rule caused by the widening[15] of the d-levels. Further widening will eventually bring μ to zero, an effect to be expected in the entire transition series. The calculated value of the bulk modulus $B_0(x)$ at the minimum is $B_0(1)$ = 2012 kbar. This compares fairly well with the observed[16] value at 300°K of 1664 kbar. The variation of μ with pressure is slow as observed by Tatsumoto et al[17,18]. The results of these authors imply that $\frac{\partial \mu}{\partial P}$ = -6.82x10^{-4}$μ_B$/kbar near the energy minimum whereas the calculated value here is $\frac{\partial \mu}{\partial P}$ = -6.4x10^{-4}$μ_B$/kbar. For the cohesive energy the calculated value is .45 Ry. per atom which is in good agreement with the observed value[19] of .312 Ry. per atom but is .138 Ry. too large. This discrepancy is possibly due to the neglect of the second integral in Eq. (11) which adds an energy of about this size to the total. The calculated overall bandwidth for the paramagnetic (μ= 0) state of α-Fe turns out to be .497 Ry. which is in close agreement with the value originally computed by Wood, namely .478 Ry. The use of Wood's value for the calculated bandwidth shifts the minimum slightly toward large r_s and improves the calculated bulk modulus. These results, plus those of Stern[7], tends to justify the scaling of Wood's band over a small range of volume. The result seems to be a very good theoretical description of α-Fe.

The description above (case 1) can be improved by estimating the value of the second integral, I_2, in Eq. (11). Although the Δρ's are not known exactly the magnitude and volume dependence of I_2 can be estimated by setting $\Delta \rho_f = \Delta \rho'' = \xi[\rho'''(\bar{\epsilon}''') - \rho_0]$. The energies $\bar{\epsilon}'''(r_s)$ can be obtained from the previous calculation and ξ is a parameter included to adjust for the error in the approximation. The integral, I_2, has been calculated in this way as a function of r_s, and ξ has been adjusted to provide an improved cohesive energy for α-Fe at r_s = 2.662. The value of ξ turns out to be .5 and the new value of F is F = .6298. The results in terms of E_{tot}, P, μ, r_s

A THEORY OF THE α→ε TRANSITION IN Fe

r_s (Bohr Units)	x	Case 1			Case 2		
		E_{tot} (Ry.)	P (kbar)	μ (Bohr Magnetons)	E_{tot} (Ry.)	P (kbar)	μ (Bohr Magnetons)
1.92	.3752	-42.2302	13898.3	.4	-42.0741	14984.6	.6
2.00	.4240	-42.5371	9764.0	.6	-42.4040	10462.4	.6
2.10	.4909	-42.8182	6200.6	.8	-42.7038	6562.1	.8
2.20	.5644	-43.0123	3841.9	1.0	-42.9075	3982.7	1.0
2.30	.6449	-43.1414	2261.8	1.0	-43.0395	2270.7	1.0
2.40	.7328	-43.2220	1239.2	1.2	-43.1190	1184.3	1.2
2.50	.8283	-43.2671	583.1	1.4	-43.1609	517.3	1.4
2.56	.8893	-43.2813	303.1	1.4	-43.1733	251.5	1.4
2.58	.9104	-43.2843	226.5	1.4	-43.1756	189.1	1.4
2.60	.9317	-43.2865	167.1	1.6	-43.1775	141.5	1.6
2.62	.9534	-43.2882	113.5	1.6	-43.1788	88.3	1.6
2.64	.9754	-43.2892	56.4	1.6	-43.1796	42.5	1.6
2.662	1.0000	-43.2895	0.0	1.6	-43.1799	0.0	1.6
2.68	1.0204	-43.2893	-41.1	1.6	-43.1797	-28.7	1.6

Table 3. Total energy and associated quantities for α-Fe in the d^7s^1 configuration.

and x are shown as case 2 in Table 3 for the d^7s^1 configuration. In this last calculation the cohesive energy is .34 Ry. per atom, $B_o(1)$ is 1583 kbar, $\frac{\partial \mu}{\partial P} = -6.72 \times 10^{-4} \mu_B$/kbar, and the bandwidth at $r_s = 2.662$ for $\mu = 0$ is .487 Ry. Note that the bandwidth is now very close to Wood's value (.478 Ry.) and the bulk modulus is much improved. (N.B. Δμ=.2 was used to minimize energies in Table 3, Δμ = .02 was used to calculate $\partial \mu / \partial P$.)

THE PARAMAGNETIC STATE

Slater[20] has pointed out that, according to Heisenberg and Van Vleck[21], paramagnetic Fe should have an energy per atom which lies an amount $\frac{1}{2} kT_c$ (T_c = Curie temperature in °K, k = Boltzmann constant) above the ferromagnetic phase. For α-Fe with T_c= 1043 °K, this amounts to .0033 Ry. per atom. The energy lines calculated for μ = 0 (in the alpha phase) by the two methods (Cases 1 and 2) described here lie at energies somewhat greater than .0033 Ry. above the energy minimum at r_s = 2.662. It will now be shown that if this energy increase were exactly .0033 Ry. per atom Fe would spontaneously demagnetize in the pressure regime of the α→ε transition. To show this it is only necessary to lower the total energy increase per atom which arises in the μ = 0 state. This can be done as if the band structure were altered so that this energy is lower than the prediction of Wood for α-Fe with μ = 0 and the same overall bandwidth. One can also accomplish this energy lowering for μ = 0 by altering the bandwidth (i.e. changing F). Whichever method is used the result is exactly the same so far as the consequent energy variation of the μ = o state with volume is concerned. This has been done for Cases 1 and 2 and the result is in Table 4. Comparing these energies with those of the ferromagnetic state one finds that the energy line of paramagnetic iron crosses that of the ferromagnetic phase at a pressure of 212 kbar and a compression of x = .9157 for case 1 and at a pressure 167.5 kbar and x = .922 for Case 2. Note that on pressure reduction (from above the point of crossing) the cross over occurs at a lower pressure of 138 kbar for Case 1 and 81 kbar for Case 2, showing a hysteris effect. From the energy curves the implied Curie temperature, $T_c(x)$, may be calculated as a function of compression for each case. This is also shown in Table 4. Note that $T_c(x)$ is the same for both cases.

These results suggest a number of conclusions about the α-ε transition in iron. First of all, since μ is not observed[3] to go to o for the alpha phase in the transition region, (as it apparently would if the μ = o state were to lie .0033 Ry. per atom above the ferromagnetic state at P = o kbar and T = o °K), then the existence of another phase with slightly lower energy at μ = o (due to small alterations of the band structure) is indicated. Because this phase should appear in the pressure regime of the observed α-ε transition, an identification with the ε phase is suggested and there is an indication that ε-Fe is paramagnetic (μ = o).

r_s (Bohr Units)	x	Case 1			Case 2		
		$E_{tot}(\mu=o)$ (Ry.)	P (kbar)	$T_c(\alpha-Fe)$ (°K)	$E_{tot}(\mu=o)$ (Ry.)	P (kbar)	$T_c(\alpha-Fe)$ (°K)
1.92	.3752	-42.2608	13713.6		-42.1062	14779.5	
2.00	.4240	-42.5632	9608.3		-42.4313	10300.8	
2.10	.4909	-42.8391	6068.3		-42.7256	6426.3	
2.20	.5644	-43.0285	3724.5		-42.9244	3862.9	
2.30	.6449	-43.1532	2173.7		-43.0519	2180.7	
2.40	.7328	-43.2297	1156.0		-43.1270	1100.0	
2.50	.8283	-43.2707	500.9		-43.1646	434.7	
2.56	.8893	-43.2824	232.2		-43.1744	180.3	
2.58	.9104	-43.2846	159.0		-43.1761	114.8	
2.60	.9317	-43.2861	92.9	126	-43.1771	57.6	126
2.62	.9534	-43.2868	33.4	442	-43.1774	8.4	442
2.64	.9754	-43.2869	-20.0	727	-43.1773	-33.6	727
2.662	1.0000	-43.2862	-70.0	1043	-43.1766	-70.4	1043
2.68	1.0204	-43.2852	-110.8	1296	-43.1756	-98.2	1296

Table 4. Total energies and pressures for paramagnetic states of Fe calculated so that the energies lie .0033 Ry. above those of α-Fe at r_s = 2.662 for Cases 1 and 2. Values of T_c for α-Fe are included.

For a large pressure range near the transition point the energy differences between the two phases remain small (of the order of the thermal energy per atom at normal temperatures which is approximately .006 Ry.). It is therefore not surprising that the transition is sluggish since a change from the bcc to the hcp structure is required and a small energy barrier can impede the transition.

The theoretical temperature dependence of the transition can be determined by comparing the free energy changes due to small temperature increases in the two phases at the point of equal free energies at $0\,°K$. At a given specific volume the specific free energy change ΔA (near $T = 0\,°K$) is given by

$$\Delta A = - \beta T^4/(\bar{c})^3 \qquad (24)$$

where $\bar{c}$ is the mean propagation velocity of the vibrational normal modes and β is a function of volume only. It is reasonable to expect that at constant volume $(\bar{c})^2$ is proportional to the bulk modulus. Since, for both Case 1 and Case 2, the bulk modulus of the ($\mu = 0$) phase at the transition point is slightly less than of the alpha phase, the free energy of the former drops faster than of the latter with increasing T and therefore becomes the phase of lowest free energy. This implies that the transition pressure, P_t, moves to lower values as temperature increases and therefore $\frac{dP_t}{dT} < 0$. This is the observed[22] trend.

At the point of transition the ε phase may be expected to appear at a volume characteristic of the transition pressure, P_t. This is a smaller volume than that possessed by the α phase at the same pressure. For Case 1 the volume change relative to v_0 is $\Delta v/v_0 = .021$ and for Case 2, $\Delta v/v_0 = .029$. These values are in reasonable agreement with the experimental value, $\Delta v/v_0 = .052$, observed at the onset[2] of the transition.

THE STRUCTURE d^8

The energy per atom of the d^8 structure has been calculated in an entirely similar way to that of $d^7 s^1$. The $\Delta_{2,3}$ widths were multiplied by $F = .6298$ as suggested by the $d^7 s^1$ calculation (Case 2). At $r_s = 2.662$, the lowest state has $\mu = 2\mu_B$ and the energy is well above (1.8 Ry.) that of the ($\mu = 0$) phase of iron in the $d^7 s^1$ configuration. The energy difference however diminishes steadily with decreasing volume. This suggests that, as pressure increases, iron is approaching the d^8 structure and loses s-electrons at the implied rate $\frac{\Delta N_s}{\Delta x} = .15$. This loss (but not the rate of loss which is a factor of 6 too small) is consistent with the observations of Pipkorn et al[3].

The tendency towards d^8 indicates increasing spherical symmetry in the d-electron distribution within a W-S cell and hence a trend toward a more isotropic distribution of nearest neighbors about an atomic site. This implies an eventual structural change from hcp (in which the distribution of nearest neighbors is highly anisotropic) to fcc. This last prediction has also been made by Olinger[23] on the basis of Engel-Brewer curves.

The trend toward a completely d-like structure is interesting from another point of view. The behavior of the d-band energy with volume depends greatly on the amount by which it is filled, since the lowest lying d-electron tends to drop continuously in total energy for a very large range of compression. This can be shown by calculating the total energy of the d^8 structure as if there were only 2 d-electrons per atom in the band ($\mu = 1\mu_B$) all other things being the same. The result is shown in Table 5 (as d^2) along with the total energy of the .8 filled band. The long continuous drop in energy for the 2 electron case suggests that in a light transition or pre-transition element (say Ca.) there is apt to be a change in configuration from d^1s^1 (if this happens to be the configuration[24] of lowest energy at P = o) to d^2. If the drop in the d^2 total energy continues over a sufficiently broad range of compression the d^2 structure could be stable at zero pressure since the minimum of the total d^2 energy could actually become lower[25] than that of d^1s^1 at the minimum of the latter. Some check on this can be made by redoing the d^7s^1 energies of Fe as if there were only 1 d-electron per atom in the d-band ($\mu = \frac{1}{2}\mu_B$). The results[26] are shown in Table 5 as d^1s^1 and confirm the possibility of a stable transition phase. Assuming an s^2 configuration at P = o does not change the qualitative result. The possibility of stable very high pressure phases is intriguing because, if they exist, it may be possible to create superdense versions of normal solids which would doubtless have very unusual properties. More detailed calculations will be done on Ca. and Sc. in the near future.

r_s (Bohr units)	$E_{tot}(d^8)$ Ry.	$E_{tot}(d^2)$ Ry.	$E_{tot}(d^1s^1)$ Ry.
1.90	-40.3115	-2.666	-2.153
2.02	-40.6194	-2.480	-2.326
2.10	-40.7846	-2.372	-2.392
2.22	-40.9789	-2.227	-2.443
2.30	-41.0775	-2.140	-2.455
2.42	-41.1886	-2.023	-2.452
2.50	-41.2455	-1.954	-2.442
2.62	-41.3121	-1.864	-2.420
2.70	-41.3530	-1.815	-2.405

Table 5. Total energies for the d^8 structure and the implied, qualitative, relationship between d^2 and d^1s^1 energies.

THE $d^6 s^2$ STRUCTURE

Since there is still evidence[27] that the $d^6 s^2$ configuration may be the one of lowest total energy for α-Fe, total energies were calculated for this configuration by the same procedure as was used for $d^7 s^1$. The results are completely different showing a jump (in α-Fe) from $\mu = 4\mu_B$ to $\mu = 0$ at a pressure of less than 200 kbar, an effect which is not observed. It was found however that the relationship between the 2.2 μ_B state of α-Fe and a state with $\mu = 0$ lying .0033 Ry. above the 2.2 μ_B state at $r_s = 2.662$, is such as to reproduce the characteristics of the α - ε transition.

REFERENCES

1. W. A. Bassett and T. Takahashi, Rev. Sci. Inst. <u>38</u>, 37 (1967).
2. P. M. Giles, M. H. Longenbach, and A. R. Marder, J. Appl. Phys. <u>42</u>, 4290 (1971).
3. D. N. Pipkorn, C. K. Edge, P. De Brunner, G. De Pasquali, H. G. Drickamer and H. Frauenfelder, Phys. Rev. <u>135</u> No. 6A, A1604 (1964).
4. R. N. Keller and A. C. Mitchell, Sol. State Comm. <u>7</u>, 271 (1968).
5. J. H. Wood, Phys. Rev. <u>126</u>, 517 (1962).
6. J. F. Cornwell, D. M. Hum and K. G. Kong, Phys. Lett. <u>26A</u> No. 8, 365 (1968).
7. F. Stern, Phys. Rev. <u>116</u> No. 6, 1399 (1959). Dr. Stern has kindly provided the information in the Appendix most of which is not included in his 1959 paper.
8. Bouckaert, Smoluchowski and Wigner, Phys. Rev. <u>50</u>, 58 (1936).
9. For example see M. F. Manning, Phys. Rev. <u>63</u>, 190 (1943).
10. The radius of a sphere, centered on a nucleus, which has the volume per atom.
11. Here "H-F energy" is used in a restricted sense since correlation effects are included in the energy calculations.
12. The method is given further on in the text.
13. R. E. Watson, Phys. Rev. <u>119</u>, 1934 (1960).
14. See, for example, K. J. Duff and T. P. Das, Phys. Rev. B <u>3</u>, 192 (1971).
15. As the d-bands widen the energy can be lowered by the increased pairing of spins. The maximum value of $\mu(\mu=3\mu_B)$ is therefore not achieved below a certain value of r_s.
16. M. W. Guiman and D. N. Beshers, J. Phys. Chem. Soc., <u>29</u>, 541 (1962).

17. E. Tatsumoto, T. Kamigaichi, H. Fujiwara, Y. Kato and H. Tange, J. Phys. Soc. Japan **17**, 592 (1962).

18. E. Tatsumoto, H. Fujiwara, H. Fujii, N. Iwata and T. Okamoto, J. Appl. Phys. **39**, 894 (1968).

19. Computed from the data in, Thermodynamic Properties of the Elements by Stull and Sinke (American Chemical Society, 1956).

20. J. C. Slater, Phys. Rev. **49**, 537 (1936).

21. See, for example, J. H. Van Vleck, Electric and Magnetic Susceptibilities, (Oxford, the Clarendon Press 1932).

22. F. P. Bundy, J. Appl. Phys., **36**, 616 (1965).

23. B. W. Olinger, PhD Thesis, Univ. of Chicago (1970).

24. This is assumed merely for the purpose of constructing a hypothetical case. The best description of the lowest energy configuration for solid Ca., at P = o, surely lies between d^1s^1 and s^1p^1. The less d character in the ground state the more certain is the result obtained in the ensueing calculation in the text.

25. This is because the lowest lying s-electron begins to rise in energy at a much larger volume than does the lowest lying d-electron.

26. The d^1s^1 energies have been lowered by a constant amount so that at r_s = 2.7.the difference between d^2 and d^1s^1 is .59 Ry./atom. This difference is greater than the sum of the transition energy $4s^2$ (1S) → $3d^2$ (3P) of the Ca. free atom (.442 Ry.) plus the cohesive energy per atom of Ca. (.14 Ry.). In normal Ca. the energy difference between the ground state and the d^2 configuration should be <.59 Ry. because the d^2 energies should at this point be below the free atom value.

27. V. P. Tsvetkov, A. V. Kalenichenko and A. P. Ganzha, Bull. Acad. Sci. U.S.S.R., Phys Ser. U.S.A., **36** No. 2, 287 (1972).

Acknowledgment

This work was supported by The Naval Ordnance Lab Independent Research Funds.

A THEORY OF THE α→ε TRANSITION IN Fe

APPENDIX - DETAILS NOT COVERED IN THE TEXT

A - Scalability of n(η)

The assumption, $\Delta n(\eta)$ = constant, implies that the ratio, $(\varepsilon_2-\varepsilon_1)/\Delta$, between the separation of any two levels and the total bandwidth is a constant with compression. Stern has calculated selected energy levels in α-Fe (μ = o) as a function of volume and of electronic configuration. In the table below are shown values of $(\varepsilon_2-\varepsilon_1)/\Delta$ as a function of r_s for several points of high symmetry in the Brillouin zone based on Sterns unpublished work for the $d^{7\cdot1}s^{\cdot9}$ configuration. These results provide some indication that n(η) may be scaled with volume for Fe at least in the small volume range of interest here.

Δ (Ry.)	r_s (Bohr Units)	$(H_{25'}-H_{12})/\Delta$	$(P_4-P_3)/\Delta$	$(\Gamma_{12}-\Gamma_{25'})/\Delta$	$(N_3-N_2)/\Delta$	$(N_3-N_1)/\Delta$
1.237	2.3	.854	.369	.360	.987	.979
.706	2.66	.865	.405	.362	.963	1.000
.374	3.1	.882	.406	.369	.952	1.000

B - Free Atom Energies and Functions

$d^6 s^2$ Configuration. A very good ionization energy can be calculated for the down spin d-electron of the Fe^{++} free atom (^5D) using a potential derived from Watson's[13] Argon core and radial d-functions. Therefore his d-functions were used to obtain ρ_o in this configuration. Watson's exchange interaction between 2 d-electrons (-.093348 Ry.) is too large in magnitude relative to the value (-.0589 Ry.) computed from observed energy levels which differ in spin configuration. The calculated exchange potential was consequently multiplied by (.0589/.093348) to account for this.

The free atom s-energies computed with Watson's functions were too high, possibly because of neglected correlation effects. To account for this, the calculated s-core exchange interaction was multiplied by 1.75. By this method the calculated s-energies become exactly the observed values and the total energy of the 8 valence electrons in the d^6s^2 ground state (^5D) is -42.8388 Ry.

$d^7 s^1$ Configuration. For this configuration the d-d exchange energy was forced to the observed value of -.06369 Ry. through multiplication of Watson's d-d exchange interaction energy by the factor (.06369/.093348). This did not however provide a good energy for the down spin electrons of Fe^+ (^4F) due to the spreading out of the d-functions to accommodate an extra member. To account for this the calculated potential $v_d(r)$ due to one Watson d-function was set equal to $(1 + \exp(-\gamma r))$ and γ was calculated so that $v_d(r)$ could be regenerated from the exponential expression. In subsequent computations of the down spin energy of Fe^+, γ, was steadily reduced through multiplication by a constant factor, σ, (to simulate the spreading of the d-functions) until the correct ionization energy of the down spin d-functions was achieved. This led to $\sigma = .89909$ and the calculated values of ρ_o for the d^7s^1 configuration are based on $v_d(r) = (1 + \exp(-\sigma\gamma r))$. The d-d coulomb interaction energy $V_o(\infty)$, computed on this basis (the weighted average for the up and down spin electrons) is 1.6 Ry. This is in excellent agreement with the value (1.60876 Ry.) required to have the total energy of the d^7s^1 configuration the observed amount above the d^6s^2 ground state. The ratio $V_o(r_s)/V_o(\infty)$ was computed with $\sigma = .89908$ but $V_o(\infty)$ was taken as 1.60876 Ry. to insure that the d^7s^1 energies would approach the correct value relative to the d^6s^2 ground state at $r_s = \infty$. The ratio $J_o(r_s)/J_o(\infty)$ ($J_o(\infty) = -.06369$ Ry.) was found to have the same dependence on r_s as $V_o(r_s)/V_o(\infty)$. Values of the latter ratio for various r_s are given in the table below along with values of the integral I_2/ξ^2. The integral I_2 is given by

$$\xi^2 \tfrac{1}{2} \int_0^{r_s} \int_0^{r_s} \left\{ [\rho(\overline{\epsilon'''}, r_1) - \rho_o(r_1)] \frac{2}{r_{1,2}} [\rho(\overline{\epsilon'''}, r_2) - \rho_o(r_2)] \right\} d\tau_1 \, d\tau_2$$

in which $\rho(\overline{\epsilon'''}, r)$ is obtained by solving $\nabla^2 \psi_d + [\overline{\epsilon'''}(r_s) - v'(r)]\psi_d = 0$ for the energies $\overline{\epsilon'''}(r_s)$ obtained from Case 1 in the text with $\mu = 0$.

r_s (Bohr units)	$V_o(r_s)/V_o(\infty)$	$(I_2/\xi^2) \times 10^2$ (Ry.)
1.9	.9319	1.053
2.0	.9438	.846
2.1	.9537	.733
2.2	.9619	.696
2.3	.9686	.712
2.4	.9741	.761
2.5	.9786	.823
2.6	.9823	.876
2.7	.9854	.902

The s-core potential for $d^7 s^1$ was obtained in the same way as that for $d^6 s^2$.

A THEORY OF THE $\alpha \to \epsilon$ TRANSITION IN Fe

$\underline{d^8\ \text{Configuration}}$. The calculations for d^8 were done in the same way as those for $d^7 s^1$ except that an extrapolated value for $J_o(\infty)$ was used. The value is $-.06841$ Ry.

C - The Interpolation Scheme for Changes in Spin Configuration

The interpolation scheme employed in Eqs. (16) through (19) was checked by using it to calculate the energy differences between the ^5D, ^3P, and ^1I states of the free atom in the $d^6 s^2$ configuration and between the ^4F and ^2G states of Fe^+ in the d^7 configuration. For $d^6 s^2$ the observed values are $^5D \to ^3P = .167$ Ry. and $^5D \to ^1I = .267$ Ry. The calculated values (by interpolation) are respectively .179 Ry. and .239 Ry.

For d^7 the transition energy for $^4F \to ^2G$ is .127 Ry. The calculated value is .111 Ry.

The small differences between the interpolated and experimental energies are probably due to the slight changes in the d-d Coulomb correlation energy in the different states. These changes are not accounted for and may not exist to any appreciable extent in the solid state.

COMMENTS ON "A THEORY OF THE HIGH PRESSURE $\alpha \rightarrow \epsilon$ PHASE TRANSITION AND OF POSSIBLE HIGHER PRESSURE TRANSITIONS IN Fe"

Lynn M. Barker

Sandia Laboratories, Shock Wave Phenomena Division

We have been doing some shock experiments on iron in which we measure the free surface motion with the new high resolution VISAR instrumentation system. Some of our preliminary findings may be of interest with regard to Pastine's paper.

The configuration just prior to impact in one of our experiments is shown in the inset of the figure. The measured free surface velocity history is characterized by (A) an elastic precursor, (B) the so-called plastic I wave, (C) the plastic II, or phase transition wave, and (D) a previously unresolved final increase in free surface velocity.

The phase transition stress at the top of the plastic I wave is somewhat ill-defined because of curvature in the profile, especially at the shorter propagation distances. Nevertheless, we observe a stress decay of the order of 8 kbar between 3 and 19 mm propagation distance. Also, from free surface velocity profiles like this one, it has been possible to start mapping out the release stress-volume path. We find that the ϵ phase is substantially stable on release down to a stress of about 100 kbar, whereas it is well-known that the α phase is stable on compression to about 130 kbar. Thus, in the 100 to 130 kbar region, both phases can co-exist. It turns out that both phases do co-exist for a short time with a boundary between the two phases near the free surface of the shocked specimen. This phenomenon gives rise to the final abrupt increase in velocity (D) after the arrival of the phase transition wave (C). The velocity increase (D) results from a wave reflection off the α iron - ϵ iron interface near the free surface. The observed stress hysteresis in the phase transition suggests that the equilibrium phase transition pressure may be in the neighborhood of 115 kbar.

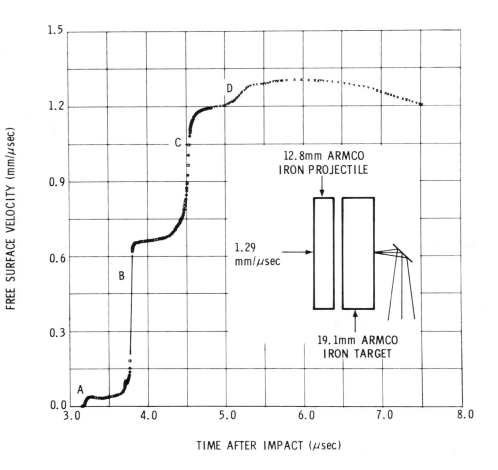

REMANENT MAGNETIZATION AND STRUCTURAL EFFECTS DUE TO SHOCK IN
NATURAL AND MAN-MADE IRON-NICKEL ALLOYS

Peter J. Wasilewski[1,2] and Arthur S. Doan, Jr.[1]

[1]NASA/Goddard Space Flight Center, Greenbelt, MD

[2]George Washington University, Washington, D.C.

ABSTRACT

Natural and man-made iron alloys are remagnetized when subjected to dynamic deformation. In nature the dynamic remagnetization is manifest in meteorite impact at planetary surfaces or in space, while the man-made material is remagnetized during laboratory experiments. There is a broad range of remagnetization, from the $\alpha \rightarrow \varepsilon \rightarrow \alpha$ transition in BCC iron alloys to the ferromagnetic products resulting from shock melting of paramagnetic iron silicates. Distinction is made between the shock effect <u>senso stricto</u> and the thermal effects associated with shock in porous and polymineralic material such as soil and rock. Micrographs illustrating shock effects and post-shock thermal effects in meteoritic metal and shock lithification of lunar samples are presented. Lunar samples can be classified on the basis of their magnetic hysteresis properties into three groups related to mode of origin. Demonstration is made of the use of magnetic measurements to classify shock-formed products, and to evaluate the effects of shock metamorphism.

Shock impact increases coercive force (H_c) in iron and iron-nickel alloys. The H_c values are related to domain wall motion, and are not related to stable remanence components. H_c values for shocked alloys with greater than 10% nickel are less than for quenched or annealed alloys. The coercive force in high-pressure regions is greater by a factor of 2 than in the low pressure region for samples shocked in the free surface mode. The saturation isothermal remanence and shock remanence in iron shocked to between 2.5 and 15.5 kb show systemmatic increase in intensity which is qualitatively related to the increasing twin density. The thermal effects associated with high peak pressures are shown to reduce coercivity in iron.

INTRODUCTION

Planetary surfaces have been subjected to extensive modification by meteorite impact. Meteorites which survive earth impact contain evidence of several episodes of impact deformation. This evidence is available upon metallographic examination of the recovered meteorites and ranges from simple mechanical twinning to complete recrystallization at the extremes. All meteorites and lunar samples contain measurable magnetic remanence which was acquired outside the influence of the terrestrial magnetic field. The origin of meteoritic remanence is unknown at present, and from our preliminary studies we feel that impact had a significant influence on the modification of any primary remanence components and indeed may have been responsible for much of the measurable remanence in meteorites. Lunar samples present a different problem since the evidence of deformation noted in meteorites cannot be found in lunar samples, even though it is clear that lunar breccia samples were formed by impact processes. There is at present no clear picture of how the lunar samples acquired their remanent magnetization. Thus, a great deal of interest has been generated in an attempt to understand the effects of cratering and shock lithification at planetary surfaces. We are evaluating shock as a remagnetization process in iron and iron-nickel alloys, in porous ferromagnetic dispersions, and in shock-melted iron silicates. Evidence to date indicates that shock is a significant remagnetization mechanism and that the effects of shock can be studied by magnetic measurements. The visible effects of shock deformation in meteorite metal phases, the role of shock as a lithification process at the lunar surface, the classification of lunar material based on magnetic hysteresis measurements, the effects of shock melting, and the change in coercivity and demagnetization stability of iron and iron-nickel alloys are considered here.

BACKGROUND

Magnetic _remanence_ is the state of magnetization in a specimen after any thermal, thermochemical, thermophysical, or dynamic process has acted on the specimen. A ferromagnetic specimen cooled through its Curie point in the presence of an ambient field will acquire a magnetic remanence whose vector sense coincides with the ambient field vector and whose intensity is a function of grain size, grain shape, and composition. The magnetic vector changes due to thermal or dynamic processes which take place in the presence of an ambient field can be used to study the process. Shock remagnetization, in the present context, refers to any modification of intensity or sense of the original remanent magnetization due to the influence of the shock event. Separation must be made between shock senso stricto and the thermal effects associated with the shock. Magnetic remanence mechanisms in FeNi alloys differ from conventional mechanisms observed to take place in ferrites and other oxides. Six new mechanisms have been designated for the FeNi alloys (1):

Mechanism	Designation	Identification
Martensitic thermal	MTRM	Cooling thru M_s-rate dependent
Reverse thermal	RMRM	$\alpha \rightarrow \gamma \rightarrow \alpha_1$
Recrystallization	RCRM	Anneal of prior strained metal
Shock transition (first order)	SRRM	$\alpha \rightarrow \varepsilon \rightarrow \alpha$
Transition point (secnd order)	TPRM	$\Delta T_c / \Delta P$
Duplex anneal	DARM	$\alpha \rightarrow \alpha + \gamma; \gamma \rightarrow \alpha + \gamma$

All of the above mechanisms can be operative at specific stages of shock metamorphism, shock metamorphism being the thermophysical alteration of an original sample as a consequence of shock impact. Based on our experimental studies and the implications from published literature (2) on shock studies, the following shock remagnetization mechanisms have been designated (3):

1. first-order reversible crystallographic transitions in body-centered cubic iron-nickel alloys,

2. second-order Curie temperature transitions in face-centered cubic iron-nickel alloys,

3. shock-induced uniaxial anisotropy due to magnetoelastic coupling of magnetic vectors to the shock wave,

4. shock melting of iron containing silicates,

5. subsolidus reduction and FeO decomposition,

6. partial thermoremanence due to post-shock temperature,

7. total thermoremanence due to post-shock temperature,

8. production of a superparamagnetic distribution of iron

which is sensitive to surface temperature fluctuation, and

9. thermal effects in metal and alloy phases.

METALLOGRAPHIC OBSERVATIONS

Meteorites contain both BCC and FCC iron-nickel. Iron meteorites are classified according to the geometrical dimensions of the Widmanstatten pattern (4):

Structural type	Band width (mm)	Median Ni content (wt.%)	Median P content (wt.%)
H	50	5.6	0.24
Og-H	3 -50	6.4	0.18
Og	1.5-3	7.1	0.22
Om	0.5-1.5	8.3	0.23
Of	0.2-0.5	8.1	0.06
Off	0.2	13.0	0.75
Off-D	0.2	10.0	0.16
D	0.1	16.2	0.10

The chondrite meteorites are classified on the basis of their chemistry and textural relationships (5).

Chemical Groups	Fe/SiO_2	Fe^o/Fe	Fa	
E	0.77	0.80	0	Fe - total iron
C	0.77			Fe^o - metallic iron
H	0.77	0.63	18	Fa - fayalite mole%
L	0.55	0.33	24	in olivine solid
LL	0.49	0.08	29	solution

The chemical criterion are sufficient for class grouping and the petrologic classes are based on presence or absence of glass, the metallic minerals, overall texture, matrix texture, and carbon and water content, as well as homogeneity of the ferromagnesium silicates. BCC metal in chondrites contain 6-7% Ni at the center and 5-6% Ni near the edges of the grains, and FCC metal contains 25-35% Ni at the center and 45-55% at the edges. The FCC iron contains generally more Ni at the center and edges as grain size decreases. The Vickers micro hardness (VHN) for meteoritic phases are:

BCC iron-nickel ∿ 150-250
FCC iron-nickel ∿ 250-500
phosphide ∿ 1000
carbide ∿ 1300

indicating an impedence mismatch during interphase shock traversal. In impactites which contain metal spheres, it has been shown that (6):

> Ni increases as sphere size decreases;
> Fe decreases as sphere size decreases;
> P decreases as sphere size decreases;
> Co increases as sphere size decreases.

The composition of spherules in impactites range from nearly pure iron to 80 wt.% nickel alloys.

Meteorites, being spheroidal-shaped or irregular-shaped bodies, will contain complex deformation patterns upon post-impact examination. However, in portions of any large iron meteorite, it is possible to find extensive regions where the angular distribution of deformation bands can be used to pick the shock direction (7). Rarefaction shocks and shock interactions can also be observed, particularly when FCC or phosphide areas are in the path of a planar wave. In chondritic meteorites metal grains are usually irregular in shape, and are dispersed in a silicate matrix. Intense shock waves will melt FeS and some silicates, and it has been domonstrated that blackening of chondrite meteorites will take place when they are shocked, due to the formation of finely-dispersed troilite, formation of dense glass and micro-crystalline silicates (8). It is possible to find a broad range of shock effects in the same chondrite meteorite. The myriad variety of shock effects seen in metallic and chondritic meteorites are not seen in the lunar sample metal; one reason being that they were not looked for, the other reason being that little or no meteoritic metal is found as a component of impact breccia. Purely mechanical effects are noted in deformed meteorites:

1. slip bands and band rotations,
2. deformation twinning,
3. faulting with offset,
4. twins which fault phosphides and twins which show no faulting, and
5. metal fracturing and shearing.

In reheated meteoritic metal the textures and structures produced are numerous:

1. recrystallized edges of deformed monocrystals,
2. polycrystalline BCC FeNi due to recrystallization,
3. $\alpha + \gamma + Ph$ polycrystal,
4. $\alpha + \gamma$ duplex structures,
5. quench martensite,

6. precipitate decoration on twinning and shear zones,
7. homogenization of composition gradients,
8. addition of secondary α and γ to metal and sulfide peripheries,
9. production of zoned structures, and
10. fractured phosphides rounded off by diffusion processes.

The wide variety of effects associated with dynamic deformation of natural materials are summarized in Plates I and II. Plate I illustrates the range of mechanical effects noted in BCC and FCC iron-nickel in meteorites. Plate II illustrates the range of thermal effects noted in shock reheated chondrite meteorites (A to F). Shock lithification at the lunar surface is deomonstrated in G, H and I; and in J a chondrule which may have been formed by shock melting is illustrated. In K and L, the profuse distribution of metal and sulfide spherules found in impactites is illustrated.

COERCIVE FORCE

Magnetic hysteresis properties of natural ferromagnetic dispersions and metal and alloy spheres have been measured (Table I - Lunar Sample Data due to Dr. T. Nagata and Dr. F. Schwerer). Coercive force (H_c), remanent coercive force (H_R), saturation remanence (I_R), and saturation magnetization (I_S) were measured and the ratios R_I ($=I_R/I_S$) and R_H ($=H_R/H_c$) were computed. There is a relationship between R_I and R_H for all natural materials, dependent on the grain size distribution, the grain shape distribution and the composition of the ferromagnetic components (9). The ferromagnetic components responsible for the magnetic properties of natural materials range from cubic and rhombohedral iron-titanium oxides in terrestrial samples to FeNiCo alloys in the lunar samples. Redox modifications and thermophysical alterations such as shock metamorphism, alter the hysteresis properties in predictable directions. For example, shock lithification, shock melting and fragmentation associated with meteorite impact at the lunar surface will produce group B or C (Table I, bottom) material. Thermal metamorphism of group B or C material will reduce the paramagnetic component by adding multidomain iron due to subsolidus reduction, and will destroy the superparamagnetic component responsible for the reduced coercivity and large R_H values. Thus the reheated material will fall into Group A.

The H_R value is the backfield which reduces remanence to zero. In basaltic rocks and nickel spheres, both H_c and H_R increase with shock and plastic deformation respectively, but the H_c increases relatively more than H_R so that the R_H value decreases. The H_R value is more time consuming to measure, but

Table I. Magnetic hysteresis properties for natural materials, nickel and alloy spheres.

	SAMPLE	H_C	H_R	R_H		$R_J(10^{-2})$
BASALTS	CH21-001	540	800	1.5		
	NSU	100	230	2.3	- unshocked	
	NSS	200	300	1.5	- shocked	
	FUJI-2	80	275	3.4	- reduced	
IMPACTITE	Monturaqui	90	225	2.5		
CHONDRITES	Modoc	75	1100	14.6		
	Claytonville	80	1900	23.0		
	Atwood	110	1710	15.5		
	Beenham	30	700	23.0		
	Calliham	165	270	1.6		
	Clovis	80	200	2.5		
	Allende	110	320	2.9		
CHONDRULES	Allende AA	65	130	2.0		
	Allende 01	175	550	3.1		
	Chainpur	75	735	9.8		
	Bjurbole	60	2700	45		
Ni SPHERES	Ni^{II}ANNEAL	4.95	180	36		
	Ni^{II}COLDWORK	25.5	330	12.9		
ALLOYSHPERE	52100	51	1055	20.7		
		54	870	16.1		
		45	900	20.0		
LUNAR	A. 14053	20	80	4.0		1.9
	14313	27	180	6.6		1.6
	14311	17	140	8.7		0.6
	12053	8	76	9.5		0.4
	10024	45	160	3.6		1.0
	B.					7.^
	14047	26	350	13.4		4.4
	14259	19	300	15.7		4.0
	14161	26	430	16.5		4.4
	14301	27	450	16.7		0.93
	15301	20	400	20.0		4.7
	12070	22	450	20.5		4.8
	C. 15498	78	770	9.9		8.8
	10048	50	520	10.4		7.2
	15601	38	450	11.8		7.3
	10084	36	460	12.8		7.2

but is more meaningful than H_c. For example, a quenched 9% nickel alloy has an H_c value of 4.6 Oe while the H_R value is 320 Oe. Isothermal remanence in low applied fields is associated with domain wall motion, the increase in H_c with nickel content (see Figure 1) is the result of the quench transformation structure. The stable remanence, which will persist through time, and which will not be altered by nominal handling is not associated with the H_c field. Stable remanence is associated with high shape anisotropy and any unidirectional pinning mechanisms. Shock reversal of martensite which results in a net shape anisotropy, as has been observed in the experiments by Rohde, et al. (10), or any constrain which causes a net orientation of martensite plates, will result in stable acquired remanence which is not necessarily parallel to the ambient field.

Values of coercive force for iron and iron-nickel alloys subject to various metallurgical treatments, plastic deformation and shock loading are presented in Figure 1. H_c values for annealed alloys are shown as open circles (11). A continuous curve can be drawn through the points or two linear trends can be considered. We are in the process of obtaining high purity iron-nickel alloys to redetermine the H_c values for annealed and quenched BCC alloys. It is clear that H_c increases with added nickel. Cold rolled (99% reduction) iron and iron-nickel alloys were studied by Hirsch and Eliezer (12). H_c values for iron are greater than for iron-nickel up to about 25% nickel, measured at frequencies between 2 H_z and 0.02 H_z. The frequency dependence of H_c shown by the open diamonds in Figure 1 suggest significant domain relaxation in deformed alloys. There is a significant difference in the coercivity values for massively transformed and martensitically transformed alloys, which is also reflected in magnetic remanence and remanence demagnetization stability (see Figures 2a and 2b).

Data for iron shocked to peak pressures up to 750 kb following various pre-shock treatment are indicated in the inset at the upper left of Figure 1. Iron H_c values increase with increasing pressure, but the effect of the temperature rise during the 750 kb and 300 kb shocks is in evidence. As an example, the H_c values for iron shocked to 164 kb and 750 kb can be identical. The effects is further exemplified for samples twinned at -196°C prior to shocking, the 190 kb shock results in an H_c value of 10.4 Oe while the 750 kb shock produces a sample with an H_c value of 3.7 Oe.

The dotted line with the "thin foil" label connects H_c values for thin foils made from the low pressure region of samples shocked to 164 kb in the free surface mode. The open square for the 8% alloy labeled 164 is the H_c value for the high pressure

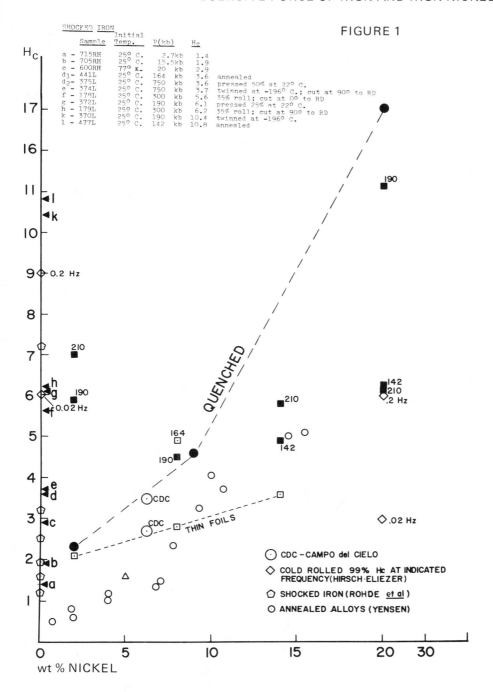

FIGURE 1

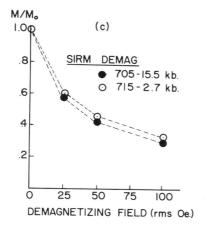
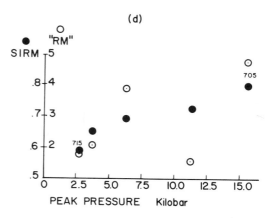
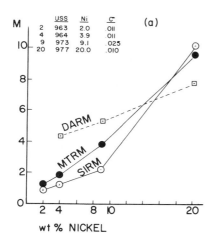
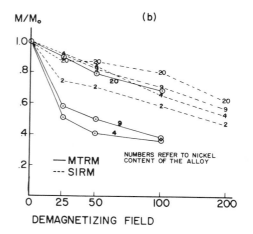

Figure 2. INTENSITY

(a) Magnetic remanence vs. wt.% nickel in iron-nickel alloys for three remanence states: MTRM - martensitic thermal remanence; SIRM - saturation isothermal remanence, DARM - duplex anneal remanence.

(b) Alternating field demagnetization curves for MTRM and SIRM in iron-nickel alloys.

(c) Alternating field demagnetization curves for shocked Ferrovac E iron.

(d) SIRM intensity for samples of Ferrovac E iron shocked between 2.7 kb and 15.5 kb, and shocked state remanence intensity value after 100 Oe demagnetization.

region of the same sample(3). The amplitude of the Plastic I wave, which is the pressure reached in the low-pressure region, decreases with added nickel in BCC iron-nickel alloys (13). Along the "thin foil" line, the highest pressure was experienced by the 2% alloy; the lowest by the 14% alloy. It is also interesting to note that the H_c value for the low-pressure region of the shocked 14% alloy is less than the value for an "annealed" alloy of similar composition. With added nickel (>10%), the H_c of annealed and quenched alloys increase rapidly compared to shocked alloys of similar composition. The solid solution effect of nickel and the transformation substructure appear to provide more impedance to domain wall motion than shock induced substructure. To date only trends are available; since the sampling is limited, the samples are of different purities, and the shock ranges are limited.

SHOCKED IRON

Johnson and Rohde (14) demonstrated that the twin density in Ferrovac E iron increases monotonically with shock stress to 15.5 kb. For Specimen 715 RH shocked to 2.7 kb, the H_c is 1.4 Oe, while for Specimen 705 RH shocked to 15.5 kb, the H_c is 1.9 Oe, indicating an impedance to domain wall motion with increasing twin density. The remanent magnetization values, for samples shocked between 2.7 kb and 15.5 kb, after 100 Oe peak alternating field demagnetization are indicated in Figure 2d by open circles. The same specimens were given saturation isothermal remanence (SIRM) in a 7.5 K Oe field (black circles in Figure 2d). Specimens were machined to identical dimensions from the same area of each shocked disk so that a comparison of coercivity and remanence is valid. The demagnetization spectra of SIRM are similar in shape but the stability of SIRM in the 2.7 kb specimen is slightly greater than in the 15.5 kb specimen.

Since the twinning increases monotonically over the range of pressures evaluated, there should be a correlation between twin density, SIRM intensity and coercivity. Qualitatively these correlations are established. Twinning is related to the extent of prior substructure since in prior deformed metal, fewer twins are necessary to complement slip in accomodating the imposed strain rate. Prestrained iron and annealed iron which is then shocked should show quite different magnetic effects. This is implied in the observations of stress relaxation made by Holland (15). The activation energies for strain aging of stress relaxation are comparable to activation energies for C and N diffusion in iron. Time-dependent magnetization changes should occur in strain-aged iron if C and N diffusion is important.

According to Giles, et al. (16), the $\alpha \to \varepsilon$ transition starts at P = 130 kb and is not complete until about 170 kb. The $\varepsilon \to \alpha$ reverse transition exhibits considerable hysteresis beginning at ∼80 kb and finishing at ∼45 kb. At high peak pressures the sluggish nature of the transition will not be important, but if an iron sample is shocked between 130 and 170 kb there should be detectable magnetic effects, since the sample will not be completely transformed, and reversal of $\varepsilon \to \alpha$ will take place in the presence of untransformed α. Since the ε phase is apparently paramagnetic and the α phase is ferromagnetic, an experiment could be done which might help to evaluate (a) the role of the transition <u>per se</u>, and (b) the role of the mechanical anisotropy induced by the planar shock pulse. Iron can be given a thermal remanence prior to shock loading in any chosen direction. The sample can then be arranged so that the thermal remanence vector is perpendicular to the shock direction.

In the assembly, an external field can be applied which is greater than the 0.5 Oe geomagnetic field, in the plane of the thermal remanence vector, but at some angle, say, 90° to the thermal remanence vector. Prior to loading, the external field will induce a magnetization component which cannot override the thermal remanence, and can be eliminated by a demagnetizing field of the order of the applied field. This can be quantitatively evaluated prior to setting up the experiment. During the shock event there are two competing effects,(a) the $\alpha \to \varepsilon \to \alpha$ transition, and (b) the shock-induced anisotropy. The anisotropy can be evaluated independently by loading to peak pressures below the transition. If $\alpha \to \varepsilon$ is a ferromagnetic to paramagnetic transition complete demagnetization of the prior induced thermal remanence component should take place and during the $\varepsilon \to \alpha$ transition a new stable magnetic component should be induced whose vector sense will coincide with the applied external field. Since the remanence vectors can be determined and their stability assessed, a fairly complete picture of the magnetic effects associated with the shock transition can be made.

THE ROLE OF NICKEL

Studies to date (1) demonstrate that (a) the acquisition of martensitic thermal remanence (MTRM) on quenching iron-nickel alloys depends on the amount of nickel in the alloy, (b) the demagnetization stability depends on the amount of nickel in the alloy which implies a relation between the coercivity spectrum and the transformation structures, (c) martensitic transformations impart shape anisotropy if not macroscopically, at least locally within the alloy, (d) there is a relationship between magnetic hysteresis and the transformation microstructure, (e) there is considerable magnetic remanence hysteresis associated with the M_s and reverse A_s transformations.

The role of structure contrast between massively and martensitically transformed alloys in determining the intensity and stability of MTRM and SIRM can be seen in Figure 2 (a and b). The MTRM and SIRM intensities for 2, 4 and 9% alloys fall on approximate linear trends related to increasing nickel content. We have not evaluated alloys between 10% and 20% nickel. As can be seen in Figure 2b, the demagnetization stability of MTRM in the 4% and 9% alloys are similar, in striking contrast to the stability of the 20% alloy. The stability of SIRM appears to increase systematically with increasing nickel, no contrast noted between massive and martensitic alloys. Notably the demagnetization spectrum for any alloy cannot be explained in terms of the low measured coercivities of alloys which are of the order of 10 Oe, since demagnetization does not take place in fields of several hundred Oe.

The microhardness of iron-nickel alloys increases with added Ni in annealed alloys (<10% nickel). For quenched alloys the hardness increases more rapidly than in the annealed alloys. The yeild strength, defined as the flow stress at 0.2% offset, for quenched and annealed iron-nickel has been shown by Speich and Swann (17) to be related to (a) the solid solution hardening effect of nickel and (b) the contribution due to the defect structure. The influence of the change from random dislocations to a cell structure and higher density is noted for MTRM but not for SIRM. The yield strength (YS) for meteorites having nickel contents between 7.4 and 11.0 wt.% nickel (18) are in good agreement with the results of Speich and Swann (17). There is some shock hardening and precipitation hardening in meteorites. Canyon Diablo a quenched poly-crystalline sample has a YS of 62.9 K psi which is reduced to 60.0 K psi on annealing, while the single-crystal Odessa meteorite has a YS of 46.6 K psi. Both fall near the YS curves for quenched (Canyon Diablo) and annealed (Odessa) alloys. The role of nickel in quenched alloys can be summarized (Table II):

TABLE II

Nickel (wt.%)	0	10	20
Transformation	equiaxed	massive	martensitic
Dislocations	random	cell structure	
Coercivity	2.3	4.6	17.1
MTRM Intensity	increase→	—	
SIRM Intensity	increase→	—	
Demagnetization Stability MTRM	increase→	—	
Demagnetization Stability SIRM		continuous increase→	

In the shocked alloys there is coercive force anisotropy in low applied fields, but no apparent anisotropy in fields of 16 K Oe. The coercivity increases with added nickel and also for increasing peak shock pressure. At present we do not have the samples to decipher the competing roles of shock pressure, nickel concentration, and initial alloy state, on coercivity and remanence.

MAGNETIC HYSTERESIS CLASSIFICATION

Three groups of lunar samples have been defined on the basis of magnetic hysteresis measurements. Group A includes igneous rocks and thermally-metamorphosed samples, including those with notable subsolidus reduction. Group B contains Apollo 14 breccia samples, and lunar fines. Group C contains lunar fines and breccia samples.

The following relationships exist for the hysteresis parameters for the three groups:

$$H_R(C) > H_R(B) > H_R(A)$$
$$R_I(C) > R_I(B) > R_I(A)$$
$$R_H(B) > R_H(C) > R_H(A)$$

The R_H value has been used quite effectively for most ferromagnetic dispersions, but in the absence of a quantitative estimate of the superparamagnetic component, the value becomes non-definitive, since it depends on H_c, which is reduced according to the superparamagnetic component. Shock lithified lunar samples fall into Groups B and C. If the rocks in Groups B and C are thermally metamorphosed, they will fall into Group A. An interrupted sequence of magnetic changes are associated with shock lithification and fragmentation at the lunar surface. The data for the three groups of lunar samples are plotted in Figure 3.

Fayalite powder (Fe_2SiO_4) is paramagnetic under normal conditions. When the powder was shocked to approximately 200 kilobars, the product, iron spheres plus glass, were highly ferromagnetic. This is clear evidence during shock melting of the coalescence of iron in the melt, presumably due to the reduced state of the melt, since Fe^o is not soluble to any extent in glass. Shock melting in the lunar environment is certain to produce iron metal from glass or iron silicates. Shock melting is also considered to be important in the formation of chondrite meteorites (19).

The thermal demagnetization curve for the iron spheres embedded in glass is shown in Figure 4. Iron cooled through the

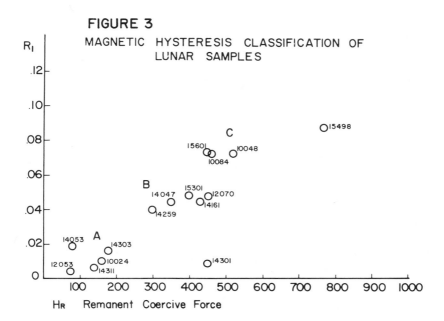

FIGURE 3
MAGNETIC HYSTERESIS CLASSIFICATION OF LUNAR SAMPLES

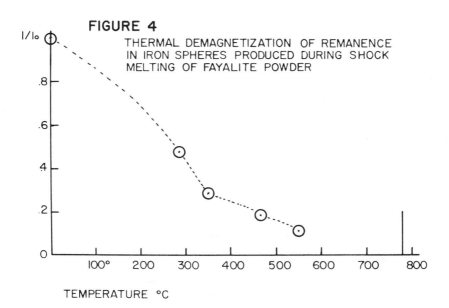

FIGURE 4
THERMAL DEMAGNETIZATION OF REMANENCE IN IRON SPHERES PRODUCED DURING SHOCK MELTING OF FAYALITE POWDER

the Curie point (~780°C) would, on thermal demagnetization, yield a continuously-decreasing curve out to the Curie point. The curve in the figure is a two-step curve, indicating that a two-component assemblage is present which is probably not likely. Instead, the first step in the demagnetization is attributed to annealing of the complex shock structure in the iron after which the "iron" component remains.

CONCLUSIONS

Explosive shock or meteorite impact are significant remagnetization processes. The mechanisms of remagnetization associated with the dynamic processes depend on the peak shock pressure, the nature of the shocked materials and the behavior of the shock in the material. Magnetic measurements can be used to classify products formed during a shock process, and magnetic measurements can be used to investigate the process itself because of the special characteristics of remanent magnetization vectors.

The magnetic coercive force increases more rapidly in quenched and annealed iron-nickel alloys as nickel is added than it does in the alloys which have been shocked. The coercive force values are associated with low field domain wall motion and do not relate to stable magnetic remanence. For low-level shocks in iron where it was shown that the twin density increased monotonically with pressure, a qualitative relationship between shock remanence stability, saturation isothermal remanence intensity, and twin density. Six new remanence mechanisms have been designated for iron-nickel alloys. The vector remanence properties which we are concerned with are not normally considered in metallurgical research but might have promising applications. Research in progress will establish the role of nickel, metallurgical treatment and shock on remanence properties in BCC iron-nickel alloys.

ACKNOWLEDGMENTS

One of us (P.J. Wasilewski) gratefully acknowledges NASA Grant NGR-09-010-080 for support of this research. Special thanks are extended to Dr. M.D. Fuller of the University of Pittsburgh and Dr. F.C. Schwerer of the U.S. Steel Research Labs for providing access to their research facilities. Special thanks are extended to Dr. Speich, Dr. Leslie, Dr. Rose, Dr. Rohde and Dr. Sclar for providing samples used in this study. Mrs. Inman has helped extensively with specimen preparation, and we thank her for her able assistance.

REFERENCES

1. Wasilewski, P.J., presentation to the Third Lunar Science Conference, March, 1973.
2. Grady, D.E., J. Appl. Phys. 43, 1942, 1972.
   Grady, D.E., G.E. Duvall, and E.B. Royce, J. Appl. Phys. 43, 1949, 1972.
   Graham, R.A., J. Appl. Phys. 39, 437, 1968.
   Graham, R.A., D.H. Anderson, and J.R. Holland, J. Appl. Phys. 38, 223, 1967.
   Wayne, R.C., J. Appl. Phys. 40, 15, 1969.
   Bartel, L.C., J. Appl. Phys. 40, 661, 1969.
   Bartel, L.C., J. Appl. Phys. 40, 3988, 1969.
   Fuller, M.D., M.F. Rose, and P. Wasilewski, to be published, 1973.
   Royce, E.B., J. Appl. Phys. 37, 4066, 1966.
   Christou, A., and N. Brown, J. Appl. Phys. 42, 4160, 1971.
3. Wasilewski, P.J., The Moon, to be published, first issue, 1973.
4. Goldstein, J.I. and A.S. Doan, Jr., Geochim. Cosmochim. Acta 36, 51, 1972.
5. van Schmus, W.R. and J.A. Wood, Geochim. Cosmochim. Acta 31, 747, 1967.
6. Bunch. T.E. and W.A. Cassidy, Contrib. Mineral. Petrol. 36, 95, 1972.
7. Smith, C.S., Trans. Met. Soc. AIME 212, 574, 1958.
8. Fredriksson, K., P. De Carli, A. Aaramae, Space Research, Proc. Intern. Space Sci. Symp., Washington, 1962, North Holland Publ. Co., Amsterdam, 1963.
9. Wasilewski, P., Earth and Planet. Sci. Lett., in press, 1973.
10. Rohde, R.W., J.R. Holland and R.A. Graham, Trans. Metall. Soc. AIME 242, 2017, 1968.
11. Yensen, J., Am. Inst. Elec. Eng. 39, 396, 1920.
12. Hirsch, A.A. and Z. Eliezer, IEEE Trans. Mag. MAG-6, 732, 1970.
13. Fowler, C.M., F.S. Minshall and E.G. Zukas, in Response of Metals to High Velocity Deformation, ed. P.G. Shewmon and V.F. Zackay, Interscience Publ., New York, 1961.
14. Johnson, J.N. and R.W. Rohde, J. Appl. Phys. 42, 4171, 1971.
15. Holland, J.R., Acta. Met. 15, 691, 1967.
16. Giles, P.M., M.H. Longenbach and A.R. Marder, J. Appl. Phys. 42, 4290, 1971.
17. Speich, G.R. and P.R. Swann, J. Iron and Steel Inst. 203, 480, 1965.
18. Knox, R., Geochim. Cosmochim. Acta 27, 261, 1963.
19. Wlotzka, F., Meteorite Research, ed. P. Millman, D. Reidel, Holland, 1969.

Plate I: Micrographs of Polished Surfaces of BCC and FCC Iron-Nickel. 1% Nital Etch.

   a. Mechanical twins in the low pressure region and comples transition structure in the high-pressure region of an Fe-2% Ni alloy shocked to 210 Kb in the free surface mode. Free surface is at the top of the micrograph. Horizontal bar to the right marks the transition.
   b. BCC iron-nickel in Kandahar chondrite containing mechanical twins and slip bands. Twin curvature is associated with a second deformation. Average VHN (50 gm) is 196. (The bar is 0.1 mm.)
   c. BCC iron-nickel in Allegan (type H5) chondrite. The only structures are straight mechanical twins associated with low-level shock; no thermal effects. (The bar is 0.05 mm.)
   d. BCC iron-nickel in Dalgety Downs chondrite. Complex mechanical twinning due to moderate shock level deformation. (The bar is 0.05 mm.)
   e. Complex microstructure resulting from recrystallization and plastic deformation of BCC iron-nickel in Renazzo chondrite. (The bar is 0.0125 mm.)
   f. and g. FCC iron-nickel in Dalgety Downs chondrite. Slip bonds and shear displacements post date the thermal episodes and responsible for diffusion which produced the composition gradients responsible for the zoned etching pattern. (The bar is 0.0125 mm.)
   h. Fine mechanical twins due to shock in Campo del Cielo meteorite. There appear to be some activated slip systems which are apparently responsible for the discontinuous propagation of some twin sets. Metal is BCC with approximately 6.6% nickel. (The bar is 0.05 mm.)
   i. Faulting displacement of phosphide grain intersected by mechanical twins in Campo del Cielo meteorite. (The bar is 0.0125 mm.)
   j. Void formation along a grain boundary in the Campo del Cielo meteorite. Macroscopic evidence demonstrates shear and plastic flow along discontinuous bands. Slip bands show marked change in orientation on either side of the voids indicative of crystallographic difference at the grain boundary. (The bar is 0.1 mm)
   k. Fine structure of slip bands in Campo del Cielo meteorite. (The bar is 0.0125 mm.)
   l. Offset faulting of an FCC iron-nickel grain along a continuous shear zone in the Dalgety Downs chondrite. Metal grain first developed composition gradients due to diffusion at high temperature; subsequent offset faulting was associated with reheating which produced recrystallization along the faulted faces and added secondary metal to the grain edges. (The bar is 0.05 mm.)
   m. Fracture of BCC metal showing voids and metal structure at fracture tip. Dalgety Downs chondrite. (The bar is 0.0125 mm.)

PLATE I

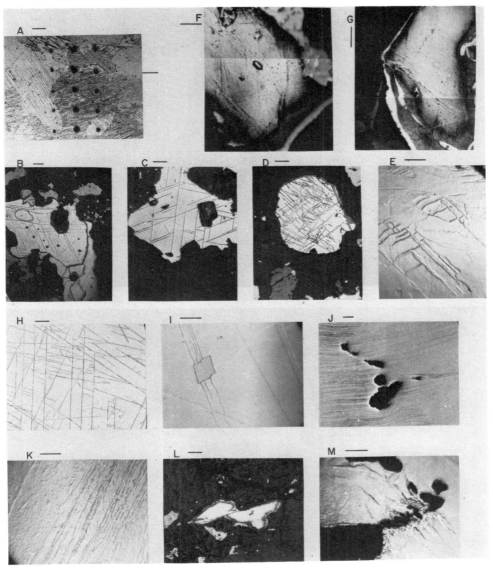

Plate II: Micrographs of Polished Surfaces of Meteorites (Post Shock Thermal Effects), Lunar Samples (Shock Lithification), and Impactites (Sphere Formation). Meteorites etched with 1% NITAL.

    a. Polycrystalline BCC FeNi due to reheating in the Stalldallen (H5) chondrite. The VHN (25 gm) is 116 $\pm$ 5 (The bar is 0.05 mm.)

    b. Recrystallized FCC+BCC+Phosphide invaded by oxide along grain boundaries at upper left and lower right. Freemont Butte (L4) chondrite.

    c. Reheated grain in Kyushu (L6) chondrite, showing diffusion aided decomposition of single crystal. VHN (25 gm) ranges from 175 to 213. (The bar is 0.025 mm.)

    d. Martensitically transformed BCC grain in Farmington (L5) chondrite. The VHN (25 gm) values 187 $\pm$ 31 show considerable variation.

    e. and f. Partially-recrystallized BCC iron-nickel in New Concord chondrite. Twins show some decoration by precipitates. The serrated advance of apparently recrystallized material is controlled in part by precipitates. (The bar is 0.0125 mm.)

    g. Welded breccia, Apollo 16 (61175). Angular shapes of grains, welded grain edges, and the glass sphere containing a metallic sphere all indicate a shock lithification process produced the sample. (The bar is 0.05 mm.)

    h. Breccia Sample 15255 (Apollo 15). Large glass fragment containing metallic spheres. Glass was subsequently fractured after shock lithification. (The bar is 0.05.)

    i. Shock welded Apollo 15 breccia (15287). (The bar is 0.05 mm.)

    j. Metallic grains in Allegan chondrule (285) illustrating the metal, silicate disposition in these small spherical bodies which may have been produced during shock melting. (The bar is 0.05 mm.)

    k. and l. Spherical metal, metal-sulfide, and sulfide in Monturaqui impactites. (The bar is 0.05 mm.)

PLATE II

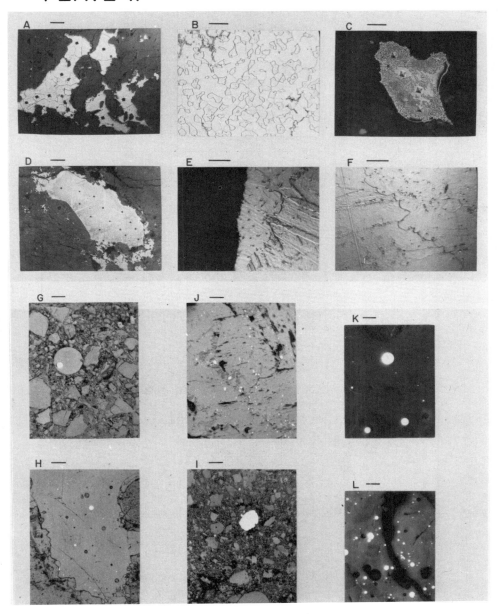

COMMENT ON SHOCK-INDUCED MELT TRANSITIONS IN ALUMINUM AND BISMUTH

J. R. Asay

Sandia Laboratories, Albuquerque, New Mexico 87115

We have recently measured wave profiles for shock wave propagation near equilibrium melt states in aluminum and bismuth. These data represent the first measurements of wave profiles obtained for states of partial melting under shock loading and have allowed some preliminary conclusions regarding melting rates on the submicrosecond time scale of the experiment. The initial objective of the experiments was to determine whether shock-wave compression in the solid-liquid mixed phase region can be described with equilibrium thermodynamics, i.e., with the assumption that melting begins instantaneously for states in the mixed phase region.

Bismuth and aluminum are representative of materials for which the melt temperature either decreases or increases with pressure. In bismuth, the melt temperature decreases from its atmospheric pressure value of 271°C to 183°C at the triple point of 17 kbar connecting the two solid phases (solids I and II) and the liquid phase. When solid I bismuth is preheated to a temperature near the atmospheric melt temperature and shocked to a high pressure, the Hugoniot can intersect the melt boundary. At the point of intersection, there is a discontinuity in isentropes which gives rise to a two-wave structure for the shock if melting begins instantaneously for states in the solid-liquid mixed phase region. This effect is similar to that observed in polymorphic transitions under shock loading and allows a determination of whether melting occurs on the submicrosecond time scale of the experiment.

We have performed a number of plate impact experiments[1] on bismuth for initial temperatures ranging from 170°C to 240°C using quartz gauges[2] to measure the transmitted wave profiles at the rear surface of the specimens. In some of the experiments the specimens were shock-loaded to states only in the solid I-liquid mixed phase region, and in others they were shocked to stresses in excess of the triple point. The measured profiles were then compared with analytic solutions based on a three phase equation of state for bismuth[3] (solid I-liquid-solid II). Generally, we found that the measured profiles did not show a definite two wave structure and were much more dispersed than the profiles predicted from equilibrium thermodynamics. This indicates that melting is not instantaneous for shock wave propagation in bismuth.

In aluminum, the melt temperature increases with increasing pressure so that a two-wave structure does not develop when the material is shocked across the melt boundary. In this case, however, a two-wave structure occurs in the release wave profile under equilibrium conditions, if the material is shocked to a state below the melt boundary and then released to an equilibrium state in the solid-liquid mixed phase region.

Energy states near the melt temperature in aluminum were obtained by shocking porous specimens with an initial density of approximately 60% of solid density. Incipient melting occurs at a shock stress of about 75 kbar for this material. We have measured nearly complete release wave profiles for initial stresses ranging from 45 to 73 kbar, using an interferometer technique.[4] The highest stress attained results in an equilibrium release state approximately 40% across the solid-liquid mixed phase region.

The measured release wave profiles were compared with profiles calculated with a hydrodynamic computer code which assumes equilibrium thermodynamics. The calculated solutions indicated that a two-wave structure should occur in the release wave profile for impact stresses greater than about 50 kbar. However, the measured profiles do not show a structure which can be attributed to melting under equilibrium conditions. The release profiles show much more dispersion near the foot of the wave if the final release state is well into the solid-liquid mixed phase region.

These results do not preclude the conclusion that some degree of melting occurs on the time scale of shock wave experiments, but simply show that melting is not <u>instantaneous</u> in either aluminum or bismuth. This is contrary to the common assumption used to treat wave propagation near melt states. However, it should be recognized that our experiments have been confined to regions close to the melt boundary where the rate of transformation associated with melting might be expected to be small.

References

1. S. Thunberg, G. E. Ingram and R. A. Graham, Rev. Sci. Inst. 35, 11 (1964).

2. G. E. Ingram and R. A. Graham, "Quartz Gauge Technique for Impact Experiments," in 5th Symposium on Detonation, Office of Naval Research, Report ACR-184 (1970).

3. J. N. Johnson, D. B. Hayes and J. R. Asay, to be published.

4. L. M. Barker and R. E. Hollenbach, J. Appl. Phys. 43, 4669 (1972).

RAPID MELTING OF ALUMINUM INDUCED BY PULSED ELECTRON BEAM EXPOSURE

F. B. McLean, R. B. Oswald, Jr., D. R. Schallhorn and
T. R. Oldham

Harry Diamond Laboratories
Washington, D.C. 20438

In view of the discussions concerning the speed at which phase transitions occur under dynamic loading conditions we think it appropriate to report preliminary results of an experimental study of the material dynamic response of aluminum exposed to a high fluence, low energy, pulsed electron beam. The energy depositions were sufficient to induce melting and some partial vaporization of the surface regions of the target samples and enabled us to probe the melting transition under rapid thermal loading conditions. We compared the experimental results with calculations based on two equilibrium equation of state (EOS) models for metals in the melt regime. The first model is the simple Mie-Grüneisen (M-G) scheme for solids extrapolated into the liquid phase. The second model--the so-called GRAY EOS recently developed at Lawrence Livermore Laboratories[1]--is a three phase EOS which provides a more detailed and thermodynamically complete description of metals in the melt region. The essential result of the comparison was that after the onset of melting the simple M-G model departs appreciably from experiment, whereas the GRAY EOS provides quite acceptable agreement with the experimental results. The results indicate that melting occurs on the time scale of the deposition times (70 nsec) and that the melting process can be described by an equilibrium (but thermodynamically complete) EOS model.

The experimental studies consisted of measuring the dynamic response of 6061 aluminum to a pulsed 185 keV electron beam for fluences ranging from 15 to 50 cal/cm^2, resulting in peak energy depositions of 450-1500 cal/g. The experimental condition produced a mass blow-off which was dominated by mass in the melt state. By conservation of momentum the expanding melt material produces an impulsive load which is coupled to the remaining solid portion of the sample via the transmission of a progagating large amplitude stress wave into the solid region. The material response was characterized by simultaneous (for each electron pulse) measurements of total impulse, transmitted stress, blow-off velocities and mass loss.

The predicted response to the rapid heating was calculated with a recently developed hydrodynamic computer code, the RIP code[2], in conjunction with the two EOS models. A time dependent energy deposition profile was read as input. To describe the stress wave propagation through the cold portion of the samples an appropriate mechanical constitutive relation was used. For the present calculations we employed a strain rate dependent model[3] designed to accurately describe plastic flow in metals.

The Mie-Grüneisen EOS relates the thermal pressure directly to the internal energy density via the Grüneisen parameter, $\gamma_s(V)$, which is assumed to be dependent only upon the specific volume. No thermodynamic distinction is normally made between the solid and liquid phases nor are the details of the melting transition considered. In the GRAY formulation material near normal density and at temperatures above the onset of melting is described by a scaling law EOS which is based on the following assumptions: (1) the entropy of melting is a constant independent of pressure; (2) the temperature dependence of the specific heat in the liquid is a universal curve, scaled on the melting temperature, and (3) the pressure dependence of the melting temperature is given by a modified Lindeman law. For a detailed description of the GRAY EOS, the reader is referred to the original literature.[1] Here, we point out only that immediately above the onset of melting the scaling law EOS for the liquid takes the form of a modified M-G theory, in which the effective "average" Grüneisen parameter $\Gamma(V,T)$--defined as the ratio of the thermal pressure to the thermal energy density--is temperature dependent as well as volume dependent. For aluminum $\Gamma(V,T)$ increases from the solid Grüneisen value $\gamma_s(V)$ at the onset of melting to a value approximately 40% greater than $\gamma_s(V)$ at the completion of melting. Thereafter, $\Gamma(V,T)$ decreases slowly, dropping below $\gamma_s(V)$ at a temperature about four times the melting temperature. The implication is that for constant volume conditions at temperatures immediately above melting the thermal pressures calculated by GRAY are of the order of 40% greater than calculated by the M-G model.

Fig. 1 is illustrative of the experimental results[4] and of the comparisons with calculations. Here we plot specific impulse (impulse per unit area) vs the incident electron fluence. Each of the experimental data points shown represents the average of about ten exposures. The impulse calculated using the GRAY EOS is in excellent agreement with the measured values over the entire range of the experimental data. In contrast, the predicted impulse based on the M-G EOS is significantly lower than the measured values over the entire fluence range from 18 to 50

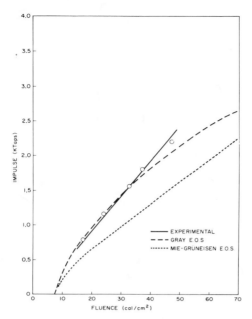

Fig. 1. Comparison of calculated and measured impulse produced by the exposure of 6061 Al to the pulsed electron beam.

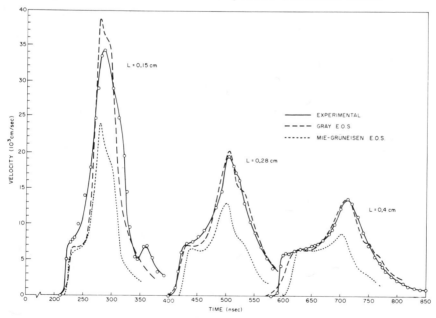

Fig. 2. Comparisons of calculated and measured rear surface velocity histories for three thicknesses of 6061 Al exposed to a fluence of 36 cal/cm^2.

cal/cm^2. Since the impulse reflects the magnitude of the thermal pressures generated in the deposition regions, the experiments clearly confirm the physical validity of the scaling law model in the melt regime. Further evidence is presented in Fig. 2 where also the changing nature of the stress wave as it propagated through the sample is shown. Here we compare the measured and predicted rear surface velocity histories for three different sample thickness 0.15, 0.28, and 0.4 cm resulting from an exposure to a fluence of 36 cal/cm^2. These histories are good examples of the elastic-plastic response of aluminum produced by the pulsed electron beam exposures. The velocities computed with the M-G EOS are significantly lower than experimental values, again an indication that the M-G scheme under-predicts the generated thermal pressures. On the other hand, the velocity histories calculated using the GRAY EOS agree in excellent fashion with the measured histories for each of the three thicknesses. The good agreement for the thicker samples, for which appreciable attenuation and broadening of the plastic peak has occurred, also gives support to the validity of the strain rate dependent constitutive relation employed in the calculations.[3]

In summary our results confirm GRAY's prediction that the thermal pressure in the melt regime can be described by an effective "average" temperature dependent Grüneisen parameter which for aluminum at the completion of melting is of the order of 40% greater than the solid Grüneisen value. As a corollary, it appears that the thermal pressures generated during the pulsed thermal loading conditions of the present experiments can be described by equilibrium melting processes.

REFERENCES

1. E. B. Royce, "GRAY, a Three Phase Equation of State for Metals," Lawrence Livermore Laboratory, Rept. UCRL-51121 (1971). See also R. Grover, J. Chem. Phys $\underline{55}$, 3435 (1971); D. A. Young and B. J. Adler, Phys. Rev. $\underline{A3}$, 364 (1971).

2. R. H. Fisher, G. A. Lane and R. A. Cecil, "RIP, A One-Dimensional Material Response Code," Systems, Science and Software, Rept. 3SR-751 (1972).

3. H. E. Read, J. R. Triplett and R. A. Cecil, "Dislocation Dynamics and the Formulation of Constitutive Equations for Rate-Dependent Plastic Flow in Metals," Systems, Science and Software, Rept. DASA-2638 (1970). This is essentially the same model as discussed by H. E. Read at this conference as applied to beryllium.

4. A complete report of the investigation will be submitted for publication later. The discussion presented here is a preliminary report of our major findings.

MICROSCOPIC MECHANISMS OF DISLOCATION DRAG

A. V. Granato

Department of Physics and Materials Research Laboratory

University of Illinois, Urbana, Illinois   61801

ABSTRACT

Dislocation drag effects are reviewed. It is found that drag effects are well enough understood to conclude that dislocation generation rates, rather than dislocation velocities, provide the rate-limiting step under high strain-rate conditions.

I.   INTRODUCTION

Dislocation drag effects have always played an important role in ultrasonic attenuation. Recently, they have been found to play a decisive role in the plastic deformation of superconductors, where dislocations are thought to achieve high speeds. The question naturally arises as to whether or not drag effects are of importance under shock-loading conditions. To help assess this question we first survey in Section II what is expected theoretically from drag effects. In Section III, a limited selection of experimental results are examined to determine to what extent the theoretical expectations are realized. This is a semi-quantitative survey only, as these questions are considered in more detail by others in this session of the conference. Finally, in Section IV, the significance of drag effects for plastic flow questions in general and for shock-loading conditions in particular, is discussed.

II. DISLOCATION DRAG MECHANISMS

Known agents providing dislocation drag are: 1. phonons,
2. electrons,
3. radiation
and 4. point defects.
Phonon drag is thought to be of importance in all materials at high temperatures. At low temperatures in metals when the phonons are removed, electron drag remains. When the electrons are removed, as in superconductors or insulators, radiation drag remains for oscillating dislocations. For dislocations moving slowly enough for diffusion of point defects to occur, point defect drag effects appear but these are not of interest for high speed effects and will not be discussed further.

A. Phonon Drag

The earliest estimate of phonon drag by Leibfried (1) in 1950 remains one of the simplest of, and about as accurate as, any of the many succeeding more sophisticated calculations. Leibfried supposed that for a plane lattice sound wave incident on a dislocation, all of the energy in a strip of width of the order of atomic dimensions would be scattered. The scattering cross section per unit length σ is then $\sigma = b$ where b is the Burgers vector. The force per unit length acting on the dislocation is then given by

$$F \sim \sigma E \qquad (1)$$

where E is the phonon energy density. E has the dimensions of energy/vol. or force/area which when multiplied by σ gives the force/length F acting on the dislocation. For a dislocation in an isotropic flux of phonons, there is no net force. However, if it moves through this flux with speed v, a fraction v/c of the waves now appear in the forward direction rather than the backward so that the net force/length is given by

$$F = g\sigma E v/C = b\sigma = Bv \qquad (2)$$

where g is a geometrical factor arising from averages which must be taken over a distribution of angles between incident phonons and the dislocation. Leibfried estimated this factor to be $g \sim 1/10$. Eq. 2 defines the drag coefficient B to be

$$B = g\sigma E/C \qquad (3)$$

1. **Radiation Scattering.** The first mechanism for which a detailed calculation of the scattering cross section σ was given was for reradiation scattering by Eshelby (2) (1949). If a phonon incident on a dislocation has a shear stress component on the slip plane in the slip direction, then the phonon drives the dislocation into forced oscillation. The oscillating dislocation radiates a cylindrical wave. The calculation is similar to that for an electron in an electromagnetic field. Eshelby found that the scattering cross section σ_R for this process (sometimes known as the flutter process) goes as

$$\sigma_R \sim d^2 \omega^3 \tag{4}$$

where d is the dislocation oscillation amplitude and ω is the frequency. For high frequencies, the dislocation motion is inertia-limited and

$$d \sim \left[M(\omega) \omega^2 \right]^{-1} . \tag{5}$$

In Eq. 5 the dislocation mass depends logarithmically on frequency. The resulting frequency dependence from Eqs. 4 and 5 is approximately given by (3,4)

$$\sigma_R \simeq 2b(\omega_D/\omega)^{\frac{1}{2}} , \tag{6}$$

where ω_D is the Debye frequency. Eshelby's calculation is a continuum elasticity calculation which should be valid for $\omega \ll \omega_D$. Eshelby also calculated a drag arising from a thermoelastic effect in the same article. For this it was assumed that macroscopically defined concepts such as thermal conductivity could be applied locally in the field of a dislocation in regions small compared to phonon (and electron) mean free paths. The calculated drag is smaller than the scattering drag, and this approach will be discussed further later.

2. **Strain-field Scattering.** The second known scattering mechanism arises from non-linear elastic effects. The material close to the dislocation line is finitely strained, and scatters sound waves because of changes in the elastic constants and density in that region. This mechanism was first discussed by Nabarro (1951)(5) and has subsequently been considered further by many authors. The simple calculation below, again using continuum elasticity theory valid for $\omega \ll \omega_D$, illustrates the main features involved. This calculation is different from, but equivalent to, most of those which have appeared in the literature. The wave equation

$$\nabla^2 \mu - \frac{\rho}{C} \frac{\partial^2 \mu}{\partial t^2} = 0 \tag{7}$$

is to be solved, where the density ρ and the elastic constant C of the medium vary in the strain field of the dislocation. The elastic constant (and density) may be determined from $C = C_o + \delta C$, where

$$\delta C = \frac{\partial C}{\partial \tau_i} \tau_i , \qquad (8)$$

where C_o is the elastic constant for an unstrained medium and τ_i is the stress field of the dislocation. Then, letting

$$\mu = \mu_o + \mu_1 \qquad (9)$$

where μ_o represents a plane wave incident phonon and μ_1 a cylindrical wave scattered phonon, the scattered wave is given by

$$\nabla^2 \mu_1 + k_o^2 \mu_1 = k_o^2 \frac{\partial C}{\partial \tau_i} \frac{\tau_i}{C} \mu_o \qquad (10)$$

where k_o is the propagation vector $k_o = \omega C_o$ for the unstrained medium, and the quantities on the right hand side of Eq. 10 are all known. The solution of the Poisson Eq. 10 is simple by the Born approximation. The difficult part is to know what to use for $\partial C/\partial \tau_i$ and to carry out all the necessary averages over phonon directions and polarizations. The coefficients $\partial C/\partial \tau_i$ are in fact just the third order elastic constants. In practice, most calculations have been made using estimates of average values of these effects. A simple estimate is obtained, for example, by choosing C to be the shear modulus G and the stress τ_i to be a hydrostatic pressure P in $\partial \tau / \partial \tau_i$ but the dislocation shear stress otherwise. Then $\partial C/\partial \tau_i$ is represented by $\partial G/\partial P \simeq 2\gamma$, where γ is Gruneisen's constant. In this way, one finds

$$\sigma_S \simeq \frac{\gamma^2}{8} \frac{\omega}{\omega_D} b \qquad (11)$$

From Eqs. 6 and 11, for $\omega = \omega_D$ and $\gamma = 2$, one finds $\sigma_R/\sigma_S \simeq 4$. However Nabarro (4) estimates that $B_R/B_S \simeq 1/3$ because of the different way that the geometrical factor g enters for the two effects.

3. <u>Subsequent Calculations of Phonon Drag</u>. The reradiation scattering σ_R as given in Eq. 6 increases with decreasing ω. This increase continues until the dislocation resonant frequency

$$\omega_o \simeq \pi C/L , \qquad (12)$$

where L is a dislocation segment length between pinning points, is reached. Below this, the dislocation displacement d in Eq. 4 is tension limited, and $\sigma_R \sim \omega^3$. The resonance is sharp, and limited by radiation damping (6). For a random distribution of loop lengths, the resonance is broadened out somewhat and shifted to lower frequencies. For screw dislocations above the resonant frequency, an additional scattering is expected for phonons incident on dislocations at angles for which the phonon phase matches that of traveling waves on the dislocation (7).

There are numerous articles (8-17) offering further elaboration of the strain-field scattering mechanism. These papers provide further discussion of questions involving techniques of averaging, estimates of the non-linear strain field parameters and the influence of kinks on the dislocation lines. There is not space available here to review this work, but most of these have been reviewed recently by Nabarro (4), Brailsford (18), Hikata, et. al. (19) and Anderson and Malinowski (20). Loss mechanisms postulated to arise from the lattice structure, most of which lead to a non-linear dependence on velocity, have also been reviewed by Nabarro (4).

A calculation of a different kind was offered by Mason (21). Mason considered the phonons as a "gas" in a "box" provided by the extent of the dislocation strain field. A moving dislocation strain field then provides a shear strain rate which stirs the gas up and the energy loss and effective viscous drag force can be computed by standard continuum fluid dynamic calculations in terms of the viscosity of the phonon gas. By kinetic theory the viscosity, related to momentum transport, and the thermal conductivity, related to energy transport, can both be expressed in terms of the mean free path of the gas. By eliminating the mean free path between these relations and expressing the viscosity in terms of the thermal conductivity, Mason gave an expression for the drag in terms of the thermal conductivity. His result depended inversely on the square of a cut-off radius, which was taken to be of atomic dimensions.

This procedure had successfully been applied earlier by Mason to compute the interaction between ultrasonic waves and electrons in solids. The results obtained in this way agree exactly with detailed atomistic calculations (22) for the regime of $\lambda \gg \ell e$, where λ is the sound wavelength and ℓe is the electronic mean free path. This is the regime where kinetic theory should apply. However, this method fails to give correct results for the regime $\lambda \ll \ell e$. Indeed it has been pointed out by many, most recently by Brailsford (18), that this technique is inapplicable for dislocations since the dominant Fourier wavelength of the dislocation strain field is of the order of atomic dimensions so that $\lambda \ll \ell p$, where ℓp is the phonon mean free path. An extended discussion of this is given by Brailsford (18), who finds that the long wavelength Fourier components of the dislocation strain field do indeed lead to a viscous drag of the type described by Mason, but that this part of the drag is always smaller than the direct scattering type because the important Fourier components are of short wavelengths. Presumably, the thermoelastic drag calculation of Eshelby mentioned earlier arising from heat flow between hotter and colder strained regions of the crystal resulting from a moving dislocation is a similar type calculation to that of Mason's for heat transport instead of momentum transport.

In summary, phonons are expected to produce drag by two mechanisms, reradiation and strain field scattering. The reradiation cross section has an underdamped resonance at ω_o, falling as ω^3 for lower frequencies and as $\omega^{-\frac{1}{2}}$ for higher frequencies. The strain field scattering is linear in frequency, and both are expected to be comparable at $\omega = \omega_D$ or $T = \theta$, where θ is the Debye temperature. The resulting drag should be linear in velocity with a drag coefficient $B \sim 10^{-4}$ in cgs units at $T = \theta$.

4. <u>Expected Temperature Dependence for Experimental Results</u>. Drag effects can be observed through the measurement of the drag coefficient B in ultrasonic attenuation measurements, etch pit velocity measurements at high velocities, and cross sections σ may be determined from thermal resistance measurements.

The temperature dependence of B arises through that for the factors $\sigma(T)$ and $E(T)$ in Eq. 3. For high temperatures, $\sigma = \sigma(\omega_D)$ and $E = 3NkT/V$ where V is the volume per atom, so that $B \sim T$.

For low temperatures, $E \sim T^4$ and $\sigma_S \sim \omega$ for the strain field scattering. Defining the dominant phonons at temperature T to have a frequency $\nu = 3.8kT/h \simeq 10^{11}T(Hz/K)$, $\sigma_S \sim T$. For reradiation scattering $\sigma_R \sim \omega^{-\frac{1}{2}} \sim T^{-\frac{1}{2}}$. Thus, one expects the low temperature dependences of B for reradiation scattering (B_R) and strain-field scattering (B_S) to be

$$B_R \sim T^{7/2} \quad \text{and} \quad \tag{13}$$

$$B_S \sim T^5, \text{ for } T \ll \theta \quad .$$

Thermal conductivity measurements should offer a means for measuring $\sigma(\omega)$, as such measurements can be made with relatively high precision. Using the kinetic theory relation

$$K \sim \frac{1}{3} Cv\ell \quad , \tag{14}$$

where K is the thermal conductivity, C the specific heat, v the velocity of sound and ℓ the mean free path, together with

$$\ell^{-1} = \Lambda \tau \tag{15}$$

where Λ is the dislocation density, the frequency dependence may be inferred from the temperature dependence of K. Since $C \sim T^3$ at low temperatures, one then expects

$$\begin{aligned} K_R &\sim T^{7/2}, \text{ and} \\ K_S &\sim T^2, \text{ for } T \ll \theta \end{aligned} \tag{16}$$

for the temperature dependence if the thermal conductivity is limited by the dislocation resistivity.

B. Electron Drag

Electrons are also scattered by dislocations. There is no flutter here as the frequencies are too high. The interaction between the displacement field and the electrons is well known from ultrasonic attenuation studies (22).

The first theory offered for this effect was given by Mason (23) and is completely analogous to the phonon viscosity mechanism already discussed. Mason predicts that the electron drag is proportional to the conductivity. Huffman and Louat (24) have also estimated an electron drag in terms of the Boltzmann transport equation. Their theory also predicts a drag proportional to the conductivity. Mason's theory has been criticized (25,26,27) on the grounds that an electron gas concept is not valid for electronic mean free paths much larger than the typical Fourier wavelengths of a dislocation which are of atomic dimensions. Brailsford (27) has pointed out an error in the Huffman-Louat theory. When their theory is corrected for this error, it predicts a value of B_e in agreement with that of Holstein (28) and Kravchenko (29) from a theory based on electron scattering produced by the deformation potential of the strain field of a moving dislocation. This theory predicts that B_e is independent of both conductivity and temperature, and proportional to the electronic density Ne, i.e.

$$B_e \sim N_e \tag{17}$$

The magnitudes of B_e expected are typically of order 10^{-5} in cgs units, or about one order of magnitude smaller than phonon drags at room temperature.

III. EXPERIMENTAL EVIDENCE

A. Phonon Drag Measurements

1. <u>Ultrasonic Measurements</u>. Ultrasonic measurements are analyzed in terms of the vibrating string model introduced by Koehler (30) and developed further by Granato and Lücke (31). Under the influence of an oscillating shear stress at low frequency, the dislocations oscillate as strings with the maximum displacement limited by the dislocation tension and the segment length L, as illustrated in Fig. 1. At high enough frequencies, a dislocation viscous drag is felt and the displacement is limited by the drag force instead. If a pinning point is added the low frequency displacement is greatly reduced, while the high frequency displacement is affected only in the neighborhood of the pinning points, which are widely separated.

An example of the effect of adding pinning points is shown in Fig. 2. These are measurements of the ultrasonic damping in Cu (32) as it is affected by pinning points introduced by cobalt gamma irradiation. One sees that the low frequency damping is strongly reduced, but the high frequency damping is hardly affected by the pinning points. At high frequencies, the attenuation is simply proportional to Λ/B, and is independent of the difficult to determine dislocation segment lengths and their distribution function. By estimating the dislocation density Λ from etch pit counts, the drag constant was found from these measurements to be in reasonable agreement with Leibfried's estimate. In this technique the accuracy is limited by the measurement of Λ and the need to separate the dislocation component of attenuation from the total attenuation. The latter is done by assuming that light irradiations affect only the dislocation component.

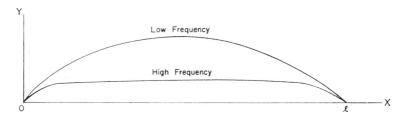

Fig. 1. - Schematic Dislocation Displacement $y(x)$ as a Function of Coordinate x for Low Frequencies and High Frequencies. At Low Frequencies the Displacement is Limited by Tension Forces. At High Frequencies, the Displacement is Limited by Viscous Forces.

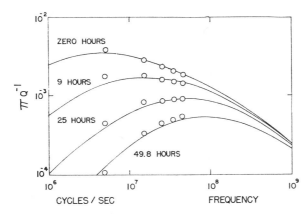

Fig. 2. - Decrement as a Function of Frequency for Several Times During Cobalt Gamma Irradiation in a 6000 Curie Source. The Solid Curves Are Theoretical (After Stern and Granato.)

Ultrasonic measurements of attenuation and velocity in copper by Alers and Thompson (33) showed that B is linear in temperature between liquid nitrogen and room temperature. This is again in agreement with the Leibfried theory and shows furthermore that $B_p \gg B_e$, where B_p and B_e are the phonon and electron drag, respectively.

The accuracy in these early measurements was not high (estimated at about a factor of four by Stern and Granato). In addition, subsequent measurements by different workers on the same materials showed disagreements of about one order of magnitude.

A strong attempt to improve the accuracy of ultrasonic measurements was made by Fanti, et. al. (34) for NaCl and LiF. Many systematic errors were identified and accounted for. Ultrasonic attenuation and velocity were measured at two frequencies as a function of cobalt gamma irradiation. Measurements of attenuation and velocity at one frequency are equivalent to measurements of attenuation as a function of frequency if the segment length distribution is known. Measurements at two frequencies provide a check of the assumed random segment length distribution function. The check showed that the actual distribution is not the commonly assumed random distribution but was close enough so that the estimated accuracy was to better than a factor of two. The resulting drag limited velocities in NaCl as determined by Eq. 2 are shown in Fig. 3 as a straight line. The line is extrapolated to higher stresses to compare with the velocities as determined in etch pit

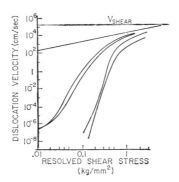

Fig. 3. - Dislocation velocities as a function of applied stress in NaCl. The straight line is a linear extrapolation of ultrasonic velocities to higher stresses. The curved lines are the etch-pit measurements of Gutmanas et. al. (After Fanti, et. al.)

measurements by Gutmanas, et. al., (35). A number of conclusions may be reached from these results.

a. Since velocities determined ultrasonically are orders of magnitude higher than velocities as deduced from etch pit measurements at low stresses, the etch pit measurements do not measure instantaneous velocities in this stress regime, but only average velocities. The dislocations are free to oscillate at relatively high speeds between pinning points. This was first noted by Baker (36).
b. Since the deduced drag constant is of the order predicted by Leibfried, the scattering cross section is of order of atomic dimensions.
c. The viscous drag determined ultrasonically is strictly linear in velocity. If this were not so, then the observed dependence on radiation dose and frequency would not be obtained.
d. At the yield stress of soft crystals at room temperatures, relativistic velocities (relative to sound speeds) cannot be achieved due to phonon drag. This answers an old question in dislocation dynamics concerning the possibility of inertial effects in plastic flow.
e. The same mechanism responsible for ultrasonic attenuation also limits dislocation etch pit velocities at high speeds.
f. Different mechanisms operate in different ranges. At low stresses, dislocation etch pit velocities are determined by interactions with obstacles. In an intermediate range, viscous drag limits the speeds. For high stresses, relativistic effects limit speeds to sound velocities.

Since the mechanisms are different, there is no hope of finding a single relation which describes the results over the entire stress range.

g. For stresses greater than about 10K gm/mm^2 or 1Kbar, viscous drag effects become ineffective in limiting dislocation velocities at room temperature in NaCl. Also, since phonon drag effects do not depend strongly on temperature or material properties, this order of magnitude of stress applies for most circumstances.

From these room temperature measurements we are not able to infer which mechanism of scattering is the more effective. Also, the accuracy of ultrasonic and etch pit measurements is not yet sufficiently precise to distinguish between the low temperature dependences predicted in Eqs. 13.

2. <u>Thermal Conductivity Measurements</u>. We should expect to be able to distinguish reradiation scattering from strain-field scattering in thermal conductivity measurements, according to Eqs. 16. Systematic measurements of the effect of dislocations on the thermal conductivity of LiF down to about 2°K were made by Sproull, et. al., (37). They found that the thermal resistivity was approximately proportional to the dislocation density. The temperature dependence was approximately T^2 as expected for static strain-field scattering, but the magnitude of the effect was two to three orders of magnitude greater than that predicted for strain field scattering. This magnitude however is that expected for reradiation scattering; the ratio of the two from Eqs. 6 and 11 being of the order of 10^3 for this temperature range.

It seems to have been implicitly assumed by most that the reradiation mechanism was somehow inoperative in this temperature range, and these measurements inspired heroic efforts to find ways of increasing the estimated magnitude of the strain field scattering (12-17). If, in fact, Eq. 11 did underestimate strain-field scattering by 2 - 3 orders of magnitude, then the resulting estimated drag constant for ultrasonic attenuation and etch pit measurements would be 2 - 3 orders of magnitude greater than those observed, which are in agreement with Leibfried's estimate. This fact was not noticed.

3. <u>Direct Observation of Reradiation Scattering at 10 - 100 MHZ</u>. Reradiation scattering was directly measured (38) in LiF in the 10 - 100 MHZ range. Thin walls of dislocations were put in LiF crystals by shear and signals from the dislocation walls were observed. These are seen in Fig. 4 as the small echoes between the large echoes from the specimen end faces. The dislocations in the wall are each forced into oscillation by the impressed MHZ sound wave. Each dislocation sends out a cylindrical sound wave,

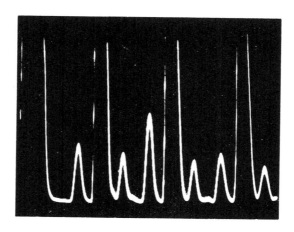

Fig. 4. - Photograph taken on a Matec ultrasonic comparator screen which shows the rectified reradiated wall echoes for a deformed LiF crystal versus display time. (After Schwenker and Granato)

but they are all in phase and combine to yield by Huygen's principle a macroscopic plane sound wave which is then detected by a quartz transducer. The proof that these are reradiated waves and not strain field scattered waves is obtained from orientation and radiation studies. The signal is only obtained when there is a resolved shear stress in the dislocation slip planes. Also the signal disappears when the specimen is lightly irradiated so as to immobilize the dislocations. The frequency dependence of the radiated wave was found to agree with that of Eq. 4 when account was taken of the effect of damping on the dislocation displacement amplitude.

4. <u>Thermal Conductivity of LiF in the .03 to 1°K Range</u>.
Using the same dislocation wall specimen configuration, Anderson and Malinowski (20) showed that reradiation scattering is far more important than strain-field scattering for the thermal conductivity of LiF. Their results for one specimen are shown in Fig. 5 as a function of cobalt gamma irradiation. The conductivity results have been normalized to the values obtained for the part of the specimen containing no dislocation wall where the conductivity was limited by specimen boundary scattering. When a dislocation wall is introduced, the conductivity drops to about half its original value. Immobilizing the dislocations by pinning restores the conductivity at low temperatures or frequencies. As the dislocation radiation is increased, the cross-over temperature moves to higher frequencies. The pinned dislocations remain in the specimen, but produce no detectable resistivity, showing that the reradiation scattering is far greater in magnitude. These measurements cover

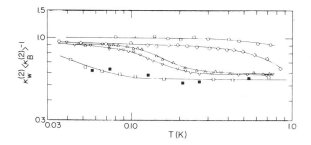

Fig. 5. - Reduced thermal conductivity for the wall region of sample 2, $K_w^{(2)} \langle K_B^{(2)} \rangle^{-1}$, vs temperature: □ - deformed sample; ■ - deformed sample, remeasured; ▽ - 800-R total irradiation; △ - 1000-R total irradiation; ◇ - 26 000-R total irradiation; ○ - 136 000-R total irradiation. (After Anderson and Malinowski)

the range from 4×10^9 HZ to 10^{11} HZ, and show that reradiation scattering is still effective at 1°K. The remaining conductivity is presumably due to that carried by the phonon modes having no resolved shear stress on the slip planes. Indeed, separate ballistic phonon measurements which resolve the different phonon modes by time-of-flight analysis showed that only those modes with resolved shear stress components in the slip plane were reduced by the dislocation wall.

B. Electron Drag Measurements

As noted earlier, electron drag is expected to lead to a drag coefficient of order of 10^{-5} independent of temperature. Ultrasonic measurements on aluminum by Hikata et. al. (39) are shown in Fig. 6. The results can be analyzed as the sum of a temperature dependent phonon component which is linear in temperature at high temperature, and a temperature independent electron component ~ 10^{-5}. The method used by Hikata, et. al., is to measure the change in attenuation with an applied bias stress, assuming that the dislocation segment length distribution is random.

Since the electronic drag depends only on electronic density, it is difficult to test Eq. 17. One possible method becomes available using superconductors, however, since the number of electrons which can be scattered by dislocations is reduced in the superconducting state by reducing the temperature. No direct ultrasonic measurements of B_e have yet been made. The electronic drag can be measured however by an indirect method using the amplitude-dependent component of the attenuation. Amplitude

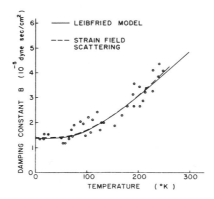

Fig. 6. - Damping parameter B as a function of temperature. (After Hikata, Johnson and Elbaum)

dependence arises from dislocation breakaway from pinning points with increasing stress (31). It was found by Tittman and Bommel (25) that a larger stress is required to produce the same degree of breakaway in the normal state than in the superconducting state. This occurs because the dislocations are unable to exert as large a depinning force when their displacement is drag-limited rather than tension-limited. Hikata and Elbaum (40) used an explicit relation (31) for the drag-dependent depinning force to analyze their amplitude dependent measurements to obtain a value of B_e in the normal state in agreement with the calculation of Holstein and Kravchenko.

A possibility for determining the electronic density dependence of B_e arises from analysis of the temperature dependence of the flow stress changes which occur when superconducting materials are switched between the normal and superconducting states. For these macroscopic flow measurements, it has been found that the drag coefficient plays a decisive role. For superconductors in the normal state, the phonon drag is negligible and only electron drag and radiation drag remain. In the superconducting state at low temperatures, only radiation drag remains, and it has been shown (41) that this drag is weak enough so that all dislocation segments in superconductors are underdamped. This permits inertial effects to operate, allowing dislocations to overcome obstacles at lower stresses than for the normal state. A survey of this effect will not be given here, as Dr. Galligan will discuss it later in the conference. For our present purposes, we only note that the stress change is predicted by the inertial model to be proportional to the density of superconducting electrons. Measurements illustrating this effect are shown in Fig. 7. The solid line is data for lead by Suenaga and Galligan (42), the dashed line is data for indium by Alers, et. al. (43) (closed squares) and by Hutchinson and

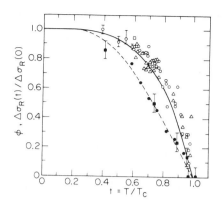

Fig. 7. - The temperature dependence of $\Delta\sigma$ for Pb (open symbols) according to Ref. 42 and for In (closed symbols) according to Ref. 43 (squares with error bars) and to Ref. 44 (circles). The dashed line is Muhlschlegel's calculation of the superconducting electron density according to the BCS theory for weak superconductors, and the solid line is $1 - t^4$.

Pawlowicz (44), (closed circles). The temperature dependence of the density of superconducting electrons is different for indium than for lead since the energy gap is different.

IV. SIGNIFICANCE OF DRAG EFFECTS FOR PLASTIC FLOW

A. The Strain-Rate Equation

The significance of the fundamental relation of dislocation dynamics

$$\overset{\circ}{\varepsilon} = \Lambda b v \qquad (18)$$

derived by Seitz and Read (45) from the Orowan (46) relation was appreciated with the work of Gilman and Johnston (47). They showed that v and Λ could be measured in many cases. Also they showed that many different kinds of plastic flow effects (flow at constant strain rate, creep, stress relaxation, etc.) could be related in terms of Eq. 18 so that it is clear that a theory giving the stress dependence of v and the strain dependence of Λ leads to a theory of plastic flow.

B. Plastic Flow at Low Dislocation Velocities

The large values of the drag constants normally found support a view of plastic flow as a quasistatic type of process in which inertial effects play no role. The dislocation velocity becomes the rate-limiting process and theories proceed by specifying the rate at which dislocations overcome obstacles by thermal fluctuations. The strain rate is then written as

$$\overset{\circ}{\varepsilon} = \Lambda bv = \Lambda bv_o \exp\left[-U(\sigma)/hT\right] \qquad (19)$$

where $U(\sigma)$ is the activation energy required to overcome an obstacle at a stress σ. The effects of long-range stresses are incorporated into $U(\sigma)$.

C. Plastic Flow at High Dislocation Velocities

Recently we have become aware of many plastic flow phenomena which occur at high dislocation velocities. Some of these are:
1. plastic flow in superconductors. This effect is unintelligible in terms of conventional plasticity theories.
2. etch pit observations at high velocities,
3. fast slip band formation, as observed in copper-aluminum alloys by Schwarz and Mitchell(48),
4. ultrasonic attenuation and
5. shock loading.

The first three of these are high velocity, but not high strain-rate effects, while the last two are both high velocity and high strain-rate effects.

But it has been shown (49) that for high velocities, including the viscous drag regime, dislocation velocities are not the rate-limiting factor in the strain-rate, but instead the dislocation generation rate is controlling. For example, for a velocity $v \sim 10^5$ cm/sec, a dislocation which travels a distance $d \sim 1$mm has a lifetime $\tau \sim 10^{-6}$ sec and the mobile dislocation density Λ in Eq. 18 is just the number which move simultaneously in this time interval. Thus

$$\Lambda = \overset{\circ}{\Lambda}\tau = \overset{\circ}{\Lambda}d/v \qquad (20)$$

so that the explicit dependence on velocity in the strain rate Eq. 18 cancels out to give

$$\overset{\circ}{\varepsilon} = \overset{\circ}{\Lambda}bd \qquad (21)$$

Eq. 21 is simply another way of writing Eq. 18 for the high velocity regime which emphasizes that the rate-limiting factor is the dislocation generation rate.

D. Plastic Flow Under Shock Loading Conditions

Dislocation drag effects appear to be well enough understood to conclude that they will not be effective under shock loading conditions. The largest effect is phonon drag, which is not strong even near the melting temperature. For strong materials, obstacles may still be of some significance for the lower shock loading stresses, but for

$$\sigma \gg \sigma_y, \quad v \to c \tag{22}$$

where σ_y is the yield stress and c is the sound velocity. The strain rate may then be written as

$$\dot{\varepsilon} = \Lambda bc \tag{23}$$

or

$$\dot{\varepsilon} = \Lambda b \dot{d} \tag{24}$$

where

$$d = c\tau \tag{25}$$

Eq. 24 is to be preferred over Eq. 23, as it emphasizes the need for a theory of dislocation generation rates. It is also valid for $v < c$. It may be anticipated that $\dot{\Lambda}$ should be a strong function of applied stress and internal stress concentrations with a weak, if any, dependence on temperature. With these relations one has for $b \sim 3 \times 10^{-8}$ and $c \sim 3 \times 10^5$ that $\dot{\varepsilon} = 10^{-2}\Lambda$. The representative value of $\dot{\varepsilon} = 10^5 \text{sec}^{-1}$ requires an instantaneous mobile dislocation density of 10^7cm^{-2}. For $\tau \sim 10^{-6}\text{sec}$, $d \sim 3\text{mm}$ and $\dot{\Lambda} \sim 10^{13} \text{cm}^{-2}\text{sec}^{-1}$ while for $\tau \sim 10^{-9}$ sec, $d \sim 3 \times 10^{-4}\text{cm}$ and $\dot{\Lambda} \sim 10^{16} \text{cm}^{-2}\text{sec}^{-1}$.

<u>Acknowledgment</u>: This work was supported by the National Science Foundation.

REFERENCES

1. G. Leibfried, Z. Phys. $\underline{127}$, 344 (1950).
2. J. D. Eshelby, Proc. R. Soc. $\underline{A197}$, 396 (1949). See also Proc. R. Soc. $\underline{A266}$, 222 (1962).
3. A. V. Granato, Phys. Rev. $\underline{111}$, 740 (1958).
4. F. R. N. Nabarro, Theory of Crystal Dislocations, (Oxford U. P., Oxford, England, 1967), p. 505.
5. F. R. N. Nabarro, Proc. R. Soc. $\underline{A209}$, 278 (1951).
6. J. A. Garber and A. V. Granato, J. Phys. Chem. Solids $\underline{31}$, 1863 (1970); in Fundamental Aspects of Dislocation Theory (Special publication 317) J. A. Simmons, R. deWit, and R. Bullough, eds. (National Bureau of Standards, U.S.A. 1970) p. 419.
7. T. Ninomiya, in Fundamental Aspects of Dislocation Theory (Special publication 317) J. A. Simmons, R. deWit and R. Bullough, eds. (National Bureau of Standards, U.S.A. 1970) p. 315.
8. P. G. Klemens, Proc. Phys. Soc. (London) $\underline{A68}$, 1113 (1955) See also Solid State Physics, ed. by F. Seitz and D. Turnbull (Academic Press, N. Y. 1958) Vol. 7, p. 1.
9. J. Lothe, Phys. Rev. $\underline{117}$, 704 (1960). See also J. App. Phys. $\underline{33}$, 2116 (1962).
10. J. M. Ziman, Electrons and Phonons, Clarenden Press, Oxford (1960) ch. vi, p. 4.
11. F. R. N. Nabarro and J. M. Ziman, Proc. Phys. Soc. Lond. $\underline{78}$, 1512 (1961).
12. P. Carruthers, Rev. Mod. Phys. $\underline{33}$, 92 (1961).
13. H. Bross, A. Seeger and P. Gruner, Annln. Phys. $\underline{11}$, 230 (1963). See also H. Bross, A. Seeger, and R. Haberkorn, Phys. Stat. Sol. $\underline{3}$, 1126 (1963).
14. P. Gruner and H. Bross, Phys. Rev. $\underline{172}$, 583 (1968).
15. K. Ohashi, J. Phys. Soc. Japan $\underline{24}$, 437 (1968).
16. A. Seeger and H. Engelke, in Dislocation Dynamics. ed. by R. Rosenfeld, G. T. Hahn, A. Bement, Jr., and R. I. Jaffe (McGraw Hill, N. Y. 1968), p. 623.
17. P. P. Gruner, in Fundamental Aspects of Dislocation Theory (Special Publication 317), J. A. Simmons, R. deWit and R. Bullough, eds. (National Bureau of Standards, U.S.A. 1970) p. 363.
18. A. D. Brailsford, J. Appl. Phys. $\underline{43}$, 1380 (1972).
19. A. Hikata, J. Deputat, and C. Elbaum, Phys. Rev. $\underline{B6}$, 4008 (1972).
20. A. C. Anderson and M. E. Malinowski, Phys. Rev. $\underline{5B}$, 3199 (1972).
21. W. P. Mason, J. Acoust. Soc. Am. $\underline{32}$, 458 (1960) J. Appl. Phys. $\underline{35}$, 2779 (1964).

22. A. B. Pippard, Phil. Mag. 46, 1104 (1955).
23. W. P. Mason, Phys. Rev. 97, 557 (1955); Appl. Phys. Letters 6, 111 (1965); Phys. Rev. 143, 229 (1966)
24. G. P. Huffman and N. Louat. Phys. Rev. 176, 773 (1968). See also Phys. Rev. Letters 19, 518 (1967) and Phys. Rev. Letters 19, 774(E) (1967).
25. B. R. Tittmann and H. E. Bommel, Phys. Rev. 151, 178 (1966).
26. C. Elbaum and A. Hikata, Phys. Rev. Letters 20, 264 (1968).
27. A. D. Brailsford, Phys. Rev. 186, 959 (1969).
28. T. Holstein, in appendix in ref. 25.
29. V. Ya Kravchenko, Fiz. Tverd. Tela 8, 927 (1966)[Sov. Phys. Solid State 8, 740 (1966)].
30. J. S. Koehler, in *Imperfections in Nearly Perfect Crystals* (Edited by W. Shockley, J. H. Holloman, R. Maurer and F. Seitz), p. 197, Wiley, N. Y. (1952).
31. A. V. Granato and K. Lucke, J. Appl. Phys. 27, 583 (1956).
32. R. M. Stern and A. V. Granato, Acta Met. 10, 358 (1962).
33. G. A. Alers and D. O. Thompson, J. Appl. Phys. 32, 283 (1961).
34. F. Fanti, J. Holder, and A. V. Granato, J. Acoust. Soc. Amer. 45, 1356 (1969).
35. E. Y. Gutmanas, E. M. Nadgornyi, and A. V. Stepanov, Sov. Phys. Solid State 5, 743 (1963).
36. G. S. Baker, J. Appl. Phys. 33, 1730 (1962).
37. R. L. Sproull, M. Moss and H. Weinstock, J. Appl. Phys. 30, 334 (1959).
38. R. O. Schwenker and A. V. Granato, Phys. Rev. Letters 23, 918 (1969); J. Phys. Chem. Solids, 31, 1869 (1970).
39. A. Hikata, R. A. Johnson and C. Elbaum, Phys. Rev. Letters 24, 215 (1970); Phys. Rev. B2, 4856 (1970).
40. A. Hikata and C. Elbaum, Phys. Rev. Letters 18, 750 (1967).
41. A. V. Granato, Phys. Rev. Letters 27, 660 (1971); Phys. Rev. B4, 2196 (1971).
42. M. Suenaga and J. M. Galligan, Scr. Met. 5, 63 (1971).
43. G. A. Alers, O. Buck, and B. R. Tittman, Phys. Rev. Lett. 23, 290 (1969).
44. T. S. Hutchison and A. T. Pawlowicz, Phys. Rev. Letters 25, 1272 (1970).
45. F. Seitz and T. A. Read, J. Appl. Phys. 12, 470 (1941).
46. E. Orowan, Proc. Phys. Soc. (London) 52, 8 (1940).
47. W. G. Johnston and J. J. Gilman, J. Appl. Phys. 30, 129 (1959).
48. R. B. Schwarz and J. W. Mitchell, Bull. Am. Phys. Soc. 17, 285 (1972).
49. W. S. deRosset and A. V. Granato, in *Fundamental Aspects of Dislocation Theory* (Special Publication 317), J. A. Simmons, R. deWit and R. Bullough, eds. (National Bureau of Standards, U. S. A. 1970), p. 1099.

Discussion by Frank A. McClintock
            Massachusetts Institute of Technology

   I think that strain-rates at high dislocation velocities are determined by dislocation generation rates, and not by dislocation velocities. Therefore, I don't think one needs to worry about abrupt increases in stress which would follow from assumptions about constant dislocation densities. I don't know of any data relevant to the questions you raise, but almost any kind of a dislocation nucleation theory should lead to a slowly varying (logarithmic) dependence of the strain rate on stress.

Discussion by T. Vreeland, Jr. and K. M. Jassby
            California Institute of Technology

   We wish to point out two areas of disagreement between experimental observations and the theoretical predictions of B. The first has to do with the magnitude of B at room temperature. Brailsford's calculation of energy dissipation through the process of strain field scattering gives $B \approx 4 \times 10^{-3}$ cgs for copper at room temperature. Both direct and indirect measurements give $B \approx 2 \times 10^{-4}$ cgs, a factor of 20 smaller. The second disagreement is with the temperature dependence of B near room temperature. You indicated that a linear B vs. T was found. Our direct measurements show a linear B vs. T below the Debye temperature which changes to a temperature-independent B above the Debye temperature in copper, aluminum and zinc. We note that the decrement measured in internal friction experiments is not a very sensitive measure of this temperature dependence near the Debye temperature, and therefore the direct measurements of B should give a better indication of B vs. T.

Author's Reply

An important contribution of Brailsford's theory is the unified way in which the thermoelastic damping, phonon viscosity, phonon scattering and reradiation damping are all treated within the same framework. This permits an assessment of the relative influence of the various effects which should be valid whatever estimate is made of the multiplying constants which give the absolute value of the total drag. For relatively detailed theories such as Brailsford's, it seems inevitable that there will arise difficulties in making accurate estimates of the absolute value of the drag. It is indeed true that Brailsford's estimate is about a factor of 20 larger than measured values and also larger than Leibfried's value by about the same amount. Brailsford tried to avoid some of these difficulties by expressing his result in terms of thermal resistivities arising from strain field scattering, thereby eliminating the parameters which are difficult to estimate. However, we now know that the measured thermal resistivities do not arise from strain field scattering, so that this method will lead to an overestimate of the drag. Hikata, Deputat and Elbaum used a different estimate of average sound velocity from that used by Brailsford to obtain an estimate of the drag from Brailsford's theory which is in much closer agreement to the measured values and to the Leibfried's value.

Before this meeting, the only measurements of the temperature dependence of B for $T > \theta$ that I knew of were those by V. B. Pariiskii and A. I. Tret'yak (Sov. Phys. Sol. St. $\underline{9}$ (1968) 1933) for etch pit measurements in KBr. They found that B increased at least as fast as linearly in the range of T/θ between about 1.5 and 2.5. Whatever problems a constant B for $T > \theta$ may make for theorists, the fact that B doesn't increase strongly at high temperatures means that viscous drag effects will not be of great importance at stresses and temperatures typical of shock loading conditions.

EXPERIMENTAL MEASUREMENT OF THE DRAG COEFFICIENT[*]

T. Vreeland, Jr. and K.M. Jassby

W.M. Keck Laboratory of Engineering Materials

California Institute of Technology, Pasadena, Calif.

The drag coefficient is related to the dissipative viscous force which acts on a dislocation in motion. The magnitude of the drag coefficient for a dislocation of known Burgers vector is determined by measurement of the viscous force at a known dislocation velocity, or by measurement of the energy dissipation brought about by the viscous force. We discuss here these measurements and explore the special conditions which make possible the determination of the drag coefficient.

DISSIPATIVE AND NON-DISSIPATIVE FORCES

When the dissipative viscous drag force predominates over all other forces which retard dislocation motion, a constant resolved shear stress, τ, will produce a terminal dislocation velocity, v, such that the driving force per unit dislocation length, τb, is equal to the viscous force per unit length, Bv, where B is the drag coefficient and b is the Burgers vector. Thus

$$B = \frac{\tau b}{v} . \qquad (1)$$

The most direct determination of B is the measurement of v for a known τ and b under conditions where eq. 1 applies. This approach is complicated by the existence of non-dissipative forces which also act on the moving dislocation. Interactions between the moving dislocation and nearby surfaces, other dislocations, and point defects all give rise to non-dissipative forces, as do Peierls forces, inertial forces and curvature forces. These forces will usually vary as the dislocation moves through the

[*]This work was supported by the U.S. Atomic Energy Commission and the California Institute of Technology.

crystal, and the variation may have a characteristic wave length, λ. It is then useful to consider the mean dislocation velocity, $\bar{v}$, over a distance large compared to λ. In many cases the non-dissipative forces will have the effect of reducing the driving force on the dislocation. Then $\bar{v}$ will be less than the terminal velocity, v, that would be attained for the same applied resolved stress in the absence of non-dissipative forces. In one important case, i.e., the growth of slip bands, the forces on the leading dislocation due to those following may make $\bar{v}$ larger than v.

A linear, or viscous, relationship between $\bar{v}$ and the applied resolved stress may result from the influence of drag mechanisms which are non-dissipative. For example, $\bar{v}$ vs. τ measurements in the diamond cubic lattice[1] have shown a linear relationship to be followed. Thermal activation over the Peierls barrier, a non-dissipative process, is thought to be responsible for determining the linear $\bar{v}$ vs. τ relationship. Viscous drag forces also act in this case, but they do not directly influence the $\bar{v}$ vs. τ relationship and the drag coefficient cannot be obtained from this data. The direct determination of B using eq. 1 requires that the non-dissipative forces be small compared to the dissipative force (or that they be accurately known and taken into account). The non-dissipative forces do not similarly complicate the determination of B from measurements of energy dissipation (i.e. the internal friction method).

THE DIRECT METHOD

A calculation of the stresses and velocities involved in direct measurements of B is useful to show the importance of accounting for non-dissipative forces. Drag coefficients of the order of 10^{-4} cgs are typical. Taking $b = 2.5 \times 10^{-8}$ cm, eq. 1 gives $v = 250$ cm/sec/bar. We see immediately that we must be prepared to either measure high dislocation velocities or conduct the experiments at very low stresses. The flow stress of the crystal is a fair measure of the effective non-dissipative stresses, and as discussed above, the applied stress must be greater than these stresses for the direct method to apply. For this reason the applied resolved shear stress in the direct method will usually be greater than one bar. High dislocation velocities are therefore necessary, and continuous observation of high velocity dislocations is not currently possible. The stress pulse technique, in which mean velocities are deduced from dislocation displacement observations is the most direct method available.

Stress pulse techniques have been reviewed elsewhere.[2] The stress pulse must be of short enough duration to stop the dislocations before they move out of the crystal. Longitudinal and torsional stress pulses in rods have been used for measurement of the drag coefficient. As the high velocity dislocations are not continuously observed, the entire stress-time history between dislocation observations must be known. Careful attention must therefore be given to possible wave reflections which could cause multiple stress pulses in the specimen.

An example of the use of a torsional stress pulse and Berg-Barrett topography for dislocation displacement measurements is shown in fig. 1. The topograph of an (0001) surface of zinc in fig. 1a was taken after scratching the surface to introduce basal edge dislocations. Kodak-type R X-ray film was used with an exposure time of 12 min. A higher resolution film or plate would require a longer exposure and therefore a longer time between the introduction of the dislocations and the application of a stress pulse. This time is held to a minimum in the experiments to minimize climb and impurity pinning of the fresh dislocations. Figure 1b shows a topograph taken after a torsional stress pulse was applied to the (0001) end surface of the cylindrical crystal. This topograph was made using a Kodak high-resolution plate and required a 9 hour exposure. The stress pulse caused the dislocation displacements which are seen to vary linearly with radius. The applied resolved shear stress amplitude varied linearly with radius, and a knowledge of the torsional pulse amplitude and the duration permits us to relate the displacements of fig. 1b to dislocation velocity and the radial position to stress amplitude (maximum values in fig. 1b were 6.6×10^3 cm/sec and 25 bar). The leading dislocation line of fig. 1b then represents the velocity vs. stress curve. The slope of this curve, together with a knowledge of b (whose direction is confirmed by topographs using $\{10\bar{1}3\}$ reflections for which $\bar{g} \cdot \bar{b} = 0$) gives B using eq. 1. A linear velocity-stress relationship is observed in fig. 1b at stress levels as low as one bar, which indicates that non-dissipative forces were either very small or were linearly dependent on $\bar{v}$. The temperature dependence of B determined in these experiments[3] is the opposite of that expected when the $\bar{v}$ vs. τ curve is controlled by thermally activated processes. This observation, and the agreement between the B values determined in the direct experiments and in the internal friction experiments leads us to believe that we have succeeded in making the non-dissipative forces negligible compared to the Bv forces.

When a crystal is hardened by discrete obstacles, the drag coefficient may be found from $\bar{v}$ vs. τ measurements at stresses larger than about twice the critical resolved stress. Figure 2 shows an example of this in zinc where the discrete obstacles were forest dislocations introduced by second-order pyramidal slip.

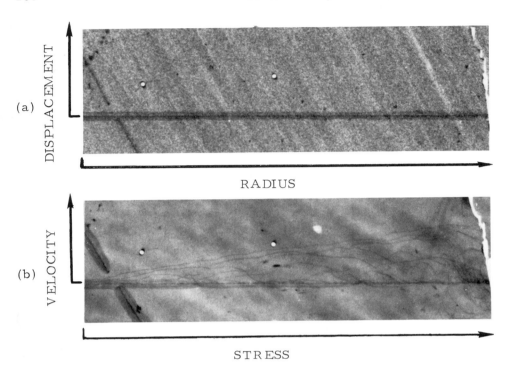

Fig. 1. Berg-Barrett topographs of an (0001) end surface of a zinc cylinder, (10$\bar{1}$3) reflection, CoKα, 40 Kv, 7 ma, 26.5 X. Center of cylinder is at the left edge of the topograph. a) After scratching, b) After application of a torsional stress pulse at 66°K.

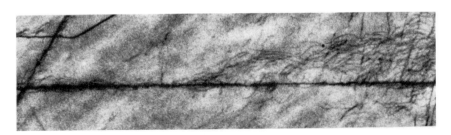

Fig. 2. Berg-Barrett topograph of an (0001) surface of zinc containing a forest dislocation density of about $7 \times 10^4/cm^2$, after scratching and applying a torsional stress pulse, (10$\bar{1}$3) reflection, 40X.

The critical stress to move basal dislocations through the forest with a dislocation density of $7 \times 10^4/cm^2$ is about 3 bar, and at stresses of about 6 bar the $\bar{v} - \tau$ relationship becomes linear (extrapolating to the origin) with the same slope found in crystals with a much lower forest density (and a critical stress less than 1 bar). This behavior, with a transition from obstacle controlled velocity to viscous drag controlled velocity with increasing stress was predicted by Frost and Ashby.[5] A similar transition has been observed in the $\bar{v} - \tau$ relations for screw dislocations on the second-order pyramidal slip planes of zinc.[6] At resolved stress levels below about 20 bar, Lavrentev et. al. found the $\bar{v}$ vs. τ curve to be non-linear, and the stress at the transition is likely to be Peierls stress. We have recently confirmed this transition for both screw and edge oriented dislocations on the second-order pyramidal system of zinc using longitudinal stress waves which produce a single short duration stress pulse in the crystal. Dislocations are observed after the pulse is applied using Berg-Barrett X-ray topography. Individual dislocations in a slip band are resolved in the topograph shown in fig. 3. The dislocation interaction forces may be calculated knowing the dislocation positions, so that an estimate of their importance compared to the viscous force may be made.

Attempts have been made to obtain the drag coefficient from measurement of the strain rate vs. stress relation at strain rates above about $10^2/sec$. The product of mobile dislocation density and the average dislocation velocity may be related to the strain rate. Hence, a knowledge of the mobile dislocation density is needed to determine the average dislocation velocity. Only an estimate of the mobile dislocation density can be made, since it cannot be measured under the test conditions. Therefore, reliable estimates of the dislocation velocities in these tests cannot be made. Nagata and Yoshida[7] estimated that the mobile dislocation density at plastic strain rates between $7 \times 10^2/sec$ and $2 \times 10^3/sec$ was equal to the initial density determined from etch pit counts ($\rho = 5 \times 10^6/cm^2$), and calculated $B = 2.4 \times 10^{-4}$ from the slope of the linear strain rate vs. τ relationship they found at small plastic strains in copper at room temperature. We have determined a value of $B = 2.0 \times 10^{-4}$ for copper from torsional stress pulse tests at room temperature.

INTERNAL FRICTION MEASUREMENTS

Dislocations are assumed to be strongly pinned by a network of discrete obstacles (usually dislocation nodes) and more weakly pinned at intermediate locations by point defects in the Granato-Lücke theory[8] which describes dislocation-induced energy loss mechanisms in internal friction experiments. Each dislocation,

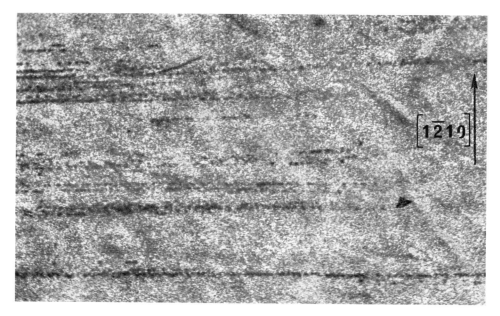

Fig. 3. Berg-Barrett topograph of an (0001) surface of zinc after application of a 7 μsec duration compression stress pulse along [1$\bar{2}$10], (10$\bar{1}$3) reflection, 135X.

excited by an externally applied oscillating stress field, vibrates between its pinning points. Two distinct cases have been considered: (i) during each cycle of stress, the dislocation breaks away from the weaker pinning points, but remains pinned at its network lengths, and (ii) the dislocation remains pinned at each discrete obstacle throughout the stress oscillation. In the former case, called amplitude-dependent internal friction, (see also Ref. 9) dislocation damping is derived from both hysteretic effects attributed to dislocation breakaway from intermediate pinning points and phase lag effects induced by linear or viscous damping. The energy loss (decrement) is a function of the amplitude of the exciting stress. In the latter or amplitude independent case, (see also Ref. 10) only phase lag losses are considered, and the decrement is a function of the oscillating frequency of the applied stress.

In both experimental situations, dislocation-induced losses can be separated from those losses attributed to direct interaction between the exciting stress wave and the crystal by carrying out the experimental measurements both before and after neutron irradiation of the crystal.[11] The strong pinning of dislocations by irradiation induced defects should make the dislocation induced losses negligible compared to the other losses. Highly perfect crystals are not employed for these experiments because they do not contain a sufficient density of mobile dislocation line to induce a measurable decrement. Both cases demand low resolved shear stresses and consequently small dislocation velocities (of the order of 10cm/sec.

In the most simple dislocation model, edge and screw dislocations are not treated separately, but rather "averaged" values are implied for the various parameters which enter the theory. Granato and Stern[12] have shown that when dislocations of more than one orientation contribute to energy absorption, a much broader peak in the curve for decrement as a function of exciting frequency results than in the situation where only one type of dislocation is active. In the former case, accurate analysis of experimental measurements precludes treatment on the basis of the simpler model. In one internal friction experiment in copper,[12] where a broad peak was observed in the decrement vs. frequency curve, the measurements were analyzed by considering contributions from both edge and screw oriented dislocations. In this case the magnitude of B deduced was in good agreement with that derived from direct measurements.[13]

The energy loss from a network of vibrating dislocations is normally characterized by two geometric parameters, the average dislocation loop length and the total length of mobile dislocation line, values of which must be estimated for each particular crystal in order to interpret the decrement measurements and hence evaluate B. Under certain experimental conditions the decrement is independent of the average loop length.[11] Two experimental techniques have recently been developed which enable one to determine B independently of knowledge of these two paremeters. The first technique[14] applies to the measurement of B below the superconducting transition temperature in superconducting metals.*

---

*Hikata and Elbaum, discussing this method, assumed that conduction electrons provide the only significant source of dislocation damping at temperatures near absolute zero and hence the value of B in the superconducting state is negligible compared to that in the normal state. Recent experimental work carried out by the present authors has indicated the presence of a residual phonon damping in copper at $4.2°K$, contrary to this assumption.

The second technique,[15] which employs a bias stress, is applicable to all materials over an extended temperature range. Both methods provide more promising solutions to the problem of measuring B at low dislocation velocities, by removing the uncertainties inherent in estimation of the average dislocation loop length and the total length of mobile dislocation line. However, B cannot be deduced independently of the dislocation effective mass with these techniques. This latter quantity must be estimated from theoretical considerations and the value of B will depend on the accuracy of this estimation.

REFERENCES

1. S. Schafer, Phys. Status Solidi, 19, 297 (1967).
2. T. Vreeland, Jr., Techniques of Metals Research, Volume II, Part 1, R.F. Bunshah, ed., Interscience Publishers, 1968, p. 341.
3. T. Vreeland, Jr., and K.M. Jassby, Mat. Sci. Eng. 7, 95 (1971).
4. N. Nagata and T. Vreeland, Jr., Phil. Mag. 25, 1137 (1972).
5. H.J. Frost and M.F. Ashby, J. Appl. Phys. 42, 5273 (1971).
6. F.F. Lavrentev, O.P. Salita, and V.L. Vladimirova, Phys. Stat. Sol. 29, 569 (1968).
7. N. Nagata and S. Yoshida, J. Met. Soc. Japan 32, 385 (1968).
8. A. Granato and K. Lücke, J. Appl. Phys. 27, 583 (1956).
9. D.H. Rogers, J. Appl. Phys. 33, 781 (1962).
10. A. Granato, in Dislocation Dynamics, McGraw-Hill, New York, 1968, p. 117.
11. G.A. Alers and K. Salama, in Dislocation Dynamics, McGraw-Hill, New York, 1968, p. 211.
12. R.M. Stern and A.V. Granato, Acta Met. 10, 358 (1962).
13. K.M. Jassby and T. Vreeland, Jr., Phil. Mag. 21, 1147 (1970).
14. A. Hikata and C. Elbaum, Trans. Jap. Inst. Met. Suppl. 9, 46 (1968).
15. A. Hikata, R.A. Johnson, and C. Elbaum, Phys. Rev. 2, 4856 (1970).

Discussion by Frank A. McClintock
        Massachusetts Institute of Technology

   The assumption of constant mobile dislocation densities, along with a phonon drag and a limiting dislocation velocity, leads to the strain rate dependence illustrated in Fig. 2 of the discussion of the paper by McClintock. Can you suggest data or theory giving a more plausible relation to be used in calculating the dynamic growth of holes in high strength materials at strain rates of the order of 10^6/sec to 10^9/sec?

Authors' Reply

   Earlier in this conference, Dr. O. E. Jones and Prof. P. P. Gillis discussed experiments and theories relating dislocation dynamics to shock wave mechanics, and Prof. A. V. Granato emphasized the importance of dislocation multiplication at high strain rates. I believe the conclusions are that either supersonic dislocation velocities are occurring (see paper by J. Weertman), or very significant dislocation multiplication takes place in the shock wave experiments. Unfortunately there are no direct measurements of dislocation velocity at the high stress levels of interest to your calculations nor are there any measurements of the mobile dislocation density.

THE ROLE OF DISLOCATION DRAG IN SHOCK WAVES

PETER P. GILLIS

University of Kentucky

JAMES M. KELLY

University of California

INTRODUCTION

Various theories of dislocation mobility have been developed in recent years having the object of leading to an understanding of the physics of dislocation damping in metallic crystals. Models have been based on thermal activation mechanisms[1,2] on dislocation-thermal phonon interactions[3,4] (these using either phonon viscosity or phonon scattering), on dislocation-conduction electron interaction,[4,5] on overcoming the Peierls barrier[6,7] or on empirical relationships between dislocation velocity and stress[8,9]. To a certain extent differences between these models can be summarized as to whether or not they predict a limiting dislocation velocity. The dislocation-thermal phonon interaction based on phonon viscosity, for example, predicts a linear relation between dislocation velocity and applied shear and does not include a limiting dislocation velocity. It is well known that on the basis of linear elastic theory, the energy of a dislocation line becomes infinite as its velocity approaches the shear wave velocity of the material which would suggest that the dislocation velocity could not exceed this value. However the use of linear elasticity in association with the very high stress fields predicted at speeds close to the shear wave speed is highly questionable and many workers have taken the attitude that if driving forces are available the dislocations will respond by moving at appropriate velocities which could include supersonic velocities.

The resolution of these basic differences could be met by experiments involving direct measurement of the stress dependence of dislocation velocity at velocities close to the shear wave speed. This requires the production of correspondingly high driving forces. Techniques to measure dislocation velocities directly are available only at low velocities; typically, near the low rates of the original Johnston-Gilman experiments on LiF[9]. Further developments have been made including a torsional stress wave bar by Pope et al[10] which allows a considerable increase in strain rate but this device is limited in the maximum stress it can produce and is not adequate for polycrystalline metals. The torsion bar has, however, considerable advantages over other techniques as, for example, the Hopkinson bar in that there are no dispersion effects and that the measured stress and strain are in the form of true stress and true strain. This advantage is shared by the plane shock wave test and in this test the stresses and strain rates which can be generated are virtually unlimited by comparison.

The wave referred to in the literature as the plastic shock is the second component in the double wave structure which exists in most metals for a certain range of impact pressures. The first, wave referred to as the elastic precursor has the essentially constant velocity of sound waves. The second wave has a velocity which increases with the pressure level. The upper limit of the range of the double wave structure is the pressure at which the second wave speed equals that of the elastic precursor, providing no phase transformation effects intervene.

In the earliest plane wave experiments two parameters that could be determined were the Hugoniot elastic limit (or stress level) associated with the elastic precursor wave and the dynamic compressibility (or bulk modulus) associated with the following plastic wave (or shock wave in the terminology used here). A comprehensive review of relevant theoretical and experimental work prior to 1957 is given by Rice, McQueen and Walsh[11], and subsequent research along these lines has been reviewed by Duvall[12] and more recently by McQueen et al.[13].

More complex theories of material behavior have led to the extraction of much additional information from such experiments. The analysis of precursor-amplitude decay with specimen thickness has been suggested by several investigators[14-16] as a fruitful technique for the evaluation of proposed constitutive equations.

A significant development in this area is due to Rhode[17] who seems to have made the first systematic study of precursor amplitude dependence upon temperature. Rhode attempted to reconcile several postulated material equations with his experimental results.

There are, however, considerable difficulties associated with the measurement of precursor amplitudes. It has been shown[18] that dislocation multiplication in the region behind the wave front has the effect of producing an immediate drop in the stress behind the wave. As a consequence, the precursor wave profile is extremely narrow and the stress maximum that is recorded depends very strongly on the resolution of the sensor. Furthermore, since dislocation multiplication can vary considerably from grain to grain it is found that the reproducibility of these measurements is poor.[16]

Thus, Kelly and Gillis[19] proposed an alternative approach to the evaluation of the drag parameters that could be obtained from the study of shock thickness. A system of equations governing the kinematics of propagation of one-dimensional strain waves of finite deformation in an elastic, viscoplastic material was developed and a simple dynamical condition was obtained by assuming steady-state propagation of the wave. Under this assumption a method was presented which allows the principal features of the wave to be calculated with great accuracy by elementary numerical techniques.

The use of plane wave methods, however, involves some substantial disadvantages from the analytical point of view. These are associated with the non-linear compressibility of the material and the heat produced by both the high levels of pressure and the plastic deformation. The material response must be characterized as both elastic and plastic since elastic deformations are of the same order as plastic. The dilatational elastic deformations are very sensitive to non-linear effects due to lattice compaction; the plastic deformations, to volume-preserving dislocation motions; and there is much uncertainty concerning deviatoric elastic deformations.

The initial wave generated by impact on a plane face of a target splits generally into two waves, an elastic precursor and a plastic shock. Although the precursor is described as elastic it is nevertheless strongly affected by plastic deformations particularly by the plastic strain rate immediately behind the wave, which has the effect of reducing its magnitude with propagation distance. The plastic shock is strongly influenced by elastic deformations within its structure. As it travels, the plastic shock develops, more or less rapidly, a steady shape that persists until interactions with unloading waves from free surfaces alter it. This steady wave which develops is a balance between dissipative effects that tend to smooth the wave and convective effects due to the non-linearity in the elastic response that tend to sharpen the wave. In effect, the more non-linear the elastic stress, strain relation becomes at high stresses, the more tendency there is for higher stresses to overrun lower stresses in front. When dissipative influences are small, a true shock develops.

Thus, dislocation drag effects have the following influences. They cause the elastic precursor to decay, they retard the formation of a steady shock and they increase its thickness. From the experimental point of view the last two effects have both advantages and disadvantages. The major disadvantage is that at high shock pressures non-linear elastic effects dominate the drag effects in which we are interested. On the other hand is the compensating advantage that a steady-state wave is rapidly formed at high pressure and can be maintained for some time. The mathematical problem associated with identifying drag parameters from experimental data is considerably less severe when a steady-state wave can be assumed that when a completely transient situation has to be modelled.

KINEMATIC AND CONSTITUTIVE ASSUMPIONS

Symmetry arguments apply on a scale that is large compared to the average grain size of a polycrystalline specimen. These require that, on this scale, the particle displacements in the specimen material be in the direction of propagation of the shock, for the usual plane-wave experiment. The situation is then described as a state of uniaxial strain. For large compressions uniaxial strain produces extremely large shearing strains, as indicated schematically in Fig. 1(a). The role of dislocations in the shock are to relieve these shearing strains and to transform the crystal structure to a state more closely approximating uniform hydrostatic compression, as in Fig. 1(b). The role of dislocation drag is to determine how fast the dislocations can move to relieve the shearing strain and, therefore, what thickness the shock will have and what its structure will be.

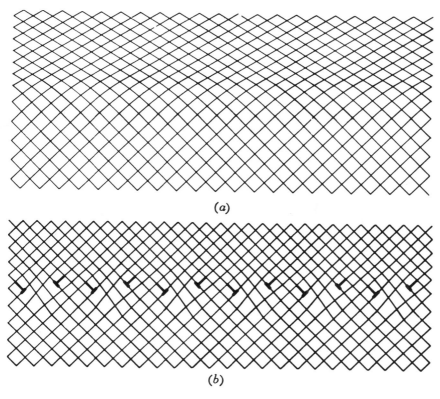

Figure 1. (a) Direct uniaxial compression of the schematically represented lattice produces large shear strains. (b) Dislocation motions can substantially relieve the shear strains thus producing a hydrostatically compressed lattice. (Taken from C.S. Smith[20]).

The more usual description of this situation is in terms of stresses. As the stress begins to increase with the arrival of the shock the initial response of the material tends to be elastic. Denoting the shock stress by σ and the transverse stress by σ_t these are initially related by $\sigma_t = \sigma \nu/(1-\nu)$ where ν is Poissons ratio. By symmetry σ and σ_t are principal stresses and thus the initial maximum shear stress is $\tau = (\sigma - \sigma_t)/2 = (\sigma/2)(1-2\nu)/(1-\nu)$, the ordinary value for uniaxial elastic strain. However, this shear stress tends to produce dislocation motions. These in turn produce plastic strains that relieve the driving shear stress. The final state towards which the deformation proceeds in one of hydrostatic pressure* in which $\sigma_t = \sigma$. In this process the stress σ builds up to a rather large value, the shock stress, while the shear stress τ, after an initial buildup, relaxes toward zero.

Returning to the kinematic variables, denote the shock strain by ε and the transverse strain by ε_t. The latter is, of course, zero according to the symmetry arguments. These strains are decomposed into elastic and plastic portions according to several different schemes[21] but it is usual to assume that no change of volume is associated with the plastic deformation. Using superscripts e and p for elastic and plastic respectively the two conditions above can be written as: $\varepsilon_t^e + \varepsilon_t^p = 0$ and $\varepsilon^p + 2\varepsilon_t^p = 0$. Thus, the elastic strains can be expressed in terms of the two variables ε and ε^p: $\varepsilon^e = \varepsilon - \varepsilon^p$ and $\varepsilon_t^e = \varepsilon^p/2$.

It is usual to assume that the stresses are related to the elastic strain through Hookes law although the stiffness constants can be taken as functions of the pressure $p = -(\sigma + 2\sigma_t)/3$. In practice only the variation of bulk modulus with pressure is known with any certainty so the more useful form for Hookes law is: $\sigma = (K + 4G/3)\varepsilon - 2G\varepsilon^p$, $\sigma_t = (K - 2G/3)\varepsilon - G\varepsilon^p$; thus $\tau = 2G(\varepsilon - 3\varepsilon^p/2)$ and $p = -K\varepsilon$. Here K and G denote the bulk and shear moduli respectively. Two significant aspects of these relations are that the final plastic strain in the fully relaxed material behind the shock is $\varepsilon^p = 2\varepsilon/3$ and that the shock thickness is governed by how fast this plastic strain can accumulate. The first statement implies that the elastic strains are not negligible in comparison to the plastic strains as is frequently assumed in plastic deformation analyses of many other situations. The second statement implies that for crystalline materials dislocation dynamics govern the shock structure.

It is observed experimentally that shock thickness decreases with increasing shock intensity. In terms of the foregoing discussion, for large enough shock stresses the maximum shear stress quickly becomes large enough to provide an enormous dislocation

---

*In practice this state is finally approached at an ever-decreasing rate so that the final state is usually taken as some shear stress small enough to produce no further significant dislocation flux. For the purpose of discussion, however, it is simpler and qualitatively correct to consider hydrostatic compression as the final state.

flux that quickly relaxes the material to its hydrostatically compressed state. High intensity shocks have thus been successfully analysed on the basis of rate insensitive constitutive equations. However, it is usually necessary in these analyses to employ what is asserted to be merely a computational device, euphemistically called artifical viscosity As lower shock intensities are considered, inviscid relationships become progressively less satisfactory. The analysis of the elastic precursor wave, in particular, requires a rate sensitive constitutive equation.

This brings up the main difficulty in trying to describe the role of dislocation drag in shock waves. It seems as if no specific consideration of dislocations needs to be included in an adequate analysis of a shock. <u>Any</u> rate sensitive relation will suffice. There will always be a sufficient number of adjustable parameters to enable calculation of wave profiles that correspond to experimental measurements. Plastic strain rate relations based on dislocation concepts are no exception. Usually only an average Burgers vector can be estimated a priori with any degree of accuracy and the several other parameters involved are adjusted to fit the experimental results. Thus, in the development of quantitative relationships the title of this paper must be reversed and we can attempt to describe the role of shock wave experiments in assessing dislocation drag.

As previously noted, dislocation behavior is related to the propagation of shock waves in two ways: decay of the elastic precursor wave and the exact shape of the plastic wave profile. The former can be more easily formulated in concise mathematical form and is treated first.

PRECURSOR ATTENUATION

A complete derivation of the theory of precursor decay can be found elsewhere[14,16]. For the purpose of the present discussion we merely note that the elastic precursor wavefront is considered to propagate as a jump discontinuity in stress, denoted by $[\sigma]$, and associated kinematic variables including the plastic strain rate, $[\dot{\epsilon}^P]$. Here the superposed dot is used to denote material time derivative. Because the precursor stress level is fairly low away from the impacted face of the specimen, non-linear effects in the bulk compressibility are neglected. In the undisturbed material ahead of the precursor both stress and strain rate are zero. Thus, the magnitudes of the stress and the strain rate just at the precursor equal the respective jumps. According to the linearized theory these are related by:

$$d[\sigma]/dx = -(G/c)[\dot{\epsilon}^P] \qquad [1]$$

An initial stress of magnitude σ_0 applied to the impacted surface of the specimen decays with propagation distance x as prescribed by Eq. [1]. Here c is the elastic wave speed obtained from $c^2 = (K + 4G/3)/\rho_0$ where ρ_0 is the initial density of the material.

SHOCK PROFILE

The development of the relationships which govern the structure of the steady-state shock is somewhat more involved than that for the precursor decay. In contrast to the precursor, analysis, it is essential here to consider the elastic response of the material to be non-linear. A relatively simple non-linear theory has been developed in which the non-linearity has been achieved through the use of equations linear in the logarthmic strains and their conjugate stresses but non-linear in Cauchy stress and Eulerian strain. This theory leads to relatively simple equations for the steady-state shock structure. Then any specific form for the dislocation drag can be used to compute a profile which can be compared to the measured form. It is also possible to take into account the influence of temperature rise and this has been done in a recent paper.[22] In the interests of brevity the derivation that includes temperature effects has been omitted here.

The theory makes use of the deformation analysis developed by Lee[23]. In this theory, the elastic and plastic strains are defined through introduction of an intermediate unstressed configuration that is obtained from the elastic, plastic configuration by a process of purely elastic unloading. Thus, the deformation gradient of the elastic, plastic configuration with respect to the initial configuration is given by matrix multiplication of the plastic deformation gradient and the elastic deformation gradient.

In the plane wave problem, the principal axes are known and constant and it is convenient to develop the equations in terms of principal values of the deformation gradients. We denote the

THE ROLE OF DISLOCATION DRAG IN SHOCK WAVES 295

principal stretch ratios by λ_i, λ_i^e, λ_i^p, $i = 1, 2, 3$ and we have:

$$\lambda_i = \lambda_i^e \lambda_i^p, \quad (i \text{ not summed}) \qquad [2]$$

It is convenient to introduce logarithmic strains $\epsilon_i = \ln \lambda_i$, $\epsilon_i^e = \ln \lambda_i^e$, $\epsilon_i^p = \ln \lambda_i^p$ for which:

$$\epsilon_i = \epsilon_i^e + \epsilon_i^p \qquad [3]$$

In terms of these, the rate of work per unit mass $\dot{W}$ is given by:

$$\rho_0 \dot{W} = \sigma_1 \lambda_2 \lambda_3 \dot{\lambda}_1 + \sigma_2 \lambda_3 \lambda_1 \dot{\lambda}_2 + \sigma_3 \lambda_1 \lambda_2 \dot{\lambda}_3 \qquad [4]$$

Here ρ_0 is the mass density of the material in the undeformed configuration and σ_1, σ_2, σ_3 denote the principal stresses with respect to the deformed configuration. The density ρ of the elastic, plastic configuration is related to ρ_0 through:

$$\rho_0 = \lambda_1 \lambda_2 \lambda_3 \rho \qquad [5]$$

In view of Eqs. [3] and [5], Eq. [4] takes the form:

$$\rho_0 \dot{W} = (\rho_0/\rho)(\sigma_i \dot{\epsilon}_i^e + \sigma_i \dot{\epsilon}_i^p) \quad (i \text{ summed}) \qquad [6]$$

from which we obtain the plastic work $\dot{W}^p$ and the recoverable work $\dot{W}^e$ in the form:

$$\rho_0 \dot{W} = \rho_0 \dot{W}^e + \rho_0 \dot{W}^p; \quad \dot{W}^e = \sigma_i \dot{\epsilon}_i^e/\rho, \quad \dot{W}^p = \sigma_i \dot{\epsilon}_i^p/\rho \qquad [7]$$

Suitable stress-strain relations for the elastic response are immediately suggested by analogy with the usual derivation of infinitesimal stress, strain relations. We assume the existence of an elastic strain energy u per unit mass having the form:

$$\rho_0 u = \tfrac{1}{2} K e^2 + G[(e_1^e)^2 + (e_2^e)^2 + (e_3^e)^2] \qquad [8]$$

where $e = \epsilon_1 + \epsilon_2 + \epsilon_3$ is the dilatation and $e_i^e = \epsilon_i^e - e/3$ are the principal deviatoric elastic strains. We assume that there is no plastic change in volume so that $\epsilon_1^e + \epsilon_2^e + \epsilon_3^e = \epsilon_1 + \epsilon_2 + \epsilon_3 = e$. Since $\rho_0 \dot{W}^e = \rho_0 \sigma_i \dot{\epsilon}_i^e/\rho$ we have $\rho_0 \sigma_i/\rho = \partial(\rho_0 u)/\partial \epsilon_i^e$ from which we obtain:

$$\rho_0 \sigma_i/\rho = Ke + 2G(\epsilon_i^e - e/3) \qquad [9]$$

The three Eqs. [9] can be written in the more convenient form:

$$\sigma_i + p = 2G (\rho_0/\rho)(\varepsilon_i^e - e/3) \qquad [10]$$

and:

$$p = -K(\rho/\rho_0) \ln(\rho_0/\rho) \qquad [11]$$

where $p = -(\sigma_1 + \sigma_2 + \sigma_3)/3$.

In plane shock waves the strain tensor reduces to a single nonzero component which may be taken to be ε_1 and the stress state can be described by the pressure p and the first component of the stress deviator $\sigma_1 + p$, the transverse stresses $\sigma_2 = \sigma_3$ being obtained if necessary from $\sigma_1 + 2\sigma_2 = -3p$. For convenience, we therefore drop the subscript 1 so that hereafter σ and ε refer to σ_1 and ε_1.

Equation [5], conservation of mass, reduces to:

$$\varepsilon = \ln(\rho_0/\rho) \qquad [12]$$

and the plastic strain $\varepsilon^p = \varepsilon - \varepsilon^e$ is then easily obtained in the form:

$$\varepsilon^p = (\tfrac{2}{3}) \ln(\rho_0/\rho) - (\rho_0/\rho)(\sigma + p)/2G \qquad [13]$$

Generally, the plastic strain cannot be related to the state of stress in a unique manner through a material constitutive equation. However, in the present case, the specific kinematical conditions associated with waves of one-dimensional strain do relate the plastic strain directly to the stress state. It is this unusual feature of the governing equations which is the basis of the present analysis.

For the one-dimensional steady-state wave, conservation of momentum, takes the particularly simple form:

$$\sigma\Big|_{x_2} - \sigma\Big|_{x_1} = \rho_0 V_0^2 [(\rho_0/\rho_{x_2}) - (\rho_0/\rho_{x_1})] \qquad [14]$$

where V_0 is the speed of the wave with respect to the undeformed material points and x_2, x_1 are the Lagrangian coordinates of any two points. If x_1 is taken at a distance sufficiently far in front of the plastic wave, it may be identified with the Hugoniot elastic limit if the function defining $\dot{\varepsilon}^p$ includes a static yield stress.

The response of a material having a static yield stress to the initial loading by the wave is wholly elastic up to the stress level σ_h referred to as the Hugoniot elastic limit. If we denote the corresponding density by ρ_h, then from Eq. [13] with ϵ^p set equal to zero we get:

$$(\rho_0/\rho_h)(\sigma_h + p_h) = (4G/3)\ln(\rho_0/\rho_h) \qquad [15]$$

The quasi-static yield is given by:

$$\ln(\rho_0/\rho_h) = \pm Y/2G \qquad [16]$$

where Y is the yield stress in simple tension. Furthermore, from the foregoing Eq. [11] and [15]: $(\rho_0/\rho_h)\sigma_h = (K + 4G/3)\ln(\rho_0/\rho_h)$ which with equation [16] gives:

$$(\rho_0/\rho_h)\sigma_h = \pm (K/2G + \tfrac{2}{3})Y \qquad [17]$$

Thus Eqs. [16] and [17] define the necessary Hugoniot elastic wavefront quantities in terms of Y.

On the other hand, if $\dot{\epsilon}^p \neq 0$ for all $(\rho_0/\rho)(\sigma + p) \neq 0$, then $\sigma|_{x_1}$ may be taken to be zero. The two forms of the momentum equation are:

$$\sigma - \sigma_h = \rho_0 V_0^2 [(\rho_0/\rho) - (\rho_0/\rho_h)] \qquad [18a]$$

or:

$$\sigma = \rho_0 V_0^2 [(\rho_0/\rho) - 1] \qquad [18b]$$

The corresponding deviatoric stresses are:

$$(\sigma + p) = \sigma_h + \rho_0 V_0^2 [(\rho_0/\rho) - (\rho_0/\rho_h)]$$
$$- (\rho/\rho_0) K \ln(\rho_0/\rho) \qquad [19a]$$

or:

$$(\sigma + p) = \rho_0 V_0^2 [(\rho_0/\rho) - 1] - (\rho/\rho_0) K \ln(\rho_0/\rho) \qquad [19b]$$

We note that the value of $\rho_0 V_0^2$ which appears in the foregoing equations must be at least equal to K for a plastic wave to exist and for a smooth plastic wave must be no greater than $(K + 4G/3)$. Above this value, the plastic wave overtakes the precursor elastic wave whose velocity is given by $[(K + 4G/3)/\rho_0]^{\frac{1}{2}}$ and a discontinuous plastic wave is formed.

We thus have three equations for the four unknowns σ, p, ρ_0/ρ and ϵ^p. The fourth relation needed to complete the set of equations is the relationship between $\dot{\epsilon}^p$ and the drag parameters. Since the equation for the plastic work $\dot{W}^p$ can, under the assumption of isochoric plastic strain and uniaxial total strain, be expressed as:

$$\rho_0 \dot{W}^p = (3/2)(\rho_0/\rho)(\sigma + p)\dot{\epsilon}^p \quad [20]$$

it is clear that in order to ensure that $\dot{W}^p$ be independent of elastic behavior the dependence of $\dot{\epsilon}^p$ on stress should be though the quantity $(\rho_0/\rho)(\sigma + p)$ the deviatoric or shearing stress.

Various common forms for the dependence of $\dot{\epsilon}^p$ on stress and on dislocation drag parameters have been investigated and reported elsewhere[19,22]. As previously noted, there are usually enough disposable parameters associated with such relations to be able to obtain an adequate fit to any limited sample of experimental data. In the section that follows we attempt to rationalize the existence of apparently contradictory forms.

DISLOCATION DRAG PARAMETERS

The role of dislocation drag is introduced through the specification of a functional dependence of the plastic strain rate on the other variables and some set of parameters. Broadly, such specifications fall into two classes[24] having the general forms:

$$\dot{\epsilon}^p = \nu \exp\{-(Q - V\tau)/kT\} \quad [21]$$

and:

$$\dot{\epsilon}^p = \Phi b N v \quad [22]$$

Equation [21] is a thermally activated model for the plastic strain rate in which ν is a frequency factor, Q an activation energy, V an activation volume, τ the shear stress and kT has its usual meaning. All but the last of these can be functions of the variables or else fixed parameters characteristic of the material. Here the energy barrier Q is the quantity that relates directly to dislocation drag. Equation [22] is a geometrical description of the plastic strain in which Φ is an orientation factor, b is a Burgers vector, N is dislocation density and v is average dislocation velocity. As above, all of these can be functions of the variables or else fixed parameters. Some drag parameters will be associated with the dislocation velocity. Within these general forms a further distinction can be drawn between rate equations which have an upper limit or maximum rate that can be achieved, and those which do not.

It is observed experimentally that at constant temperature dislocation density depends mainly upon strain and dislocation velocity upon stress. Frequently, therefore, the isothermal representations used for these quantities in calculations are of the form $N = N(\epsilon^P)$, $v = v(\sigma)$. For example, a linear dependence of density on strain, $N = N_0 + \alpha \epsilon^P$ where α is a multiplication coefficient, is often encountered. However, one of the chief advantages in the interpretation of precursor decay is that the plastic strain rate jump, $[\dot\epsilon^P]$ in Eq. [1], is from zero to some value corresponding to the initial mobile dislocation density. Therefore, it is not necessary to consider any variation of either density or velocity with plastic strain. Postulated velocity expressions are more numerous and more varied. They include: the power law $v = v_0 (\tau/\tau_0)^n$ where the velocity v_0 corresponds to the stress τ_0 and n is a material parameter; the Gilman relation $v = v_\infty \exp\{-D/\tau\}$ where v_∞ is the limiting dislocation velocity and D is a drag stress that is sometimes taken to be strain dependent: $D = D_0 + H\epsilon^P$; the Newtonian viscous relation resulting from the power law when the exponent is unity, but customarily written $v = b\tau/B$ where B is the quantity most often alluded to by the expression dislocation drag.

Before proceeding further we note that in the foregoing relations dependence upon plastic strain are preferred to total strain on purely heuristic grounds since experimental scatter generally far exceeds the small differences between total and plastic strains. Secondly, the stress dependences are usually interpreted as relating to some resolved shear stress components although the experimental evidence concerning the effect of the mean normal stress (hydrostatic pressure) on dislocation drag[25,26] is subject to conflicting interpretations at present.

The many functional forms postulated for the stress dependence of average dislocation velocity can be attributed to many possible causes, including different interpretations of identical experimental results[27]. The dominant cause, however, is likely to be differences in the prevalent mechanism of dislocation drag from one situation to another and from one material to another. For example, Ashby[28] recently published deformation information for eleven different materials as maps in stress, temperature coordinates indicating the boundaries between six regimes in which different deformation mechanisms dominate. Each map is for a specific grain size and deformation rate which somewhat limits its usefulness but the point is that different deformation mechanisms do occur.

In plastic shock waves dislocation glide is undoubtedly a principal deformation mechanism. Even for this one process, however, different defect interactions can be responsible for dislocation drag under different conditions. The experimental techniques that have mainly been used to study dislocation drag or, its inverse, mobility are: (1) direct observation by means of selective etching of the position of a dislocation before and after a stress pulse is applied[8,9]; or the same by X-ray topography[10] (2) interpretative analysis of ultrasonic attenuation measurements[29] (3) interpretative analyses of macroscopic mechanical tests[30]. The last of these three is recognized by many investigators as being uncertain at best; self-contradictory at worst[31].

Two limiting types of behavior emerge from the experimental studies. One is a linear dependence of velocity on stress which has been observed both by means of ultrasonic attenuation measurements and by velocity measurements. The other type of behavior is non-linear, the velocity being an exponential function of the stress. We believe that at high stresses a limiting dislocation velocity must be approached because dislocations cannot exceed sonic speeds in ordinary stress fields. If this saturation effect is included, two functions have been proposed to give analytic descriptions of velocities in terms of stress. Following Gillis, Gilman and Taylor[32] we let v_s be the saturation velocity. Then the functions are:

$$v/v_s = (v_s B/2b\tau) \{[1 + (2b\tau/v_s B)^2]^{\frac{1}{2}} - 1\} \qquad [23]$$

and:

$$v/v_s = \exp\{-D/\tau\} \qquad [24]$$

The forms of Eqs. [23] and [24] are illustrated in Fig. 2 in which z is a parameter proportional to the applied stress. Although the two forms appear similar in the figure there is an enormous difference between the two at velocities less than 0.1 (v/v_s) since in this region v/v_s in one is linear with z while in the other v/v_s changes extremely rapidly with z. That is, Eq. [24] has a 'yield stress' below which the velocity is rather small and above which it quickly rises to large values. Furthermore, the absolute stress dependence will not generally be the same for the two equations because of different apparent saturation velocities and different constants of proportionality relating to the stress.

Linear viscosity is characteristic of pure metals and salts in which dislocations glide easily at all temperatures, and of covalent crystals at high temperatures. Non-linear viscosity is observed in impure crystals and in pure covalent ones at low temperatures. Most real crystals will obey neither Eq. [23] nor Eq. [24] exactly because they contain a mixture of bonding types. Taylor has suggested a formulation for describing such heterogeneous substances and this was amplified by Gillis, Gilman and Taylor[32] along the following lines.

In the analysis of ultrasonic attenuation measurements it is postulated that a segment of a dislocation line is held fast at either end by obstacles, and vibrates about its equilibrium position in response to the acoustic waves. The vibration amplitude is very small so the segment never moves far enough to encounter other obstacles. On the other hand, the distance that a dislocation must travel for direct velocity measurements to be made is large. Therefore, the moving dislocation must pass a large number of obstacles during a velocity measurement in an impure crystal.

Consider an array of homogeneous obstacles in an otherwise clear lattice. Denote by L the average spacing of these obstacles as seen by dislocations traversing glide planes through the lattice and let f be the fraction of the length occupied by obstacles. Assuming that dislocation acceleration times are negligible, the time spent by a dislocation in moving a distance L will thus be the sum of the time, t_1, required to transverse the distance L (1 - f)

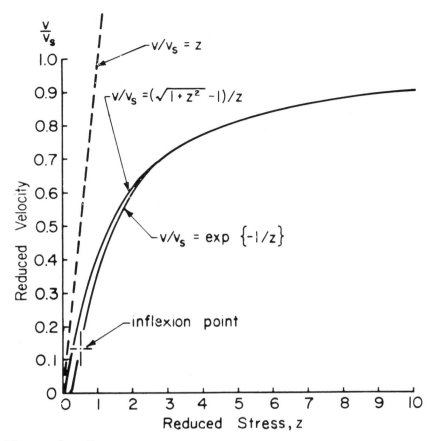

Figure 2. Comparison of the forms of the functions given in Eqs. [23] and [24] relating dislocation velocity to stress. Also included is a dashed line indicating simple linear dependence. (Taken from Ref. 32).

through clear lattice plus the time, t_2, that is required for the dislocation to traverse obstacles. The average dislocation velocity is therefore:

$$v = L/(t_1 + t_2) \quad [25]$$

It is assumed that the dislocation velocity for steady-state motion through clear lattice is given by Koehler's theory[33] of acoustic attenuation:

$$v_1 = b\tau/B \quad [26]$$

and that the viscous damping coefficient is of the form:

$$B = B_0 / [1 - (v_1/c_s)^2] \qquad [27]$$

to account for relativistic effects when the clear lattice velocity approaches the shear wave velocity, c_s, given by: $c_s^2 = G/\rho$. The combination of Eqs. [26] and [27] leads directly to [23] with $v_s = c_s$.

It is further assumed that the average velocity for dislocation motion past an obstacle is described by the theory of Gilman[7] so that it has the form of Eq. [24]:

$$v_2 = v^* \exp \{ - D/\tau \} \qquad [28]$$

Here v^* is some limiting velocity, and D is a characteristic drag stress.

When the obstacles are relatively far apart the average velocity can be obtained by inverting Eq. [25] and adding the reciprocals of v_1 and v_2. This leads to the expression:

$$(v^*/v) = \exp \{ 1/x \} + mx (v^*/c_s)/$$
$$[(1 + m^2 x^2)^{\frac{1}{2}} - 1] \qquad [29]$$

Here $m = 2bD/c_s B$, a normalized ratio of drag parameters and $x = \tau/D$, a reduced stress. How the average dislocation velocity depends upon the parameters m, x and (v^*/c) was described in some detail previously[32].

This description is briefly summarized in Fig. 3 which shows the stress values x_1 and x_2 at contours on which v_1 and v_2 in x vs. m space for various values of (v^*/c). For very low stress v_2 is always the dominant factor in determining average velocity unless no dislocation unpinning occurs. For $c < v^*$ and sufficiently small m there are two stresses at which $v_1 = v_2$. Thus, v_2 is the dominant factor at very low and at very high stresses but v_1 can become dominant in an intermediate stress regime. However, there is some limiting value for m, dependent upon (v^*/c), and for larger values of m v_2 will dominate at all stress levels. For $v^* = c$ there is at most one stress for which $v_1 = v_2$ and the limiting value of m is unity. Finally, for $c < v^*$ there will always be a unique stress at which $v_1 = v_2$ and this will occur at a small value of x except when (v^*/c) is in the range 1-10. At this stress the linear viscosity mechanism becomes the dominant dislocation drag mechanism and remains so for higher stresses.

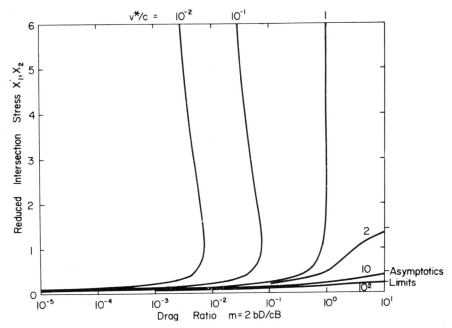

Figure 3. Values of reduced stress, at which $v_1 = v_2$, versus m for various ratios of v^*/c. (Taken from Ref. 32).

DISCUSSION

While the foregoing analysis tends to rationalize the existence of apparently contradictory dislocation velocity functions it does not tell us which to select in a given situation. One must simply make a choice and then attempt to calculate wave behaviors that satisfactorily match experimental results. The difficulty is, as stated previously and repeated for emphasis, that any one will do if the number of disposable parameters associated with it is large enough. This statement is exemplified by the work of Taylor. He and Rice published[34] the first thorough experimental study of elastic precursor decay in which they measured the precursor amplitudes in specimens of different thicknesses. In their initial paper they correlated this data well with calculations based upon a standard linear relaxing solid. Later Taylor published[14] an equally good correlation based upon Gilman's dislocation model.

One source of this difficulty is that the decay rate predicted by all of these equations is extremely rapid. We showed elsewhere[16] that these relations gave different results at distances very close to the impact surface of the specimen (less than 2 mm). At that time, however, difficulties in distinguishing elastic from plastic waves at such correspondingly short separations prevented useful experiments on ultrathin specimens. More recently[35] improved experimental techniques are opening this area to careful investigation.

A very clever alternative approach to studying the precursor decay with specimen thickness has been used by Rhode[17]. He has tested conveniently thick specimens at various temperatures in order to assess the temperature dependence of dislocation drag. However, alternative interpretations of his results are also possible[36].

To our knowledge there have been only two attempts made to correlate precursor decay data using a strain rate equation in which there were no adjustable parameters. Both failed in exactly the same way. Jones and Mote[37] report experiments on copper monocrystals in which dislocation mobilities were independently and directly determined[38] and initial dislocation densities could be reasonably estimated as 10^6 cm^{-2}. Calculations based on the linear viscosity dislocation drag relation using the predetermined drag coefficient could only be reconciled to the experimental results by using a value for initial mobile dislocation density somewhat larger that 10^8 cm^{-2}. Subsequent extension of this work[39] to W, NaCl and LiF monocrystals gave the same, unexpectedly high value for initial density required to fit calculations to experiment, approximately 10^8 cm^{-2}.

The second attempt[40] involved polycrystalline tantalum specimens for which dislocation mobilities and initial density produced by the pre-test annealing procedure, were determined by interpretative analysis of uniaxial tension tests over a broad range of deformation rates[41] and based upon the Gilman relation. Although the analysis indicated an initial density of 10^6 cm^{-2} it was again required to assume a value of 10^8 cm^{-2} to obtain agreement with the results of precursor decay experiments.

In the tantalum polycrystal experiments rise times of the elastic precursor waves were of the order of 100 nsec. It was argued semi-quantitatively[40] that dislocation multiplication could, in fact, increase the density by two orders of magnitude during this short period. Indeed it was estimated to be possible during the even shorter, 10 nsec, rise times of the copper monocrystal precursors. However, if this argument is correct then Eq. [1] does not rigorously apply to the precursor decay. In that case some higher order theory would be required, perhaps propagation of a discontinuity in particle acceleration rather than velocity, and this would obviate the simplicity of the governing equation.

The general role of dislocation drag in shock waves is to increase the thickness of the shock when other factors remain constant. Conversely, dislocation mobility tends to decrease shock thickness. A more precise statement of this effect is that shock thickness decreases with the plastic strain rate which is proportional to dislocation flux. This last statement may eventually provide some illuminating insights into the limiting behavior of dislocations at high stress. In plane wave experiments it is consistently observed that shock thickness decreases with impact pressure. This implies that either dislocation flux continually increases or that non-linear elastic effects override the drag viscosity. If we could separate these compressibility effects and the relative contributions to dislocation flux from density and velocity it would be possible to assess the response of the dislocations to extremely high levels of stress.

REFERENCES

1. H. Conrad: in Chemical and Mechanical Behavior of Inorganic Materials, A. W. Searcy, D. V. Ragone and U. Columbo, eds., Wiley-Interscience, 1970, p. 285.
2. J. E. Dorn and J. D. Mote: in High Temperature Structures and Materials, A. M. Freudenthal, B. A. Boley and H. Liebowitz, eds., McMillan, 1964, p. 95.
3. W. P. Mason: Proc. Fifth U. S. Nat. Congr. Appl. Mech., ASME, N. Y., 1966, p. 361.
4. A. Kumar, F. E. Hauser and J. E. Dorn: Acta Met., 1968, vol. 16, p. 1189.
5. V. Y. Kravchenko: Sov. Phys. - Sol. St., 1966, vol. 8, p. 740.
6. E. W. Hart: Phys. Rev., 1955, vol. 98, p. 1775.
7. J. J. Gilman: Aust. J. Phys., 1960, vol. 13, p. 327; J. Appl. Phys., 1965, vol. 35, p. 3195; Ibid., 1968, vol. 39, p. 6068.
8. W. G. Johnston and J. J. Gilman: J. Appl. Phys., 1959, vol. 30, p. 129.
9. D. F. Stein and J. R. Low: J. Appl. Phys., 1960, vol. 31, p. 362.
10. D. P. Pope, T. Vreeland, Jr., and D. S. Wood: Rev. Sci. Instr., 1964, vol. 35, p. 1351.

11. M. H. Rice, R. G. McQueen and J. M. Walsh: Solid State Physics, 1958, vol. 6, p. 11.
12. G. Duvall: in Stress Waves in Anelastic Solids, H. Kolsky and W. Prager, eds., Springer, 1964, p. 20.
13. R. G. McQueen, S. P. Marsh, J. W. Taylor, J. N. Fritz and W. J. Carter: in High Velocity Impact Phenomena, R. Kinslow, ed., Academic Press, 1970, p. 293.
14. J. W. Taylor: J. Appl. Phys., 1965, vol. 36, p. 3146.
15. J. N. Johnson and W. Band: J. Appl. Phys., 1967, vol. 38., p. 1578.
16. J. M. Kelly and P. P. Gillis: J. Appl. Phys., 1967, vol. 38, p. 4044.
17. R. W. Rhode: Acta Met., 1969, vol. 17, p. 353.
18. J. M. Kelly: Int. J. Solids Structures, 1971, vol. 7, p. 1211.
19. J. M. Kelly and P. P. Gillis: J. Appl. Mech., 1970, vol. 37, p. 163.
20. C. S. Smith: Trans. AIME, 1958, vol. 212, p. 574.
21. R. J. Clifton: in Shock Waves and the Mechanical Properties of Solids, J. J. Burke and V. Weiss, eds., Syracuse Univ. Press, 1971, p. 73.
22. P. P. Gillis and J. M. Kelly: J. Mech. Phys. Solids, 1970, vol. 18, p. 397.
23. E. H. Lee: Proc. Fifth U. S. Nat. Congr. Appl. Mech., ASME, N. Y., 1966, p. 405.
24. F. Seitz and T. A. Read: J. Appl. Phys., 1941, vol. 12, p. 470.
25. J. E. Hanafee and S. V. Radcliffe: J. Appl. Phys., 1967, vol. 38, p. 4284.
26. W. L. Haworth, L. A. Davis and R. B. Gordon: J. Appl. Phys., 1968, vol. 39, p. 3818.
27. P. P. Gillis and J. J. Gilman: J. Appl. Phys., 1965, vol. 36, p. 3370.
28. M. F. Ashby: Acta Met., 1972, vol. 20, p. 887.
29. A. V. Granato and K. Lucke: J. Appl. Phys., 1956, vol. 27, p. 789.
30. R. W. Guard: Acta Met., 1961, vol. 9, p. 163.
31. Agenda discussion: p. 479 in Dislocation Dynamics, A. R. Rosenfield, G. T. Hahn, A. L. Bement and R. L. Jaffee, eds., McGraw-Hill, 1968.
32. P. P. Gillis, J. J. Gilman and J. W. Taylor: Phil. Mag., 1969, vol. 20, p. 279.
33. J. S. Koehler: in Imperfections in Nearly Perfect Crystals, W. Shockley, J. H. Hollomon, R. Maurer and F. Seitz, eds., John Wiley, 1952, p. 197.

34. J. W. Taylor and M. H. Rice: J. Appl. Phys., 1963, vol. 34, p. 364.
35. D. R. Curran: in Shock Waves and the Mechanical Properties of Solids, J. J. Burke and V. Weiss, eds., Syracuse Univ. Press, 1971, p. 121.
36. P. P. Gillis: Scripta Met., 1970, vol. 4, p. 533.
37. O. E. Jones and J. D. Mote: J. Appl. Phys., 1969, vol. 40, p. 4920.
38. W. F. Greenman, T. Vreeland, Jr., and D. S. Wood: J. Appl. Phys., 1967, vol. 38, p. 3595.
39. J. N. Johnson, O. E. Jones and T. E. Michaels: J. Appl. Phys., 1970, vol. 41, p. 2330.
40. P. P. Gillis, K. G. Hoge and R. J. Wasley: J. Appl. Phys., 1971, vol. 42, p. 2145.
41. K. G. Hoge and P. P. Gillis: Met. Trans., 1971, vol. 2, p. 261.

Discussion by T. Vreeland, Jr. and K. M. Jassby,
    California Institute of Technology

   This discussion is intended to emphasize a very important influence of the drag coefficient - namely the dissipation of energy during plastic deformation. A large majority of the irreversible work done by the stresses applied to the boundaries of a crystalline material is dissipated as heat through the action of viscous drag on the moving dislocations. It is interesting to note that the energy dissipated is not usually sensitive to the magnitude of B at low to moderate strain rates because B does not significantly influence the stress level required to maintain the deformation rate. Because of the viscous drag, dislocations do not continue to accelerate after they break away from an obstacle, instead they reach a drag limited velocity after moving a few atomic distances. Without a viscous drag, the kinetic energy of a dislocation would continue to build up and be available to overcome local obstacles. (Large kinetic energies are postulated for dislocations in superconductors where B is expected to be very small. This postulate is used to explain the drop in flow stress at the normal to superconducting transition and is discussed in the paper by J.M. Galligan.)

Discussion by Frank A. McClintock
    Massachusetts Institute of Technology

   The assumption of constant mobile dislocation densities, along with a phonon drag and a limiting dislocation velocity, leads to the strain rate dependence illustrated in Fig. 2 of the discussion of the paper by McClintock. Can you suggest data or theory giving a more plausible relation to be used in calculating the dynamic growth of holes in high strength materials at strain rates of the order of 10^6/sec to 10^9/sec?

Author's Reply

The comments by Prof. Vreeland and Jassby are quite appropriate and the question by Prof. McClintock very provocative. In response to the latter the following discussion seems pertinent.

The concept of a limiting dislocation flux being incorporated into fairly simple representations of material response was suggested by us (Acta Met., 1972, vol. 20, p. 947) and named ideal viscoplasticity, by analogy with the simple inviscid representations of ideal plasticity. Such models can provide plausible theories for rate-sensitive material behavior. Prof. McClintock's Fig. 2 is essentially such a model, and is used by him as the basis for the analysis of the dynamic growth of holes.

On the other hand, the ideas put forward in the paper by Prof. Grenato indicate that the generation of mobile dislocations becomes the mechanism controlling the plastic strain rate when dislocation velocity is high. For constant velocity v the mean free path $\varkappa$ of a dislocation segment is related to its lifetime τ in the mobile population by: $\varkappa = v\tau$. The plastic strain rate relation $\dot{\epsilon} = bNv$ can thus be rewritten $\dot{\epsilon} = b\varkappa(N/\tau)$, as pointed out previously by Argon (Materials Science & Engineering, 1968/69, vol. 3, p. 24). Obviously the factor in parentheses is the generation rate of mobile dislocations. However, our interpretation is that the two forms of the strain rate equation are equivalent. One may be more useful than the other in a given situation but both say the same thing. Although we believe these two views to be compatible we recognize that an important problem exists in estimating what happens to the mobile dislocation density at high rates and large stresses. This factor is prominent in both formulations.

Prof. Weertman in his paper presented another alternative view of high strain rate dislocation behavior that contradicts the concept of ideal viscoplasticity. However, we believe that no experimental evidence exists to support the notion that supersonic dislocations are of substantial physical significance.

Plastic Deformation at Low Temperatures and Drag Mechanisms

J. M. Galligan, Institute of Materials Science
University of Connecticut, Storrs, Conn.
and
M. Suenaga, Brookhaven National Laboratory
Upton, New York

Introduction

    Plastic deformation at low temperatures occurs through the motion of dislocations and this motion of dislocations is resisted by processes involving other defects, lattice vibrations and electrons.[1] In the first of these resistive mechanisms, the dislocation is taken as being stationary, while the second and third resistances involve a viscous drag and, thus, a velocity. At low temperatures the phonon resistance disappears, in the same way as the specific heat, so that the major viscous resistance to dislocation motion will be an electron viscosity. Various attempts have been made to distinguish between barrier (defect) models of resistance to dislocation motion and viscous drag effects, but only recently has it become possible to make some headway on this problem. In order to study the interaction of dislocations with electrons, a method of instantaneously changing the electron drag must be found. Secondly, the velocities of the dislocation must be high if there is to be an interaction with the electrons of the metal. The first condition can be obtained by exploiting the properties of superconductors,[2] i.e. when a metal becomes superconducting, persistent currents are generated and, once these are established, they are not easily destroyed, since the slowing down involves all the electrons at once.[2] This is an extremely unlikely process which requires an energy Δ (the Bardeen, Cooper, Schrieffer energy gap) from a single phonon. This can be related to the viscous drag on a dislocation, since an interaction between a dislocation and electrons would involve the destruction of

persistent currents. Also, it is important to emphasize that
the switching time from the normal state to the superconducting
state is, typically, of the order of microseconds[2], which is
faster than any dislocation transit time.

An early indication that dislocation motion could be influenced by the pairing of electrons, i.e. superconductivity,
came from the experiments of Tittman and Bömmel[3], who
observed that the dislocation damping (in an ultrasonic
attenuation experiment) changed dramatically when a lead crystal
was cooled below the superconducting transition temperature.
The second major impetus came from an experiment conducted by
Kojima and Suzuki[4], who noted that the plasticity of lead and
niobium crystals was influenced dramatically by switching such
crystals from the normal to the superconducting state. This
result suggests strongly that dislocations in an ordinary tensile test are moving at very high velocities, or barriers must
change height when the transition occurs from the normal to the
superconducting state. In the first case one would expect the
difference in stress for dislocation motion in the normal state
compared to the superconducting state, $\Delta\sigma_{N-S}$, to vary as the
B.C.S. energy gap. In contrast to this, a change in barrier
height might be expected to vary as a typical, thermally
activated process. Some experiments which bear on these two
possible types of processes in a solid, are outlined below
as well as a simple model of dislocation motion. It is concluded that these experiments can be consistently interpreted
by use of a simple string model of a dislocation moving in a
viscous medium.

II. Experimental

The experiments were performed on a variety of materials,
so that it is of interest to emphasize that while we use
particular examples to illustrate the situation, the results
are of general validity. The experiments which we shall discuss
were undertaken in lead and lead alloys in the following manner:
Crystals were plastically deformed at specific temperatures
(below the superconducting temperature), while the material
was in the normal state. (The applied magnetic field, H, was
greater that H_c, the critical field for superconductivity).
During the plastic deformation process the material was switched
to the superconducting state (H → 0) and a stress difference
$\Delta\sigma_{N-S}$ was observed. The experiment can also be carried out in
the stress relaxation mode, which we have mainly used, and one
can do this for a variety of temperatures, fig. 1. Secondly,
one can vary the effect by changing the stress for plastic
deformation. The temperature dependence in that case is
essentially the same as that found at lower stress, which

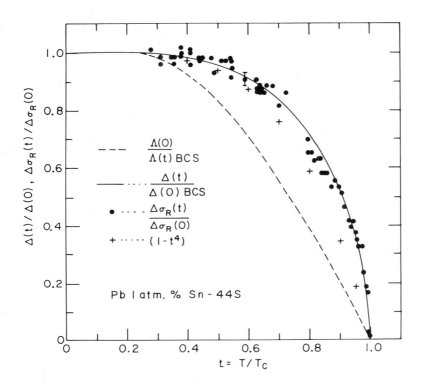

Fig. 1. The temperature dependence of the difference in stress, in a stress-relaxation experiment, as a function of deformation temperature. The sample is first deformed in the normal state, then the elongation is stopped and the specimen is switched to the superconducting state. The switching time is shorter than the speed of sound, so that the stress change is instantaneous.

indicates the same basic mechanism is responsible for the process. In addition, the magnitude of the effect, fig. 2, is quite large (~10% in the stress relaxation case) and some observers have found the change in flow stress to amount to as much as 40%. This shows that the state of the electrons plays a major role (at low temperatures) in dislocation processes.

A second major finding, fig. 3, is that the magnitude of the effect varies with the concentration of solute, c, in a rather simple way.

$$\Delta \sigma_{N-S} \sim c^{1/2}$$

In addition, there is no systematic variation with the valency of the solute, as expected in the case of a change in activation barrier. For example, solutes of the same valence as the solvent show as large an effect as solutes with valences as different as ± 2 units. A final experiment compares $\Delta \sigma_{N-S}$ with the lattice parameter change, Δa, (for constant solute type) for various concentrations of solute, fig. 4. In this case it is found that

$$\Delta \sigma_{N-S} \sim \frac{da}{dc}$$

Discussion of Results

The observed temperature dependence $\Delta \sigma_{N-S}(T)$, fig. 1, shows the following:

a. The effect saturates at low temperatures. This would not be expected for a thermally activated process.

b. The effect only appears below the critical temperature.

c. The temperature dependence $\Delta \sigma_{N-S}(T)$ follows the temperature dependence of the energy gap for a superconductor. We take these as conclusive arguments that the process involves a viscous drag. Secondly, the observed influence of solutes on $\Delta \sigma_{N-S}$ is not what one expects for a change in potential, which would be associated with a change in activation barrier. More specifically, if the barrier involves a solute through the difference in electrostatic charge between solute and solvent, then solutes from the same column in the periodic chart as solvents should have $\Delta \sigma_{N-S} = 0$. This is not observed.

If dislocation motion is impeded by an electron viscosity then this process can be looked at in the following simple manner: If a dislocation experiences a decrease in viscosity, during transit between obstacles, fig. 5, then it can overshoot

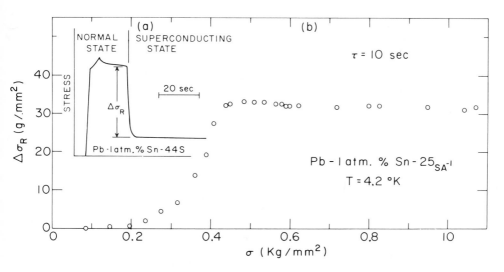

Fig. 2. The observed difference in stress-relaxation for the normal state and the superconducting state, as a function of the applied stress.

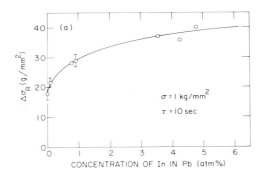

Fig. 3. The measured difference in stress, between the normal state and the superconducting state, as a function of solute concentration. This observed dependence of $\Delta\sigma_{NS}$ upon the concentration of solute, is what one expects for a random distribution of obstacles.

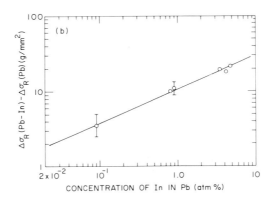

Fig. 4. The measured difference in flow stress, for constant number of solute atoms, as a function of the size difference of the solute and solvent in lead base alloys.

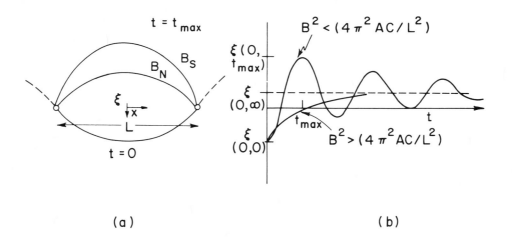

(a)   (b)

Fig. 5

the barrier if it acts as an underdamped oscillator. The equation of motion for such a situation can be written as[5],

$$A \frac{\partial^2 \xi}{\partial t^2} + B \frac{\partial \xi}{\partial t} - C \frac{\partial^2 \xi}{\partial x^2} = b\sigma$$

where ξ is the displacement of a dislocation, $A = \pi\rho b^2$ is the effective dislocation mass per unit length, B is the viscosity coefficient, and $C \approx Gb^2$ is the dislocation line tension. Note that ρ is the density of the material, G the shear modulus and b the Burger's vector. The equation is subject to the following boundary conditions:

i.) $\partial\xi/\partial t = V_0(x)$ at $t = 0$

ii.) $\xi = \xi_0(x)$ at $t = 0$

iii.) $\xi = 0$ at $x = \pm L/2$ for all t

where $\xi_0(x)$ is a given function of x which describes an initial dislocation segment and $V_0(x)$ is the initial velocity. A particular solution is

$$\xi(x_1,t) = A' \{\exp(-Bt)\cos(wt)\cos(L) - (1 - \frac{4x^2}{L^2})\}$$

where $A' = \frac{\sigma L^2}{4Gb}$, $w = \pi(C/A)^{1/2}/L$ and

$\beta = B/2A$. The function $\xi(o,t)$ is shown in fig. 5 which illustrates the motion of the dislocation as a function of time. It is important to note that the dislocation moves past the static position, $\xi(o,\infty)$ in the underdamped case and this increases the line tension on the pinning points such that breakaway can occur. The only requirement (other than an underdamped oscillator) is that B for the superconducting state be less than B for normal state. With this in mind we can proceed to calculate a critical stress for depinning, as a function of the applied stress, σ, and we perform the calculation for the superconducting state and the normal state, and find the difference in stress for this situation:

$$\Delta\sigma_{NS} \approx \frac{U}{16b^2} C^{1/2} (AC)^{-1/2} [1 - B_s(t)/B_N]$$

where U is the activation barrier for the depinning process, B_s and B_n are the viscosities for the superconducting state and the normal state and t now stands for T/T_c.

This model makes three major predictions

a. $\Delta\sigma_{N-S}$ is given by $\left(\frac{da}{dc}\right)$ at c=a constant which is what the data shows

b. For any particular alloying element $\Delta\sigma_{NS}$ should vary as $c^{1/2}$, as shown by the data.

c. The temperature dependence of $\Delta\sigma_{NS}$ should follow the temperature dependence of the B.C.S. energy gap, Δ, which is shown in fig. 1.

In conclusion we have presented measurements which show that electrons do affect dislocation motion, this effect is a consequence of the high velocities involved and we have presented a model which explains all the known results.

References

1. F. R. N. Nabarro, Theory of Crystal Dislocations, The Clarendon Press, Oxford (1967).

2. E. A. Lynton, Superconductivity, John Wiley (New York).

3. B. R. Tittman and H. E. Bommel, Phys. Rev. 151, 178 (1966).

4. H. Kojima and T. Suzuki, Phys. Rev. Letters 21, 290 (1968).

5a. J. S. Koehler, Imperfections In Nearly Perfect Crystals, Wiley, New York (1952).

b. A. V. Granato, Phys. Rev. Letters 27, 660 (1971).

c. M. Suenaga and J. M. Galligan, Phys. Rev. Letters 27, 721 (1971).

DISLOCATION MECHANICS AT HIGH STRAIN RATES

J. Weertman
Departments of Materials Science and Geological
Sciences and the Materials Research Center,
Northwestern University, Evanston, Illinois 60201

ABSTRACT

In this paper a high strain rate is defined as high if the dislocations that produce the plastic strain must move at velocities of the order of, or larger than the shear wave velocity a_3. Except possibly in the initial portions of stress strain curves dislocation velocities are much smaller than a_2. Only in shock loaded specimens should very fast dislocation velocities exist. An approximate analysis is made to show how supersonic dislocations in a shock wave interface (the Smith interface) can produce a hydrostatic stress behind the interface.

WHAT IS A HIGH VALUE FOR A PLASTIC STRAIN RATE?

Crystalline material is deformed plastically at strain rates in the range 10^{-16} to 10^{-13} s^{-1} (geologic strain rates); 10^{-12} to 10^{-8} s^{-1} (glacier flow strain rates); 10^{-8} to 1 s^{-1} (ordinary stress-strain and creep tests); 10 to 10^3 s^{-1} (impulse loading tests); and at virtually infinite strain rates (shock wave tests). The plastic strain rate $\dot{\varepsilon}$ is given by the well known equation

$$\dot{\varepsilon} = \alpha \rho b v \qquad (1)$$

where b is the magnitude of the Burgers vector of a dislocation, ρ is the density of the mobile dislocations, v is the average velocity of the dislocation, and α is a dimensionless constant ($\alpha \sim 1$). The average velocity v of a dislocation thus is of the order of

$$v = \dot{\varepsilon}/\alpha\rho b \tag{2}$$

The stress field of a dislocation is essentially independent of the value of the dislocation velocity if v is much smaller than the shear wave velocity a_2. Only if v is of the order of, or larger than a_2 is the stress field of a moving dislocation radically different from the stress field of a stationary dislocation[1]. Because the properties of a dislocation change markedly at $v \sim a_2$ it is reasonable to expect that the dislocation mechanics of plastic deformation may be qualitatively different at high strain rates. A plastic strain rate thus reasonably can be defined to be "high" if it satisfies the inequality

$$\dot{\varepsilon} > \alpha\rho b a_2 \tag{3}$$

The dislocation density in a metal can be expected to be of the order of or larger than 10^{10} m^{-2}. A high strain rate thus is greater than about 10^4 s^{-1} since $b \approx 3 \times 10^{-10}$ m, $a_2 \approx 3$ km s^{-1}, and $\alpha \approx 1$.

The strain rates that are produced in various impulse type loading tests are not larger than about 10^3 s^{-1} (see, for example, the papers of Hockett and coworkers[2-6] or of Nagata and Yoshida[7,8]). Such tests are considered in the literature, with justification, to be high strain rate tests because the strain rates are orders of magnitude larger than those employed in ordinary stress-strain tests. However by the criterion given by Ineq. [3] these are not high strain rate tests. Dislocation theories developed to account for the results obtained at relatively small strain rates can be applied to tests in which the strain rate does not exceed 10^3 s^{-1}. The properties of the dislocations have not changed qualitatively in such tests because v is small compared with a_2. At strain rates in which Ineq. [3] is satisfied the properties of a fast moving dislocation in a test specimen are qualitatively different from the properties of a slowly moving dislocation in a low strain rate test. Dislocation theories of plastic deformation developed for low strain rate tests obviously must break down when dislocations move at velocities compared to a_2.

Transient Effect

It should be noted that dislocations may have to move at a fast velocity at the start of even a low strain rate test if the initial mobile dislocation density ρ is relatively small. Only after the density ρ has increased in magnitude because of dislocation multiplication do the dislocations move at a slow velocity in a low strain test. The stress-strain curves of materials with a very low initial dislocation density should exhibit an upper yield point effect. A high initial stress is required to move the few dislocations at a high velocity; a low stress later is required to move the many dislocations at a low velocity.

The relationship between stress and dislocation velocity can be obtained experimentally with the use of impulse loaded samples that contain relatively few mobile dislocations. Figures 1 through 4 show results that were obtained at Northwestern University on dislocations in lead[9], aluminum[10], potassium[11], and iron[11]. The plots shown in Figs. 1 through 4 demonstrate that the dislocation velocity is proportional to the stress in the velocity region: $0.01 a_2 < v < 0.3 a_2$. At higher velocities an increase in the velocity requires a proportionately greater increase in the stress (that is, the velocity-stress curve bends over at the higher velocities). The linear relationship between dislocation velocity and stress at the lower velocities is to be expected if phonon or electron drag mechanisms are responsible for the velocity limitation (see the references cited in ref. 10). The bend-over of the velocity-stress plot at higher velocities also can be explained qualitatively by invoking these damping mechanisms[12].

Stress-Strain Curve

A theoretical stress-strain curve can be derived if the dislocation velocity is controlled by a phonon or electron damping mechanism. The dislocation velocity in the linear velocity region is given by

$$v = B^{-1}(\tau - \tau_i) \qquad (4)$$

where B is the damping constant (a typical value of B from Figs. 1 through 4 is $B \approx 30 \; \mu N \; s \; m^{-2}$), τ is the applied stress, and τ_i is the internal stress. If the dislocations are evenly spaced an approximate value for the internal stress is

$$\tau_i = \gamma \mu b \rho^{\frac{1}{2}} \qquad (5)$$

where γ is a dimensionless constant of the order of 1/10.

The rate of change of the dislocation density with time t is given by

$$d\rho/dt = \gamma^* \rho^m v \qquad (6)$$

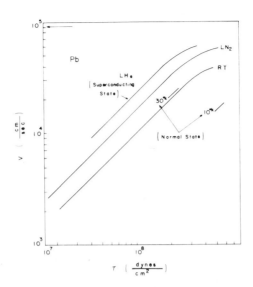

Fig. 1. Log-log plot of dislocation velocity versus stress for lead at liquid helium temperature, 10°K, 30°K, liquid nitrogen temperature, and room temperature. Data of Parameswaran and Weertman[9]. Upper arrow indicates the value of the sound velocity at liquid helium temperature. (Note that 10^7 dynes/cm^2 = MN m^{-2}.)

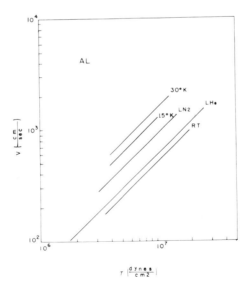

Fig. 2. Log-log plot of dislocation velocity versus stress for aluminum at various temperatures. Data of Parameswaran, Urabe, and Weertman[10]. (Note that 10^7 dynes/cm^2 = 1 MN m^{-2}.)

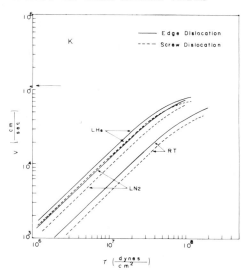

Fig. 3. Log-log plot of dislocation velocity versus stress for potassium at various temperatures. Upper arrow indicates value of the sound velocity at liquid helium temperature. Unpublished data of Urabe and Weertman[11]. (Note that 10^7 dynes/cm^2 = 1 MN m^{-2}.)

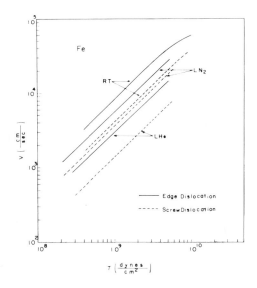

Fig. 4. Log-log plot of dislocation velocity versus stress for iron at various temperatures. Unpublished data of Urabe and Weertman[11]. (Note that 10^9 dynes/cm^2 = 100 MN m^{-2}.)

if no dislocation annihilation occurs. The exponent m equals 3/2 if the density of dislocation sources is of the order of the dislocation density divided by the mean spacing of the dislocations; m equals zero if the density of dislocation sources is a constant. The value of m thus should lie in the range $0 < m < 3/2$. The quantity γ^* is a dimensionless constant of order of 1/10 if m = 3/2 and is approximately equal to the density of dislocation sources if m = 0.

By combining Eqs. (1), (4), (5), and (6) the following equation can be found that relates stress with strain at a constant plastic strain rate

$$\tau = (B\dot{\epsilon}/\alpha b) / \left[\rho_0 + (2-m)\gamma^*\alpha^{-1}b^{-1}\epsilon \right]^{1/(2-m)} + \gamma\mu b \left[\rho_0 + (2-m)\gamma^*\alpha^{-1}b^{-1}\epsilon \right]^{1/(4-2m)} \quad (7)$$

where ρ_0 is the initial dislocation density and ϵ is the plastic strain. (This derivation is essentially the same as that given by Johnston and Gilman[13] and by Hahn[14]. The use of a linear relationship between dislocation velocity and stress is not essential to the derivation.)

Equation (7) predicts that if m = 3/2 stress is linearly related to strain at large strains and that $d\tau/d\epsilon = \gamma\gamma^*\mu/2\alpha \sim \mu/100$. In copper single crystals and polycrystalline copper a linear relation is found with a slope of this magnitude in low temperature tests carried out to a total strain of about 0.2 at strain rates up to 10^3 s^{-1} (see Figs. 3 and 4 of ref. 7 and Fig. 5 in ref. 15). A similar relationship is found in the same strain range in aluminum (see Fig. 5 of ref. 2) at strain rates up to 227 s^{-1}. However at larger strains (obtained in a Cam Plastometer [2-6]) the stress-strain curve of aluminum becomes parabolic (see Fig. 5 of ref. 2), a result that, according to the theory of Eq. (7), implies the occurrence of dislocation annihilation.

DISLOCATION BEHAVIOR AT HIGH STRAIN RATES

Consider now the behavior of dislocations in a high strain rate test, where high strain rate is defined by Ineq. (3). The strain rates considered thus are greater than about 10^4 s^{-1}. Such large strain rates are encountered only in shock wave test. Cyril Stanley Smith[16,17] has pointed out that it is reasonable to expect that a shock wave traveling through a crystal contains a dislocation interface. If there were no dislocation interface the crystal lattice on either side of the shock front would appear as shown in Fig. 5a. The lattice behind the shock front would undergo very large compressive strains in the direction of propagation of the shock. Shear stresses would be set up on planes whose normals

DISLOCATION MECHANICS AT HIGH STRAIN RATES 325

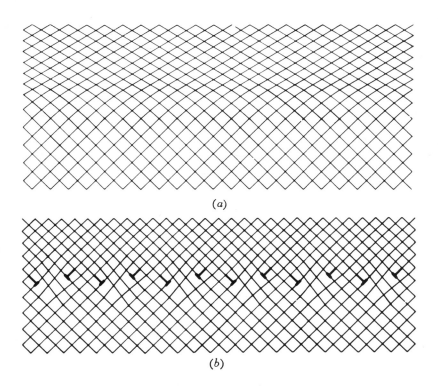

Fig. 5. Shock wave traveling through a lattice. (a) Compression only in the direction of motion of shock wave. (b) Hydrostatic compression behind shock wave that contains a Smith dislocation interface. Figure is taken from paper of Cyril Stanley Smith[16].

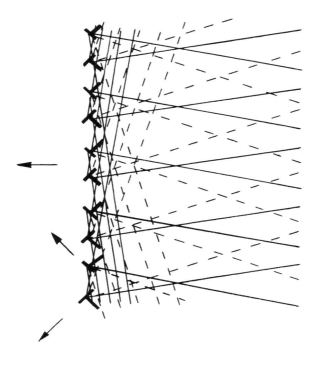

Fig. 6. Sound waves radiated by supersonic dislocations in a Smith interface. Solid lines: longitudinal sound waves; dashed lines transverse sound waves.

are not in the direction of propagation that would be equal to the theoretical strength of the material. A relaxation of the uniaxial compressive stress to a hydrostatic stress thus would be expected. If a moving dislocation interface (the Smith dislocation interface) exists in the shock front, as shown in Fig. 5b, hydrostatic stress will prevail behind the shock front. A hydrostatic stress cannot set up shear stresses on any crystalline plane.

An analytical treatment of dislocations on a Smith interface does not exist in the literature. It is the purpose of the remainder of this paper to develop an analytical treatment for these dislocations.

The number of dislocations in the Smith interface can be found from the difference in the lattice parameter on either side of the shock front. Let ρ_0 be the density of material into which a shock front propagates and let ρ_1 be the density of material behind the shock front (see Fig. 5b). Let the Burgers vectors of the two symmetrical sets of dislocations shown in Fig. 5b make the angle φ with the normal of the Smith interface. The total number n of dislocations in a unit length of the cross section of a Smith interface is

$$n = (b \sin \varphi)^{-1} \left[1 - (\rho_0/\rho_1)^{1/3} \right] \qquad (8)$$

(It should be pointed out that in order to permit hydrostatic pressure to exist behind the shock front a second group of dislocations, oriented perpendicular to the dislocations shown in Fig. 5b, must exist in the Smith interface. This second group permits relaxation of the non-hydrostatic deformation in the direction perpendicular to the plane of Fig. 5b. The number of dislocations in this second group also is given by Eq. (8). More complicated arrangements of dislocations in the Smith interface are, of course, possible.)

The velocity u_{so} of propagation of the shock front measured in the coordinate system in which the material ahead of the shock front is at rest is given by[18]

$$u_{so}^2 = [(P_1 - P_0)/(\rho_1 - \rho_0)](\rho_1/\rho_0) \qquad (9)$$

where P_1 is the pressure exerted across any plane behind the shock front that is perpendicular to the direction of propagation of the shock wave and P_0 is the pressure when the plane is ahead of the front. In the coordinate system in which material behind the shock front is at rest the shock front propagates at a velocity u_{s1} given by[18]

$$u_{s1}^2 = (P_1 - P_0)/(\rho_1 - \rho_0) \qquad (10)$$

The velocities u_{so} and u_{s1} may be compared with the transverse sound velocity a_2 and the longitudinal sound velocity a_1. The sound velocities must satisfy the equation

$$a_1^2 - (4/3)a_2^2 = \partial P/\partial \rho \tag{11}$$

where ρ is the material density at pressure P. The shear wave velocity a_2 usually has a value of the order of $a_1/2$. The longitudinal sound velocity in the material ahead of the shock wave thus has a value given by the equation

$$(a_1)_0^2 \sim (3/2)(\partial P/\partial \rho)_{\rho=\rho_0} \tag{12}$$

and in the material behind the shock front has the value given by the equation

$$(a_1)_1^2 \sim (3/2)(\partial P/\partial \rho)_{\rho=\rho_1} \tag{13}$$

The term ρ in Eqs. (12) and (13) is now the material density rather than the density of dislocations. The sound velocity $(a_1)_1$ is larger than the sound velocity $(a_1)_0$ because the laws for the finite deformation of an elastic solid rather than the laws for the infinitesimal deformation of a linear elastic solid apply when the pressure is so high that the pressure difference $\rho_1 - \rho_0$ becomes of the order of, or larger than $0.1\rho_0$.

In shock wave experiments (see, for example, results listed in Table VIII of ref. 18 or in the appendices of ref. 19) the shock wave velocities u_{so} and u_{s1} usually have values that satisfy the relationships

$$u_{so} > (a_1)_0 \tag{14}$$

and

$$(a_2)_1 \approx (1/2)(a_1)_1 < u_{s1} < (a_1)_1 \tag{15}$$

Thus the shock velocity has a supersonic velocity with respect to the unshocked material and a transonic velocity with respect to the material behind the shock front.

The derivations of the stress field and the strain field of a moving dislocation that are given in the literature generally employ the equations of a linear elastic solid. Thus the velocities of sound ahead of the dislocation and behind it have the same value. For the moving dislocations in the Smith interface of Fig. 5b the sound velocities in these two regions do not have the same value.

Consider the approximation in which $(a_1)_0 = (a_1)_1$ and the

Smith interface is moving at a supersonic or transonic velocity. If the Smith interface moves at the velocity u_{so} each (edge) dislocation in the Smith interface moves at the velocity

$$v = u_{so}/\cos\varphi \tag{16}$$

Assume that the values of u_{so} and φ are such that v is a supersonic velocity. Each dislocation in the interface must radiate shear waves and longitudinal waves (see Fig. 9 of ref. 1). These waves are radiated in directions into the region behind the interface.

Consider the average value of the tensile stresses p_{xx}, p_{yy}, and p_{zz} produced by all the radiated shear and longitudinal waves. (The x direction is the direction of propagation of the interface, and y direction is perpendicular to both the x direction and the direction of the dislocation lines of Fig. 5b, and the z direction is perpendicular to both the x and y directions.) Over a plane whose dimensions are large compared with the spacing of the dislocations it is simple to show from the stresses in the radiated sound waves (see Section IV-B-2 of ref. 1) that the average values of p_{xx}, p_{yy}, and p_{zz} are given by

$$p_{xx} = 0 \tag{17}$$

$$p_{yy} = p^°_{yy} = (2nb\mu a_2{}^2/v^2) \left\{ [\tan(\theta_1 - \varphi) + \tan(\theta_1 + \varphi)] [-\beta*_1 \sin\varphi + \alpha^2\cos\varphi] + [\tan(\theta_2 - \varphi) + \tan(\theta_2 + \varphi)] [-\alpha^4\beta*_2{}^{-1}\sin\varphi - \alpha^2\cos\varphi] \right\} \tag{18}$$

$$p_{zz} = \nu\, p_{yy} \tag{19}$$

where μ is the shear modulus, ν is poisson's ratio, $\beta*_1{}^2 = (v/a_1)^2 - 1$, $\beta*_2{}^2 = (v/a_2)^2 - 1$, $\alpha^2 = 1 - v^2/2a_2{}^2$, $\sin\theta_1 = a_1/v$, and $\sin\theta_2 = a_2/v$. The average value of the shear stresses is equal to zero.

The group of dislocations in the Smith interface that are perpendicular to those shown in Fig. 5b produce an average stress p_{xx} that is given again by Eq. (17), a stress $p_{zz} = p^°_{yy}$, and a stress $p_{yy} = \nu p^°_{yy}$. The total stresses produced by the two groups of perpendicular dislocations thus are

$$p_{xx} = 0$$

$$p_{yy} = (1+\nu)p^°_{yy} \tag{21}$$

$$p_{zz} = (1+\nu)p^{o}_{yy} \tag{22}$$

where p^{o}_{yy} is the value of p_{yy} given by Eq. (18). In the region in front of the Smith interface the stresses are zero in value.

The average stresses given by Eqs. (20), (21) and (22) are only those produced by the dislocations. To these stresses must be added the stresses that exist in the "shock" wave before hydrostatic relaxation occurs (that is, the stresses that exist in Fig. 5a when no dislocations are in the shock front). Let these additional stresses be p^{*}_{xx}, p^{*}_{yy}, and p^{*}_{zz}. If linear elasticity theory were valid, then $p^{*}_{yy} = p^{*}_{zz} = [\lambda/(\lambda+2\mu)]p^{*}_{xx}$ where λ is the other Lamé constant. For hydrostatic conditions to prevail behind the interface the stresses p_{xx}, p^{*}_{xx}, etc. must satisfy the equation

$$p_{xx} + p^{*}_{xx} = p_{yy} + p^{*}_{yy} = p_{zz} + p^{*}_{zz} = -P_{1} \tag{23}$$

From Eq. (23) the following relationship is found

$$p^{o}_{yy} = -\left[2\mu/(1+\nu)(\lambda+2\mu)\right]P_{1} \tag{24}$$

By combining Eqs. (18) and (24) the value of n can be found for a given value of the angle φ (or vice versa).

The average particle velocity in the displacement velocity field of a moving dislocation is equal to zero. The dislocations in the Smith interface do not produce a particle velocity behind the shock wave. The average particle velocity, which is equal to $u_{so} - u_{s1}$ and is in the x direction is that produced by the shock wave before hydrostatic relaxation occurs. The average particle velocity component in the z or y direction is equal to zero.

In a real shock wave the approximation $(a_1)_o = (a_1)_1$ no longer holds. However the analysis that was just developed should still describe qualitatively the stresses produced by the supersonic dislocations.

It should be emphasized that the shear and longitudinal waves radiated by the supersonic dislocations in the Smith interface will produce very rapidly fluctuating stresses behind the interface. The stresses given by Eqs. (20), (21), and (22) are only average values of these fluctuating stresses.

ACKNOWLEDGEMENT

This work was supported by the Advanced Research Projects Agency and the National Science Foundation through the Materials Research Center of Northwestern University.

ADDED NOTE

Figure 6 shows how a rich texture of planar sound waves are radiated in the shocked region behind the Smith interface by the supersonic dislocations. The group of dislocations that are oriented perpendicular to the dislocations shown in Fig. 6 radiated a similar set of sound waves. The radiated sound waves transport away energy released when forces, produced by the (resolved) shear stress of the shock wave, act on moving dislocations.

REFERENCES

1. J. Weertman and J. R. Weertman: Chapter 9 in Dislocation Theory: A Treatise, F. R. N. Nabarro, ed., Marcel Dekker, New York (in press).

2. J. E. Hockett, Trans. TMS-AIME, 1967, vol. 239, p. 969.

3. J. E. Hockett, Appl. High Speed Testing, vol. VI: The Rheology of Solids, p. 205, Insterscience, New York, 1967.

4. H. J. McQueen and J. E. Hockett, Met. Trans., 1970, vol. 1, p. 2997.

5. J. E. Hockett and N. A. Lindsay, 1971 J. Phys. E: Scientific Instruments, vol. 4, p. 520.

6. J. E. Hockett and H. J. McQueen, Second International Conf. on the Strength of Metals and Alloys, vol. III, p. 991, American Society for Metals, 1970.

7. N. Nagata and S. Yoshida, Trans. Japan Inst. Metals, 1972, vol. 13, p. 332.

8. N. Nagata and S. Yoshida, Trans. Japan Inst. Metals, 1972, vol. 13, p. 339.

9. V. R. Parameswaran and J. Weertman, Met. Trans., 1971, vol. 2, p. 1233.

10. V. R. Parameswaran, N. Urabe, and J. Weertman, J. Appl. Phys., 1972, vol. 43, p. 2982.

11. N. Urabe and J. Weertman, unpublished.

12. J. Weertman, Physics of Strength and Plasticity, A. S. Argon, ed., p. 75, M. I. T. Press, Cambridge, 1969.

13. W. G. Johnston and J. J. Gilman, J. Appl. Phys., 1959, vol. 30, p. 129.

14. G. T. Hahn, Acta Met., 1962, vol. 10, p. 727.

15. E. W. Billington and A. Tate, Proc. Roy. Soc. (London), 1972, vol. 327A, p. 23.

16. C. S. Smith, Trans. Met. Soc. AIME, 1958, vol. 212, p. 574.

17. C. S. Smith, The Mechanism of Phase Transformations in Metals, p. 291, Institute of Metals, London, 1956.

18. M. H. Rice, R. G. McQueen, and J. M. Walsh, Solid State Physics, F. Seitz and D. Turnbull, eds., vol. 6, p. 1, Academic Press, New York, 1958.

19. R. Kinslow, editor, High-Velocity Impact Phenomena, Academic Press, New York, 1970.

Discussion by T. Vreeland, Jr. and K. M. Jassby
California Institute of Technology

We note two differences between our observations and those you reported: 1) Our data for B in aluminum at room temperature gives a value about 2.8 times smaller than your value. 2) Our data in copper does not show the marked increase in B at low temperatures that you find in aluminum. If we are measuring an intrinsic interaction (of dislocations with phonons and conduction electrons) B should be the same at a given temperature in your aluminum samples and in ours. We also expect that the temperature dependence in aluminum and copper should be similar. We have no explanation for the differences at low temperature, but we hope to make some low temperature measurements in aluminum.

A MICRODYNAMICAL APPROACH TO CONSTITUTIVE MODELING OF SHOCK-INDUCED DEFORMATION

H. E. Read

Systems, Science and Software

La Jolla, California 92037

INTRODUCTION

Until several years ago, the existing theories of stress wave propagation were based on purely empirical approaches, in which the underlying physical processes of plastic flow were not considered [1]. Recently, however, the rapid advancements, both theoretical and experimental, in the knowledge of dislocation behavior have deepened our understanding of the microscopic processes that occur in metals during rapid plastic flow [2]. Correspondingly, there has been an increasing interest in relating the macroscopically observed response of shock-loaded metals to dislocation behavior at the microlevel [3]. This has resulted in a new (microdynamical) approach for describing rapid plastic deformation in which an attempt is made to reflect the salient features of the basic dislocation mechanisms in the constitutive description. Such an approach has been used to explain stress wave propagation phenomena in a number of metals [4-8]. In all of these studies, the emphasis has been on the initial loading process; virtually no attempts have been made to apply the microdynamical approach to the unloading process.

The present work is an attempt to develop a microdynamical framework for describing the uniaxial strain deformation of metals during stress wave propagation. A constitutive model is formulated by considering the basic physical processes that influence the motion of a dislocation during low temperature plastic flow. The model is applied to S-200 beryllium, using quasi-static strain hardening data, low-to-medium strain rate data, stress wave profiles, and a finite-difference computer code. The

THEORY

Consider the uniaxial strain deformation of a macroscopically homogeneous and isotropic polycrystalline metal and define the only non-vanishing strain component ϵ in terms of the density as $\dot{\epsilon} = \dot{\rho}/\rho$, where the dot denotes time differentiation following the material. Also, consider $\dot{\epsilon}$ to be decomposable into an elastic component $\dot{\epsilon}_e$, which obeys an incremental form of Hooke's law, and a plastic component $\dot{\epsilon}_p$. If there is no net plastic dilatation, the maximum shear stress τ, which occurs on planes inclined at 45° to the direction of ϵ, obeys the following relation:

$$\dot{\tau} = \mu(\dot{\epsilon} - \frac{3}{2} \dot{\epsilon}_p) \qquad (1)$$

where μ, the shear modulus, may vary with the state of deformation. Finally, the principal stress component σ in the direction of ϵ can be written in the form:

$$\sigma = p + \frac{4}{3} \tau \qquad (2)$$

where p, the hydrostatic pressure, will be taken to depend on the total strain ϵ. To develop an expression for the plastic strain rate, it will be assumed that most of the plastic flow occurs on planes of maximum shear stress. From dislocation theory, the tensor component of the plastic shear strain rate on these planes, $\dot{\gamma}_p$, can be defined in terms of the Burgers vector b, the mobile dislocation density N_m, and the mean dislocation velocity $\bar{v}$ by the equation:

$$\dot{\gamma}_p = bN_m \bar{v} \qquad (3)$$

Since $\dot{\gamma}_p = \frac{3}{4} \dot{\epsilon}_p$ in uniaxial strain, Eq. (3) may be recast in the form:

$$\dot{\epsilon}_p = \frac{4}{3} bN_m \bar{v} \qquad (4)$$

Lack of knowledge of the mobile dislocation density N_m poses one of the outstanding problems in dislocation theory today. There is no technique by which direct experimental measurements of this quantity can be made, and aside from some work by Gilman [3], virtually no theoretical progress has been made in analytically characterizing it. Because of the great uncertainty in N_m, an empirical approach is adopted here in which N_m is taken to depend on the plastic strain in the following manner:

$$N_m = \begin{cases} (N_{mo} + a_1 \epsilon_p^n) \exp(-a_2 \epsilon_p) & \text{, for initial loading} \\ N_{m\infty} + (N_m^* - N_{m\infty}) \exp(-a_3 \bar{\epsilon}_p^2) & \text{, for reverse loading} \end{cases} \quad (5)$$

In the above expression, N_{mo} is the initial mobile dislocation density, N_m^* denotes the final value of N_m at the end of the initial straining, $\bar{\epsilon}_p$ is the reverse plastic strain, $N_{m\infty}$ is the limiting value of N_m during reverse straining, and a_1, a_2, a_3 and n are arbitrary constants.

To derive an expression for the mean velocity $\bar{v}$, consider the motion of a generic dislocation along a glide plane in the crystal lattice. Following Conrad [9], the internal stress field due to various obstacles on and near the glide plane is assumed to have the form shown in Fig. 1. Large obstacles, such as precipitates and second-phase particles, which cannot be overcome by thermal activation produce the long range stress field τ_μ. It is assumed here that τ_μ is nominally periodic with amplitude τ_A^μ and has a vanishing mean value. Local perturbations in the long range stress field result from smaller obstacles, such as forest dislocations and the Peierls-Nabarro stress, which can be overcome by the thermal motion of the lattice. Upon assuming that the time required for acceleration of the dislocation is negligible, the mean velocity of the dislocation over the wavelength λ of the internal stress field is given by the expression:

$$\bar{v} = \frac{\lambda}{t_1 + t_2} \quad (6)$$

where t_1 denotes the time spent by the dislocation awaiting thermal activation over the short-range obstacles, and t_2 is the time consumed in gliding between these obstacles.

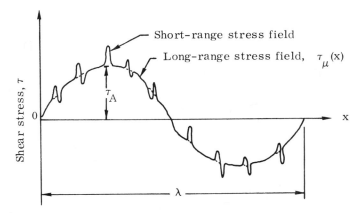

Figure 1. Internal stress field encountered by a dislocation.

Either one of two expressions may be used to describe the thermal activation time t_1. In the first, an empirical relation of the form:

$$t_1 = \frac{\lambda}{v_o} \left(\frac{\tau - \tau_A}{\tau'}\right)^{-m} \tag{7}$$

can be derived by considering that, at strain rates where thermal activation dominates, $\bar{v}$ generally depends on the quantity $(\tau - \tau_A)$ through a power law relation [10]. In the above equation, v_o, m and τ' are arbitrary parameters that depend, in general, on the plastic strain and temperature. An alternate expression for t_1, which explicitly contains the temperature, can be obtained by using an Arrhenius-type rate equation to describe the thermal activation process. If it is assumed that most of the time for thermal activation is spent overcoming the most difficult obstacle located at, or near, the peak of the long range stress field [9], t_1 can then be described by the relation:

$$t_1 = \frac{1}{\omega} \exp\left(\frac{H}{kT}\right) \tag{8}$$

where ω denotes the vibration frequency of the dislocation, H is the activation energy associated with the most difficult obstacle, k is Boltzmann's constant, and T represents the absolute temperature. In cases where deformation only at a fixed temperature is of interest, or where insufficient experimental data are available for establishing the explicit thermal dependence in Eq. (8), use of Eq. (7) for t_1 is generally preferable.

After the dislocation has overcome a short range obstacle through thermal activation, it moves along the glide plane under the action of a driving force, $b[\tau - \tau_\mu(x)]$. The time t_2 required to travel between the short range obstacles over the length λ is:

$$t_2 = \int_0^\lambda \frac{dx}{v(x)} \tag{9}$$

where $v(x)$ is the instantaneous glide velocity at location x. For dislocation velocities substantially below the relativistic range, the glide velocity generally depends linearly on the driving force, i.e.,

$$b[\tau - \tau_\mu(x)] = Bv(x) \tag{10}$$

Following Taylor [11], the above expression can be generalized to allow for relativistic effects at high dislocation velocities by taking:

CONSTITUTIVE MODELING OF SHOCK-INDUCED DEFORMATION

$$B = B_o(1-v^2/c_s^2)^{-1} \tag{11}$$

where c_s is the shear wave velocity and B_o denotes the drag coefficient for $v \ll c_s$. Upon solving Eqs. (10) and (11) for v, and introducing the result into Eq. (9), we obtain:

$$t_2 = \frac{1}{c_s} \int_0^\lambda \frac{dx}{\sqrt{\left[\frac{\alpha}{\tau-\tau_\mu(x)}\right]^2 + 1} - \frac{\alpha}{\tau-\tau_\mu(x)}} \tag{12}$$

where $\alpha = B_o c_s / 2b$. To evaluate the above integral, the dependence of τ_μ on x must be specified. For simplicity, an alternating-square wave variation with amplitude τ_A was adopted*, and for this case the evaluation of Eq. (12) leads to the following expression for t_2:

$$t_2 = \frac{\lambda}{2c_s} \left\{ \frac{\tau-\tau_A}{\sqrt{\alpha^2+(\tau-\tau_A)^2}-\alpha} + \frac{\tau+\tau_A}{\sqrt{\alpha^2+(\tau+\tau_A)^2}-\alpha} \right\} \tag{13}$$

An expression for $\dot\epsilon_p$ may be obtained by combining either Eq. (7) or Eq. (8) with Eqs. (4), (6), and (13). Selecting Eq. (7), the resulting constitutive equation may be cast into the form:

$$\dot\epsilon_p = \frac{4}{3} \left(\frac{\phi \psi}{\phi + \psi} \right) \tag{14}$$

where

$$\phi = \left(\frac{\tau-\tau_A}{\beta} \right)^m \tag{15}$$

$$\psi = 2bc_s N_m \left\{ \frac{(\tau-\tau_A)}{\sqrt{\alpha^2+(\tau-\tau_A)^2}-\alpha} + \frac{(\tau+\tau_A)}{\sqrt{\alpha^2+(\tau+\tau_A)^2}-\alpha} \right\}^{-1} \tag{16}$$

and $\beta = \tau'(bv_o N_m)^{-1/m}$. Some of the qualitative features of this constitutive relation are illustrated in Fig. 2, where the dependence of ϵ_p on τ is shown for a given value of plastic strain.

APPLICATION TO S-200 BERYLLIUM

An attempt was made to apply the preceding constitutive model to room temperature S-200 beryllium. This metal has a nominal unstrained density $\rho_o = 1.850$ gm/cm^3, and exhibits a small degree of elastic aniso-

---

*A sinusoidal function would be more physically realistic, but it introduces complexities in integration that have not yet been overcome.

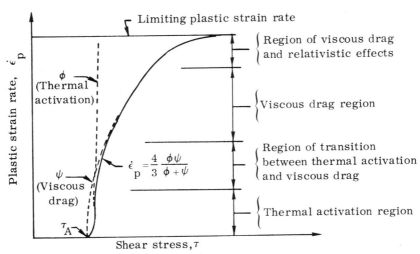

Figure 2. General dependence of the plastic strain rate on the shear stress for a given value of plastic strain according to the present constitutive model.

tropy, which was neglected here. The shear modulus μ and the elastic shear wave velocity c_s for this metal, determined from ultrasonic measurements [12], were found to have the values: μ = 1453 kbars and c_s = 8.86 x 10^5 cm/sec. For beryllium, the Burgers vector has the value b = 2.29 x 10^{-8} cm. The shock hydrostat was used to specify the pressure p for the small compressions considered herein. On this basis, the following expression for p was determined for pressures up to 30 kbars from the results of shock wave experiments on S-200 beryllium given in Ref. 12:

$$p = A_1 \theta + A_2 \theta^2 \qquad (17)$$

Here, $\theta = \rho/\rho_o - 1$, A_1 = 1114 kbars and A_2 = 3784 kbars.

Since there are insufficient data on S-200 beryllium to accurately specify some of the constants and functions in the expression for the plastic strain rate, it was necessary to use approximate methods of evaluation. For lack of better information, τ_A was simply taken as the quasi-static flow stress. Quasi-static stress-strain curves for S-200 beryllium determined under uniaxial stress conditions [12] were used to develop an expression for the dependence of the flow stress on the plastic strain. The transformation of the strain hardening data from uniaxial stress to uniaxial strain conditions was performed according to a procedure described in Ref. 13. In this manner the following representation of τ_A was developed which provides a reasonably accurate description of the quasi-static flow stress of S-200 beryllium during initial loading and reverse loading, where the material exhibits a Bauschinger effect:

$$\tau_A = \begin{cases} \tau_0 \sqrt{1 + c_1 \epsilon_p} & \text{, for initial loading} \\ \tau_\infty [1 - \exp(-c_2 \sqrt{\bar{\epsilon}_p})] & \text{, for reverse loading} \end{cases} \quad (18)$$

In this expression, $\bar{\epsilon}_p$ denotes, as before, the reverse plastic strain, and τ_o, τ_∞, c_1 and c_2 are constants which have the following values: $\tau_o = 1.35$ kbars, $\tau_\infty = 2.20$ kbars, $c_1 = 75$ and $c_2 = 17$.

The parameter m and β were determined from the data given in Refs. 14 and 15, which describe the dynamic behavior of S-200 beryllium for initial loading at strain rates from 3×10^{-3} to 2×10^{3} sec^{-1}. The strong temperature dependence of the shear stress exhibited in these experiments indicates that thermal activation is the rate-controlling dislocation mechanism over the range of strain rates considered. When thermal activation dominates, and viscous drag effects are negligible, Eq. (14) simplifies to:

$$\dot{\epsilon}_p = \frac{4}{3} \left(\frac{\tau - \tau_A}{\beta} \right)^m \quad (19)$$

where, as noted earlier, $\beta = \tau'(bv_o N_m)^{-1/m}$. Without explicitly specifying τ' and v_o, β and m were selected to provide a fit of Eq. (19) to the data in Refs. 14 and 15. In this manner, N_m remains defined by Eq. (5) given earlier, and the dependence of τ' and v_o on ϵ_p is implicitly defined. On using the first of Eqs. (18) to represent τ_A, an accurate fit of Eq. (19) to the data was obtained by taking m = 6.25, and allowing β to depend quadratically on ϵ_p. To include reverse loading, this result was generalized by assuming m remains fixed and β varies with plastic strain as follows:

$$\beta = \begin{cases} \beta_o + \beta_1 \epsilon_p^2 & \text{, for initial loading} \\ \beta^* & \text{, for reverse loading} \end{cases} \quad (20)$$

where $\beta_o = 0.125$ kbar-sec$^{1/m}$, $\beta_1 = 55$ kbar-sec$^{1/m}$, and β^* is the value of β at the end of initial loading. Fig. 3 shows a comparison between results predicted by Eq. (19), with m = 6.25 and β given by the first of Eqs. (20), and the corresponding experimental data [14, 15].

The constants a_1, a_2, a_3, n, N_{mo} and $N_{m\infty}$ in the expression for N_m given in Eq. (5) were evaluated from numerical studies of the stress

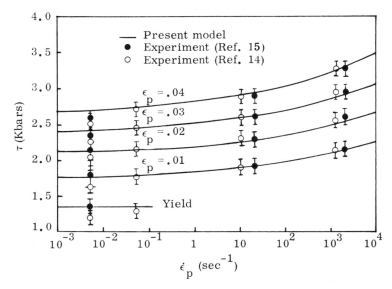

Figure 3. Dynamic behavior of S-200 beryllium at low-to-medium strain rates.

wave propagation experiments on S-200 beryllium reported in Ref. 14. In these experiments, fused silica flyer plates were impacted on stationary targets consisting of S-200 beryllium backed by fused silica. Particle velocity histories at the beryllium-fused silica interface in the targets were recorded with a laser interferometer. The numerical studies were performed with the one-dimensional, finite difference RIP code [16], using the present constitutive model of S-200 beryllium with tentative values for the constants listed above, and the constitutive model of fused silica from Ref. 17. To accomplish this, it was necessary to assume a reasonable value of B_o for S-200 beryllium, and for this purpose B_o was taken to have the value 2×10^{-13} kbar-sec, typical of many metals. The constants a_1, a_2, n and N_{mo} were determined from the loading sides of the wave profiles for Shots 260 to 264 given in Ref. 14, and the unloading sides were used to evaluate a_3 and $N_{m\infty}$. By iterating on the constants until an acceptable agreement was obtained between the experimental and calculated particle velocity histories, the following values of the constants were determined: $a_1 = 0.75 \times 10^{12} \text{cm}^{-2}$, $a_2 = 0$, $a_3 = 1 \times 10^4$, $n = 2$, $N_{mo} = 2 \times 10^6 \text{cm}^{-2}$ and $N_{m\infty} = 0.75 \times 10^6 \text{cm}^{-2}$.

The experiments from Ref. 14 also indicate that for stresses up to 24 kbars the release waves in S-200 beryllium travel slower than would be predicted by assuming the shear modulus retained its initial value of 1453 kbars throughout the deformation. By allowing the shear modulus to decrease linearly with the instantaneous maximum strain ϵ_m in the

CONSTITUTIVE MODELING OF SHOCK-INDUCED DEFORMATION

following manner:

$$\mu = \mu_o - \delta |\epsilon_m| \quad , \quad \text{for} \quad |\epsilon_m| \leq \frac{\mu_o}{\delta} \quad . \tag{21}$$

good agreement was achieved between the calculated and observed release wave velocities. In the above equation, $\mu_o = 1453$ kbars, $\delta = 4 \times 10^4$ kbars and μ is set to zero if $|\epsilon_m|$ becomes greater than $\frac{\mu_o}{\delta}$. The above form of the dependence of μ on strain precludes any healing effect in the shear strength.

The extent to which the present constitutive model, with the final values of the constants listed above, describes the loading profiles for Shots 260 to 264 from Ref. 12 is illustrated in Figs. 4 and 5. In Fig. 4, the influence of propagation distance on the amplitude of the elastic precursor and on the smearing of the plastic wave is shown for stress waves of about 11.5 kbars intensity. Fig. 5 shows the effect of stress intensity on the plastic wave profile for waves that have traveled equal distances.

As a proof test of the constitutive model described above, numerical calculations were performed with the RIP code for Shots 267 and 268 in Ref. 12. These plate impact experiments produced initial stresses of 24 kbars in the S-200 beryllium targets, and were designed to provide insight into stress wave attenuation. The results of the numerical cal-

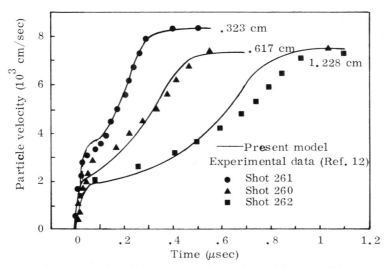

Figure 4. Particle velocity histories at the beryllium-fused silica interfaces, showing the effect of propagation distance on loading wave profiles.

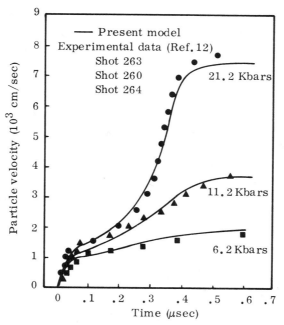

Figure 5. Particle velocity histories at the beryllium-fused silica interfaces, showing the effect of peak stress intensity on loading wave profiles.

culations are depicted in Figs. 6 and 7, where the calculated and observed particle velocity histories at the beryllium-fused silica interface in the targets are shown. As these figures reveal, the agreement between the numerical results and the experimental data is excellent.

CONCLUSION

An attempt has been made to develop a microdynamical approach for describing the uniaxial strain deformation of metals during stress wave propagation. Although much progress has been made recently in understanding dislocation behavior at high strain rates, a number of gaps in our knowledge, and in our ability to measure some of the important dislocation parameters, still exists. In regard to stress wave propagation, virtually nothing is currently known about the basic dislocation processes during the rapid reverse loading from a strained plastic state that take place during the passage of an unloading wave. Because of this, and other uncertainties, it is necessary, in pursuing a microdynamical approach to modeling shock-induced deformation, to empirically describe those aspects of the model which present dislocation knowledge is unable to provide

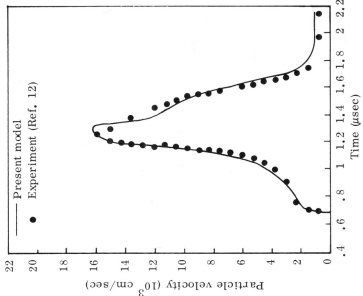

Figure 7. Particle velocity histories at the beryllium-fused silica interface for Shot 268.

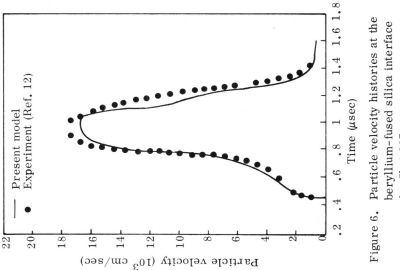

Figure 6. Particle velocity histories at the beryllium-fused silica interface for Shot 267.

guidance in characterizing. As a result, such a "microdynamical" approach still results in a phenomenological model, but it attempts to reflect through its development from dislocation concepts, the salient features of the basic underlying physical processes. As future advances are made in our knowledge of dislocation dynamics, and in the ability to measure the basic dislocation parameters, it should be possible to refine various aspects of the general model developed herein, without substantially altering the overall framework, thus leading to a more fundamental theory of dynamic plasticity based on the microstructural processes.

ACKNOWLEDGMENT

The author is indebted to Mr. R.A. Cecil for his invaluable assistance in developing the numerical methods for treating the constitutive model in the RIP code and for performing the numerical calculations. This work was supported by the Defense Nuclear Agency under Contract DASA01-70-C-0055.

REFERENCES

1. Hopkins, H.G., in Stress Waves in Anelastic Solids, Ed. by H. Kolsky and W. Prager, Springer-Verlag, 1964.

2. Klahn, D., A.K. Mukherjee and J.E. Dorn, Proc. 2nd Int. Conf. Strength Metals and Alloys, Amer. Soc. Metals, 1970.

3. Gilman, J.J., Applied Mechanics Reviews, 21, (1968) 767.

4. Taylor, J.W., J. Appl. Phys., 36 (1965) 3146.

5. Butcher, B.M., and D.E. Munson, in Dislocation Dynamics, Ed. by A.R. Rosenfield, et al., McGraw-Hill, 1968.

6. Johnson, J.N., and L.M. Barker, J. Appl. Phys., 40 (1969) 4321.

7. Jones, O.E., and J.D. Mote, J. Appl. Phys., 40 (1969) 4920.

8. Johnson, J.N., O.E. Jones and T.E. Michaels, J. Appl. Phys., 41 (1970) 2330.

9. Conrad, H., Mater. Sci. Eng., 6 (1970) 265.

10. Smidt, F.A., Jr. and A.L. Bement, in Dislocation Dynamics, Ed. by A.R. Rosenfield, et al., McGraw-Hill, 1968.

11. Gillis, P.P., J.J. Gilman and J.W. Taylor, Phil. Mag. (1969) 187.

12. Christman, D.R., and F.J. Feistmann, "Dynamic Properties of S-200 Beryllium", Matls. Struc. Lab., General Motors Corp., Rept. DNA 2785F, 1972.

13. Read, H.E., and R.A. Cecil, "A Rate Dependent Constitutive Model of Shock-Loaded S-200 Beryllium", Systems, Science and Software, Report DNA 2845F, 1972.

14. Green, S.J., and F.L. Schierloh, "Uniaxial Stress Behavior of S-200 Beryllium, Isotropic Pyrolytic Boron Nitride, and ATJ-S Graphite at Strain Rates to 10^3/sec and 700°F", Matls. Struc. Lab., General Motors Corp., Rept. MSL-68-11, 1968.

15. Kumble, R.G., F.L. Schierloh, and S.G. Babcock, "Mechanical Properties of Beryllium at High Strain Rates", Matls. Struc. Lab., General Motors Corp., Rept. MSL-68-8, VI, 1968.

16. Fisher, R.H., and H.E. Read, "RIP - A One-Dimensional Material Response Code," Systems, Science and Software, Rept. DNA 2993F, 1972.

17. Barker, L.M., and R.E. Hollenbach, J. Appl. Phys., 41 (1970) 4028.

WAVE PROPAGATION IN BERYLLIUM SINGLE CRYSTALS*

L. E. Pope and A. L. Stevens

Sandia Laboratories

Albuquerque, New Mexico 87115

ABSTRACT

Transmitted wave profiles are presented for beryllium single-crystals that were shock loaded along several crystallographic directions to activate selectively each of the primary, secondary and tertiary slip systems. A definite elastic precursor with a risetime of a few nanoseconds was exhibited by each shock-loaded single crystal, independent of orientation. For single crystals impacted along directions such that secondary and tertiary slip systems were activated, elastic precursor spikes followed by regions of stress relaxation were observed. The phenomenon of multiple plastic wave propagation was observed in a crystal shock loaded along an axis of nonsymmetry.

The resolved dynamic shear strength on basal, prism and pyramidal planes for conditions of uniaxial strain are calculated and compared to quasi-static values. The elastic responses of the single crystals are compared to the ill-defined elastic precursors observed in shock-loaded polycrystalline beryllium, and it is concluded that the elastic response of polycrystalline beryllium is a consequence of the grains in aggregate. Spallation of beryllium single crystals verifies that the dynamic fracture planes are the same as those observed quasi-statically; specifically, fracture occurs on basal, type II prism and pyramidal planes.

---

*This work was supported by the U. S. Atomic Energy Commission.

1. INTRODUCTION

In the past few years several experimental studies have been made of the elastic-plastic wave propagation in polycrystalline beryllium at stress levels up to 40 kbar.[1-5] In these studies the observed transmitted wave profiles typically exhibited dispersive elastic wave fronts in addition to dispersive plastic waves which are commonly observed in many metals. The characteristics of the elastic waves generally were not consistent from material to material. For example, the risetime of the elastic precursor of a beryllium powder-metallurgy product (S-200) varied from 40 to 80 nsec,[4] and the risetime of the precursor of a wrought high purity beryllium was approximately 100 nsec.[2] Both of these observations are in contrast with the observed wave profiles exhibited by a high-purity beryllium powder-metallurgy product (HP-10), which had no elastic precursor at all in the "as-pressed" condition, but did have a precursor with a risetime of only a few nsec in a highly textured state.[5] In all of these examples the risetimes of the elastic waves was not explained by tilt at the impact interface.

The dispersive behavior of the elastic wave in polycrystalline beryllium may be a consequence of the grains in aggregate, such as the thermal and/or acoustic anisotropy resulting from the random orientation of grains. Alternately it may be due to grain boundaries or to a low yield strength on basal planes with substantial work hardening. The explanation of this behavior will have a significant influence on the development of a constitutive model to explain the dynamic response of beryllium. A rate-dependent constitutive model is presently available for S-200 beryllium[6] which works well in the plastic region. As will be discussed later, however, rate-dependence may not be the complete explanation for the dispersion of the plastic wave. In addition, the apparent agreement between the measured response and the calculated response in the elastic region is fortuitous, since the risetime of the calculated elastic precursor is a consequence of the numerical scheme used in the code.

It is apparent that a complete understanding of the dynamic response of beryllium in the elastic region will not be obtained by studying polycrystalline material exclusively; for this reason the present single crystal study was undertaken. Until this time, the only data available concerning shock wave propagation in beryllium single crystals consisted of two plate-impact experiments.[1] These experiments were designed such that the shock propagation direction was either perpendicular or parallel to the c-axis and, therefore, activated secondary and tertiary slip systems (slip occurred on prism and pyramidal planes, respectively).

In this paper we report the results of plate-impact experiments on beryllium single crystals of several different crystallographic orientations. For each orientation special attention is directed toward the elastic response. The orientations were chosen to activate selectively basal $\{0001\}$ $<11\bar{2}0>$, prism $\{10\bar{1}0\}$ $<11\bar{2}0>$ and pyramidal $\{11\bar{2}2\}$ $<11\bar{2}3>$ slip systems, and to observe the phenomenon of multiple plastic wave propagation predicted theoretically[7] for shock propagation directions other than 0° or 90° with respect to the c-axis. Shock wave transit times are measured, and Lagrangian shock velocities are calculated. In addition, for some orientations experiments were designed to produce spallation.

2. PLATE-IMPACT EXPERIMENTS

Beryllium material characterization and plate-impact specimen preparation are discussed in this section. The experimental configuration, which is more complex than the standard plate-impact experiment,[8] is illustrated, and the special considerations taken in measuring shock wave transit times are described. Also, the slip systems activated by shock waves propagating along several crystallographic directions are analyzed.

2.1 Specimen Characterization and Preparation

The beryllium single crystals were obtained from The Franklin Institute Research Laboratories. As specified by The Franklin Institute, the starting material was a Pechiney CR grade 1.5-inch diameter rod on which three floating zone passes were taken to make the large single crystals. Two boules were grown, and semiquantitative emission spectrographic analysis was performed on a section taken from each boule. The impurities detected are listed in Table I. Specimens were cut from the central 6-inches of a 12-inch rod using a carborundum wheel. Plate specimens 1- to 1.5-inches in diameter by nominally 0.188-inch thick were cut such that the plate normal (the shock propagation direction) was parallel, respectively, to an a-axis, a c-axis, 45 degrees from a c-axis, and 33 degrees from a c-axis. All orientations except the c-axis were determined to within ±0.5 degree by the back reflection Laue technique; the c-axis crystals were oriented to within ±5 minutes of arc by a special x-ray, lapping technique.[9] The surfaces of all crystals had a 600 grit finish, and the surfaces of each specimen were parallel within 0.002-inches. The "as-received" specimens were lapped flat within 4 light bands; special hand lapping procedures on nylon blocks were used on c-axis specimens to avoid pitting.

Table I. Chemical Composition[+]

Element	Be	Si	Ni	Mg	Pb	Fe	Cu	Cr
at %, boule 1	Bal	.0005	.05	.005	.005	.05	.005	.0005
at %, boule 2	Bal	.0005	.05	.0005	N.F.	.05	.0005	.0005
Elements checked, not found				Sb, As, Ba, Bi, B, Cd, Nb, Sa, Ge, Au, Mo, P, Pt, K, Sr, Te, W, V, Zr, Al, Ca, Co, Mn, Ag, Na, Sn, Ti, Zn				

[+]W. B. Coleman Co., Philadelphia, Pennsylvania

2.2 General Configuration

The plate-impact configuration is shown in Fig. 1. For each experiment a hollow foam projectile was faced with an instrumented quartz gauge assembly.[10,11] The quartz gauge, using a 0.5-inch guard ring in the shorted configuration, was potted into an aluminum ring with insulated flying contactor pins extending through the aluminum ring to establish electrical contact just prior to impact. The quartz gauge was electrically isolated from the rest of the system. The electrical signal transmitted to the target assembly was resistively loaded for a controlled 6 to 7 volt signal to a Tektronix 585 oscilloscope. The quartz gauge essentially had a free back surface with the thickness of the gauge being selected to control the time, relative to impact, when an unloading wave was transmitted back into the specimen. For the

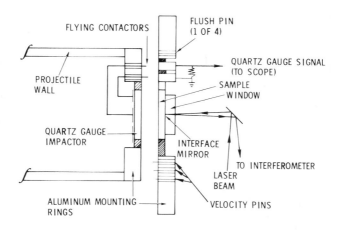

Figure 1. The experimental configuration.

c-axis sample a sapphire facing was bonded on the quartz gauge in order to achieve the required impact stress amplitude. Prior to potting the target into an aluminum mounting ring, a fused silica window with a mirror plated on the front surface was epoxyed onto the back of the beryllium target. Laser interferometer instrumentation[12,13] was used to measure the particle velocity of the mirror at the interface, with the maximum velocity determined to within ±2%. Also potted into the aluminum mounting ring were four flush pins and two velocity pins, each of which were insulated from the mounting ring. A ground pin established a common ground between the aluminum rings prior to closure of the other pins. The velocity pins measured impact velocity (±.2%), and the flush pins measured the tilt between the aluminum ring in the projectile facing and the target mounting ring. The quartz gauge directly measured the tilt of impact between the gauge and the target.

The oscilloscopes recording the quartz gauge signal were triggered from one of the velocity pins, which closed 2 μsec prior to impact. The fiducial circuitry and the 519 Tektronix oscilloscopes which recorded the interferometer signal were triggered from one of the flush pins. A fiducial was placed on both the quartz gauge record and the interferometer record, and a one-nanosecond counter recorded the real time between these fiducials during an experiment. The shock wave transit time through the target specimen was established by correlating the quartz gauge impact record to the transmitted wave profile measure by the interferometer. Wave transit time was measured to within ±6 nsec (±1.7% error in wave speed), by accounting for the transit times of signal cables, the light path length from the interface mirror to the interferometer, and the photomultiplier tube transit time.

2.3 Shock Wave Propagation

Here we briefly summarize shock wave propagation in plate-impact experiments, then we apply the concepts to shock wave propagation along different crystallographic directions in beryllium single crystals and determine the slip systems that will be activated. The method for comparing dynamic and quasi-static shear strengths is discussed.

In this study each specimen was subjected to carefully controlled planar impact, incident uniformly and simultaneously, over one of its faces. The high velocity impact generates plane longitudinal compressive waves which propagate from the impact interface into both the target and impactor materials. A single compressive wave propagates through the impactor, since the quartz gauge and sapphire impactor materials remain in the elastic region. However, for the target material the dynamic yield strength is exceeded and two compressive waves propagate through the beryllium. The first, the elastic precursor wave, travels at the longitudinal sound speed and loads the beryllium to its yield strength. The second, the plastic wave, travels at the bulk sound speed and loads the beryllium from the yield strength up to the impact stress. For the a- and c-axis loading a state of uniaxial strain is developed, and maintained for a finite time, in the central region of the specimen. In this region of uniaxial strain the only nonzero strain component is parallel to the shock propagation direction.[14] For more general loading directions additional strain components exist.[7] Due to interfacial mechanical impedance mismatching, the compressive waves reflect as rarefaction (unloading) waves from the beryllium-fused-silica interface and from the free surface of the quartz gauge (or from the sapphire-quartz-gauge interface), respectively. In addition, compressive waves are transmitted into the fused quartz window, which is thick enough that the compressive waves do not reach its back surface during the observation time. The thickness of the impactor is also selected to predetermine the time that the unloading wave from the impactor is transmitted into the beryllium target. When twice the transit time in the impactor is less than the transit time of the beryllium target, a tensile stress is developed in the specimen. This tensile stress will cause spall if it is large enough in magnitude.

Before discussing the expected results for shock loading in different crystallographic directions, it is helpful to review the deformation response of beryllium. It is well known that the primary, secondary and tertiary slip systems in beryllium are $\{0001\} <11\bar{2}0>$, $\{10\bar{1}0\} <11\bar{2}0>$ and $\{11\bar{2}2\} <11\bar{2}3>$, respectively.[9,15-18] These three slip systems are illustrated in Fig. 2 where the slip plane and Burgers vector for each slip

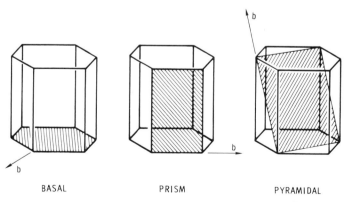

Figure 2. The three slip planes in beryllium; the Burgers vectors are identified.

system are identified. The shear stresses required to cause basal, prismatic and pyramidal slip are approximately 0.1, 0.7 and 10 kbar, respectively, under quasi-static loading at room temperature. It is the relatively large differences in shear stress required to activate slip on the three planes that allow each slip system to be activated selectively.

We consider first the state of uniaxial strain that is developed in the central region of a c- or an a-axis specimen in plate-impact experiments. In this region of uniaxial strain the only nonzero component is parallel to the shock propagation direction. Thus, for shock loading along the c-axis (perpendicular to basal planes) shear stresses are not generated on either basal or prism planes. Slip will occur, however, on pyramidal planes, and a large yield strength is expected. For shock loading along an a-axis only shear stresses on prism and pyramidal planes are nonzero. The ease with which prism slip is activated by comparison to pyramidal slip suggests that yielding, for this orientation, is controlled by slip on prism planes. A yield strength significantly less than that for shock loading parallel to a c-axis is expected. The HELs (the dynamic yield strengths in uniaxial strain) for shock loading parallel to an a-axis and a c-axis have been measured previously as 4 and 40 kbar, respectively.[1] For shock loading along a direction 33 or 45 degrees from a c-axis, shear stresses are developed on basal, prism and pyramidal planes. However, for shock propagation directions greater than a few minutes of arc from the c-axis, basal slip is activated.[9,19] For large enough angles pyramidal slip is absent, which is the case for 33 and 45 degree crystals. Also, at 33 and 45 degrees, it has been shown that the resolved shear stress on the basal plane is more than twice the maximum resolved shear stress on prism planes.[19] Thus, shock loading along a direction 33 or 45 degrees from the c-axis activates basal slip, and a yield stress less than that for prism slip is expected.

In order to compare dynamic yield data from shock experiments with quasi-static data, it is necessary to calculate the resolved shear stress on appropriate planes in the direction of the Burgers vector. The approach is based on the general framework laid down by Johnson, Jones and Michaels,[20] and assumes that a true uniaxial strain load condition is developed in the elastic region for all load directions. The error introduced by this assumption is small, since for beryllium the maximum departure from pure longitudinal wave propagation in the elastic region is less than 1%.[21] The elastic constant data of Smith and Arbogast[22] are used as the elastic moduli in evaluating the following expressions. For shock propagation parallel to the c-axis, each of the six individual $\{11\bar{2}2\}$ pyramidal planes are subject to a shear stress τ, along the $<11\bar{2}3>$ directions, given by[23]

$$\tau = \sigma B^2 r(1 - C_{13}/C_{33}) = 0.434\sigma \tag{1}$$

where $1/B^2 = 1 + r^2$, $r = c/a = 1.57$ for beryllium, C_{13} and C_{33} are elastic moduli, and σ is the measured HEL.*

For shock propagation parallel to an a-axis, the shear stress τ on the $\{10\bar{1}0\}$ prism planes in the $<11\bar{2}0>$ directions is given by[23]

$$\tau = \sigma(1 - C_{12}/C_{11}) \sin \beta \cos \beta = 0, \pm.393\sigma \tag{2}$$

where C_{12} and C_{11} are elastic moduli, $\beta = 0, \pm\pi/3$ depending on which of the three independent prism planes are being considered, and σ is the measured HEL.

For shock propagation along an axis where the angle between that axis and the c-axis is greater than 0 degrees (c-axis) and less than 90 degrees (a-axis), the shear stress τ on the basal plane $\{0001\}$ in an arbitrary $<11\bar{2}0>$ direction is given by[19]

$$\tau = \sigma(1/R_{11})\sin(\varphi + \beta)[\sin\theta\cos\theta(R_{11} - R_{33})$$

$$+ (\cos^2\theta - \sin^2\theta)R_{13}] \tag{3}$$

where

$$R_{11} = C_{11}\sin^4\theta + C_{33}\cos^4\theta + 2(C_{13} + 2C_{44})\sin^2\theta\cos^2\theta ,$$

$$R_{13} = (C_{11} - C_{13} - 2C_{44})\sin^3\theta\cos\theta + (C_{13} + 2C_{44} - C_{33})\cos^3\theta\sin\theta ,$$

$$R_{33} = (C_{11} + C_{33} - 4C_{44})\sin^2\theta\cos^2\theta + C_{13}(\cos^4\theta + \sin^4\theta) ,$$

---

*$\sigma = Y\left(\dfrac{1-\nu}{1-2\nu}\right)$, where σ is the measured HEL in uniaxial strain, Y is the static yield stress in one-dimensional stress, and ν is Poisson's ratio.

the C_{ij} are elastic moduli, θ is the angle between the shock propagation direction and the c-axis (here 33 and 45 degrees), $\beta = 0, \pm\pi/3$ depending upon which arbitrary $<11\bar{2}0>$ direction is being considered, φ is the angle between an arbitrary $<10\bar{1}0>$ direction and the projection of the shock propagation direction in the basal plane, and σ is the measured HEL. The angle φ was calculated from back reflection Laue photographs to be 20.4 and 17.8 degrees for the 33 and 45 degree crystals, respectively. Equation (3) then reduces to

$$\tau = -.283\sigma, \; .155\sigma, \; .439\sigma \tag{4}$$

for $\theta = 33$ degrees and to

$$\tau = -.334\sigma, \; .152\sigma, \; .486\sigma \tag{5}$$

for $\theta = 45$ degrees.

It is not uncommon that the measured HEL (σ) is a factor of two larger than the HEL calculated from static yield strength data. This arises because the measured HEL is not a unique quantity; it depends on both the target thickness and the impact stress. It is often observed that, as the propagating distance increases, the measured HEL decreases toward the static value of the material.

3. RESULTS AND DISCUSSION

3.1 Transmitted Wave Profiles

The plate-impact data for beryllium single crystals impacted parallel to four different crystallographic directions are listed in Table II. Each crystal is identified by the shock propagation direction; 33 and 45 degrees are measured relative to the c-axis. It is noted that the shock velocity of the elastic precursor agrees with the ultrasonic wave speed for all orientations (the ultrasonic wave speed for the 33 degree crystal was not measured).

The transmitted wave profile for each of these four

Table II

Plate-Impact Data

	c-axis	33 Degree	45 Degree	a-axis
Ultrasonic Wave Speed* (mm/μsec)	13.99		13.52	12.75
Elastic Wave Speed+ (mm/μsec)	14.29	13.82	13.43	12.74
Plastic Wave Speed+ (mm/μsec)	8.41	11.86 6.15	8.51	8.34
Projectile Facing	Sapphire	X-Cut Quartz	X-Cut Quartz	X-Cut Quartz
Target Thickness (mm)	2.977	4.554	4.746	4.752
Projectile Velocity (mm/μsec)	.471	.318	.193	.128
Max. Particle Velocity (mm/μsec)	.400	.183		.063
Free Surface Velocity (mm/μsec)			.183	
Load Stress (kbar)	68.6	24.2	14.0	9.6

*±.6%

+±1.7%

orientations is shown in Fig. 3. The ordinate is the particle velocity of the mirror at the beryllium-fused-silica interface or the free surface velocity for the 45 degree crystal; the abscissa is time. The profiles are displaced in relative time along the abscissa. The vertical arrows indicate the positions on the plastic waves at which the plastic wave speeds listed in Table II were calculated. The dashed portion of the c-axis profile is only a suggested profile; in this region the frequency response of the 70045 RCA photomultiplier tubes was exceeded. Consequently, the maximum of the precursor spike is not known, but it could not have exceeded 72.2 kbar (the impact stress). The time span covered by the dashed portion of the profile is 7 nsec. The crystal thickness for each orientation is listed. We remind the reader that slip is occurring on basal planes for both the 33 and 45 degree crystals, on prism planes for the a-axis crystal, and on pyramidal planes for the c-axis crystal.

Several features are readily apparent from Fig. 3: (1) Each single crystal has a definite elastic precursor wave which has a

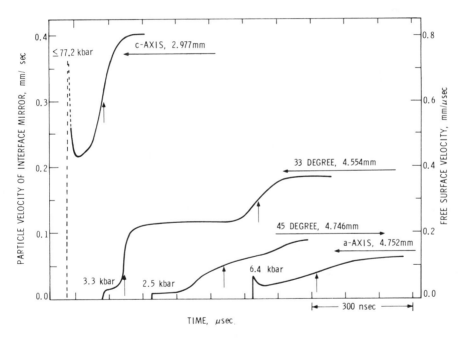

Figure 3. Particle or free surface velocity vs. time profiles of single beryllium crystals. The specimen identification refers to the shock propagation direction; the bold arrows under the identifications indicate which ordinate is applicable for each profile.

risetime of a few nsec. (2) The HELs of the primary, secondary, and tertiary slip systems differ considerably as do the quasi-static shear strengths. (3) The measured HEL of the 33 degree crystal is larger than that of the 45 degree crystal; slip occurs on basal planes for both of these crystals. It will be shown later that the resolved yield stress on the basal plane is nominally the same. (4) The a- and c-axis crystals, where secondary and tertiary slip systems operate, respectively, show the same features, that is, a large amplitude precursor spike (of nominal magnitude of the impact stress) followed by a region of stress relaxation prior to the arrival of the plastic wave. (5) The 33 degree crystal has a characteristically different structure behind the elastic precursor, i.e., two plastic waves obviously exist.

Several of the features listed above require further discussion. The phenomenon of multiple plastic wave propagation is a consequence of the plastic anisotropy of beryllium and has been predicted theoretically by Johnson.[7] The 33 degree crystal result provides the first experimental verification of this prediction. The first plastic wave, a quasi-longitudinal wave, travels at near the elastic precursor speed, while the second plastic wave, a quasi-transverse wave, travels at a speed less than the shear wave speed. It is possible that the dispersion in the plastic wave in polycrystalline beryllium, previously attributed to rate-effects,[6] may be due in part to this multiple plastic wave phenomena.

Consider the typically observed long risetime of the elastic precursor in shock-loaded polycrystalline beryllium. The measured HEL for the 45 degree single crystal is of the same magnitude as the HEL of polycrystalline beryllium.[2,23] Thus, the ramp of the elastic precursor in polycrystalline beryllium is not explained by a low yield strength on basal planes with substantial work hardening. Furthermore, it was observed that all shock-loaded single crystals exhibited a definite elastic precursor with a few nsec risetime. Thus, the elastic response of shock-loaded polycrystalline beryllium is not a result of the intrinsic response of beryllium to shock waves. It is concluded, therefore, that the long risetime of the elastic precursor in shock-loaded polycrystalline beryllium is a consequence of the grains in aggregate.

A plausible explanation of the elastic response of polycrystalline beryllium is that "locked-in" thermal stresses effectively lower the observed yield strength. Armstrong and Borch[24] calculated that residual thermal stresses, developed in beryllium due to thermal anisotropy across grain boundaries, are a few kbar when beryllium is cooled from the melt and are greater than 1 kbar when beryllium is cooled from 1300°F. In addition, the change in the structure of the precursor with texturing[5] supports this residual stress hypothesis. The "as-pressed" material, which has a nearly random orientation of grains, has a very long elastic ramp. The same material in a highly textured state (the density of basal planes nearly parallel to the specimen surface is 8 times that found in the "as-pressed" material) has an elastic precursor of 6.8 kbar with a risetime of a few nsec. With texturing, greater macroscopic thermal compatibility would exist. This would produce a resultant decrease in "locked-in" thermal stresses and a larger measured yield strength, as observed.

The existence of the precursor spike and stress relaxation behind the precursor for the a- and c-axis crystals makes a direct comparison between dynamic and quasi-static shear strengths difficult. Equations 1-3 are used to estimate the maximum elastic shear stress on prism and pyramidal planes for the a- and c-axis crystals and to calculate the resolved shear stress on the basal plane for the 33 and 45 degree crystals; the results are listed in Table III along with approximate quasi-static values. However, in the stress relaxation region, the amount of accumulated plastic strain is not known, and resolved shear stresses cannot be calculated. The slight difference in resolved shear stress for basal slip as measured on the 33 and 45 degree crystals is explained by the difference in impact stress for these crystals.

The explanation of the existence of a precursor spike followed by stress relaxation for secondary and tertiary slip systems in beryllium single crystals is based on an initial small dislocation density with subsequent dislocation motion and multiplication. It is hypothesized that the beryllium single crystals initially have

Table III

Resolved Shear Stresses (kbar)

	Basal	Prism	Pyramidal	Impact Stress
Quasi-static	.1	.7	10	
33 Degree	1.4			24.2
45 Degree	1.2			14.0
A-axis		2.5		9.6
C-axis			≤33.5	68.6

a low density of grown-in dislocations on prism and pyramidal planes, and all deformation introduced during specimen preparation occurs on basal planes. This is due to the large stresses required, relative to basal slip, to activate slip on prism or pyramidal planes. In plate-impact experiments a large impact stress is applied on a nsec time scale (less than 6 nsec), and the a- and c-axis crystals respond elastically at impact, even though the yield strengths are exceeded. Initially, the precursor spike has an amplitude equal to the impact stress, but with propagation distance the amplitude of this spike decreases. Consider an arbitrary plane within the target which is perpendicular to the shock propagation direction.

At the arrival of the precursor spike at this plane the mobile dislocations are accelerated, and dislocation multiplication processes are initiated. These processes relax the stress toward the equilibrium elastic limit. The exact magnitude to which the stress relaxes depends on the dislocation processes which occur and on the arrival time of the following plastic wave. The profiles in Fig. 3 show the structure of the precursor spike and stress relaxation for a propagation distance equal to the specimen thickness. Propagation through a larger distance would result in a decrease in amplitude of the precursor spike and perhaps also the stress plateau in the stress relaxation region.

3.2 Spallation

Additional plate-impact experiments, designed to produce spall, were conducted on a-axis, c-axis, and 45 degree crystals. Previous observations show that spall damage in brittle materials is of a different character from fracture produced by quasi-static load application, because the duration of application of the impulsive load is too short to permit extensive crack propagation. For this reason spall damage is characterized by a distribution of small, independent cracks inside the sample. Complete separation of the body into disjoint parts results only if the number and size of these cracks is such as to exhaust the cross-sectional area on some plane.

A soft recovery system was used to catch the samples intact and eliminated all secondary impacts large enough to modify the spall character. Recovered specimens were sectioned with a spark cutter, and the sectioned surfaces were polished for metallographic examination. The reader is reminded that the cleavage planes observed quasi-statically are basal $\{0001\}$, type II prism $\{11\bar{2}0\}$ and pyramidal $\{10\bar{1}2\}$.

The a-axis crystal was loaded to 10 kbar with a 444 nsec pulse (0.050-inch quartz-gauge impactor). Several sections were examined. All spall cracks were essentially parallel to the specimen surface and were independent of the orientation of the section plane. It is concluded that the spall plane is a type II prism plane, for shock propagation parallel to an a-axis. This is expected because the type II prism plane is perpendicular to the shock propagation direction and experiences the largest tensile stress.

In single crystals where the c-axis is parallel to the shock propagation direction, the basal plane experiences the largest tensile stress. Quasi-statically, the stress for basal cleavage is much less than the stress for cleavage on either type II prism or pyramidal planes.[25] The initial experiment loaded a c-axis single crystal to 7 kbar with a 340 nsec pulse (a 0.040-inch fused-quartz impactor); spall did not occur. A second c-axis crystal was loaded to 11.5 kbar with a 256 nsec pulse (a 0.030-inch fused-quartz impactor), and spall did occur. Even at this much larger stress, spall was observed only over 50% of the projected spall plane area. The observed spall cracks were parallel to the specimen surface, and the fracture plane was identified as the basal plane.

The a- and c-axis experiments show that the dynamic stress required to cause spall on basal planes is nominally the same as that to cause spall on type II prism planes. This larger-than-expected spall strength on the basal plane is not understood. Quasi-static measurements[25] show the stress for cleavage on type II prism planes to be about 6 times larger than the stress for cleavage on the basal plane. This is still a factor of 4 less than the dynamic stress required to cause spall on either basal or type II prism planes. This may indicate that the dynamic fracture process is strongly time dependent, i.e., the amount of damage observed may be dependent upon the time required to initiate cracks at defects and to propagate cracks along the cleavage planes. Further, the process may be dominated by the time required to initiate the cracks. This is particularly plausible if shear stresses play a role in the crack nucleation process. For example, the c-axis crystals were stringently oriented such that very low shear stresses were developed on the basal planes which were the observed spall planes. The ratio of dynamic- to quasi-static-fracture strength is 24:1. In contrast, the a-axis crystals were less stringently oriented, thus resulting in higher shear stresses on the fracture planes and only a 4:1 ratio of dynamic- to quasi-static-fracture strength.

The 45 degree crystal was loaded to 14 kbar with a 444 nsec pulse (0.050-inch quartz gauge impactor). A sufficiently large tensile stress was developed so that complete spall occurred; the recovered sample had separated into two pieces. Examination of the spall surface indicated it was rough and noncoplanar, with planes of several different crystallographic orientations. In order to identify the fracture planes the recovered specimen was mounted on a goniometer, and the fractured surface was illuminated by a laser beam incident parallel to the shock propagation direction. Reflected beams were observed, each coming from a number of planes having the same orientation. The angles between the reflected beams were used to identify the fracture planes as both basal $\{0001\}$ and pyramidal $\{10\bar{1}2\}$. It was not possible to identify whether or not fracture occurred on type II prism planes. The spall surface was also examined with a scanning electron microscope. A stereo pair of micrographs showed that a 5×10^{-3} cm^2 projected area on a plane perpendicular to the shock propagation direction contained as many as 5 different fracture planes, the largest of which was 10^{-3} cm^2. No evidence of ductile fracture was observed.

4. SUMMARY

Plate-impact experiments using instrumented impactors were conducted on beryllium single crystals of several different orientations. The shock propagation directions (a c-axis, an a-axis, and an axis 33 and one 45 degrees from a c-axis) were chosen to activate selectively each of the tertiary $\{11\bar{2}2\}$ <$11\bar{2}3$>, secondary $\{10\bar{1}0\}$ <$11\bar{2}0$>, and primary $\{0001\}$ <$11\bar{2}0$> slip systems. Transmitted wave profiles were measured and correlated in time to the impact records. Shock wave transit times were measured to within ±6 nsec which resulted in a ±1.7% uncertainty in shock wave speeds. The elastic wave speed agreed with the ultrasonic wave speed for all orientations.

The measured HEL for the three slip systems shows a large variance in magnitude as do the quasi-static shear strengths. Equations, valid in the elastic region, are presented for calculating the resolved shear strength on basal, prism and pyramidal planes for each of the shock propagation axes. For those crystals where the secondary and tertiary slip systems were activated, an elastic precursor spike of near the impact stress amplitude followed by a region of stress relaxation was observed prior to the arrival of the plastic wave. This response is explained by an initial low mobile dislocation density on prism and pyramidal planes with subsequent dislocation motion and multiplication. The beryllium initially responds elastically and relaxes toward the elastic yield strength, with the relaxation being controlled by dislocation processes.

A definite elastic precursor with a risetime of a few nsec was observed for all single crystals examined. Comparing the well-defined elastic precursor in single crystals to the ill-defined elastic precursor previously reported for shock-loaded polycrystalline beryllium, it is concluded that the polycrystalline response is a consequence of the grains in aggregate and is not due to the intrinsic response of the beryllium within a grain. It was also shown that the polycrystalline response is not due to a low yield strength in basal slip with substantial work hardening. It is suggested that the elastic response of shock-loaded polycrystalline beryllium is explained by "locked-in" thermal stresses which lower the measured yield strength.

ACKNOWLEDGMENTS

The authors are grateful to J. N. Johnson for his critical review of the manuscript. During the course of the work helpful discussions were held with J. N. Johnson and L. W. Davison. We also wish to thank A. G. Beattie for measuring the ultrasonic wave speeds. The capable technical assistance of G. T. Holman in building and instrumenting the experiments is appreciated.

REFERENCES

1. J. W. Taylor, "Stress Wave Profiles in Several Metals," <u>Dislocation Dynamics</u>, edited by A. R. Rosenfield, G. T. Hahn, A. L. Bement, Jr., and R. I. Jaffee, McGraw-Hill Book Co., New York, 1968.
2. D. R. Christman, N. H. Froula, and S. G. Babcock, "Dynamic Properties of Three Materials, Vol. 1: Beryllium," Report MSL-68-33, General Motors Corporation, Materials and Structures Laboratory, Nov. 1968. Also reported, in part, by D. R. Christman and N. H. Froula, AIAA J. $\underline{8}$, 477 (1970).
3. N. H. Froula, "The Hugoniot Equation of State of S-200 Beryllium to 1000°F," Report MSL-68-16, General Motors Corporation, Materials and Structures Laboratory, July 1968.
4. D. R. Christman and F. J. Feistman, "Dynamic Properties of S-200-E Beryllium," Report MSL-71-23 (DNA 2785F), General Motors Corporation, Materials and Structures Laboratory, February 1972.
5. A. L. Stevens and L. E. Pope, "Wave Propagation and Spallation in Textured Beryllium," this proceedings.
6. H. E. Read and R. A. Cecil, "A Rate-Dependent Constitutive Model of Shock-Loaded S-200 Beryllium," Report DNA 2845F, Systems, Science and Software, April 1972. Also reported by H. E. Read, "A Dislocation Dynamics Approach to Formulating Constitutive Models of High Strain Rate Plastic Deformation in Metals: Application to Beryllium," this proceedings.
7. J. N. Johnson, J. Appl. Phys. $\underline{43}$, 2074 (1972).

8. C. H. Karnes, "The Plate Impact Configuration for Determining Mechanical Properties of Materials at High Strain Rates," *Mechanical Behavior of Materials Under Dynamic Loads*, edited by U. S. Lindholm, Springer-Verlag New York, New York, 1968.
9. G. H. London, V. V. Damiano, and H. Conrad, Trans. Met. AIME $\underline{242}$, 979 (1968).
10. R. A. Graham, F. W. Neilson, and W. B. Benedick, J. Appl. Phys. $\underline{36}$, 1775 (1965).
11. G. E. Ingram and R. A. Graham, "Quartz Gauge Technique for Impact Experiments," *Fifth Symposium on Detonation*, Office of Naval Research, Report ACR-184, 1970.
12. L. M. Barker, "Fine Structure of Compression and Release Wave Shape in Aluminum Measured by the Velocity Interferometer Technique," *Behavior of Dense Media under High Dynamic Pressure*, Proc. I.U.T.A.M. Symposium, Gordon and Breach, New York, 1968.
13. L. M. Barker and R. E. Hollenbach, J. Appl. Phys. $\underline{43}$, 4669 (1972).
14. A. L. Stevens and O. E. Jones, J. Appl. Mech. $\underline{39}$, 359 (1972).
15. H. T. Lee and R. M. Brick, Trans. ASM $\underline{48}$, 1003 (1956).
16. R. I. Garber, I. A. Gindin, and Yu V. Shubia, Soviet Physics-Solid State $\underline{5}$, 315 (1963).
17. H. Conrad and I. Perlmutter, "Beryllium as a Technological Material," Air Force Material Laboratory Technical Report AFML-TR-65-310, Nov. 1965.
18. G. L. Tuer and A. R. Kaufman, "Ductility of Beryllium as Related to Single Crystal Deformation and Fracture," *The Metal Beryllium*, edited by D. W. White, Jr., and J. E. Burke, The American Society for Metals, 1955.
19. J. N. Johnson, Sandia Laboratories, private communication.
20. J. N. Johnson, O. E. Jones, and T. E. Michaels, J. Appl. Phys. $\underline{41}$, 2330 (1970).
21. J. N. Johnson, J. Appl. Phys. $\underline{42}$, 5522 (1971).
22. J. F. Smith and C. L. Arbogast, J. Appl. Phys. $\underline{31}$, 99 (1960).
23. L. W. Davison and J. N. Johnson, "Elastoplastic Wave Propagation and Spallation in Beryllium: A Review," Sandia Laboratories Report SC-TM-70-634, Sept. 1970.
24. R. W. Armstrong and W. R. Borch, Met. Trans. $\underline{2}$, 3073 (1971).
25. J. Greenspan, "Ductility Problems," Chapter 9, *Beryllium: Its Metallurgy and Properties*, edited by H. H. Hausner, University of California Press, 1965, p. 240.

SHOCK-INDUCED DYNAMIC YIELDING IN LITHIUM FLUORIDE SINGLE CRYSTALS

Y. M. Gupta and G. R. Fowles

Shock Dynamics Laboratory, Physics Department

Washington State University, Pullman, Washington

INTRODUCTION

The study of solids at stresses where material rigidity cannot be ignored is important for an understanding of dynamic failure and thus learning about the complete constitutive relation for a solid. A number of solids in this stress range display stress relaxation which is described by a time dependent constitutive relation. The rate dependence is characterized by a decay of the elastic wave as a function of propagation distance. The early work in the area of stress relaxation involved polycrystals but the recent work has been generally on single crystals. This includes copper (Jones and Mote, 1969), sodium chloride (Murri and Anderson, 1970), tantalum (Gillis et al, 1971), lithium fluoride (Asay et al, 1972) and tungsten (Michaels, 1972). The use of single crystals make for a more meaningful theoretical analysis as the slip planes are well defined and the complications due to grain boundaries are avoided. The usual procedure has been to treat the solid as elastic-plastic-relaxing solid and use the constitutive relation proposed by Duvall (1963):

$$dP_x - a^2 d\rho = -F \cdot dt \qquad (1)$$

where P_x is the longitudinal stress, a is the elastic wave speed in Eulerian coordinates, ρ is the density and F is the relaxation function characterizing the time dependent behavior of the solid. The relaxation function F for an isotropic material is given as twice the shear modulus multiplied by the plastic strain rate. Dislocation dynamics can then be used to infer an understanding of the dynamic yield behavior.

The present status of dynamic yielding is not very clear and there are various suggestions to explain the experimental observations, but these are speculative. It is important that the parameters controlling the dynamic behavior be well understood for not only characterizing the material but also for a better understanding of dislocation behavior in dynamic loading. The work of Asay et al (1972) and Asay and Gupta (1972) have established some preliminary results about point defects (Mg++ ions) and their influence on dynamic yield in LiF. In this paper we report the study of shock propagation in lithium fluoride (LiF) single crystals and the influence of point defects and their aggregates on dynamic behavior of LiF. The present work is a continuation of the preliminary work reported above with better material characterization. An attempt to correlate the dynamic behavior with quasi-static yield behavior for the conditions employed is presented. A comparison with dielectric loss data is also made.

MATERIAL CHARACTERIZATION TECHNIQUES AND RESULTS

In the present work, two types of crystals moderately and strongly doped were used. (Gupta, 1973). The moderately doped crystals were purchased from Harshaw Chemical Co. (Solon, Ohio) and are denoted as material H. The strongly doped category were purchased from Dr. F. Rosenberger (Dept. of Physics, University of Utah) and are classified as material U.

A quantitative spectrographic analysis with absolute accuracies of 15% to 20% was done for Mg++ ions. (Gupta 1973). This was checked by an independent technique involving atomic absorption within the quoted errors. The infra-red examination did not reveal any OH$^-$ ions. Material H contained 120 ± 25 ppm of Mg++ ions and Material U had 600 ± 70 ppm of Mg++ ions.

The heat treatments were chosen so that no new dislocations were produced by thermal stresses and the rate of cooling did not introduce quenching strains. However the temperatures and times employed are such that maximum possible change in the distribution of the impurities is obtainable. (Dryden, 1965) The heat treatments were as follows.
Air Quench: The samples were put in a lava box and heated to 400°C and kept there for approximately 12 hours. After this time the cover was taken off and the crystals (sitting on a lava base 3 1/2" x 2 1/4" x 1/8" in size) were put out in the room. The samples attained room temperature in approximately six to seven minutes. This procedure was well reproducible.
Anneal II: The samples after quenching to room temperature were put in the lava box and annealed at 150°C for seven hours and then slowly cooled in the oven at an initial rate of 15°C to 16°C which

later dropped to 8°C per hour. In every case the temperature was monitored at the specimens by a thermocouple and this cooling rate was found to be very consistent in all the annealing procedures. Anneal III: This was the same as Anneal II except that annealing time was 70 hours. The cooling rate was the same as above.

The dislocation density measurements were made on the {100} faces of the crystals by the etch pit techniques developed by Gilman and Johnston (1962). Both the acidic and neutral etches were employed in the present technique. Dislocation density measurements were made for the as-received crystals of both U and H materials. The effect of heat treatment on the dislocation densities was also estimated by successive polishing and etching. The results reported in Table I were obtained by averaging over several pieces within the crystals. Seven to eight measurements were made on each piece.

TABLE I -- Dislocation Densities

Material	Dislocation Density $\times 10^{-4}$/cm	Dislocation Density in Sub-grain Boundaries $\times 10^{-4}$/cm	Comments
H(as-received)	4 - 12	1 - 5	Dislocations immobile
U(as-received)	2 - 10	1 - 5	Dislocations immobile

The heat treatments employed here did not affect the dislocation density within the range reported in Table I. However the dislocations tended to be more mobile in the air-quenched specimens for the H material as shown by neutral etch. For all other specimens they were immobile. This is in agreement with earlier work of Asay and Gupta (1972).

For a given concentration it is important to know the distribution of impurities due to different heat treatments. Dielectric loss measurements provide an indirect technique for this. The subject of distribution of atomic defects has recently been discussed in a review by Hartmanova (1971). The method consists of making the dielectric loss measurements in an applied A.C. field with electrodes provided by plating the samples with gold. The various LiF samples were cleaved to approximate dimensions of 1 cm x 1 cm x 1 mm. These were then lapped and polished. The parallelism was better than 5 μm. In the case of the annealed samples, gold was plated after annealing and preparation of the samples. For the air-quenched samples, the samples were first prepared to the desired degree and then gold plated. These were then air-quenched in such a manner that measurements could be made within one to two hours of quenching. (This was made possible through the kind cooperation of Drs. J. R. Asay and G. A. Samara of Sandia Laboratories.) The capacitance and conductance measurements were made at Sandia Laboratories through the cooperation of Dr. G. A. Samara. The Tan δ values can then easily be computed (Gupta 1973). The dipole concentration was obtained by the method used by Grant and Cameron (1966). The results are presented in Table II.

TABLE II -- Dipole Concentration for Different Specimens[a]

Material	Treatment	$10^3 \times$ [Tanδ]	$10^{-18} \times$ Conc. of Dipoles (cm^{-3})	Conc. of[b] Dipoles (ppm)
H	A.Q.	5.0	5.5	95 (90%)
	A.Q. (Room temp. anneal)	1.25	1.37	24 (23%)
	Anneal II	0.9	1.0	17 (16%)
	Anneal III	0.68	0.75	13 (12%)
U	A.Q.	6.8	7.5	130 (22%)
	Anneal II	4.2	4.6	80 (13%)
	Anneal III	3.8	4.2	73 (12%)

[a]The measurements were made at room temperature between frequencies of 10 and 100,000 Hz.

[b]The numbers in parenthesis in the column give the percentage of Mg++ ions present as dipoles.

Quasi-static yield stress measurements under compression were made for the different crystals and the various heat treatments used. Rectangular crystals ranging from 2.5 mm to 6.0 mm in lateral dimensions and from 7 mm to 12 mm in height were compressed in an Instron machine modified for compression. The anvil displacement rate for all measurements was 5×10^{-4} cm/sec. This gives the strain rate as approximately (4 to 7) $\times 10^{-4}$/sec. The yield stresses for the H and U crystals for the different heat treatments are shown in Table III.

In Fig. 1(a) we have plotted the critical resolved shear stress (given as half the yield stress along <100> compression for a NaCl type structure) for air-quenched crystals as a function of Mg++ concentration. Our data also includes materials which have not been mentioned earlier for the sake of brevity. The data of Reppich (1972) and Dryden et al (1965) have also been shown. It should be mentioned that Reppich's data are for crystals cooled

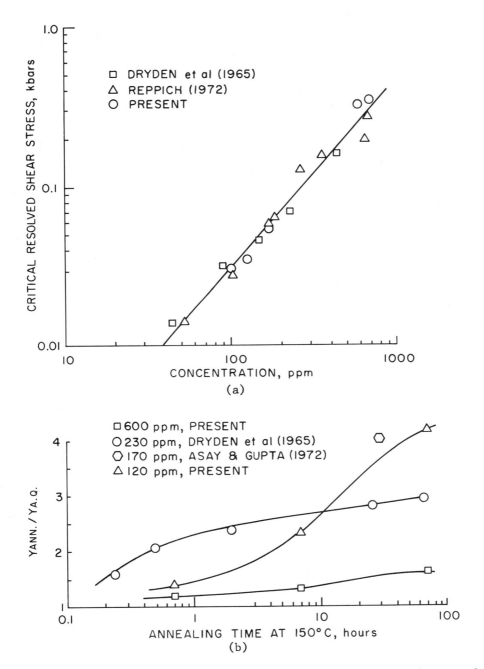

Fig. 1 (a) Critical resolved shear stress of air-quenched crystals plotted as a function of impurity concentration. (b) Ratio of yield stress of annealed and air-quenched crystals plotted as a function of annealing time.

from 500°C at a rate of 50°C/minute. However this is fairly close to our cooling rate. The line drawn to fit the entire data has a $C^{1.16}$ dependence.

TABLE III -- Yield Stress Data

Material	Yield Stress, kbars[a,b,c]			
	A.Q.	Anneal I[d]	Anneal II	Anneal III
H	.07 ± .005 (8)	.096 ± .008 (4)	.16 ± .016 (4)	.29 ± .02 (4)
U	.66 ± .04 (5)	.82 ± .08 (2)	.88 ± .04 (4)	1.07 ± .04 (5)

[a] The applied strain rate was $(4-7) \times 10^{-4}$/sec.
[b] The number in parentheses refers to the number of measurements made for each sample.
[c] Limits of error denote the standard deviation.
[d] Same as Ann. II and Ann. III except annealing time was 0.7 hrs.

The ratio of the flow stress of annealed crystals to that of the air-quenched crystals ($Y_{Ann.}/Y_{A.Q.}$) is plotted in Fig. 1(b) along with the data of Dryden et al (1965) for LiF containing 230 ppm of Mg. One point from the work of Asay and Gupta (1972) is also shown. Our results indicate that the change in flow stress due to annealing is large for the H material. The change for the U material is not quite so large. The results are analogous to the results obtained in dielectric relaxation measurements. Asay and Gupta (1972) have reported a Mg++ concentration of 130 ppm for material II based on Asay's semi-quantitative analysis. (Asay,1971) The same material now referred to as As II was found to contain 170 ppm of Mg++ by a quantitative analysis. Examination of the data near and beyond 10 hours of annealing time indicates an increase in $Y_{Ann.}/Y_{A.Q.}$ with impurity concentration with a maximum inferred at about 175-200 ppm. Beyond this the $Y_{Ann.}/Y_{A.Q.}$ decreases with impurity. At shorter times of annealing it is hard to draw any conclusions. The maximum occuring within 175-200 ppm agrees well with the applicability of the linear C-dependence for air-quenched crystals up to 200-250 ppm of Mg++ concentration. (Gupta, 1973)

SHOCK WAVE EXPERIMENTS

These experiments were the main theme of this work and consisted of studying shock wave propagation along the <100> crystallographic direction in LiF. Shock waves were produced by impacting LiF crystals with 6061-T6 aluminum projectiles travelling at an average velocity of 0.343 mm/μsec. An initial impact stress of 29 kbar is attained if a non-linear elastic equation of state is assumed for LiF (Asay, 1971). The gas gun used has been described by Fowles et al (1970). Stress time profiles were measured by bonding quartz gauges to the back of the specimens (Graham et al, 1965). This method is well suited for LiF due to the close mechanical impedances of the two materials. Experiments were done for materials H and U and the three different heat treatments. Further experimental details are omitted and can be seen in the work of Gupta (1973). The experimental results are summarized in Table IV. The rate of stress drop behind the wavefront ($-\partial P_x/\partial t$) is also shown in Table IV.

The shock wave structure in LiF consists of two waves, an elastic wave followed by a plastic wave. While the plastic wave amplitude is nearly constant (about 26 kbars in quartz), within the scatter of projectile velocity data, the elastic wave amplitude is a decreasing function of propagation distance. Figure 2 shows the elastic wave amplitudes in LiF plotted as a function of propagation distance for the different cases. The amplitudes in LiF were obtained from the quartz amplitudes by an impedance method described in the work of Asay and Gupta (1972). All the specimens except H(A.Q.) displayed a rapid decay in precursor amplitude for propagation distances less than 1mm. The decay beyond 1mm is comparatively small. The precursor amplitudes for a given sample thickness increase considerably with annealing for material H. For the U material this increase is smaller, in fact U(Ann. II) and U(Ann. III) are nearly the same. The reasons for the low precursor amplitude for the 3.05mm thick U(Ann. III) crystal are not known. Since the crystal was too thick for bending and the tilt for the shot was good, the possibility of a small crack in the specimen is speculated. The material H displays a greater change in precursor amplitudes with heat treatment as compared to material U in agreement with the dielectric loss results shown in Table II. The precursor amplitudes in Fig. 2 also increase with increasing yield stress but this is true only for the directionality of the increase. The H(Ann. III) material is fairly close to the U(Ann. II and Ann. III) materials in precursor amplitudes while their yield stresses are apart by a factor of three. The same is true for H(Ann. II) and U(A.Q.) which have yield stresses different by a factor of four. Figure 2 indicates the extrapolation to elastic impact stresses for all cases except H(A.Q.).

Material H(A.Q.): Five specimens ranging from 0.5mm to 3.2mm were shot. The decay in the precursor amplitude with propagation distance was very small. The stress drop behind the wavefront was nearly the same except for shot 71-052. Shot 71-052 showed the experimental reproducibility to be excellent as seen in Table IV from the results of the two specimens shot. The extrapolation to the elastic impact stress for this material is not obvious. The behavior of this material is very similar to that observed by Asay (1971) for the very pure LiF.

Material H(Ann. II): Three specimens ranging from 0.5mm to 3.0mm were shot. The stress drop behind the wavefront was small compared

TABLE IV -- Summary of Shock Experiments

Shot #	Sample	Sample Thickness (mm)	Projectile Velocity (mm/μsec)	Projectile Tilt (millirad)	Precursor Quartz (kbar)	Precursor LiF (kbar)	Stress Minima in Quartz (kbar)	$(-\frac{\partial P_x}{\partial t})_h$ (kbar/μsec)
72-058	H(A.Q.)	0.5	0.337	0.3	4.01	4.3	3.6	190.0
71-052	H(A.Q.)	1.4	0.35	0.7	3.44	3.69	3.2	50.0
71-052	H(A.Q.)	1.4	0.35	0.7	3.49	3.75	3.1	57.0
71-057	H(A.Q.)	1.9	0.345	0.1	3.7	3.98	2.6	190.0
71-057	H(A.Q.)	3.2	0.345	0.1	3.5	3.76	2.6	89.0
72-059	U(A.Q.)	0.35	0.343	0.3	21.1	23.1	15.7	260.0
72-026	U(A.Q.)	1.46	0.349	0.2	9.86	10.68	8.1	200.0
72-026	U(A.Q.)	2.8	0.349	0.2	8.93	9.66	7.9	100.0
72-060	H(Ann.II)	0.5	0.334	0.3	17.65	19.2	17.57	28.0
72-029	H(Ann.II)	1.5	0.348	0.1	9.0	9.75	7.37	103.0
72-029	H(Ann.II)	3.0	0.348	0.1	7.39	7.98	5.1	192.0
72-046	U(Ann.II)	0.43	0.348	0.3	20.54	22.55	16.4	226.0
72-027	U(Ann.II)	1.5	0.35	0.2	13.83	15.06	7.56	466.0
72-044	U(Ann.II)	3.06	0.335	0.3	12.2	13.24	6.2	318.0
72-038	H(Ann.III)	0.6	0.342	0.2	19.21	21.05	15.62	160.0
72-015	H(Ann.III)	1.5	0.33	0.5	12.37	13.44	7.16	333.0
72-015	H(Ann.III)	3.0	0.33	0.5	10.5	11.38	5.61	323.0
72-054	H(Ann.III)	0.46	0.334	0.1	21.06	23.1	14.6	280.0
72-033	U(Ann.III)	1.57	0.35	0.2	14.02	15.26	7.4	480.0
72-033	U(Ann.III)	3.05	0.35	0.2	8.82	9.54	5.82	228.0

DYNAMIC YIELDING IN LiF SINGLE CRYSTALS

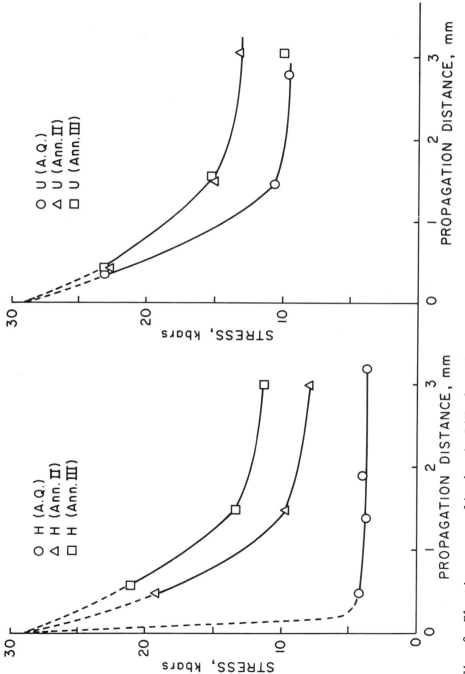

Fig. 2 Elastic wave amplitudes in LiF plotted as a function of propagation distance.

to H(Ann. III), U(A.Q.), U(Ann. II) and U(Ann. III). The $(-\partial P_x/\partial t)$ value of 28.0 kbar/μsec for the thin crystal was the smallest of all observed. All of this discrepancy cannot be attributed to an experimental flaw since the stress drops for the thicker crystals were also small compared to the other materials.

Material H(Ann. III): Three specimens ranging from 0.6mm to 3.0mm were shot for this material. While the precursor amplitude decay is slower compared to H(Ann. II), the curves are qualitatively similar. The rate of stress drop behind the wavefront is considerably larger for the two thicker crystals as compared to the thin crystal.

Material U(A.Q.): The three specimens shot ranged from 0.35mm to 2.8mm. The initial rapid decay is followed by a smaller decay. The rise time of the elastic wave for the thick specimens was larger than for the thin specimen despite the good tilt. The stress drop behind the wavefront decreased with increasing propagation distance. This is particularly interesting for the thicker specimens since the change from 200 kbar/μsec to 100 kbar/μsec in stress drop cannot be attributed to tilt as the two were shot in the same target.

Materials U(Ann. II) and U(Ann. III): These will be treated together due to the similar behavior. Three specimens ranging from approximately 0.45mm to 3.05mm were shot for each material. The behavior of both materials up to a crystal thickness of 1.5mm is nearly identical with respect to precursor amplitude and stress drop behind the wavefront. For the 3.05mm thick crystals the reasons for the different behavior are not clear, as already mentioned.

DISCUSSION OF PRESENT RESULTS AND COMPARISON WITH EARLIER WORK

Figure 2 reveals increasing precursor amplitudes with quasi-static yield stress in general. There are two interesting features. (1) At the high impurity concentrations the time of annealing causes no change in precursor amplitudes. (2) For long annealing times, the differences in precursor amplitudes due to impurity concentration are small. Asay et al (1972) have concluded a strong dependence on defect concentration and the observation of a minimum decay rate for 100 ppm of Mg^{++} concentration. They have observed the largest precursor amplitudes for a material with quasi-static yield stress of 0.32 kbar (hereafter referred to as Material II following Asay et al). Direct comparisons with the work of Asay et al (1972) are hard to make since the thermal state of the materials used in their study is not well defined. We can obtain some idea by making comparisons with the work of Asay and Gupta (1972) who have also studied the same material II with a well defined thermal heat treatment. Both the above mentioned works quote a figure of 130 ppm of Mg^{++} based on Asay's (1971) work. On subsequent quantitative measurement we obtained a concentration of 170 ppm for Mg^{++} ions in Material II. This is well within the error estimate of ±50% quote by Asay (1971). Asay and Gupta (1972)

observed a nearly constant precursor amplitude of 5.6 kbars for an air-quenched specimen. The precursor amplitude for a 2mm thick specimen which had been annealed for 30 hours at 150°C was found to be 17.34 kbars. This is higher than any of our amplitudes for that thickness. This permits the following important conclusions with respect to our data.
(1) The air-quenched crystals display higher precursor amplitudes with increasing concentration of Mg++ and thus an increasing yield stress. The precursor decay after the first millimeter of propagation is very small.
(2) The conclusions regarding the annealed crystals are a little complicated. The results of Asay and Gupta (1972) and the present results regarding precursor amplitudes suggest a minimum in precursor decay either at 170 ppm or possibly up to 200 ppm if the yield stress and dielectric loss work is an indication. This minimum decay rate seems to be strongly correlated to the impurity concentration beyond which all Mg++ ions cannot be quenched as dipoles. The prediction of a minimum is in agreement with the work of Asay et al (1972).
(3) The difference in the data for materials I, II and III(a) in the work of Asay et al (1972) could be due to the different thermal processes of various suppliers.

An interesting result is obtained if we consider the $Y_{Ann}/Y_{A.Q.}$ for 30 hours of annealing from our Fig. 1(b) and the value reported by Asay and Gupta (1972). The ratios for the material H(120 ppm of Mg++), II(170ppm of Mg++) and U(600ppm of Mg++) are 3.6, 4.0 and 1.5 respectively. The ratios of precursor amplitude of annealed to air-quenched crystals for 2mm thickness in the same order are 3.0, 3.1 and 1.4 respectively. Our values of 3.0 and 1.4 are slightly higher since they are for 70 hours annealing. The two sets of numbers compare favorably. The crystals with highest $Y_{Ann}/Y_{A.Q.}$ appear to give the highest precursor amplitude beyond the first millimeter of decay. This information can be used to infer the precursor amplitude in annealed crystals at larger distances.

CONCLUSIONS

Briefly, the important results and conclusions in this work are as follows.
(1) The stress decay in LiF under dynamic loading is strongly influenced by both Mg++ impurities and heat treatments. For a given defect concentration the impurity clustering reduces the rate of elastic wave attenuation.
(2) There appears to be a correlation between precursor amplitudes beyond 1mm propagation distance and quasi-static yield stress measurements. This correlation can predict the minimum rate of stress decay for precursor amplitudes as obtained by Asay et al (1972).

(3) Quasi-static yield stress is strongly dependent on Mg++ concentrations and heat treatments. A linear dependence with impurity concentration is experimentally demonstrated for air-quenched LiF up to 200 ppm Mg++ concentration.

(4) Dielectric loss measurements for the low impurity material (120 ppm) are in good agreement with earlier studies. The studies on high impurity material (600 ppm) show the quenching of only 22% of Mg++ ions as dipoles with a small change on annealing.

REFERENCES

Asay, J. R., Ph.D. Thesis, Washington State University (1971).
Asay, J. R. and Y. M. Gupta, J. Appl. Phys. 43, 2220 (1972).
Asay, J. R., G. R. Fowles, G. E. Duvall, M. H. Miles
    and R. F. Tinder, J. Appl. Phys. 43, 2132 (1972).
Dryden, J. S., S. Morimoto and J. S. Cook, Phil. Mag. 12, 379 (1965).
Duvall, G. E., in *Stress Waves in Anelastic Solids* (Springer-Verlag, Berlin, 1964).
Fowles, G. R., G. E. Duvall, J. Asay, P. Bellamy, F. Feistmann,
    D. Grady, T. Michaels and R. Mitchell, Rev. Sci. Instr. 41, 984 (1970).
Gillis, P. P., K. G. Hoge and R. J. Wasley, J. Appl. Phys. 42, 2145 (1971).
Gilman, J. J. and W. G. Johnston, in *Solid State Physics* (Academic Press, New York, 1962) Vol. 13.
Graham, R. A., F. W. Nielson, W. B. Benedick, J. Appl. Phys. 36, 1775 (1965).
Grant, R. M. and J. R. Cameron, J. Appl. Phys. 37, 3791 (1966).
Gupta, Y. M., Ph.D. Thesis, Washington State University (1973).
Hartmanova, M., Phys. Status Solidi (a) 7, 303 (1971).
Jones, O. E. and J. D. Mote, J. Appl. Phys. 40, 4920 (1969).
Murri, W. J. and G. D. Anderson, J. Appl. Phys. 41, 3521 (1970).
Michaels, T. E., Ph.D. Thesis, Washington State University (1972).
Reppich, B., Acta Met. 20, 557 (1972).

ACKNOWLEDGMENTS

This work was supported by the Air Force Office of Scientific Research, ARPA Contract No. F44620-67-C-0087 and National Science Foundation Grant No. GH 34650.

EFFECTS OF MICROSTRUCTURE AND TEMPERATURE ON DYNAMIC DEFORMATION OF SINGLE CRYSTAL ZINC*

P. L. Studt, E. Nidick, F. Uribe

University of California, Lawrence Livermore Laboratory

Livermore, California

A. K. Mukherjee

University of California, Davis, California

INTRODUCTION

Considerable progress has recently been made in the understanding of the physical processes which govern the dynamic elastic-plastic response of metals.[1-3] Some of the workers in dynamic plasticity have tried to interpret their data in terms of the various rate-controlling mechanisms from dislocation theory. Most of such work has been phenomenological in nature and used an Arrhenius-type rate equation without incorporating explicitly the substructural details of the micromechanism for flow and the pertinent rate-controlling step. The principal objective of this report is to rationalize the various probable rate-controlling mechanisms for plastic deformation and correlate them with the present experimental data. Major emphasis will be given towards the synthesis and unification of strain rate effects arising from the simultaneous operation of several mechanisms. In particular, attempt will be made to estimate theoretically plastic strain rates over wide ranges of stress and temperature, based on as realistic a model as possible for the pertinent rate-controlling steps in the substructure.

The first series of experiments consisted of gas gun flyer plate tests while the second series of experiments consisted of Hopkinson split bar tests. Gas gun flyer plate tests were used to study the influence of impact surface region microstructure on

---

* Work performed under auspices of the U.S. Atomic Energy Commission.

elastic precursor decay and elastic overshoot in a one dimensional strain state. Hopkinson split bar tests were used to study yield stress dependence on strain rate and temperature in a one dimensional stress state.

Zinc single crystals were oriented four different ways with respect to shock propagation direction. In the flyer plate experiments the shock directions were perpendicular to the first order prism plane, or the second order prism plane (along the a axis), or the basal plane (along the c axis). These orientations limit slip to various combinations of the $<11\bar{2}3>\{11\bar{2}2\}$ slip systems. In the Hopkinson split bar experiments the shock directions were perpendicular to the second order prism plane or 30° to the basal plane with an a axis direction oriented 15° from the maximum resolved shear stress direction in the basal plane. The first orientation again limits deformation to c + a slip and the second orientation to basal slip.

ONE DIMENSIONAL STRAIN TESTS

Test Technique

Conventional gas gun flyer plate test techniques were used to impact single crystal zinc targets with 2024-T4 aluminum alloy flyer plates. Three, three and one half,[4] and four inch bore diameter guns were used to generate impact shock pressures from 25 Kb to 78 Kb. Tilts were measured by means of contact pins and raster scopes. Projectile velocities were measured with shorting pins, time interval counters and elapsed time flash x-ray photos of the projectile. X-cut quartz crystals in the current mode[5,6] and 50 ohm manganin in-material gages[7,8,9] were used to measure shock pressures. The resulting voltage-time outputs were fed to a variety of oscilloscopes and photographed. The photographic records were then finally transformed into pressure-time plots. The impacted targets were recovered by deflecting the projectile and catching the target in a rag-filled container. Multiple target tests were performed in addition to single target tests to reduce the variability in the data caused by changes in projectile velocity and tilt.

Specimens

One and one half inch diameter by ~ 6 inches long zinc single crystals were purchased from commercial sources.[10,11] Nominal purity of these crystals was 99.999%. The crystals were received with the a axis or the c axis or the normal to the first order prism plane within 2° of the rod axis. The actual angle and orientation were determined by analysis of back reflection laue photographs. Target specimens were electrical discharge machined (EDM) from the rods. Surfaces were prepared several different ways. All specimens were first EDM planed to provide flat and parallel specimen surfaces. Additional preparations were (a) heavy lapping (~ 10μ alumina grit), (b) light lapping (~ 10μ alumina grit), (c) light lapping followed by chemical polishing with a 10% nitol-90% water solution, and (d) vibratory polishing (1μ diamond grit) followed by chemical polishing with the above noted nitol solution. The prepared specimen disks were then examined by x-ray diffractometry, Laue back reflection photography and, in the case of chemically polished specimens, their surfaces were examined optically for evidence of twinning. Specimens were then cored out of the disks by EDM.

Metallographic examination of cross-sections of the as-prepared specimens revealed that condition (a) caused a 9μ layer of recrystallized material to form. Beneath that layer a 13μ deep region of twins 25μ long were formed. Only twins with traces on the first order prism plane that were parallel to the basal plane were observed. The twins were very dense just beneath the recrystallized layer and reduced in density to zero at approximately 22μ beneath the surface. Back reflection laue patterns showed Debye rings consisting of discrete spots superimposed over the single crystal pattern. Condition (b) gave a very thin, ~ 1μ deep recrystallized layer and a few isolated twins beneath it. All other conditions gave no indication of surface damage.

Initial dislocation densities were determined by dislocation etch pit counting on material remaining after the specimens were cored out. A combination of dislocation etch pit counting on first order prism planes[12,13,14] and basal planes[15,16] was used to determine the dislocation densities for second order pyramidal slip and basal slip. Total initial dislocation densities were approximately 2.5×10^6 cm^{-2}, as measured on the first order prism plane. Dislocation densities on the basal plane were approximately 9×10^4 cm^{-2}. Since dislocations lying in the basal plane are not etched on the basal plane the initial second order pyramidal slip dislocation density is taken as 9×10^4 cm^{-2} and the basal slip dislocation density as 2.5×10^6. These values have been corrected from the raw count of number/cm^2 to cm/cm^3 by means of the Schoeck correlation[17] and for crystallographic affects.

Results

The shock, pressure-time traces from thirty-three single crystal zinc targets have been analyzed with respect to (a) the fractional approach of the peak shock pressure to the predicted equilibrium Hugoniot shock pressure as a function of target thickness, and (b) elastic precursor stress as a function of target thickness. McQueen[18] pointed out that the initial shock response of impacted strain-rate sensitive materials is elastic. The materials then plastically decay to the equilibrium Hugoniot. The initial shock pressure-particle velocity relationship is described by the 1-D Hugoniot where 1-D signifies no elastic or plastic strain transverse to the shock direction. The final equilibrium pressure-particle velocity state is described by the equilibrium or 3-D Hugoniot where 3-D signifies no net strain transverse to the shock direction but canceling amounts of elastic and plastic transverse strain along with plastic strain in the shock direction. The initial and final shock pressures for each target were determined graphically from Fig. 1 in the usual manner.[19] Figure 1 is derived from data from references 20, 21 for zinc and 21, 22 for aluminum 2024-T4 alloy. The difference between the 1-D (P_{1D}) and 3-D (P_{3D}) Hugoniot shock pressure ($P_{1D} - P_{3D}$) for a given flyer plate velocity was then divided into the difference between the experimental peak stress (P_{exp}) and P_{3D}. The resulting values of elastic overshoot are plotted in Fig. 2. Elastic precursor stress as a function of target thickness is plotted as Fig. 3.

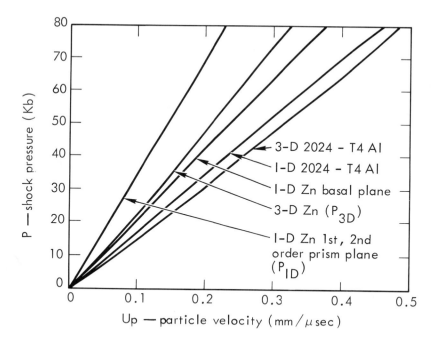

Figure 1. Shock pressure versus particle velocity.

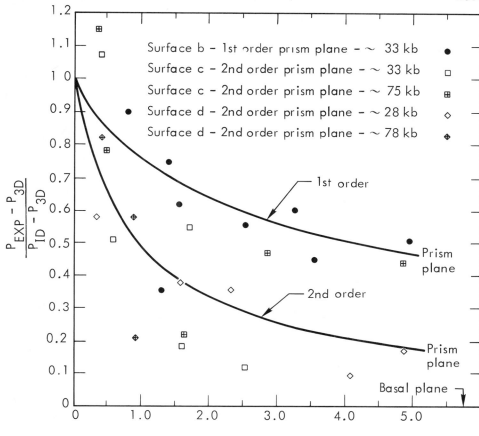

Figure 2. Fractional elastic overshoot as a function of target thickness.

ONE DIMENSIONAL STRESS

Test Techniques

Conventional Hopkinson split bar apparatus was used to compress right circular cylinders of single crystal zinc at shear strain rates from ~ 600 sec^{-1} to 2×10^4 sec^{-1}. The specific apparatus has been described elsewhere.[23] Semiconductor strain gages[24] were used to increase sensitivity of pressure measurement. Data from the input bar and output bar were analyzed as described in Ref. 25 and 26 with two exceptions. Because of the softness of basal slip specimens the difference between the stresses at the input bar and output bar interfaces of the zinc specimen could not

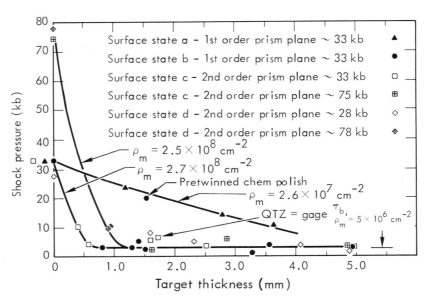

Figure 3. Elastic precursor decay.

be measured. The output bar stress measurement was therefore taken as the average stress in the specimen and the input bar stress measurement was used to calculate strain rates. Second, because the area under the output bar stress-time curve was less than 1% of the input bar stress-time curve, the former curve was neglected in calculation of the strain-time plot.

Elevated temperature tests were performed using stainless steel input and output bars protruding into a small wire element box furnace. The bars were preheated in the furnace before specimen insertion. Prior studies were made to determine soaking times necessary to minimize or eliminate temperature gradients in the specimen at time of deformation. Typical times to reach temperature from ambient and perform test were 5 to 6 minutes.

Specimens

Disks were prepared according to (d) in the section on preparing specimens for one dimensional strain tests. Right circular cylinders of nominal 0.25 in. diameter were then cored out by EDM. Initial dislocation densities were also determined in the same manner as described above. Length to diameter ratios for the specimens were varied from ~ 1 to $\sim 1.8"$.

Results

Results of the Hopkinson split bar tests are plotted as yield stress versus shear strain rate in Figs. 7 and 8. Figure 7 is for basal slip and Fig. 8 is for second order pyramidal slip. Specimens were oriented as described above in the introduction. The Schmid factor was used to transform the measured uniaxial stress and strain rate values to shear stress and shear strain rate.[27] For basal slip the relationships between uniaxial stress and strain rates and shear values are $\tau/\sigma = 0.418$; $\dot{\gamma}/\dot{\epsilon} \approx 2.39$. The influence of temperature on yield shear stress for basal slip is shown by Fig. 6. In the case of the second order pyramidal slip specimens, the two different orientations of the slip systems were taken into account by considering that the deformation mechanism is controlled by phonon damping. From Schmid factor calculations the ratios of the two shear stresses to the applied uniaxial stress are $\tau_1/\sigma = 0.4176$; $\tau_2/\sigma = 0.1043$ and the corresponding ratios of shear strain (rate) to corresponding uniaxial strain (rate) are approximately $\gamma_1/\epsilon_1 = 2.36$; $\gamma_2/\epsilon_2 = 8.89$. Since
$\dot{\gamma} = \rho_M(b^2\tau_1/(3B) + \rho_M 2b^2\tau_2/(3B))$, where ρ_M = mobile dislocation density, B = proportionality constant in $Bv = b\tau$ relationship between dislocation velocity and shear stress, b = Burgers vector magnitude, $\dot{\gamma}_1/\dot{\gamma}_2 = 2.0028$. These results indicate that γ_2 may be neglected. At shear stresses near the drag related back stress (discussed in following sections) slip systems with τ_2 probably contribute even less than estimated. Hence $\tau/\sigma \approx 0.4176$ and $\dot{\gamma}/\dot{\epsilon} \approx 2.36$.

DISLOCATION MODELING OF STRAIN-RATE

The characteristics of various types of deformation mechanisms are revealed by the differences in the trends of the dependence of the resolved shear stress for flow τ, on the absolute temperature T and the plastic shear strain rate $\dot{\gamma}$. The usual description of the various distinguishable ranges of plastic flow as due to athermal, thermally activated, diffusion-controlled and dislocation-drag controlled mechanisms is primarily a convenient procedure. In practice, all types of mechanisms are operative over all test conditions. Each mechanism, however, dominates over a certain range of temperature and strain rate and makes larger contributions to the total strain rate. It will be helpful at this stage to outline briefly the individual dislocation mechanisms that appear pertinent to describe plastic flow in zinc within the ranges of strain rate and temperatures of the present investigation. This discussion is limited to basal glide except for the mechanism of phonon damping where second order pyramidal slip is also considered.

Diffusion Controlled Mechanisms

In general deformation at temperatures above about $0.45\ T_m$ (melting point in °K) for low values of $\dot{\gamma}$ (shear strain rate) is determined by one or more of a series of diffusion controlled creep mechanisms. The thermal-activated events in all such mechanisms are vacancy-atom exchanges in a stress-induced chemical-potential gradient of vacancies or solute atoms in the grain boundaries, dislocation cores or the volume of the crystal. In general the steady-state rate, $\dot{\gamma}_s$, is given by

$$\frac{\dot{\gamma}_s KT}{DGb} = A(\tau/G)^n \qquad (1)$$

where D is the appropriate diffusivity, and A and n are constants dependent on the mechanism and pertinent substructural details, G is the shear modulus, b the Burgers vector and KT has the usual meaning of Boltzmann's constant times temperature. The various diffusion-controlled mechanisms have been summarized in a recent series of reviews[28,29] and will not be repeated here. For pure zinc, the available experimental data can be best correlated to a diffusion-controlled creep mechanism by dislocation climb as has been shown by Bird, Mukherjee and Dorn.[28] The experimental data on creep of zinc by Flinn and Munson[30] gave a value of $A \approx 3 \times 10^4$ and $n \simeq 4.5$. The shear modulus G was estimated from the relationship $G = G_o - G_T T$ where G_o and G_T are constants and equal to 5.981×10^{11} dy/cm^2 and 3.37×10^8 dy/cm^2 and T, the absolute temperature.[31] The diffusivity $D = D_o \exp(-H_D/RT)$, where $D_o \simeq 0.12$ cm/sec ($\perp$ to c axis) and $H_D = 24.3$ KCal/mole was calculated from the data of Shirn et al[32] and b was taken equal to 2.67×10^{-8} cm.

Athermal Mechanisms

Over an intermediate range of temperatures the deformation is athermal and τ decreases very gradually with an increase in T, usually in a manner that parallels the effect of T on G. Athermal mechanisms are characterized either by a continuously increasing free energy as a dislocation is forced through the crystal so that a saddle-point free energy (in Eyring's reaction-rate theory) is never attained or because the mechanism calls for a saddle-point free energy vastly greater than KT. In the latter case, sufficiently energetic thermal fluctuations will be so infrequent as to have a negligible effect on strain rate.

Disregarding statistical geometrical details, the theoretical estimate for the athermal stress levels due to operation of the various[33] athermal mechanisms, e.g., Frank-Read sources, the breaking of attractive junctions, the bypassing of individual dislocations, and long range stress fields all give about the same answer;

namely,
$$\tau_A = \beta G b \sqrt{\rho} \qquad (2)$$
where $0.2 \leq \beta < 0.5$ and ρ is dislocation density. It has been shown by Klahn, Mukherjee and Dorn[1] that the major factor responsible for strain hardening arises essentially from effects of long range back stresses rather than the formation of attractive junctions, etc. It should be emphasized that in pure and annealed single crystal metals, the initial dislocation density is often quite low and the value of the athermal stress τ_A, as given by Eq. 2 is likely to be negligible for such initial dislocation density of the order of $10^5/\text{cm}^2$ or lower.

Thermally Activated Mechanisms

At low temperatures and nominal strain rates, plastic deformation in metals and alloys is usually characterized by one or more of thermally activated mechanisms. At $T = 0$, a stress τ_0 (see Fig. 4) is needed to push dislocations mechanically past all barriers. At $T > 0$, thermal fluctuations can assist the stress in causing the dislocation to surmount barriers. Therefore, the value of τ needed to maintain a constant value of $\dot{\gamma}$ for a given substructural state, decreases with increasing T due to the more energetic thermal fluctuations at higher temperatures. For the same τ, higher temperatures give higher frequencies of successful fluctuations and therefore higher values of $\dot{\gamma}$. Above a critical

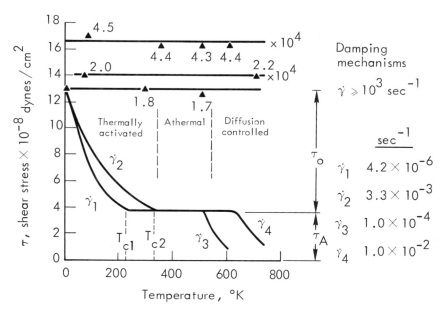

Figure 4. Shear stress as a function of temperature and strain rate.[1]

temperature T_c, which increases with increasing $\dot{\gamma}$, thermal fluctuations are sufficiently energetic to cause dislocations to surmount all of the short-range barriers responsible for the thermally activated mechanisms.

As a result of the inherent versatility of dislocation reactions, large number of dislocation mechanisms usually occur simultaneously. Occasionally one controls the deformation rate and thus becomes experimentally identifiable. This happens (1) when each dislocation undertakes a series of different mechanisms in sequence, i.e., in series, the slowest becomes rate controlling or (2) when dislocation glide is the net result of several alternate mechanisms, i.e., in parallel, and the strain rate arising from one is much greater than that from all others, it becomes rate controlling. Due to the unique differences between mechanisms, there is no single analytical approach to the analysis of data. In general, identifications of mechanisms must be based on correlations of the experimental data with predictions based on reliable models of possible mechanisms.

All approaches to the formulation of the plastic strain rate of materials are based on one or another of two equivalent geometrically justified expressions, e.g.,

$$\dot{\gamma} = \rho_m bv = NAb\nu \qquad (3)$$

where v is the average velocity of dislocations, N is the number of sites where activation can occur per unit volume, A is the average area swept out per successful activation and ν is the net frequency for successful fluctuation. The effect of τ, T and substructure on ρ_m and v or on N, A and ν is deduced from the model. A factor of major consequence in establishing the velocity v or frequency ν, is the activation free energy. Once a model is clearly visualized, it is possible to estimate the activation free energy. For the sake of brevity and emphasis on the effect of strain rate per se, we would not attempt to discuss here individually the various thermally activated mechanisms that are probable. Instead, we shall attempt to focus attention on the most probable mechanics in the present case and then proceed to show how the experimental data can be correlated on the basis of the operation of this specific mechanism of thermal activation.

Considering the present case of basal glide in pure single crystal of zinc, the most likely prototype of thermally activated mechanism is the case where glide dislocations are being thermally activated to overcome the short-range barriers imposed by forest dislocations threading through the glide planes. The dislocations of the forest are assumed to provide an array of barriers that are rigid, localized and repulsive in their interaction with the glide dislocation. One characteristic of all of such localized barriers is that the motion of dislocations past such barriers involve thermal activation of separate dislocation segments that subtend each contacted obstacle.

Typically, the free energy of activation for intersection of dislocations can be estimated from the force-displacement[34] diagram for cutting the obstacle as shown in Fig. 5. The saddle-point free energy of activation is given by

$$U_c = \int_{\tau^* b \ell}^{F_m} (x_2 - x_1) dF \quad (4)$$

where τ^* is effective stress and ℓ is the average distance between barriers. The total measured flow stress in the thermally activated region τ, is taken to be the sum of the athermal component τ_A and the thermally sensitive component τ^*. The analysis assumes that most of the athermal component is composed of long range internal back stress and the principle of superimposition is applicable here. In the absence of a precise knowledge of the exact shape of the force-displacement diagram, the simplest type of a diagram that can be considered is rectangular where $x_2 - x_1$ is uniformly equal to d. The activation energy then equals

$$U_c = \alpha \Gamma_o d - \tau^* b \ell d . \quad (5)$$

Here the parameter α represents the strength of the obstacle and Γ_o is the average dislocation line energy per unit length. This model approximates conditions that apply when undissociated glide dislocations produce jogs upon intersecting undissociated repulsive forest dislocations. This is not too unrealistic a representation of deformation of hexagonal crystals by basal glide. It should be emphasized that the rectangular representation of the force-displacement diagram is strictly for convenience and the exact value of d will depend upon the precise shape of the diagram and in specific cases can have a very small value.

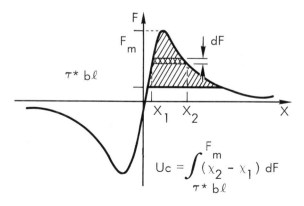

Figure 5. Force-distance diagram.

The obstacles are often assumed to be arranged in a square array where $\ell = L_s$ on a side. This leads to the Seeger[35] approximation for the shear strain rate:

$$\dot{\gamma} = \rho b \left(\frac{\nu b}{L_s}\right) L_s \exp\left\{-\frac{1}{KT}(\alpha \Gamma_o d - \tau^* b L_s d)\right\}. \tag{6}$$

As τ^* gradually approaches zero, i.e., the temperature approaches the critical temperature, there is a finite probability that the dislocation may move back to the low energy configuration on the reverse path. Since the distance moved until stopped again during such reversed motion is L_s, the effective strain rate is now given by

$$\dot{\gamma} = \rho b^2 \nu \left[\exp\left\{-\frac{1}{KT}(\alpha \Gamma_o d - \tau^* b L_s d)\right\} - \exp\left\{-\frac{1}{KT}(\alpha \Gamma_o d + \tau^* b L_s^2)\right\}\right]. \tag{7}$$

One outcome of this formulation is that the $\tau - T$ curve in the thermally activated region veers gradually and apparently asymptotically into that for the athermal region. In these instances the value of T_c cannot be accurately determined experimentally[33] because the activation energy increases rapidly as the effective stress approaches zero.

The simplifying assumption that localized obstacles form a regular square array is never valid. The obstacles are much more likely to be distributed randomly over the glide plane. It has been shown[1,33] that the neglect of this random distribution indeed led to serious errors in the past on the effect of $\dot{\gamma}$ and T on τ^* in the thermally activated region. Recent computer simulated experiments on the thermally activated motion of dislocation past localized randomly distributed barriers has shown[36,37,38] that Friedel's[39] theory for the steady state thermally activated motion of dislocations past weak obstacles gives close approximation to the statistically random situation. Hence we follow Friedel's formulation in presenting our analysis.

When the obstacles are distributed more or less randomly on the slip plane, more obstacles will be contacted as the dislocation segments, that subtend each contacted obstacles, bow out to smaller radii of curvature, as the effective stress τ^* is increased. Friedel has shown that the length of a dislocation subtended by randomly distributed weak obstacles is given by

$$\ell = \left(\frac{2\Gamma_o L_s^2}{\tau^* b}\right)^{1/3} \tag{8}$$

when there is one obstacle per area L_s^2 on the slip plane. When this value is introduced in Eq. 5 and the frequency of reversed reactions[33] is also considered, the shear strain rate is given by:

$$\dot{\gamma} = \rho b^2 \nu L_s^2 \left(\frac{\tau^* b}{2\Gamma_o L_s^2}\right)^{2/3} \exp\left\{-\frac{1}{KT}\left[\alpha \Gamma_o d - 2^{1/3}\Gamma_o d\left(\frac{\tau^* L_s b}{\Gamma_o}\right)^{2/3}\right]\right\}$$

$$\cdot \left[1 - \exp(-\frac{1}{KT})\left(\tau^* b L_s^2 + 2^{1/3}\Gamma_o d\left(\frac{\tau^* L_s b}{\Gamma_o}\right)^{2/3}\right)\right] \quad (9)$$

Viscous Drag Mechanisms

Stresses above $\tau_o + \tau_A$ (see Fig. 4) are sufficiently high to force dislocations mechanically past all barriers to their motion. Due to their low inertia, dislocations subjected to such stresses accelerate rapidly and soon reach a limiting velocity which increases linearly with the net operative drag stress τ_B, as determined by the sums of effects due to the various viscous damping processes. The net operative drag stress here will be given by the difference between the total applied stress and $(\tau_o + \tau_A)$, as shown in Fig. 4.

Damping mechanisms are also operative in the thermally activated and athermal ranges of behavior. Once a barrier has been surmounted with the aid of a thermal fluctuation, the dislocation advances to the next barrier with an average velocity that approximates that dictated by damping mechanisms. Consequently, the total time for a thermally activated excursion is the time awaiting a successful fluctuation in energy plus the time to travel to the next barrier.

All viscous drag mechanisms considered here are Newtonian and hence the drag stress τ_B, acting on a unit length of a moving dislocation is given by

$$\tau_B b = B V_{max}, \quad (10)$$

where B is the net drag coefficient. Hence,

$$\dot{\gamma} = \rho_m b V_{max} = \frac{\rho_m b^2 \tau_B}{B}. \quad (11)$$

As the dislocation velocity approaches that for sound waves, relativistic factors serve to modify the simple linear relationship between stress and dislocation velocity and Eq. 11 is no longer applicable. Some progress[40] has been made to define theoretically some of the conditions that are encountered when dislocations enter the relativistic range. There does not appear to be reliable and analysable experimental data for ultrasonic attenuation, dislocation velocity measurements by etch-pit technique, etc., or plastic strain rates for cases where the dislocation velocities begin to approach sonic velocities. Theoretical estimates for the various dislocation drag mechanisms have been recently summarized by Klahn, Mukherjee and Dorn.[1] A study of these mechanisms suggests that for zinc from room temperature to melting point, the most likely mechanism is due to phonon viscosity. It arises due to differences in the shear strain rates as the dislocations move causing a separating of the vibration frequencies of the different phonon modes. Energy loss is due to dissipation of heat from the phonons having higher temperatures to the cooler ones.

In the present investigation, the value of the net drag stress τ_B in Eq. 11 was estimated from $\tau_B = \tau_{total} - \tau_b$, where τ_b was the drag related back stress and as discussed earlier, is comparable to $(\tau_o + \tau_A)$ in Fig. 4. The value of τ_b was experimentally determined from the Hopkinson split bar data to be equal to 530 psi for basal glide. The damping coefficient B was taken to be equal to 3.2×10^{-4} dynes-sec/cm^2 from the data of Nagata et al.[41] Correlation of experimental data with Eq. 11 at the highest region of strain rates yielded a value of $\rho_m = 3.75 \times 10^8$ dislocation/cm^2. This estimated value is in harmony with the dislocation density deduced at high strain rates in the investigation of Ferguson et al.[42]

Equation 11 also applies to second order pyramidal slip. τ_b is 1.0×10^4 psi as estimated from the Hopkinson split bar data of Fig. 8 for upper yield point. B was again taken as 3.2×10^{-4} dynes-sec/cm^2. The calculated value of the mobile dislocation density is $\rho_m \approx 5 \times 10^6$ cm^{-2}. On the basis of a similar calculation, the nearly horizontal line for the lower yield point implies that ρ_m has increased to $\rho_m > 10^8$ cm^{-2}.

Elastic precursor data for second order pyramidal slip were analyzed according to equations of Johnson et al.[43] The calculated values of ρ_m and corresponding precursor decay curves are shown in Fig. 3. These mobile dislocation density values indicate several things. First, condition (a) surface damage of 22μ or less reduces ρ_m by an order of magnitude as compared to slightly or undamaged surfaces. Second, there is no difference in the steady state precursor stress (HEL) between first order or second order prism plane impact--as predicted from the equation of Johnson et al.[43] Third, τ_b, when transformed to the 1-D strain state, is very similar to the HEL. These results imply that HEL $\approx \tau_b$ and that the region near the impact surface is an important source of either mobile dislocations or dislocation barriers.

Elastic overshoot decay is also governed by high strain rate deformation mechanisms but in a much more complicated way than initial yielding. The data of Fig. 2 indicate a crystallographic orientation effect--in contrast to elastic precursor decay. The rate of approach to the equilibrium Hugoniot state is also slow. Targets were also impacted on the basal plane. These results, if plotted on Fig. 2, would be along the abscissa. The correspondence between basal plane impact results and Fig. 1 data serve to verify the accuracy of the manganin in-material gage pressure measurements.

Combined Effects of Thermally-Activated Diffusion-Controlled and Dislocation-Drag Mechanisms

All four major types of mechanisms are operative at any one time. Over limited ranges of conditions, however, one of these mechanisms becomes predominant. But the effects of the remaining mechanisms are detectable. We would now attempt to synthesize the contribution of the individual mechanisms to basal glide so as to arrive at a theoretical representation for the total applied stress--temperature--strain rate relationship and correlate it with the present experimental data. The initial forest dislocation density in the tested single crystals of zinc is $\sim 9 \times 10^4$ cm^{-2}. Nagata et al[41] showed that a dislocation density of the order of 10^5/cm^2 when applied to Eq. 2 produces a very low athermal stress level. Klahn[44] in his attempt to correlate Nagata's dislocation velocity measurements in single crystals of zinc essentially had to assume a negligible athermal stress level. Hence only those strain-rate effects ascribable to combinations of thermally activated, diffusion-controlled and drag-controlled mechanisms will be considered. For ease of analysis we assume an independent time sequence for each mechanism. Such treatment will yield a mean net dislocation velocity slightly less than the actual, due to interactions between different types of mechanisms, and particularly between the thermally activated and dislocation-drag mechanisms. But as shown in a recent survey[1] the error introduced by such treatment is indeed very small and certainly within the range of experimental scatter of data in such high strain rate region. This assumption does, however, produce a discontinuity in the numerical calculation in the region where thermal activation and dislocation drag contributes about equally to the total strain rate.

It would be worthwhile to emphasize here that the strain rate effect described is related to plastic deformation due to glide or climb of dislocations only and does not account for twinning as a rate-controlling mode of deformation. Metallographic examination of deformed specimens as well as their shape change after deformation strongly suggested that glide on basal plane was indeed the dominant mode of deformation here. Experimental investigation by Yoshida and Nagata[45] on zinc single crystals having identical orientation also confirms the operation of only basal glide at both high and low strain-rates and temperatures.

The theoretical strain-rate vs stress relationship at 293°K, 373°K, 473°K, 573°K and 673°K was calculated using Eq. 1, 9 and 11. The values for parameter in Eq. 1 and 11 have been stated already. For Eq. 9, the following values were used: line tension $\Gamma_o \simeq Gb^2/2 \simeq 3.1 \times 10^{-4}$ dynes. $\nu = 10^{13}$ sec^{-1}, relative barrier strength $\alpha = 0.4$, mean distance between barriers, $L_s = 130$ b (best fit from iterative program) and obstacle width as characterized in the force-displacement diagram, $d = 0.1$ b.

The theoretical strain-rate dependence of the flow stress is shown in Figs. 6 and 7 for the various temperatures as the solid lines and the experimental results are shown as data points. Within the scatter encountered in experimental results in such high strain rate regions, the correlation is satisfactory. For shear strain rates higher than 0.5×10^4 sec^{-1}, the dislocation drag processes dominate and become rate-controlling. At lower strain rates, thermally activated processes increasingly come into play. For constant plastic strain rates, such thermally activated mechanisms are characterized by rapidly decreasing flow stresses with increasing temperatures. As the temperature is decreased, the delay time for successful thermal fluctuations is being continually increased and consequently at really low temperatures, thermal activation plays a very little role and the plastic shear flow stress is given essentially by drag-controlled processes.

Although the present analysis incorporated the effects of diffusion-controlled dislocation climb process as a prototype of

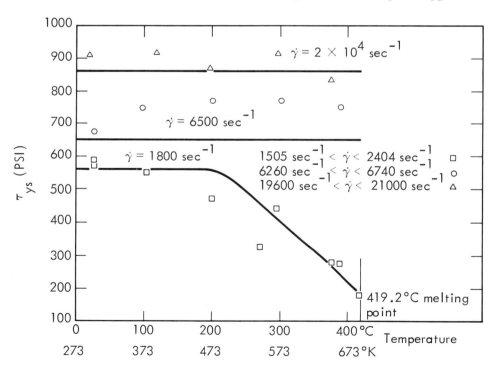

Figure 6. Shear stress vs temperature for various shear strain rates.

DYNAMIC DEFORMATION OF SINGLE CRYSTAL ZINC

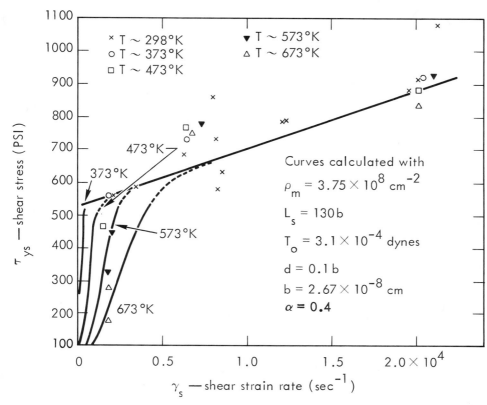

Figure 7. Shear stress vs shear strain rate for basal glide.

high temperature rate-controlling mechanism, it never did play any significant role even at 673°K. This is because the present work was conducted essentially in the region of high strain rates and there was inadequate time for successful activation over the high energy barriers that characterize such diffusion-controlled processes. The soundness and broad applicability of Eq. 1 is corroborated by the fact that this equation is in quite good agreement with low strain-rate creep data on zinc by Tegart and Sherby.[46]

The present correlation depicts that thermally activated mechanisms indeed play an important role even at the comparatively high shear strain rates of the order of 3×10^3 sec^{-1}, specially at higher temperatures. It has been commonly assumed in the past that such thermally activated process usually dominate in the low to intermediate ranges of strain rate. In order to ascertain the upper limit of operation of the thermally activated mechanism of intersection in terms of strain rate, Eq. 9 was used to estimate the critical shear strain rate, $\dot{\gamma}_c$, when the critical temperature

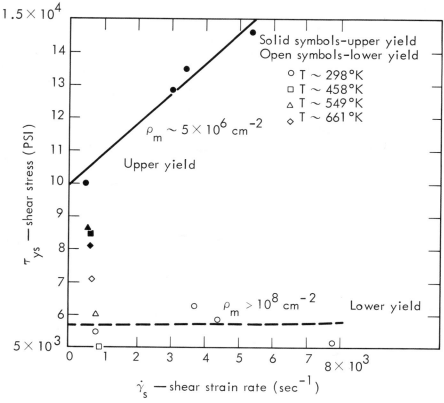

Figure 8. Shear stress vs shear strain rate for second order pyramidal glide.

τ_C was infinite. The value of τ^* was taken to be equal to 530 psi, i.e., the threshold for the dominant operation of drag-controlled processes. The estimated value of $\dot{\gamma}_C$ should provide an upper limit for the operation of the thermal activation intersection process, which is unmodified by the simultaneous presence of drag mechanism. In reality, of course, dislocation drag mechanisms will make increasingly dominant contribution at the higher strain rate region, so that the actual upper limit will be at somewhat lower strain rate value. The theoretically estimated value of $\dot{\gamma}_C \simeq 8.3 \times 10^4$ sec^{-1} as an absolute upper limit comes within one order of magnitude to the experimentally observed value of about 0.8×10^4 sec^{-1} in Fig. 7, where thermal activation is essentially superseded by dislocation drag processes. Noting that the theoretically estimated value can be lowered if one assumes a slightly lower value of ρ and L_s and that the experimentally observed value is expected to be lower anyway, due to the simultaneous operation of drag mechanism, the correlation seems satisfactory. Hence, the thermally activated mechanism of intersection effectively contributes even in the comparatively high strain rate region of $\sim 10^3$ sec^{-1}. Similarity between the upper yield data of Fig. 8 and the data of Fig. 7 implies that thermal activation mechanisms are also important in c + a slip to $\sim 10^3$ sec^{-1}.

ACKNOWLEDGEMENTS

The authors wish to acknowledge the considerable efforts of R. T. Sato, University of California at Davis, in computer calculation of the dislocation mechanisms and R. D. Breithaupt, Lawrence Livermore Laboratory, in execution of the Hopkinson split bar tests.

REFERENCES

1. D. Klahn, A. K. Mukherjee, J. E. Dorn, Proceedings of the 2nd International Conference on the Strength of Metals and Alloys, Vol. III, ASM, 951 (1970).
2. J. J. Gilman, Appl. Mech. Rev., 21, No. 8, 767 (1968).
3. A. S. Argon, Material Science & Engineering 3, 24 (1968).
4. R. J. Wasley, J. F. O'Brien, LLL Rep. No. UCRL-14543 (1965).
5. O. E. Jones, F. W. Neilson, W. B. Benedick, J. Appl. Phys. 33, 3224 (1962).
6. J. F. O'Brien, R. J. Wasley, Rev. Sci. Instr. 37, 531 (1966).
7. J. A. Charest, EGG Rep. No. EGG 1183-2L45, LLL Rep. No. UCRL-13473 (1970).
8. J. A. Charest, B. D. Jenrette, EGG Rep. No. S-59-TP (1970).
9. E. O. Williams, Sandia Lab. Rep. No. SCL-DR-70-148 (1971).
10. Crysteco, 180 East Main Street, Wilmington, Ohio 45177.
11. Aremco Products, Inc., P. O. Box 145, Briarcliff Manor, N.Y. 10510.
12. K. H. Adams, R. C. Blish, II, T. Vreeland, Jr., J. Appl. Phys. 37, 832 (1966).
13. R. C. Brandt, K. H. Adams, T. Vreeland, Jr., J. Appl. Phys. 34, 591 (1963).
14. R. C. Brandt, K. H. Adams, T. Vreeland, Jr., J. Appl. Phys. 34, 587 (1963).
15. H. S. Rosenbaum, M. M. Saffren, J. Appl. Phys. 32, 1866 (1962).
16. H. S. Rosenbaum, Acta Met. 9, 742 (1961).
17. G. Schoeck, J. Appl. Phys. 33, 1745 (1961).
18. R. G. McQueen, Metallurgy at High Pressures and High Temperatures, K. A. Gschneidner, Jr., M. T. Hepworth, N. D. D. Parlee, eds., Gordon and Breach, 108 (1964).
19. G. E. Duvall, Response of Metals to High Velocity Deformation, P. G. Shewmon, V. F. Zackay, eds., Interscience Pub., 79 (1961).
20. M. Van Thiel, A. S. Kusubov, A. C. Mitchell, eds., LLL Rep. No. UCRL-50108 (1967).

21. Group GMX-6, LASL Rep. No. LA-4167-MS (1969).
22. G. R. Fowles, J. Appl. Phys. 32, 1475 (1961).
23. R. J. Wasley, J. C. Cast, K. G. Hoge, R. D. Breithaupt, LLL Rep. No. UCRL-51098 (1971).
24. Micro Systems, Los Angeles, California, Gage No. PA3-16-350.
25. R. J. Wasley, K. G. Hoge, J. C. Cast, Rev. Sci. Instr. 40, 889 (1969).
26. F. E. Hauser, J. A. Simmons, J. E. Dorn, Response of Metals to High Velocity Deformation, P. G. Shewmon, V. F. Zackay, eds., Interscience Pub., 100 (1961).
27. R. W. K. Honeycombe, The Plastic Deformation of Metals, E. D. Arnold, Ltd., 17 (1968).
28. J. E. Bird, A. K. Mukherjee and J. E. Dorn, Quantitave Relation Between Properties and Microstructure, D. G. Brandon and A. Rosen, eds., Israel Univ. Press, 255 (1969).
29. A. K. Mukherjee, J. E. Bird and J. E. Dorn, Tran. ASM 62, 155 (1969).
30. J. E. Flinn and D. E. Munson, Phil. Mag. 10, 861 (1964).
31. G. A. Alers and T. R. Neighbours, J. Phys. Chem. Solids 7, 58 (1958).
32. G. Shirn, E. Wajda and H. Huntington, Acta Met. 1, 513 (1953).
33. J. E. Dorn, A. K. Mukherjee, LBL Report No. UCRL-19097 (1969).
34. J. S. Basinski, Phil. Mag. 4, 393 (1959).
35. A. Seeger, Phil. Mag. 46, 1194 (1955).
36. D. Klahn, D. Austin, A. K. Mukherjee and J. E. Dorn, "The Importance of Geometric Statistics to Dislocation Motion", to be published in Advances in Applied Probability.
37. R. J. Arsenault and T. Cadman, "Thermally Activated Dislocation Motion Through a Random Array of Obstacles", to be published in the Proceedings of the ASM John E. Dorn Memorial Symposium, Cleveland, Ohio, 1972.
38. P. Wynblatt "Simulation of Thermally Activated Dislocation Motion", to be published as in Ref. 37.
39. J. Friedel, Dislocations, Pergamon Press, Oxford, 224 (1964).
40. J. Weertman, J. Appl. Phys. 38, 5293 (1967).
41. N. Nagata and T. Vreeland, Jr., Phil. Mag. 25, 1137 (1972).
42. W. G. Ferguson, F. E. Hauser and J. E. Dorn, Brit. J. Appl. Phys. 18, 411 (1967).
43. J. N. Johnson, O. E. Jones, T. E. Michaels, J. Appl. Phys. 41, 2330 (1970).
44. D. H. Klahn, PhD Thesis, U.C., Berkeley, CA, LBL Rep. No. LBL-800 (1972).
45. S. Yoshida and N. Nagata, Trans. Japan Inst. Metals 9, 110 (1968).
46. W. J. Tegart and O. D. Sherby, Phil. Mag. 3, 1287 (1958).

Discussion by C. H. Karnes
Sandia Laboratories, Mechanical Response Division

It is important to point out that the use of the split Hopkinson pressure bar technique requires the assumption that the specimen and the pressure bar in the region of the strain gages be in a one-dimensional stress state before any measured data can be interpreted in terms of specimen behavior. It was stated that the pressure bar system used was 0.5 inch in diameter and that it was estimated to require about 20 μsec after loading for the required regions to reach stress equilibrium or a one-dimensional stress state. If one assumes that equilibrium is attained in 20 μsec, then the shear strains reached in that time during average strain rates of 5×10^2/sec and 5×10^3/sec are 0.01 and 0.10, respectively, which are very large compared to the strain at yield. Therefore, the data points shown in Fig. 8 for upper and lower yield stress are not interpretable in terms of the specimen behavior because the specimens and output pressure bar are in a stress state which is nowhere near one-dimensional when yielding occurs.

Another way of illustrating the problem is to consider the time after loading at which upper yield, for example, occurs. The data points for upper yield shown in Fig. 8 are for shear strain rates from 5×10^2/sec to 5×10^3/sec. If the upper yield stresses are valid as shown, then they occur at shear strains of approximately 0.002 or at times after loading of about 3 μsec for the lowest strain rates to 0.5 μsec for the highest strain rate data of Fig. 8.

The upper and lower yield phenomena represents a very short duration event at these strain rates. A short duration pulse whose risetime is significantly shorter than 5 μsec when applied to the authors' output pressure bar will be severely distorted in both amplitude and risetime because of the dispersion of the high frequency components.

It is apparent that the signals obtained by the authors from the Hopkinson pressure bar strain gages during this investigation resulting in Fig. 8 can be used only to show, in a very qualitative manner, that the upper and lower yield point phenomena exists in the material at these strain rates, but cannot, in any way, indicate the magnitudes of the yield points. Therefore, any conclusions based on "trends" observed in the data of Fig. 8 concerning the existence of various dislocation mechanisms or the values of parameters in the corresponding models are not valid.

Author's Reply

The data summarized in Figure 8 is for specimens 0.25 in. dia by 0.45 in. long. The rod speed for these specimens is 4.09 mm/μsec based on S_{11} from Huntington.[1] One wave transit time is 2.8 μsec and four transit times are 11.2 μsec. The output bar peak stress (identified as upper yield in Figure 8) occurs 12-13 μsec after it first begins to rise. Time between the beginning of roll off and the minimum stress (t_2 in Reference 2) in the input bar pressure trace (caused by reflection at the input bar-specimen interface) is 12 μsec. Offsetting the output bar stress-time trace by one wave transit time causes the input and output bar pressure traces to superimpose within 10%. Therefore, the specimens must be considered to be in a one dimensional stress state at the time upper-lower yield points are observed.

For times less than about 12 μsec, non-steady state stress-strain-strain rate conditions exist. Hence, the calculated amount of strain to the upper yield point is subject to most of the uncertainty in relating specimen strain to stress. Virtually all conventional engineering yield points are preceded by plastic deformation.[3] Further, yield points can occur after large amounts of plastic deformation.[4] As pointed out by N. Brown,[3] upper and lower yield points simply correspond to maxima and minima in the stress strain curve and do not depend on a strain definition. Interpretation of the upper yield point data in Figure 8 was limited to $\dot{\gamma} = \rho_M b v = \rho_M b^2 B \tau^*$. The result was $\rho_M \sim 10^6 \mathrm{cm}^{-2}$. Since this value is > 10 times the total initial density, it is clear significant plastic deformation did occur before the observed upper yield point.

The Hopkinson split bar technique is a constant load test at its best that involves transmission of deformation waves. Consequently it is a poor technique for microplastic yield studies and can never exactly correspond to quasistatic constant total strain rate tests. The authors did not attempt to do that.

1. H.B. Huntington, *The Elastic Constants of Crystals*, (Academic Press), 66 (1958).
2. L.D. Bertholf, "Feasibility of Two-Dimensional Numerical Analysis of The Split Hopkinson Pressure Bar System," to be published in *J. Appl. Mech.*
3. N. Brown, "Observations of Microplasticity," in *Microplasticity*, (Interscience Pub.), 59 (1968).
4. E.O. Hall, *Yield Point Phenomena in Metals and Alloys*, (Plenum Press), 5-8 (1970).

A CONSTITUTIVE RELATION FOR DEFORMATION TWINNING IN BODY CENTERED CUBIC METALS

R. W. Armstrong, University of Maryland
College Park, Maryland
and
P. J. Worthington, Central Electricity Research Laboratories
Leatherhead, Surrey, England

ABSTRACT

A constitutive relation is developed to describe the grain size, temperature and strain rate dependence of the stress required for the essentially elastic twinning of body centered cubic (bcc) metals at low temperatures. The relation is derived from a description of the bulk material strain produced by twinning and, also, from specifying on the basis of a thermal activation analysis the number of twins and the twinning nucleation rate. Two natural consequences of this development for bcc metals are that twins must be nucleated by pre-twinning slip and that twinning and brittle fracture are predicted to occur under very similar circumstances. These are normal experimental observations.

INTRODUCTION

Previous[1-3] attempts to define a stress and strain rate for twinning have been limited in scope and application. Any comprehensive constitutive relation for twinning must take account of the following observations.

The onset of deformation twinning in body centered cubic (bcc) metals is often associated with brittle fracture[4], and the stresses at which both twinning and brittle fracture occur in polycrystals are markedly dependent on the grain size, as observed, for example, in 3% silicon-iron[5,6], iron[7,8], chromium[9] and vanadium[10]. The temperature[5,11] and imposed strain rate appear to have a much smaller effect on the twinning stress than for the slip stress in most cases.

For twinning, the formation of a nucleus seems to be the critical event because any twin propagates easily once it has formed[4]. The energetics of the twinning process have been described on the basis of various models employing continuum mechanics[12], analytic dislocation configurations[4] and computer methods[13]. Atomic mechanisms for twinning have utilized partial dislocations which are geometrically arranged so as to produce an invariant shear and, more recently, by means of a disclination[14].

None of the current twinning models are able to give a full account of twinning behavior. This is because they have not been carried forward to give a constitutive relation for the process. The purpose of the present paper is to give one development for the sensibly elastic twinning of a bcc metal and to assess this relationship in regard to a number of previous measurements.

THE CONSTITUTIVE RELATION

For a particular bcc twin system, say, of $\{112\}$ <111> type, the twinning shear strain γ_{12} is given by,

$$\gamma_{12} = \frac{b_1}{b_2} = \frac{nb_1}{nb_2} = \frac{nb_1}{h} \qquad (1)$$

where b_1 is the (translation) twinning dislocation Burgers vector, b_2 is the (pole) screw component Burgers vector moving the twinning dislocation to adjacent planes, n is the number of twinning dislocations, and h is the twin thickness. The total twinning strain for a family of parallel twins within a crystal or grain volume is given by,

$$\gamma_{TOT} = N_T \cdot \gamma_{12} \cdot \frac{\bar{h}}{H} \cdot \frac{\overline{\Delta A}}{A} \qquad (2)$$

where N_T is the total number of twins having an average thickness, $\bar{h}$, measured along the crystal or grain height, H, and $\overline{\Delta A}/A$ is the average twinned area in the plane of shear (perpendicular to b_2 and containing b_1). It may be seen for a crystal which is totally twinned by one set of parallel twins that $N_T \bar{h} \overline{\Delta A}$ = HA so that $\gamma_{TOT} = \gamma_{12}$, as expected. Alternatively, it also occurs that the (volume) density of these twins, ρ_T, is implicitly specified in (2) because

$$\rho_T = \frac{N_T}{V} = \frac{N_T}{HA} \qquad (3)$$

and, also, the average volume per twin, $\bar{V}_T$, is specified by so that

$$\bar{V}_T = \bar{h}\,\overline{\Delta A} \qquad (4)$$

$$\gamma_{TOT} = \rho_T \bar{V}_T \gamma_{12} \qquad (5)$$

The total resolved twinning strain along the specimen axis for a crystal or average grain, then, is given by

$$\varepsilon_T = m' \rho_T \bar{V}_T \gamma_{12} \qquad (6)$$

where m' ≤ 0.5 is the direction cosine product obtained from an appropriate transformation of γ_{12} onto the specimen axis. m' may be evaluated, for example, by assuming that each grain undergoes the same strain as the total specimen. In this case, the product $\rho_T \bar{V}_T$ would have to be determined for each set of parallel twins and the average value used to compute m'.[15]

The rate at which the twinning strain develops can be conveniently determined from (6) by considering that ρ_T and $\bar{V}_T$ are time dependent, in which case

$$\dot{\varepsilon}_T = m'\gamma_{12} \left[\rho_T \dot{\bar{V}}_T + \dot{\rho}_T \bar{V}_T \right] \quad (7)$$

For a "penny-shaped" twin,

$$\bar{V}_T = \pi \bar{r}^2 \bar{h} \quad (8)$$

where $\bar{r}$ is the twin radius and, thus,[16]

$$\dot{\bar{V}}_T = 2\pi \bar{r}\, \bar{h}\, \dot{\bar{r}} + \pi \bar{r}^2\, \dot{\bar{h}} \quad (9)$$

Furthermore, if a polycrystal is being considered in which the volume of a hypothetical spherical grain is taken as $\pi \bar{\ell}^3/6$, then, an average value of ρ_T is determined from the measurement of an average number for N_T, also, so

$$\bar{\rho}_T = 6\bar{N}_T/\pi \bar{\ell}^3 \quad (10)$$

and

$$\dot{\bar{\rho}}_T = 6\dot{\bar{N}}_T/\pi \bar{\ell}^3 \quad (11)$$

By substitution of equations (8) through (11) into (7) and including some rearrangement of terms, the total rate of strain due to twinning is obtained in the form

$$\dot{\varepsilon}_T = \frac{6m'\gamma_{12}\bar{h}\,\bar{r}^2}{\bar{\ell}^3}\left[\bar{N}_T \left(\frac{2\dot{\bar{r}}}{\bar{r}} + \frac{\dot{\bar{h}}}{\bar{h}} \right) + \dot{\bar{N}}_T \right] \quad (12)$$

It was indicated in the INTRODUCTION that the number of twins and their nucleation rate are critical parameters to specify in assessing the influence of twinning on mechanical properties. Therefore, special attention needs to be given to evaluating $\bar{N}_T$ and $\dot{\bar{N}}_T$ in (12). A reasonable proposal for the form of N_T which is based on earlier studies[4,17] is that

$$\bar{N}_T = \bar{N}_{T_O} \exp\left[-U^*/RT \right] \quad (13)$$

where N_{T_O} is the limiting maximum number of twins that may develop in a grain volume, U^* is an activation energy for twinning, R is the gas constant and T is the absolute temperature. Although it has been reasoned that U^* should be small[17], in fact, limits on U^* have previously been given[4] as

$$C_1 \alpha [Gb_1/\tau_{12}]^2 \leq U^* \leq C_2 \alpha^3 G^2/\tau_{12}^4 \quad (14)$$

where α is the twin coherent interfacial energy, G is the shear modulus, and C_1 and C_2 are numerical constants of order unity. It is found, however, by substituting typical values for these parameters into (14) and then into (13) that U^* is too large for any twins to form by a direct stress (and temperature) assisted process, i.e. $\bar{N}_T \simeq 0$ even for an upper limiting value of $\bar{N}_{T_O} = \ell/b_2 \simeq 5 \times 10^5$. This has already been pointed out[18] for the lower limiting value of U^*. The situation is far worse for the upper limiting value. For example, substituting a reasonable applied stress for $\tau_{12} \simeq 10^8$ dynes/cm^2 for 3% silicon-iron in equation (14) gives $8.1 \times 10^{-7} \leq U \leq 2.8 \times 10^{-3}$ ergs.

It must be concluded, based on the foregoing evaluation of (14) and (13), that deformation twinning will not normally occur unless a certain amount of pre-twinning slip happens first to raise τ_{12} in (14) by a stress concentrating effect. Experimental observations on 3% silicon-iron at low temperatures[6] over a range of grain sizes have shown in fact that micro-slip precedes twinning. The well-documented results for micro-slip preceding even the brittle fracture stress of iron and steel are shown in Figure 1. The fracture stress, σ_F, which is shown for cleavage at liquid nitrogen temperature[19], 78°K, is very nearly the same as the stress required for substantial deformation twinning[20]. The fracture stress follows a Hall-Petch relation, as also does the yield stress, σ_y, which is shown according to room temperature measurements[21]. The temperature and strain rate dependence indicated in Fig. 1 for the yield stress intercept, σ_{o_y}, is so pronounced as to allow σ_y to be raised above σ_F for all grain sizes except the very smallest ones. Also shown in Figure 1 is the micro-slip stress of iron and steel as determined at room temperature by two different methods. For values of $\ell^{-1/2} \lesssim 10$, the values of σ_m were determined by direct measurement of the micro-yield stress[22], whereas for larger values of $\ell^{-1/2}$ the values of σ_m were determined from stress-strain curves[23] by an extrapolation procedure[24]. In contrast to the temperature dependence of the conventional yield stress, experimental measurements have shown σ_m to be relatively insensitive to temperature[25] and, hence, the micro-slip stress is very much smaller than the fracture stress or the twinning stress in steel.

On the basis of the preceding discussion, it should be expected that the twinning stress may be achieved from an effective shear stress for micro-slip, τ_e, as

$$\tau_{12} \simeq n_s \tau_e = C_3 \tau_e^2 \bar{\ell}/Gb_s \qquad (15)$$

where n_s is the number of slip dislocations of Burgers vector, b_s, in a slip band pile-up. Consequently, by replacement of τ_{12} in (14), it occurs that

$$C_1 \alpha (G^2 b_1 b_s / C_3 \tau_e^2 \ell)^2 \leq U^* \leq C_2 \alpha^3 G^6 b_s^4 / C_3^4 \tau_e^8 \bar{\ell}^4 \qquad (16)$$

Now, for $C_1 \simeq C_2 \simeq 5$, $C_3 \simeq 2$, $\bar{\ell} \simeq 10^{-2}$ cm, $b_s \simeq b_1 \simeq 2 \times 10^{-8}$ cm, $\alpha \simeq 2 \times 10^2$ ergs/cm^2, $G \simeq 10^{12}$ dynes/cm^2, $\tau_e \simeq 10^8$ dynes/cm^2, then $2.5 \times 10^{-13} \leq U^* \leq 3.6 \times 10^{-9}$ ergs. These values for U^* are still large, but not sufficiently so as to prevent (13) from being taken

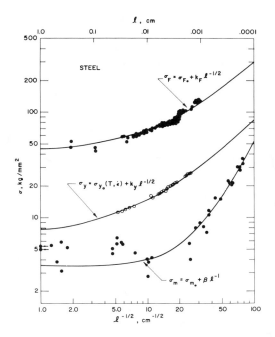

Figure 1. The micro-slip, yield, and brittle fracture stresses of steel at different grain sizes

seriously. It should also be noted that because of τ_{12} being evaluated as a concentrated stress, U^*, is an even more sensitive function of stress than resulted before and this changed dependence seems to be in the correct direction. It is normal practice in the analysis of concentrated stresses to express the effective shear stress in terms of a concentrated shear stress factor, say, k_{ST} which relates for a polycrystal, τ_e, to $\bar{\ell}$ by

$$\tau_e = k_{ST}\ell^{-1/2} \qquad (17)$$

Thus, it is found that

$$Gb_S(C_1\alpha/U^*C_3^2)^{1/4} \leq k_{ST} \leq (C_2\alpha^3 G^6 b_S^4/C_3^4 U^*)^{1/8} \qquad (18)$$

From the estimates given above, k_{ST} is found to be of the order of 10^7 to 10^8 dynes/cm$^{3/2}$ for both limiting values given in (18). This value of k_{ST} agrees very well with experimental values determined for silicon-iron[5,6].

In order to obtain the rate of twinning, $\dot{\bar{N}}_T$, advantage is taken of the stress dependence of $\bar{N}_T$, in that

$$\dot{\bar{N}}_T = (d\bar{N}_T/d\tau_e)\,\dot{\tau}_e \qquad (19)$$

Putting the limits for U^* defined in (16) into (13), then $\dot{\bar{N}}_T$ is directly obtained as

$$\dot{\bar{N}}_T = (qU^*/RT)\,\bar{N}_T\,(\dot{\tau}_e/\tau_e) \qquad (20)$$

where q is the magnitude of the exponent of τ_e in U^*, i.e. $4 \leq q \leq 8$. It also may be estimated in the evaluation of (12) that

$$(\dot{\bar{h}}/\bar{h}) \simeq (\dot{\bar{r}}/\bar{r}) \simeq (v_s/\bar{r}) \tag{21}$$

where v_s is the normalized sound velocity which is approximately 2.85×10^5 cm/sec for iron[26].

Equation (12) may be rewritten from (20) and (21) as

$$\dot{\varepsilon}_T \simeq \frac{6m'\gamma_{12} \bar{h} \bar{r}^2}{\bar{\ell}^3} \left[\frac{3v_s}{\bar{r}} + \left(\frac{qU^*}{RT}\right) \cdot \left(\frac{\dot{\tau}_e}{\bar{\tau}_e}\right) \right] \bar{N}_T \tag{22}$$

Furthermore, by estimation of the magnitude of terms in (22), it occurs that $(3v_s/\bar{r}) \gg (qU^*/RT)(\dot{\tau}_e/\tau_e)$, so that (22) actually reduces to

$$\dot{\varepsilon}_T \simeq 18m'\gamma_{12} \bar{h} \bar{r} v_s \bar{N}_T/\bar{\ell}^3 \tag{23}$$

Equation (23) is the resultant constitutive relation for the sensibly elastic twinning of a bcc metal. It appears important from (23) that the strain rate at a particular time is essentially determined by the number of twins present at that time. This result agrees with recent experimental observations on the twinning of SAE 1010 steel in compression at liquid helium temperature[20]. There is also evidence[27] that high strain rates are associated with small values of twin thickness $\bar{h}$ so that the increased values of $\dot{\varepsilon}_T$ must be accompanied by increased values of $\bar{N}_T$ which more than compensate for the reduced $\bar{h}$ values according to (23). For a fixed strain rate it is observed[5] that $\bar{h}$ increases as $\bar{\ell}$ increases. However, an increase in grain size of ~ 40x causes only the twin thickness to increase by ~ 2x, so even though it is noted that $\bar{r}$ is directly related to $\bar{\ell}$ in (23), an appreciable increase in $\bar{N}_T$ is expected as $\bar{\ell}$ increases. This is as predicted from (16), and is generally observed experimentally.

The applied shear stress for twinning, τ_T, may be determined from (23) by noting that τ_e in (15) is conventionally taken as τ_T minus a friction stress, in this case, due to the pre-yield slip occurring at τ_o, and, therefore,

$$\tau_{12} = C_3(\tau_T - \tau_o)^2 \bar{\ell}/Gb_s \tag{24}$$

By combining (13), (16), (23), (24) and utilizing the exponent, q, from (20), τ_T is obtained as

$$\tau_T = \tau_o + G \left(\frac{b_s}{C_3}\right)^{1/2} \left[\left(\frac{U^{*'}}{RT}\right) / \ln\left(\frac{\dot{\varepsilon}_{T_o}}{\dot{\varepsilon}_T}\right) \right]^{1/q} \bar{\ell}^{-1/2} \tag{25}$$

where $\dot{\varepsilon}_{T_o} = 18m'\gamma_{12} \bar{h} \bar{r} v_s \bar{N}_{T_o}$, $U^{*'} = U^*(\tau_{12}/G)^{q/2}$, and

$C_1 \alpha b_1^2 \leq U^{*'} \leq C_2 \alpha^3/G^2$. In terms of an applied stress σ_T,

$$\sigma_T = m\tau_o + mG \left(\frac{b_s}{C_3}\right)^{1/2} \left[\left(\frac{U^{*'}}{RT}\right)/\ln\left(\frac{\dot{\varepsilon}_{T_o}}{\dot{\varepsilon}_T}\right)\right]^{1/q} \bar{\ell}^{-1/2} \quad (26)$$

where m is the slip orientation factor representative of the micro-slip which produces twinning. Since twinning is a nucleation controlled event, then it should occur locally where the stress is greatest. That there is some critical energy for the onset of a significant amount of twinning is supported by the good agreement between theory and experiment on the grain size dependence of the twinning stress, previously pointed out in equations (17) and (18). The very significant stress dependence of the onset of twinning is shown in Figure 2 for a stress-strain curve obtained by compression testing[20]. For strains smaller than approximately 10^{-2}, the recorded load displacement curve exhibited very small load drops, as noted in the stress-strain curve, but these load drops were accompanied by clearly audible clicks. The strain sensitivity was insufficiently to accurately determine the stress-strain behavior at strains near to 3×10^{-3} and, hence, the dashed portion of the stress-strain curve is drawn to indicate the region where micro-slip might be expected to be predominant. At a stress level of approximately 100 kg/mm^2 appreciably sized load drops began to occur between otherwise continued elastic loading segments of the force-displacement curve. Thus, the plastic strain due to twinning resulted during the load drops which led to an accumulated strain of 0.27 when the specimen was finally unloaded. For the same grain size material tested in tension[20], brittle fracture resulted at a stress of 108 kg/mm^2 which compares favorably with Figure 2 and, also, with the value of σ_F determined from Figure 1 for steel of this grain size tested at liquid nitrogen temperature[19].

Twinning of bcc metals rarely occurs in all the polycrystal grains or, infrequently does it occur on more than one twinning system per grain. For these reasons and because τ_o is associated with the micro-slip preceding twinning, it should be expected that a lower value of m will result in (26) than for the case of homogeneous yielding by macro-slip. Thus, a smaller stress intercept, σ_{o_T}, should occur for twinning at $\bar{\ell}^{-1/2} = 0$, as compared with a σ_o value for yielding under similar circumstances by macro-slip. This is as observed[6].

DISCUSSION

The contribution to the bulk material strain of the twinned volume itself is given in equations (1) to (6). No provision has thus far been made for any accommodating strain in the untwinned matrix, say, due to emissary dislocations[28] or other mechanisms. The additional matrix strain can be an important contribution to the total deformation and this added strain should probably be proportional to the twinned volume, in which case equation (23)

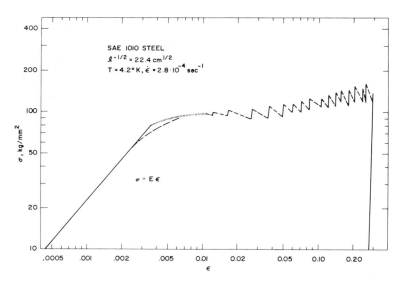

Figure 2. Stress-strain curve for the twinning of steel in compression at 4.2°K

might be modified to approximately account for it by employing a multiplying factor somewhat greater than unity on the right side of the equation. For example, for the specimen of Figure 2, approximately 25% of the total strain could be accounted for by the twinned volume of the specimen[20]. The additional strain in the matrix surrounding the twins is "locked in" as evidenced by the continued elastic loading following each twinning load drop. Equation (26) would be slightly changed for reason of adding the locked-in matrix strain by allowing for an increased value of $\dot{\varepsilon}_{T_o}$ in k_{ST}. However, the main value of equations (23) and (26) at this time should not be nearly so much in their quantitative predictions as in their ability to predict correctly the relative dependences of the twinning stress on the grain size, temperature and strain rate.

It was previously mentioned in the INTRODUCTION that a striking feature of the twinning stress was its pronounced grain size dependence, as determined by k_{ST}. The estimated value of k_{ST} of 10^7 to 10^8 dynes/cm$^{3/2}$, as calculated from equation (18), is compared in Table 1 with a number of experimental values, all of which show very good agreement with the calculated values except for niobium[29]. However, special effects due to the substructure and its attendant solute segregation may be particularly important for this material[30]. From equation (26), the value of k_{ST}/G should be approximately constant and Table 1 shows reasonable

DEFORMATION TWINNING IN BODY CENTERED CUBIC METALS

Table I: Characteristic Values (shear), based on results at 77°K, strain rate ~ 10^{-4} sec^{-1}.

Material	k_{Sy} dynes cm$^{-3/2}$	k_{ST} dynes cm$^{-3/2}$	G dynes cm^{-2}	k_{Sy}/G	k_{ST}/G	Transition $\ell^{-1/2}$ Grain Size cm$^{-1/2}$	Ref.
3% SiFe	5.6×10^7	12.2×10^7 (a)	6×10^{11}	9.0×10^{-5}	20.0×10^{-5}	18	5
V	1.1×10^7	7.1×10^7 (b)	5×10^{11}	2.2×10^{-5}	14.2×10^{-5}	6	10
Cr	3.2×10^7 (c)	21.5×10^7	9×10^{11}	3.6×10^{-5}	24.0×10^{-5}	19 (c)	9
Fe	5.8×10^7	14.5×10^7	8×10^{11}	7.2×10^{-5}	18.1×10^{-5}	10	7
Nb	0.5×10^7	very small (d)	4×10^{11}	1.3×10^{-5}	very small	3	29

(a) k_{ST} value at 20°K is 11.4×10^7 dynes cm$^{-3/2}$
(b) k_{ST} value measured at 20°K
(c) Estimated from results at higher temperatures
(d) Value at 195°K

agreement also with this prediction, even though the G values have been taken from measurements at 293°K. A small temperature dependence of k_{ST} and an even smaller strain rate dependence is directly indicated by equation (26). For 3% silicon-iron, it is calculated for the two temperatures of 293°K and 78°K, that

$$1.4 \geq [k_{ST}(78°K)/k_{ST}(293°K)] \geq 1.1$$

and for strain rates of 50 sec^{-1} and 10^{-4} sec^{-1}, that

$$1.08 \geq [k_{ST}(50 \text{ sec}^{-1})/k_{ST}(10^{-4} \text{ sec}^{-1})] \geq 1.04$$

Figure 3 shows several values of the twinning stress, σ_T, determined at various temperatures from which it may be seen that k_{ST} must be very weakly temperature dependent. Results of this type have been employed to demonstrate that the onset of major twinning occurs for each temperature at a stress level very close to the tensile fracture stress, σ_F[11]. The micro-slip stress is shown to occur below the twinning stresses. In Figure 4, the twinning-fracture stress, σ_{TF}, is shown at widely different strain rates to be definitely insensitive to this testing parameter, as indicated for the calculation from (26). The results of Figures 3 and 4 are in reasonable agreement with other experimental studies[2,10].

The importance of pre-twinning slip to obtaining a reasonable twinning rate and consequent number of twins is an essential feature of the derivation of equations (23) and (26). It has previously been proposed[31] that this was so and there is substantial experimental evidence on this point also to verify it[6,32-34]. Perhaps the most striking results are those obtained by etch-pitting

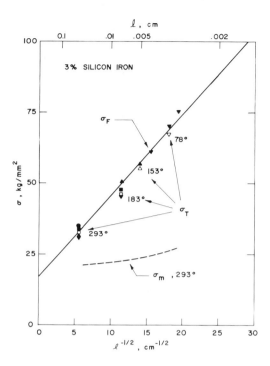

Figure 3. Twinning and brittle fracture stresses of 3% silicon-iron

dislocations in 3% silicon-iron, where it was observed that the difference in stress at which micro-slip and twinning first occurred was greater the smaller the grain size. It has proven to be difficult to observe directly slip dislocations producing twins but evidence in this direction has been given by others[35,36].

The expression for $\overline{N}_T$, equation (13), is in general agreement with the conclusion that the initiation of twinning in iron crystals is probably a thermally activated process[17]. The activation energy for twinning, U* as given in equation (16), is not a low value, however, but because of the pronounced stress dependence of U* the effects of temperature and particularly, strain rate are not very great once those stress levels are reached at which twinning may occur. The rate of twinning, $\dot{N}_T$, as given in equation (20), is also closely related to $\overline{N}_T$. By substitution in (20) of $(\dot{\tau}_e/\tau_e) = 10^{-2}$ sec^{-1}, T = 80°K, and taking those values of U* estimated following (16), then it is found that $\dot{N}_T \simeq \overline{N}_T$ at the lower limit of U* but that $\dot{N}_T > \overline{N}_T$ at the upper limit of U*. Thus, as the stress is increased during the elastic loading of a material, twins are predicted to be formed suddenly and to increase in

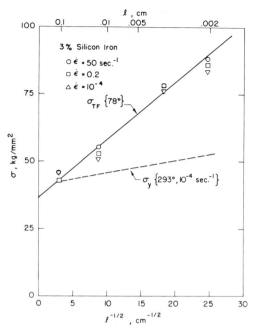

Figure 4. The twinning-fracture stress of 3% silicon-iron at various strain rates

number at a sharply increasing rate. This seems to agree well with experimental observations.

For most polycrystalline bcc metals, there is a transition grain size at which the major mode of deformation changes from slip to twinning. This transition grain size decreases with decreasing temperature[5,11] and increasing strain rate (c.f. Figures 3 and 4). The relative magnitudes of the σ_o and k values for slip and twinning seem important to understanding the transition. For example, σ_{o_y} (for slip) > $m\tau_o$ (for twinning) and $k_T > k_y$. The relatively large σ_{o_y} value is due to the need during bulk yielding of a polycrystal for slip to occur on sufficient deformation systems within nearly all the grains to ensure continuity of strain within the material. Appreciable work hardening also occurs during the pre-yield micro-straining and this contributes in a significant way to the friction stress, thus necessarily raising the applied stress necessary to cause bulk yielding[24]. Twinning occurs much less uniformly than slip and is significantly different from the slip process because of the need for pronounced stress concentrations within local regions of the microstructure. Each twin is capable of making an appreciable contribution to the bulk strain even when averaged with the essentially elastic deformation of the total specimen, especially at large grain sizes. For those metals described in Table I at 77°K, a variation in k_{ST} is observed from $k_{ST} \simeq 2k_{Sy}$ for silicon-iron to $k_{ST} \simeq 7k_{Sy}$ for vanadium. It is expected that metals with a low k would be unlikely to twin unless σ_{oy} were exceptionally large such as occurs at low temperatures and high strain rates or due to alloying, irradiation, or dislocation strengthening mechanisms whereby the "intrinsic" friction stress were increased. For a given τ_e, it occurs that increasing values of ℓ will decrease U^*, as shown in (16) and thus promote twinning as occurs for large grain size materials. An increased τ_e allows twinning to occur at finer grain sizes.

One further feature of twinning in bcc metals may also be explained by the present analysis, and this is the observation of twinning and brittle fracture often occurring at nearly the same stress level[4]. The predicted dislocation pile-up stress concentration factor for fracture, k_F, is given by Stroh[37] as $[6\pi G\gamma/(1-\nu)]^{1/2}$, where γ is the surface energy and ν is Poisson's ratio. Substituting reasonable values for the metals in Table I, it occurs that $k_F \simeq 10^8$ dynes/cm$^{3/2}$ which is very close to the twinning value, k_{ST}, previously determined. The lower $m\tau_o$ value for twinning, as compared with bulk yielding by slip, also allows then for the possibility that brittle fracture may occur at stress levels below those required for general yielding. On this basis, there is a nearly equal chance for preyield microslip to initiate twinning or brittle fracture in bcc metals at very low temperatures and high strain rates.

CONCLUSIONS

A constitutive relation for the grain size, temperature and strain rate dependence of the stress for the essentially elastic twinning of bcc metals at low temperatures has been developed. The analysis is based on an evaluation of the factors determining the twinning contribution to the bulk material strain and, also, on a thermal activation analysis of the twinning nucleation rate. The derived equations give agreement with a number of experimental observations on bcc metals, including:
  (a) The necessity of micro-slip for twin nucleation;
  (b) A pronounced grain size dependence of the twinning stress;
  (c) A relatively small dependence of the twinning stress on temperature, and an even lesser dependence on the strain rate;
  (d) The onset of twinning under conditions very much the same as for brittle fracture.

The equations also emphasize the importance of the number of twins as a controlling factor determining the material deformation rate, so that the concept of an average dislocation velocity which is used to describe the strain rate due to slip does not appear relevant to a description of the strain rate due to twinning[38].

ACKNOWLEDGEMENT

The authors wish to thank the Office of Naval Research for financial support for this research at the University of Maryland through Contract N00014-67-A-0239-0011, NR031-739.

REFERENCES

1. K. Ogawa, Phil. Mag., 11, 217 (1965)
2. G. F. Bolling and R. H. Richman, Can. Met. Quart.,5, 143 (1966)
3. R. Lagneborg, Trans.Amer.Inst.Min.Met.Engrs.,2301, 1661 (1967)
4. R. W. Armstrong, "Deformation Twinning", ed. by R. E. Reed-Hill, J. P. Hirth and H. C. Rogers, p. 356, Gordon and Breach,N.Y. (1964)
5. D. Hull, Acta Met. 9, 191 (1961)
6. P. J. Worthington and E. Smith, Acta Met., 14, 35 (1966)
7. A. S. Drachinskii, V.F. Moiseev and V.I. Trefilov, Soviet Phys. Doklady 9, 1701 (1964); V.F. Moiseev and V.I. Trefilov, Phys. Stat. Sol. 18, 881 (1966)
8. J. R. Low, Jr., "Relation of Properties to Microstructure", ASM (1954)
9. M. J. Marcinkowski and H. A. Lipsitt, Acta Met. 10, 95 (1962)
10. T. C. Lindley and R. E. Smallman, Acta Met. 11, 361 (1963)
11. P. J. Worthington, Scripta Met. 2, 701 (1968)
12. J. D. Eshelby, Proc. Roy. Soc. London, A241, 376 (1957)
13. M. J. Marcinkowski and K. S. Sree Harsha, J. Appl. Phys. 39, 6063 (1968)
14. R. W. Armstrong, Science 162, 799 (1968)
15. R. E. Reed-Hill, E. R. Buchanan and F. W. Caldwell, Jr., Trans. TMS-AIME 233, 1716 (1965)
16. J. N. Johnson and R. W. Rohde, J. Appl. Phys. 42, 4171 (1971)
17. J. Harding, Proc. Roy. Soc. London, A299, 464 (1967)
18. E. Orowan, "Dislocations in Metals", p. 116, Amer. Inst. Min. (Metall.) Engrs., N. Y. (1954)
19. N. J. Petch, J. Iron Steel Inst. 174, 25 (1953)
20. N. M. Madhava, P. J. Worthington and R. W. Armstrong, Phil. Mag. 25, 519 (1972)
21. R. W. Armstrong, I. Codd, R. M. Douthwaite and N. J. Petch, Phil. Mag. 7, 45 (1962)
22. N. D. Brentnall and W. Rostoker, Acta Met. 13, 187 (1965)
23. W. B. Morrison, Trans. ASM 59, 824 (1966)
24. C. T. Liu, R. W. Armstrong and J. Gurland, J. Iron Steel Inst. 209, 142 (1971)
25. R. Kossowsky and N. Brown, Acta Met. 14, 131 (1966)
26. R. F. Bunshah, "Deformation Twinning", ed. by R. E. Reed-Hill, J. P. Hirth and H.C.Rogers, p.390,Gordon and Breach,N.Y.(1964)

27. L. B. Pfeil, J.I.S.I., Carnegie Scholarship Memoirs, 15, 318 (1926)
28. A. Sleeswyk, Acta Met. 10, 705 (1962)
29. M. A. Adams, A. C. Roberts and R. E. Smallman, Acta Met. 8, 328 (1960)
30. A. C. Raghuram, R. E. Reed and R. W. Armstrong, Mat. Sci. & Engrg. 8, 299 (1971)
31. B. A. Bilby and A. R. Entwisle, Acta Met. 2, 15 (1954)
32. R. L. Bell and R. W. Cahn, Proc. Roy. Soc. London A239, 494 (1957)
33. R. M. Hamer and D. Hull, Acta Met. 12, 682 (1964)
34. R. Priestner, "Deformation Twinning", ed. by R. E. Reed-Hill, J. P. Hirth and H. C. Rogers, p. 321, Gordon and Breach, N.Y. (1964)
35. D. Hull, Ibid, p. 121
36. R. Priestner and W. C. Leslie, Phil. Mag. 11, 895 (1965)
37. A. N. Stroh, Advances in Phys. 6, 418 (1957)
38. W. deRosset and A. V. Granato, "Fundamental Aspects of Dislocation Theory", NBS Spec. Pub. 317, p. 1099 (1970)

MODELS OF SPALL FRACTURE BY HOLE GROWTH

Frank A. McClintock

Department of Mechanical Engineering

Massachusetts Institute of Technology

1. STATIC FRACTURE BY THE GROWTH OF HOLES

Since scanning electron micrographs show strong similarities between static and dynamic fracture by hole growth, it is worth reviewing the stages in static fracture, as they are currently understood.

Nucleation can occur in a perfect lattice at a stress related to the modulus of elasticity E by the shape of the interatomic force law. Bubble raft studies by McClintock and O'Day (1965), originally motivated by observations of spall fracture, gave values of E/10 to E/20. Estimates can also be obtained from a power-law estimate of the interatomic forces, but Lomer (1949) felt that the bubble model is more accurate for copper, at least. Since the atomic frequency is of the order of 10^{13}/sec, similar values would be expected for dynamic fracture. Such a theoretical strength was apparently obtained by McQueen and Marsh (1962). The bubble model shows that dislocations and grain boundaries reduce the strength by a factor of two. Very possibly the mosaic pattern of nucleation observed by Seaman et al. (1971) in tests on 99.999% pure aluminum at E/25 indicates nucleation at a dislocation substructure.

More commonly, hole growth nucleates at second phases, either by cracking of brittle phases or by separation of the inclusions from the surrounding matrix. Once nucleated, holes grow by plastic flow. Both the deformation and the triaxial stress applied to the surrounding matrix play a role, as indicated by the following approximate equation giving the change in semi-axis $\underline{b}$ of holes with spacing ℓ_b in material that strain-hardens according to the law

$$\bar{\sigma} = \bar{\sigma}_1 \bar{\epsilon}^n ,\qquad(1)$$

and is subject to transverse principal stress components σ_a^∞ and σ_b^∞:

$$\frac{d\ln(2b/\ell_b)}{d\bar{\epsilon}^\infty} = \frac{\sinh[(1-n)(\sigma_a^\infty + \sigma_b^\infty)/(2\bar{\sigma}/\sqrt{3})]}{(1-n)}\qquad(2)$$

Localization of void growth into a sheet which does not deform in its plane, but across which there is relative motion, is the next stage of fracture, as recently pointed out by Berg (1972). This localization should be reduced by the spatial variations in stress accompanying dynamic fracture. More diffuse fractures are possible dynamically, because the straining at one region can be continuing for a moment after a nearby region has begun to unload by hole growth.

The final stage in formation of a crack is the separation process across the localized sheet of holes. As shown recently by Nagpal, et al., (1972), the traction falls off nearly linearly with displacement, and the work per unit area during this final separation process is of the order of the yield strength Y times the spacing of the holes, ℓ.

Crack growth generally occurs by re-nucleation of holes ahead of the crack, followed by joining up. Statically, this process favors holes at either side of the crack tip, as shown by Carson (1970), and for viscous flow by Cipolla (1972). On a more macroscopic scale, the fracture criteria can be provided by the well-known equations of linear elastic fracture mechanics, which uniquely define the stress in an elastic region distant from the plastic zone but close compared to the total crack length or the distance to the next nearest boundary. This stress distribution is distorted by dynamic effects, but will not be discussed here.

2. RATE EFFECTS

Rate effects in materials can only occur in one of three ways: by inertia terms, as in wave motion; by quantum mechanical tunneling effects; and by the time required for thermal activation, including any temperature effects due to adiabatic temperature rise.

For interpreting experimental results and estimating the magnitudes of effects, it is convenient to have a number of parameters worked out in consistent units such as Table 1. The properties in Table 1 were taken to be those of the pure metal except for the tensile strength, which was taken to be that of the strongest

Table 1. Parameters for Spall by Hole Growth

Material: condition:	Aluminum pure, alloy	Copper pure, alloy	Iron pure, alloy
Young's modulus, E, ksi kbar $\approx 10^3$ atm	10,300 710	17,500 1210	30,000 2070
Poisson's ratio, ν	.33	.345	.28
Tensile strength, T.S., kbar	.47 6.1	2.2 13.8	2.1 20.6
Density, ρ, g/cm^3	2.7	8.9	7.7
Specific heat, cal/g°C	.22	.093	.11
Thermal conduct. cal/sec cm°C	.49	.92	.11

Loading parameters

Uniaxial strain modulus $E_1 = E(1-\nu)/((1+\nu)(1-2\nu))$, kbar	1050	1900	2660
Uniaxial strain wave velocity, $c_1 = \sqrt{E_1/\rho}$, cm/µsec	.62	.46	.59
Shear wave velocity, c_2, cm/µsec	.215	.225	.324
Stress/impact velocity, $\sqrt{\rho E_1}$, kbar/(cm/µsec)	1685	4100	4530
Pulse duration/flyer thickness, $2/c_1$, µsec/cm	3.2	4.3	3.4

Thermal paramaters

Temperature rise for $r_2/r_1=10$, $2\ell n10$(T.S.)/ρC, °C	90 1150	300 1850	380 2700
Melting point °K	950	1380	1830
Thermal diffusivity $\alpha = k/\rho C$, cm^2/sec	.82	1.11	.13
Quench time for a 10µ hole R^2/α, µsec	.3	.22	1.9

Viscous parameters

Viscosity, $.3kT/b^4 \Lambda c_2$, poise	5400	12,900	9800
Strain rate for viscous stress = T.S. at 10^8 disl./cm^2, µsec^{-1} (end of rate-indep. plast)	.044(.45)	.085(.29)	.15(.40)
Strain rate at 10^8 disl./cm^2 & $.5 c_2$, $\dot{\gamma} = b\Lambda .5c_2$, µsec^{-1} (end of linear viscosity)	.45	.29	.40
Viscous stress at $.5 c_2$, µ$\dot{\gamma}$ (end of linear visc.) kbar	2.4	3.7	3.9

common commercial alloy. The loading parameters are calculated on the basis of elastic collisions which should be valid when the stress pulses are only a small fraction of the modulus. In the expression for pulse duration per unit via thickness, the factor 2 arises from the total length of the pulse being set by its travel through the flier and return. This total length also governs the time of loading, for a square wave it is the time required for the reflected wave to pass the spall plane. Further discussion of rate effects due to inertia will be deferred to the last section.

Quantum mechanical effects would be more important than effects due to thermal motion only under conditions for which

$$h\nu > kT. \tag{3}$$

The frequency ν of concern here is not $\nu = 10^{13}$/sec, which holds for individual atoms, but something more like $\nu = 10^{11}$/sec, which is the vibrational frequency of the dislocations pinned against an obstacle. (Teutonico, et al. 1964) This implies that thermal motion will free a dislocation sooner than tunneling provided the temperature is greater than about 1/100 of the Debye temperature, T_d:

$$T \geq T_D/100. \tag{4}$$

Since this limit is only a few °K, quantum mechanical tunneling will not be important in these problems.

Under ordinary conditions, the strain rate effects in reasonably hard metals are quite unimportant. For instance in expressing the strain rate as a power function of stress, exponents are of the order of 60 to 120 (McGregor and Fisher, 1945); Hoggatt and Recht, 1969). On the other hand, research workers dealing with pure metals have observed effective viscosities corresponding to a linear dependence of strain rate on stress. For instance Lothe (1962) gives a dislocation phonon drag of

$$\tau b = Bv, \text{ where } B = 3kT/10c_2 b^2 \tag{5}$$

Eliminating the velocity v in terms of the strain rate $\dot{\gamma}$, by the mobile dislocation density Λ_m and the Burgers vector b

$$\dot{\gamma} = b\Lambda_m v \tag{6}$$

gives the viscosity as

$$\mu = \tau/\dot{\gamma} = .3kT/b^4 \Lambda_m c_2 , \tag{7}$$

The evaluation of these equations is illustrated in Fig. 1 for

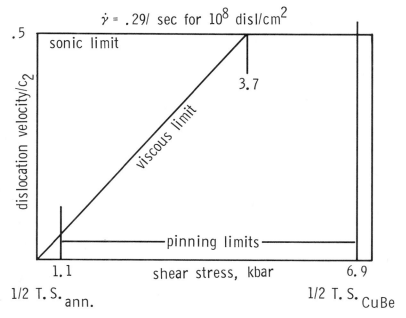

Fig. 1.

copper. Assuming a mobile dislocation density of $10^8/cm^2$, the mean dislocation velocity can be related to the strain rate by Eq. 7 and the stress calculated from the viscosity. For the annealed material, there is a small range of rate-independent plasticity, where obstacles limit the flow stress. Soon phonon drag becomes limiting. Finally inertial effects limit the dislocation velocity, and, if the mobile dislocation density remains constant by annihilation, the strain rate is also limited, giving a hole growth rate proportional to radius.

For hard beryllium copper alloy, however, there is no viscous regime. This illustrates the need to carry out studies in the regime of practical interest. Similar differences between situations are observed for aluminum, iron and their alloys.

These calculations may not leave the rate-independent theory of plasticity a clear field for hard alloys, however, as indicated in the section of Table 1 dealing with thermal behavior. Increasing the radius of a hole by a factor of ten by stretching its surface in each direction causes a local straining of $2\ln(10)$, corresponding to a strain of 4.6. As shown in Table 1, the temperature rises for the alloys are so great that melting would have to occur if the process were adiabatic. Actually these high strains are only encountered in a very thin skin, and if one estimates the quench time on the basis of the hole radius, even in that thin skin the heating only becomes important for microsecond type pulses after the holes reach a radius of the order of 5µ.

3. SIZE EFFECTS

Another factor limiting classical plasticity is the size of the hole relative to the mean free path of slip lines. The distance ℓ between pinning points on dislocations loops sets a lower bound to this dimension and can be estimated from the well-known equation for hardening:

$$\tau = Gb/\ell , \qquad (8)$$

The correlation of flow in slip bands, however, means that the slip could first be regarded as being homogeneous at a scale several orders of magnitude larger than this, say 100ℓ.

For soft and hard alloys take τ/G to be .001 and .01, respectively. If $b = 3 \times 10^{-8}$ cm,

$$100\ell = 30\mu \text{ and } 3\mu, \text{ respectively}, \qquad (9)$$

indicating that the use of homogeneous plasticity becomes doubtful for holes smaller than a micron even in hard alloys. With other solutions so difficult, however, one is tempted to push his luck.

4. INERTIA MODEL OF SPALL FRACTURE BY HOLE GROWTH

Neglecting the limitations discussed above, we can estimate the inertia effects themselves by using the equation of Hopkins (1960) for the expansion of the hole of radius $\underline{a}$ in a fully plastic regime subjected to an applied stress σ_R at the radius $\underline{R}$ midway between holes. Hopkins' Eq. 7.65 is

$$\sigma_R/\rho = (2Y/\rho)(\ln R/a) + \\ + \ddot{a}(a - a^2/R) + \dot{a}^2(3/2 + a^4/2R^4 - 2a/R) , \qquad (10)$$

A measure of the relative importance of the acceleration and the velocity terms can be obtained by first setting the radial velocity to zero and solving for the acceleration,

$$\ddot{a} = (\sigma_R - 2Y\ln R/a)/\rho(a - a^2/R) , \qquad (11)$$

and in turn setting the acceleration to zero and solving for a quasi-steady state velocity:

$$\dot{a} = \sqrt{\frac{\sigma_R - 2Y\ln R/a}{\rho(3/2 + a^4/2R^4 - 2a/R)}} \qquad (12)$$

The distance required to accelerate to steady states is of the order of

$$\dot{a}^2/\ddot{a} = \frac{a/R - (a/R)^2}{3/2 + a^4/2R^4 - 2a/R} \approx 2a/3R , \qquad (13)$$

It appears that steady state develops relatively soon, so that the radial velocity is found reasonably closely from Eq. 12. This provides an equation of state for the porosity, which could then be used with a uniaxial code, as was done by Seaman, et al (1971) with viscous hole growth models. In addition to the effects of hole growth on the porosity and average properties of the material itself, the unloading waves due to hole growth should also be taken into account. The plastic calculations outlined here should be good for high strength materials only slightly above their limit loads.

5. CONCLUSION

Order of magnitude estimates indicate that the importance of the variables such as dislocation viscosity, adiabatic temperature rise, thermal diffusivity, and mechanical inertia play markedly different roles with different alloys at different intensities of loading. The weakest assumption in these estimates is that of the mobile location density in hard alloys, about which further comment would be welcome.

6. ACKNOWLEDGEMENT

This research was supported by the Advanced Research Projects Agency and the Department of Defense under contract number DAH15-71-C-0253 with the University of Michigan. Discussions with Prof. A. S. Argon are also deeply appreciated.

REFERENCES

Berg, C.A.	1972	"Ductile Fracture by Development of Surfaces of Unstable Cavity Expansion", *Journal of Research of the National Bureau of Standards*, v. 76C, 33-39.
Carson, J. W.	1970	"A Study of Plane Strain Ductile Fracture", Ph.D. Thesis, Dept. of Mech. Engng., MIT, Cambridge, Mass.
Cipolla, R.C.	1972	"A Study of the Initiation of Ductile Fracture from Grooves", M.S. Thesis, Dept. of Mech. Engng., MIT, Cambr.,Mass.
Hoggatt, C.R. Recht, R.F.	1969	"Dynamic Stress strain Relationships Obtained from an Expanding Ring Experiment", SESA Spring Meeting, Paper 1487.
Hopkins, H.G.	1960	"Dynamic Expansion of Spherical Cavities in Metals", *Progress in Solid Mechanics*, v. 1, 86-164.
Lomer, W. M.	1949	"A Dynamical Model of A Crystal Structure, III", *Proc. Roy. Soc. A*, v. 196, 132.
Lomer, W. M.	1949	"The Forces Between Floating Bubbles and a Quantitative Study of the Bragg 'Bubble Model' of a Crystal" *Proc. Camb. Phil. Soc.*, v.45, 660.
Lothe, J.	1962	"Theory of Dislocation Mobility in Pure Slip", *Journal of Applied Physics*, v. 33, 2116-2125.
MacGregor, C. W. Fisher, J. C.	1945	"Tension Tests at Constant Strain Rates", *J. Appl. Mech.* v.12, 217-227.
McQueen, R. G., Marsh, S. P.	1962	"Ultimate Yield Strength of Copper", *J. Appl. Phys.* v. 33, 654 665.
Nagpal, V. McClintock, F.A. Berg, C. A. Subudhi, M.	1973	"Traction-Displacement Boundary Condition for Plastic Fracture by Hole Growth", *International Symposium on the Foundations of Plasticity*, Warsaw, Poland, to be published in Proceedings.

Seaman, L.　　　　　1971　　"Dynamic Fracture Criteria of
　Barbee, T. W. Jr.　　　　　Homogeneous Materials", Stanford
　Curran, D. R.　　　　　　　Research Institute, TR. AFWL-
　　　　　　　　　　　　　　TR-71-156. Menlo Park, Calif.

Teutonico, L. M.　　1964　　"Theory of the Thermal Breakaway
　Granato, A.V.　　　　　　　of a Pinned Dislocation Line with
　Lucke, K.　　　　　　　　　Application to Damping Phenomena",
　　　　　　　　　　　　　　Journal of Applied Physics, v. 35,
　　　　　　　　　　　　　　220-234.

　　See Also:

Granato, A.V.　　　　1964　　"Entropy Factors for Thermally
　Lucke, K.　　　　　　　　　Activated Unpinning of Dislocations",
　Schlipt, J.　　　　　　　　*J. Appl. Phys.* v.35, 2732-2745.
　Teutonico, L.J.

Author's Discussion on "Models of Spall Fracture by Hole Growth"

Frank A. McClintock

Department of Mechanical Engineering

Massachusetts Institute of Technology

Further comment on two points in the paper may be in order: the question of mobile dislocation density and detailed estimates of limiting stresses and growth times for voids.

The question of mobile dislocation density may be seen more clearly by re-plotting Fig. 1 as tensile strength vs. strain rate, using a logarithmic scale, as shown in Fig. 2. The engineer is interested in a phenomenal range of strain rates, from nano-seconds to hundreds of years; only the right half is shown. Consider an annealed metal. At low strain rates, obstacle and dislocation pinning limit the flow; the rate dependence is very small, with a slope of only 1-2% on a log-log scale. If the mobile dislocation density were assumed to remain nearly constant, the dislocation velocity would rise, eventually reaching a limiting

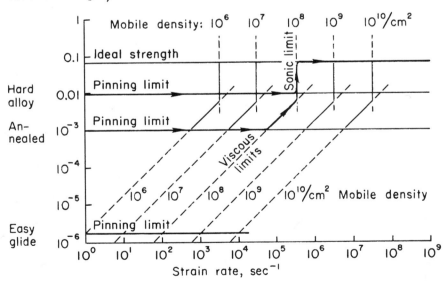

Fig. 2. Hypothetical strain rate dependence for constant dislocation density.

viscous behavior determined by phonon drag. The stress now increases linearly with strain rate. Further increases in stress would accelerate the dislocations until they approached say half the shear wave velocity, indicated by the vertical line labelled "sonic limit". Eventually, increases in stress would cause the ideal strength to be reached, at which shear can occur with no dislocation motion. Here the plot of flow stress vs. strain rate would again become horizontal, until the effects of atomic inertia were reached at frequencies of the order of 10^{13}/sec. For high strength alloys, Fig. 2 indicates that a viscous limit would never be reached whatever the mobile dislocation density, since that shifts both the viscous and sonic limits by the same amount. The assumption of a constant mobile dislocation density seems very questionable and the author would appreciate comment from others about data or more reliable estimates for the rate dependence in the high strain rate regime for strong alloys. One would expect that relatively moderate increases in strength, of the order of a factor of two occuring in several decades of strain rate, would allow the mobile dislocation density to approach the total density. This density could be estimated from either dislocation interactions at a distance ℓ or the critical stress for nucleating sources. In terms of shear strength τ, shear modulus G, and Burgers vector b it would be of the order of

$$\Lambda = \frac{1}{\ell^2} = \left(\frac{\tau}{Gb}\right)^2. \qquad (14)$$

If the rate dependence does indeed flatten out so that the flow stress only rises by a factor of 2 in 4 orders of magnitude of strain rate, then the calculations of dynamic hole growth could reasonably be based on a rate-insensitive calculation, with the mean flow stress simply elevated by an appropriate amount. For low strength materials, if there is an abrupt ramp the viscous equations might still be appropriate.

As a second point of amplification of the paper, upper limits to the time required by inertia effects in spall growth can be obtained by approximate integration of Hopkins' equations for incompressible material.

First for a hole initially at rest on the spall plane, the time required for an elastic wave appearing there to spread out to the radius R is

$$t_{load} = R/c_1. \qquad (15)$$

For typical hole spacings of $R = 5\mu$ and $c_1 = .6$ cm/μsec, the loading time is only .8 nanoseconds.

Next, for elastic growth of the hole to initial yield, from Hopkins Eq. 7.55 with $\sigma_r = 0$ at $r = a$ and $\sigma_r = \sigma_R$ at $r = R$,

$$\sigma_R/\rho = (4E/9\rho)(1 - a_o^3/a^3)(1 - a^3/R^3) +$$
$$\ddot{a}a(1 - a/R) + \dot{a}^2(3/2 - 2a/R + (1/2)a^4/R^4) \quad (16)$$

Hole growth in the elastic regime is only a fraction of the radius, so estimate the time from the acceleration term alone. First, the hole radius at initial yield, $\sigma_\phi = \sigma_\theta = Y$, for incompressible material is

$$a - a_o = a_o(Y/E)(1 - 1/2). \quad (17)$$

For a/R small compared to unity, substituting Eq. 16 into the acceleration equation gives

$$t = \sqrt{2(a - a_o)/\ddot{a}} = \left[a_o(Y/E)\frac{a_o}{\sigma_R/\rho - (4E/9\rho)(a^3/a_o^3 - 1)}\right]^{1/2} \quad (18)$$

The second term in the denominator increases from zero to $2Y/3\rho$ as the hole grows. Taking the higher value gives an upper limit to the time of initial yielding:

$$t = a_o/\sqrt{(E/\rho)(\sigma_R/Y - 2/3)}. \quad (19)$$

Comparison with the loading time of .8 nanoseconds from Eq. 15, and taking σ_R/Y twice that for static yield, gives times of the order of a fraction a/R of a nanosecond, quite negligible.

The time for elastic-plastic growth is found similarly. From Hopkins' Eq. 7.65, 7.68, and 7.71 and for $R \gg a$,

$$\sigma_R = (2Y/3\rho)(-b^3/R^3 + 1 + \ln(2E/3Y) + \ln(1 - a_o^3/a^3))$$
$$+ \ddot{a}a + 3\dot{a}^2/2 \quad (20)$$

when the plastic zone radius $\underline{b}$ is between $\underline{R}$ and $\underline{a}$:

$$1 \leq b^3/a^3 = (2E/3Y)(1 - a_o^3/a^3) \leq R^3/a^3.$$

When the plastic zone has reached R, the inner radius is

$$a^3/a_o^3 = 1 + (R^3/a_o^3)(3Y/2E). \quad (21)$$

Except for very soft, dirty metals $((R/a_o)^3 < E/Y)$, the holes will have grown by a number of radii by the time the plastic zone reaches R, so estimate the growth time from the steady state velocity:

$$t = \int \frac{da}{\dot{a}} = \int \frac{\sqrt{3/2}\, da}{\sigma_R/\rho + (2Y/3\rho)(b^3/R^3 - 1 - \ln(2E/3Y) - \ln(1 - a_o^3/a^3))} \quad (2$$

An upper limit to the time is found from a lower limit to the denominator, neglecting b^3/R^3 and a_o/a. For σ_R 10% above that for zero growth rate,

$$t < \frac{a}{\frac{2}{3}\sqrt{.1\frac{Y}{E}\frac{E}{\rho}\left(1 + \ln\frac{2E}{3Y}\right)}} \qquad (23)$$

For $Y/E = .001$ to $.01$, and $c = \sqrt{E/\rho} = .5\,\text{cm}/\mu\text{sec}$, $R = 5\mu$, and from Eq. 21

$$a = R\sqrt[3]{3Y/2E}, \qquad (24)$$

the times are less than 6 to 5 nanoseconds. Lower values of applied stress would require longer times, but are not of much practical interest because only a 10% decrease in applied stress would stop the hole growth completely.

Turning to the fully plastic regime, obtain an estimate from Eq. 12 of the original paper, with Eq. 24 for an initial value of a/R in the numerator, and a/R terms neglected in the denominator of that equation:

$$t < R/\dot{a}_{min} = \frac{R}{\sqrt{\frac{2}{3}\frac{\sigma_R}{\rho} - \frac{4Y}{9\rho}\ln\frac{2E}{3Y}}} \qquad (25)$$

Again for σ_R 10% above the limit for hole growth,

$$t \leqslant (3/2)R/\sqrt{.1(E/\rho)(Y/E)\ln(2E/3Y)} \qquad (26)$$

For $\sqrt{E/\rho} = .5\,\text{cm}/\mu\text{sec}$, $Y/E = .001$ to $.01$, and $R = 5\mu$, the times are less than 60 to 23 nanoseconds.

In conclusion, then, inertia effects are negligible (except within 10% of the limit of zero hole growth rate) when the loading times exceed 20 to 50 nanoseconds for high and low strength alloys, respectively. The limiting values of the applied stresses are typically of the order of a 4 times the yield strength, as given by the first term on the right side of Eqs. 10 or 17.

OBSERVATIONS OF SPALLATION AND ATTENUATION EFFECTS IN

ALUMINIUM AND BERYLLIUM FROM FREE-SURFACE VELOCITY MEASUREMENTS

C. S. SPEIGHT, P. F. TAYLOR, A. A. WALLACE

A.W.R.E., FOULNESS, ESSEX, UK

1.   INTRODUCTION

Reliable data on attenuation and spallation is of importance in the formulation of computer codes which are concerned with predictions of material shock behaviour. The development of a suitable model to describe spallation has been the subject of several recent publications [e.g. 1, 2], and the value of free surface velocity measurements in this connection has been demonstrated in various other papers [e.g. 3, 4]. The work presented here is based on an experimental programme carried out by the authors and, subject to the limitations of the experimental variables, a simple model for attenuation is given together with a model for spallation.

2.   EXPERIMENTAL TECHNIQUES

A capacitor gauge technique [5, 6] is used to measure the shock induced free surface velocity of Al and Be samples. The shock pulse is produced in the sample by the impact of a thin Al flier which is driven electromagnetically by a high energy capacitor bank system. A number of experimental series has been performed to include variations of target material, target thickness and flier velocity. In each case the flier, constructed from PIC ($\frac{1}{2}$H) Al foil 0.026 cm thick with an effective length and width of 12 cm x 10 cm respectively, is driven a distance of 1 cm to the target. The impact velocity, which can be continuously varied up to about 10^5 cm s^{-1}, is controlled by the capacitor bank voltage and is monitored by a transit time measurement. Direct measurement of the flier velocity, using probes, agrees to within 5% of the value deduced from the transit time measurement and the observed peak free surface velocity

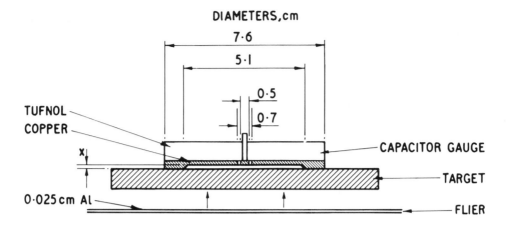

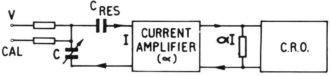

Figure 1. Practical Arrangement and Schematic Circuit for Capacitor Gauge Measuring Technique

of the thinnest Al samples. The flier and target/gauge assembly is in an environment of air at 0.1 atmospheres pressure. Two completely separate and similar facilities are used - one for toxic and one for non-toxic materials.

The experimental arrangement is indicated in Figure 1. The variable capacitor C is formed by the fixed 0.5 cm disc and the target free surface. The initial value of the gauge capacity, formed by the fixed 0.5 cm disc and the target free surface, is typically 0.3 pF and that of x is 0.06 cm. The relatively large reservoir capacitor, C_{RES}, maintains constant voltage, V, across C so that, in operation, the current I gives a direct measure of the free surface velocity, U, according to the relationship:-

$$I = VU \ (dC/dx) \qquad \ldots \ldots (2.1)$$

The gauge sensitivity dC/dx is determined from static measurement and this together with the experimental record allows U to be determined from equation 2.1.

The overall time resolution of the electrical system is a few ns and in practice the record rise time is limited by the intrinsic shock properties of the target material or by the time coherence of the shock wave at the free surface. Assessment of foil non-flatness has been made using elastic precursor measurements and here a spatial gradient in the flier of between 0.8 and 4.0 mrad is implied. (Typical precursor rise times are in the range 10-50 ns and an effective gauge diameter of 0.7 cm is taken.) Considerable effort has gone into the reduction of background noise to the extent that a discontinuity in velocity of 1 ms^{-1} is fairly easily discernible. It is interesting to note that this corresponds to a change in gauge current of only 0.3 µamps in an environment where the peak current (in the flier) is $\sim 10^6$ amps.

3. SHOCK ATTENUATION

In these experiments the flier velocity is maintained at a constant value, for each type of material, to provide an attenuation curve at a given input velocity and momentum. The target specimens were prepared from a single batch of each material. For Al both PIC ½H and PIC M material was used, and wrought ingot and 99% dense hot pressed powder (SP100C) in the case of Be.

A sample of the experimental results is shown in Figure 2 where free surface velocity (FSV) is plotted against time for targets of Al and Be over a range of thicknesses. The Al results illustrate particularly well the three characteristic regions of a typical free-surface velocity-time profile. An initial elastic precursor is observed followed by an attenuating plastic pulse and finally an oscillatory disturbance associated with spallation. The same general features are exhibited by Be but here the rise up to the peak is noticeably more ramped without any clear indication of a sharp elastic-plastic transition. It is immediately apparent that each series illustrated in Figure 2 shows considerable attenuation of the peak with target thickness accompanied by an increasing pulse width. In the case of the thinnest target the shape of the U-t profile, neglecting spallation effects, is determined essentially by the conditions at impact. These include the velocity and temperature of the flier, the shock properties of the target and flier, and the effects of gas between the flier and target. The scale of the present experiments does not allow a detailed study of the impact end of the attenuation curve and indeed the impact condition may easily be simplified in alternative experimental arrangements. The present region of interest is for targets of many flier thicknesses where the propagation time is sufficient for a number of elastic-plastic interactions to have occurred. Here, the U-t profiles appear to have a smooth and characteristic shape which is believed to be dominated by, and therefore indicative of, the stress release properties of the target material. Furthermore, the profile is found to scale for variations in width and amplitude and it is possible to construct a simple semi-analytical model for attenuation based on conservation of momentum and extraction of energy by

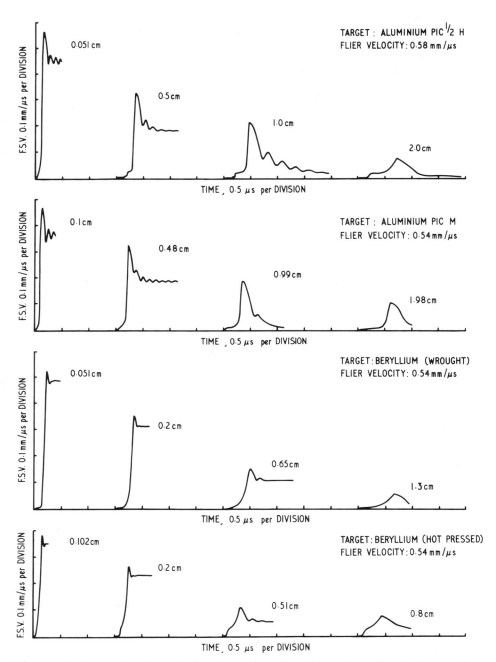

Figure 2. Free Surface Velocity-Time Profiles for Aluminium and Beryllium at Various Target Thicknesses

material absorption. The present approach is simplified by a number of assumptions to allow an easy analytical solution. There is, for example some uncertainty in accurately relating the stress components in an elastic-plastic wave to the free surface velocity, and here the variable U is retained whilst the qualitative interchangeability of stress, σ, and U is assumed. The wave profile at a depth x in the target is characterised by the amplitude $U(x)$ and arbitrarily defined width ΔT. Conservation of momentum implies and observation confirms that to a first approximation:-

$$U(x) . \Delta T = \text{Constant} = p, \text{ say,} \quad \quad \quad \quad (3.1)$$

for a given set of input conditions. Similarly there is a simple function expressing the energy in the pulse, of the form,

$$W = U^2(x) \Delta T \quad \quad \quad \quad (3.2)$$

Attenuation of $U(x)$ together with conservation of p implies deposition of energy within the material and this process is determined by a set of material properties which is usually summarized by a stress-strain diagram. In this diagram the area contained between the compression and release paths indicates the net work done on the material. Here, the simple assumption is made that this work done is proportional to the excess stress above a yield stress σy, or in terms of the present terminology:-

$$\partial W/\partial x = - \alpha (U(x) - Uy) \quad \quad \quad \quad (3.3)$$

where α is related to a material constant which expresses the energy absorbed by the material per unit volume per unit excess stress. There is an implicit assumption in (3.3) in that the wave velocity is independent of particle velocity and the degree of compression is sufficiently small to avoid the distinction between compressed and normal density. Equations (3.1), (3.2) and (3.3) give:-

$$U(x) = (U(o) - Uy) \exp(-\alpha x/p) + Uy \quad \quad \quad \quad (3.4)$$

This is the expression plotted in Figures 3 and 4 together with experimental values of $U(x)$. With the exception of the lower curve in Figure 3 the given values of α/p and Uy were obtained by a best fit procedure, and it is seen that the resultant curves provide a good description for the experimental data. The difference in the attenuation of the two Be types (Figure 4) is thought to be due to differences in material structure and also to some small porosity in the hot pressed material.

In order to test further the validity of (3.4) a series of experiments was carried out with Al using a flier with a reduced velocity i.e. reduced momentum. In equation (3.4) α and Uy remain

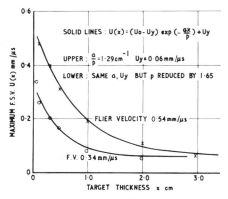
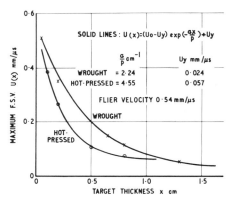

Figure 3. Attenuation of Peak in Al PIC M for two different Flier Velocities

Figure 4. Attenuation of Peak in Wrought and Hot Pressed Be

unchanged but U(o) and p now refer to the lower flier velocity and under these conditions an increase in attenuation is predicted. The result (lower curve in Figure 3) clearly demonstrates the expected dependance, and the agreement with the experimental data is very good.

The material constant α has been evaluated from the experimental results and the values, expressed in $Jcc^{-1} kb^{-1}$, are:- Al ($\frac{1}{2}$H and M), 0.20; Be (Wrought), 0.38; Be (Hot pressed) 0.78.

4.   SPALLATION

4.1   Experimental Observations

The characteristic free surface behaviour associated with spallation is illustrated in Figure 5, and actual results are given in Figure 2. Referring to Figure 5, that part of the wave up to U_2(o) is the forward wave appearing at the free surface prior to any fracture effects. At U_2(o) a discontinuity occurs and this indicates the commencement of fracture seen delayed by the transit time from the fracture plane. Following U_2(o) an oscillation occurs which appears to be the repeated reflection of a wave between the fracture plane and the free surface. Finally the free surface velocity returns to the base line in the case of incomplete fracture. (cf 0.99 cm PIC M Al result Figure 2), or in the case of complete fracture, achieves equilibrium at a constant velocity (there are a few examples of this situation in Figure 2). The period of oscillation, Δt, allows an estimate of the scab thickness, assuming the elastic wave velocity, c, is appropriate. i.e. $\Delta x = c\Delta t/2$. In practice no significant difference has been observed between

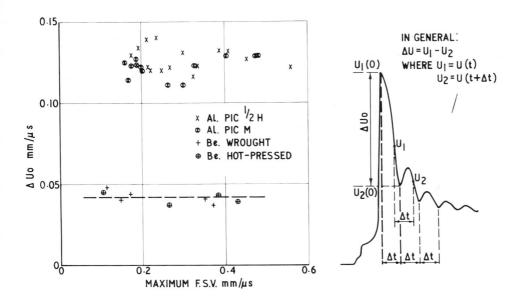

Figure 5. Experimental Values of "Pull-back" (ΔUo)

thicknesses derived in this way and those from direct measurement of micrographs. The results for Be in Figure 2 show a high dispersion of the reflected wave in the scab (compared with the Al results) and this is thought to be largely an indication of irregular fracture. Direct observation of recovered spall sections show irregular cracking at varying distances from the free surface. This irregularity may arise from the polycrystalline structure of the Be, in that the orientation of the crystals relative to the shock axis may well determine the effective strength of the material in the fracture region. The lack of any clear elasto-plastic transition and the general ramped nature of the Be results given in Figure 2 already suggests a blurring of the yield pressure by random orientation in view of the sharp transition observed in the case of single crystal results [3], [7].

The wave points $U_1(o)$ and $U_2(o)$ are coincident in time and position with the start of fracture and it therefore follows that their difference in amplitude gives a measure of the fracture stress. Measured values of $(U_1(o) - U_2(o)) = \Delta Uo$ ("pull-back") over a range of amplitudes, $(U_1(o))$, and target thicknesses are given in Figure 5 for Al and Be. Some variation in ΔUo is evident for different samples of nominally similar material but the random nature of the variation suggests a spread in the material properties coupled with some measuring error, rather than a correlatable dependence of ΔUo

on the amplitude or width of the primary stress pulse. Experimentally, therefore, ΔU_o is found to be virtually invariant for a given material within the range of the present experiments and hence the onset of fracture occurs at a critical negative stress - the incipient fracture stress σ_I. Simple arguments lead to the relation:-

$$\sigma_I = \frac{c\rho o \Delta U o}{2} \qquad \ldots \ldots (4.1)$$

where c is sound wave velocity and ρo is density. Using equation 4.1 the value of σ_I for Al (both $\tfrac{1}{2}$H and M) is 12 kb and that for Be (both wrought ingot and hot pressed powder SP100c) is 5 kb.

The records in Figure 2 for Al all indicate a reduction in the average velocity of the scab after the commencement of fracture. This provides a clear indication of a relatively long term fracture mechanism in which mechanical work is performed and it seems natural to regard this work phase as an indication of ductile fracture. An interesting feature of this phase is that its time constant is coupled with the pulse width and indeed the complete spallation process appears almost perfectly scaled in time for results of the same amplitude but different widths. For Be the picture is somewhat different in that the fracture process appears to be predominantly brittle, and to occur at a faster rate than in Al. A further interesting feature of the velocity records is the significance of the oscillation following onset of fracture. For example if complete brittle fracture occurred instantaneously then a wave of magnitude ΔU_o would be released at the fracture plane and this suggests that the observed wave is an indication of partial brittle fracture in that its amplitude is always much less than ΔU_o. Referring to Figure 5 it is seen that by generalising the method adopted for obtaining σ_I (ΔU_o) experimentally, an assessment may be made of the average stress $\sigma(\Delta U = U_1 - U_2)$ at the fracture plane during the fracture process. The variation of ΔU with t has been obtained from the experimental results and the surprising result is that, notwithstanding the curvature in the primary wave (U_1), the dependance is linear. A set of results illustrating this linear dependance is given in Figure 6 where it is also seen that the rate of fall of ΔU increases with decreasing pulse width (i.e. decreasing target thickness). Phenomenologically, it seems that the rate of fall of stress at the fracture plane is the determining factor and a just sufficient proportion of the primary wave is released from the plane to maintain a linear rate. In sections 4.2 and 4.3 a semi-analytical model is developed which provides a description of the experimental features.

4.2 <u>First Phase, Brittle Fracture</u>

The present experiments indicate that σ_I, the incipient fracture

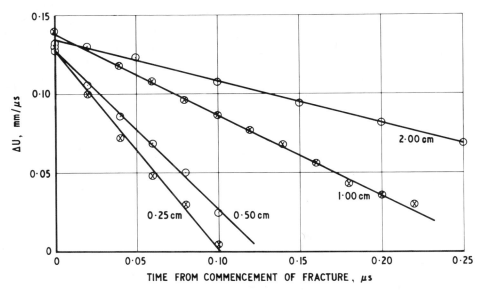

Figure 6. Variation of ΔU with Time at The Fracture Plane for Aluminium PIC(M)

strength, may be regarded as a material constant to a first approximation, although the physical significance of such a constant is not immediately apparent. A description is required which accounts both for the macroscopic uniqueness of σ_I and for the speed of stress release required to retain invariance for high speed transient loading. The solution adopted here is to regard σ_I as the stress required for propagation (macroscopic) rather than nucleation (microscopic) of cracks so that, given a sufficient number of nuclei to act as sources the rate of release by the production of free surface may be arbitrarily high for a fixed crack propagation velocity. The nuclei would be expected to arise from impurities and dislocations.

Consider a triangular wave reflecting perfectly off a free surface as shown by the full line in Figure 7. The time is chosen so that the negative peak at x just exceeds the positive amplitude by σ_I, so the situation corresponds with the onset of fracture. The negative wave continues to propagate into "virgin" material and interacts with the reducing positive forward wave to apparently produce an increasing net negative stress. However, on our interpretation of σ_I, the material releases a just sufficient amount of positive stress by the creation of free surface to ensure that σ_I is not exceeded: in the example there is a cumulative positive wave travelling with the negative wave and originating from the free surface generated between x and x + Δx. This is illustrated by $\Delta\sigma$ in the figure where the dotted line corresponds to the

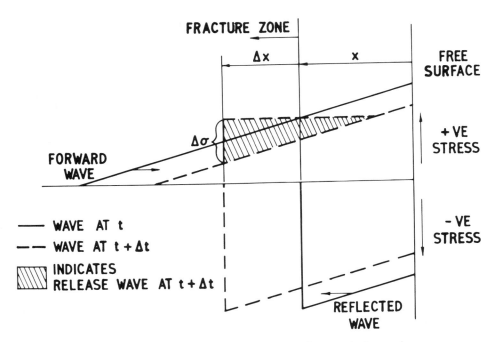

Figure 7. Illustration of Wave Interactions giving rise to Incipient Fracture

situation at $t + \Delta t$, $x + \Delta x$, neglecting for the moment any further fracture in the region Δx. It is apparent that the fracture in this mode is virtually instantaneous to ensure that the integrated release wave remains linked to the negative peak and σ_I is not exceeded. We can assume therefore that no significant deformation of the material occurs and there is no work done in the fracture. It is on these grounds that we describe this initial fracture mode as brittle and for the purposes of distinction we refer to the resultant fracture as incipient.

The degree of incipient fracture (F_I) may be represented by the amount of effective cross-sectional free surface area per unit area (A_I/A) created in the region Δx. We therefore define F_I in the form of a fracture density:-

$$F_I = A_I/A \: \Delta x = A_I/\Delta V \qquad \ldots \ldots (4.2)$$

We note the not too obvious point that before ductile flow occurs the amount of free surface in any plane is zero, since we are discussing a volume density, and there is therefore no question of an instantaneous runaway of fracture arising through a reduction in bonded area and consequential increase in stress, etc. Now, we can relate A_I to $\Delta\sigma$ since the stress throughout the creation if A_I is σ_I and therefore each element of free surface contributes a particle velocity of u (σ_I) to the wave and the total contribution averaged over an area A is:-

$$\Delta u = A_I/A \; u \; (\sigma_I) \quad \text{or} \quad \Delta\sigma = A_I/A \; \sigma_I \qquad \ldots\ldots \; (4.3)$$

within the present approximations. Equations (4.2) and (4.3) give:- $F_I = 1/\sigma_I \cdot \Delta\sigma/\Delta x$ and assuming equal wave velocities in Figure 7 for the +ve and -ve waves, it is easy to see that:-

$$F_I = 2/\sigma_I \cdot d\sigma/dx \qquad \ldots\ldots \; (4.4)$$

where $d\sigma/dx$ is now the pressure gradient on the trailing edge of the positive wave at the point of interaction. In general, this slope is not constant over the fracture zone and then it is the slope at the point of interaction that determines the local fracture density.

We may immediately comment on the experimental implications of the results thus far. A fracture zone is predicted rather than a fracture plane and the width of the zone is determined by the width of the applied stress pulse from the point designated U_2(o) down to zero. For a given material the density of incipient fracture is determined only by $d\sigma/dx$ in the applied wave and an increase in the amplitude alone merely increases the zone width. It is of course difficult to assess the brittle fracture phase experimentally in isolation from the subsequent ductile fracture. Nevertheless, it is clear from examination of recovered samples that the fracture does occur over a zone with variations in width and density according to the amplitude and gradient of the pressure pulse.

4.3 Second Phase, Ductile-Brittle Fracture

The situation prior to ductile flow is then as follows. There exists a section of material of normal density but with a distribution of free surface within it and under a negative stress of σ_I. The material surfaces of each free surface element are separating ititially at a velocity of 2u (σ_I) to create voids in the material accompanied by ductile flow, stress gradients and further fracture. We describe this latter phase as ductile-brittle on the grounds that inevitably some material flow occurs (ductile) but the possibility of further brittle fracture arising from the stresses associated with the deformation is not excluded.

It is apparent that the whole fracture zone may not be considered simultaneously due to the propagation delay described in the last section (see Figure 7). The region most advanced in time is that at x and it follows that the final fracture is most likely to occur at x and the scab thickness is thereby defined. It is for this reason that the concept of a fracture plane is an allowable one and in the present context the observations of section 4.1 remain valid.

Suppose we consider a small element δx at x, which is to include the region of final fracture rather than the region of damage (zone width). For a given material, σ_I is determined and the only variable at the start of ductile flow is the brittle fracture density determined by $d\sigma/dx$ in equation (4.4). It is very reasonable then to expect the rate of further fracture of the section δx to be determined by $d\sigma/dx$, whatever the details of the process. Furthermore, the fracture may be characterised by a reduction in bonded cross-sectional area (in contrast to the initial brittle phase) and a consequential reduction in average stress including void area, if the limiting material stress σ_I is not to be exceeded. If we add the implied assumption that σ_I is maintained constant within the material, then a curve of average stress σ at x versus time is equivalent to a fracture time curve. For this purpose we define the fracture during the ductile-brittle mode, F_D, in terms of the remaining area of contact A^1 per total area A, and $F_D = 1 - A^1/A$. For the average material stress to remain constant at σ_I, $\sigma A = \sigma_I A^1$ and therefore, we may obtain a quantitative assessment of F_D from:-

$$F_D = 1 - \sigma/\sigma_I \qquad \ldots \ldots (4.5)$$

It follows from these remarks that if we plot σ versus t for a number of experimental results, we should obtain a set of curves of identical form but with slopes proportional to $d\sigma/dx$ in the applied stress pulse. This is just what is found in practice: the curves in Figure 6 are effectively plots of $\sigma - t$ and they show the expected dependence. The constant rate of fracture for a given result appears as an experimental conclusion but it does allow the results to be conveniently summarised by a single curve of fracture rate versus $d\sigma/dx$. The method adopted for expressing the experimental results in this form is as follows. The fracture rate is obtained from equation (4.5) and $dF_D/dt = -1/\sigma_I \, d\sigma/dt$ [$\sigma = \sigma_I$ at t = 0], and

$$dF_D/dt = 1/T \text{ (T is characteristic fracture time)} \ldots (4.6)$$

Experimental values of T are readily obtained from curves of the form shown in Figure 6. The other variable $d\sigma/dx$ is more conveniently plotted as dU/dt, which is measured at the linear region in the vicinity of the point designated $U_2(o)$. The resultant points are shown in Figure 8 and it will be seen that the linear fit is fairly good. The results are in accordance with the expectation that the higher the value of $d\sigma/dx$ the greater the density of incipient fracture and for a given void growth rate the greater dF_D/dt.

There is an effective ductile-brittle growth constant included in the above interpretation which appears in the experiments as a constant relating the rate of fracture to the stress gradient (or the density of incipient fracture). It is of interest to see to what extent this constant may be related to the natural growth parameter which is the characteristic velocity associated with the internal free surface. If we assume a purely ductile phase and make the simple assumption that the free area created in the brittle

SPALLATION AND ATTENUATION EFFECTS IN Al and Be

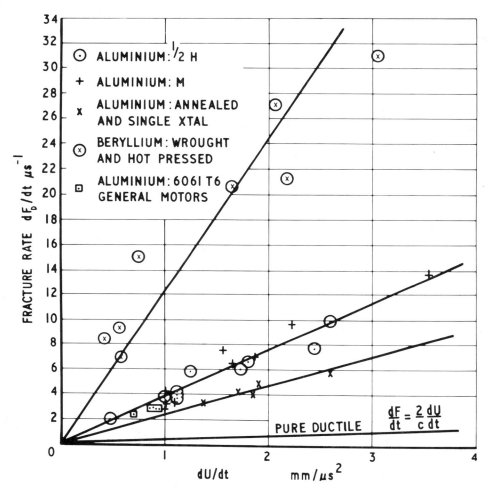

Figure 8. Rate of Fracture Versus dU/dt
(≡ Incipient Fracture Density)

mode continues to separate at the particle velocity associated with the material stress σ_I, then it can be shown that:-

$$dF_D/dt = 2/c \cdot dU/dt \qquad \ldots\ldots (4.7)$$

We note that the result is independent of the material fracture properties, given perfect ductility, so that in practice a material dependent variation in the fracture rate curves would be expected to reflect the degree of brittle fracture accompanying the ductile phase rather than a variation in σ_I, for example. The greater the brittle contribution, the greater the expected fracture rate due to the production of extra free surface as deformation occurs. The results in Figure 8 which include the purely ductile case, equation (4.7), make an interesting comparison. Although the annealed and single crystal Al results are somewhat limited in number, no significant difference in fracture rate is observed between these two cases, but both are significantly nearer the pure ductile case than PIC ½H, as expected from their higher ductility. The value of σ_I for the annealed material is virtually the same as for PIC ½H but in the case of the single crystal it is some 50% higher, and as expected this variation is not reflected in the fracture rate. No significant difference is observed in the fracture behaviour between the Be materials, but both fracture at about three times the rate of Al for a given dU/dt. Also included in Figure 8 are some results obtained by applying the present analysis to some results obtained by Isbell and Christman at General Motors [4], who used gas gun and laser interferometer techniques. The resultant wave profiles in their experiments were flat topped rather than triangular, and it is encouraging that these results show such good agreement. In addition results obtained by Christman et al [8] for OFHC copper have also been analysed and the fracture rates lie close to the Al (½H and M) curve.

5. SUMMARY

A simple attenuation model is proposed in which the momentum associated with the pressure pulse is conserved and the energy is reduced by material absorption according to an assumed linear dependence on the excess stress above a yield stress. The experimental results compare very favourably with the model and the coefficient for the energy absorbed/unit volume/unit stress is obtained. A model for the mode of fracture is proposed in which the fracture process occurs in two stages: an initial brittle phase followed by a ductile-brittle phase in which material deformation occurs. The initiation of the brittle phase is determined by the incipient fracture strength which, by virtue of the experimental evidence, is taken to be a material constant. A quantitative argument is presented which shows that the expected spatial density of the initial brittle fracture is directly proportional to the release stress gradient in the primary wave. The ductile-brittle phase can be described by a single growth parameter (for a given material) which relates the fracture rate to the degree of incipient fracture, and the growth parameter is determined essentially by the degree of brittleness of the material.

REFERENCES 1. DAVISON, L. and STEVENS, A. L., J App Phys 43, 3, 988 (1972)
2. BARBEE, T. W., Jnr, SEAMAN, L., CREWDSON, R., CURRAN, D., J Matls JMLSA, 7, 3, 393 (1972).
3. TAYLOR, J. W. DISLOCATION DYNAMICS, McGRAW HILL, NEW YORK, 1968, pp 573-589. 4. ISBELL, W. M., CHRISTMAN, D. R. MSL-69-60. MATERIALS AND STRUCTURES LABORATORY, MFRG DEVPT, GENERAL MOTORS CORP. Feb 1970. 5. RICE, M. H. Rev Sci Inst 32, 449 (1961). 6. TAYLOR, J. W., RICE, M. H., J App Phys 34, 364, 1963). 7. CHRISTMAN, D. R., FROULA, N. H. A1AA Jnl 8, 3, 477 (1970). 8. CHRISTMAN, D. R. ISBELL, W. M., BABCOCK, S. G. DASA 2501-5, MSL-70-23, Vol. V, July 1971.

EFFECTS OF METALLURGICAL PARAMETERS ON DYNAMIC FRACTURE*

W. B. Jones and H. I. Dawson
Division of Metallurgical Engineering
University of Washington
Seattle, Washington 98195

ABSTRACT

The effects of systematically varying three common metallurgical parameters on the fracture pattern caused by impact loading have been studied using optical and electron microscopy. 1) Increasing dislocation densities were produced in pure copper specimens by compression prior to impact testing. The dynamic fracture strength increased and the total amount of strain induced by the impact loading, decreased for larger prestrains. The fracture surface of the copper showed typical ductile fracture morphology. 2) Decreasing the stacking fault energy using alloys of copper with up to 7 weight percent aluminum, was found to increase the dynamic fracture strength and the twin density resulting from the impact loading. Recovered fracture surfaces showed typical morphology of twin markings. 3) The state of precipitation in two age hardenable aluminum alloys was found to affect the dynamic fracture behavior. Samples of 2024 aluminum alloy showed a continuously increasing level of spall damage with increasing aging times. Samples of 6061 aluminum alloy on the other hand displayed decreasing fracture damage for aging times up to the maximum hardness condition and increasing fracture damage for longer aging times. The fracture surface morphology was typically ductile in nature and did not vary in appearance among the samples with the different precipitation states studied.

* Work supported by the U. S. Army Research Office - Durham.

INTRODUCTION

The present understanding of the response of solids to high stresses of a short duration is the result of two different approaches. Assumption of continuous homogeneous media has led to the description of stress wave deformation utilizing the principles of continuum mechanics. The second approach has attempted to describe the same deformation on the basis of dislocation models. The utilization of either approach involves oversimplified assumptions. The elastic response of continua is fundamentally described in terms of elastic stress wave propagation, but plastic wave propagation presents a problem in that the details of plastic deformation processes are very complex. In order to describe plastic behavior using continuum mechanics, certain adjustable parameters have to be introduced. While these parameters are implicitly related to the deformation mechanisms and can be adjusted to adequately describe the results, their fundamental significance, if any, is not apparent. Continuum mechanics can produce useful fracture criteria and constituitive equations, but it cannot include the processes of dislocation motion and interactions which constitute the very mechanism by which plastic deformation occurs. Neither can continuum mechanics incorporate microstructures that are metallurgically significant into a description of plastic deformation. Dislocation models as such have been successfully developed to describe the passage of a plastic wave through a perfect crystal lattice (1). On the other hand, a description of dynamic response including fracture of engineering materials, has not been accomplished, since simple dislocation models cannot quantitatively incorporate the effect of complicated microstructures on dislocation motion.

The problems associated with a fundamental description of the effects of various metallurgical parameters on dynamic deformation have necessitated qualitative and semi-quantitative studies of these effects. Increasing numbers of such studies are being performed and some of the results obtained so far are inconclusive, or even contradictory. The present situation points to a distinct need for detailed and systematic studies of systems with well characterized microstructures. The results of such studies combined with static stress-strain data may lead to a correlation of dynamic response, microstructure, and static properties. This approach can produce information which may further stimulate the development of continuum, as well as dislocation models.

The present paper reports some preliminary results of an investigation which deals with the effects of three common metallurgical parameters, on material response to impact loading. In pure copper, the effect of dislocation density on dynamic deformation and fracture is being studied. Varying the concentration of aluminum in copper is used to study the effect that different stacking fault energies have on the response to dynamic loading. Finally, two commercial age hardening aluminum alloys are used to determine how the degree and mode of solid state precipitation affects the dynamic fracture response. Experiments are being performed in an attempt to contribute toward the development of atomistic and microstructural explanations of high strain rate deformation and fracture. The preliminary results reported here emphasize the significance and necessity of taking metallurgical parameters into account if a better understanding of the dynamic response of materials is to emerge.

EXPERIMENTAL

Dynamic loading was performed using the method developed by Fyfe (2). In this method hollow cylindrical specimens are used, which results in some differences with the well known plate impact experiment. Typical plate impact tests, as described by Karnes (3), have been used by many investigators to study effects at high strain rates. This method of high strain rate loading results in a stress pulse of short duration and is accompanied by uniaxial strain. If fracture occurs, it appears as a set of coalesced microcracks lying approximately on a plane a short distance below the outer surface of the sample. The major stress component in plate impact tests is the stress normal to the propagation direction of the wave. Early attempts to establish the existence of a critical fracture stress were unsuccessful, and several investigators later proposed that the duration of the stress pulse, as well as its amplitude, are the primary criteria in determining whether or not fracture would occur. The plate impact test is well suited for investigation of this relationship since the flyer plate velocity determines the stress wave amplitude and the stress duration is dependent on the flyer plate thickness. The time dependence of dynamic fracture has been established by several investigators using the plate impact experiment (4).

In the experiment designed by Fyfe, a second dimension is added to the state of strain by the generation of radial stress waves in thick walled hollow cylinders. Only the radial displacements are non zero, but geometric effects result in a biaxial state of strain. While the cylindrical geometry does alter the quantitative analysis of the stress history compared to that of the plate impact experiments, it does not affect the qualitative results. Figure 1 shows the specimen configuration of the cylindrical experiment, and figure 2 shows a schematic of the experimental set-up. After a specimen has been secured to the apparatus, two 15 microfarad capacitors are charged to fixed potentials in the 20 kilovolt region, storing energies of the order of 600 joules. On discharging the capacitors,

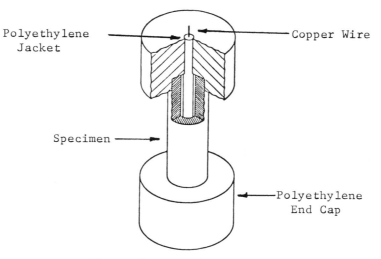

Figure 1. Specimen Configuration

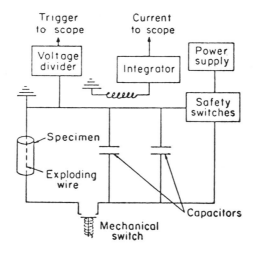

Figure 2. Experimental Apparatus

the heated copper wire (see figure 1) vaporizes and expands radially. The resulting high pressure drives the polyethylene jacket outward until it impacts against the inner surface of the hollow cylindrical specimen. The end caps serve to center the wire. Kerr Cell photography has shown that the inner boundary loading resulting from the impact is uniform both radially and along the axis of the sample (5).

Stress histories calculated by Fyfe for cylindrical specimens, using a material constitutive equation and the inner boundary loading history, are shown in figures 3 and 4. This particular pair of stress histories refers to a sample of 6061-T651 aluminum alloy, but the relative relationships are independent of the metal used. All stresses plotted are principal stresses and correspond to a cylindrical coordinate system (r,θ,z) with the z-axis coinciding with the cylindrical specimen axis. Figure 3 shows the nature of the stress state as being strongly triaxial. The significant feature of the relationship among the principal stresses is the dominance of the hoop stress ($\sigma_{\theta\theta}$) in the tensile rarefaction wave. Each of the principal stresses also varies as a function of radius in a manner which is unique to the cylindrical geometry. The hoop stress history at two radial positions within the sample are shown in figure 4. The first of two important effects to note from this figure is the decay of the compressive stress pulse as it travels outward while the amplitude of the tensile reflection increases as it travels inward. Secondly, the duration of the tensile rarefaction is a maximum where it first appears then shortens as the stress pulse travels inward.

When dynamic fracture occurs in the cylindrical samples it exhibits radial and axial symmetry on a macroscopic scale. In this experiment, as in the plate impact experiment, the fracture behavior is generally determined by both the stress amplitude and the stress pulse duration. While these variables are established in the plate impact experiment by the flyer plate velocity and thickness respectively, in the cylindrical experiment the wall thickness determines both the stress pulse amplitude and duration; because of the radial attenuation of the stress pulse, the degree of fracture damage can be made to decrease by increasing the wall thickness, for a given energy input and material. In addition, the stress wave amplitude can be varied by changing the discharge voltage of the capacitors, thus altering the inner boundary loading. If the tensile pulse, as it first appears, has sufficient intensity and duration to cause crack nucleation, growth and coalescence, a clear plane of spall fracture will be formed. If the outer radius of the specimen is large enough so that the tensile rarefaction pulse, as it first appears, has insufficient amplitude to cause fracture, it will travel inward until it has intensified to the level necessary to cause fracture. In this case, the specimen will show diffuse radial cracks in its interior with no clear spall plane (5).

Specimen Preparation

The following specimens were prepared for the three sections of this investigation. 1. 99.999% pure copper was used in the section of the investigation which involved varying the dislocation density by static pre-deformation. Short bars of the pure copper were annealed for 6 hours at 600°C in a 10^{-6} torr vacuum. Following this, the bars were staticly deformed in compression using an Instron Tensile Machine to intervals of plastic strain between 0% and 22%. Each bar was then machined to final dimensions and impact loaded immediately.

2. 99.999% pure copper was used with additions of 99.999% pure aluminum to study the effect of varying the stacking fault energy. Copper-aluminum alloys containing up to 7% by weight aluminum were prepared in a vacuum of 10^{-6} torr using an induction furnace. All samples contained aluminum in concentrations less than the solid solubility limit. Alloy samples were wrapped in titanium foil and homogenized for 24 hours at 850°C in a 10^{-6} torr vacuum. The castings were then machined to final specimen dimensions and annealed for 6 hours at 600°C under the same vacuum.

3. Alloy systems used in the study of the effect of precipitation state on dynamic fracture were two commercial alloys. 1 inch diameter bar stock of 2024-T351 and 6061-T651 aluminum alloys were machined to final dimensions and solution treated in air for one hour at 495°C and 529°C respectively and quenched in water (SHT/WQ). Samples of both alloys were aged at 211°C over times long enough to study all precipitation states. Basically three conditions were examined, solution treated and quenched with no aging (NA), critical aging (CA), and overaging (OA). After solution treatment, all samples were stored in liquid nitrogen.

Testing Procedure

The testing procedure and specimen examination varied among the three sections of this investigation. The pure copper specimens and the copper-aluminum specimens were machined to 15.85 mm cylinder length with an inner diameter of 4.75 mm and an outer diameter of 9.50 mm. A 0.84 mm solid copper conductor encased in a 2.95 mm polyethylene jacket was used for the generation the stress pulse. The specimens of the age hardening aluminum alloys were machined to a length of 58 mm and an inner diameter of 7.5 mm. The outer surface of each specimen was evenly stepped to diameters of 16 mm, 20 mm, and 24 mm. Each step length was one-third the total specimen length. Using the stepped specimens, three experiments could be done simultaneously, insuring that all were receiving identical inner boundary loading.

METALLURGICAL PARAMETERS ON DYNAMIC FRACTURE 449

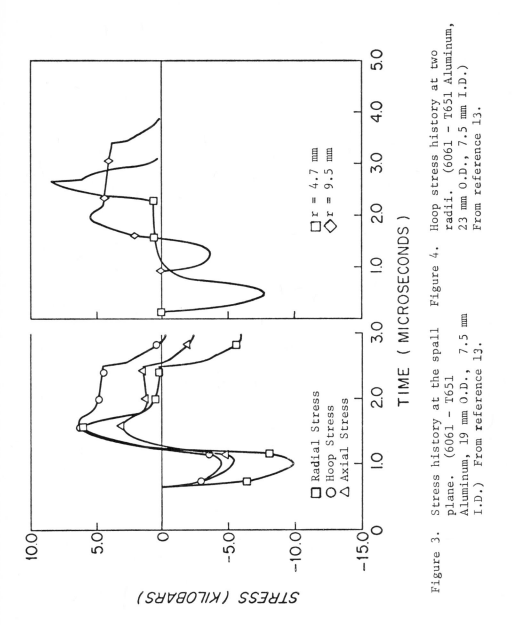

Figure 3. Stress history at the spall plane. (6061 - T651 Aluminum, 19 mm O.D., 7.5 mm I.D.) From reference 13.

Figure 4. Hoop stress history at two radii. (6061 - T651 Aluminum, 23 mm O.D., 7.5 mm I.D.) From reference 13.

Specimen Examination

After impact testing, the specimens were recovered and examined using several microscopy techniques. Recovered samples were sectioned on a plane perpendicular to the cylindrical axis of the specimens at a point near the center of each step. The surfaces thus exposed were polished and macroetched and examined with optical and scanning electron microscopy. Observation of the cracks on this surface allowed the extent of fracture and the distribution of cracks to be determined for all specimens. The fracture surfaces were also studied using the scanning microscope. Transmission electron microscopy methods were employed to study the defect structures induced by the impact loading in specimens of varying concentrations of aluminum in copper.

RESULTS AND DISCUSSION

1. Effects of Varying the Density of Pre-Existing Dislocations

In pure copper, increasing dislocation densities, caused by static deformation of up to 22% compression of the cylindrical specimens, resulted in decreasing degrees of dynamic fracture caused by an inner boundary loading of 18 kbar. Figure 5 illustrates the trend in the degree of fracture and the typical ductile fracture morphology observed. The plastic strain caused by impact testing was found to decrease with increasing pre-strain. The dynamically induced strain dropped from 14% for the annealed copper to about 1% for the samples given 12% pre-strain.

Prestrain introduces dislocation networks through which the stress wave must pass during impact testing. The dislocation tangles of the network hinder the motion of the dislocations associated with the moving stress wave and limit the amount of deformation that can take place during the passage of the stress wave. Thus the total plastic strain caused by impact loading decreases for increasing amounts of pre-strain. The experimental results indicate that pre-existing dislocations affect the fracture process as well as affecting the degree of plastic deformation. It is reasonable to assume that in this high purity copper containing no second phase particles, crack nucleation takes place by a coalescence of several moving dislocations associated with the stress pulse. The extent to which crack nucleation takes place would be dependent on the degree to which moving dislocations are allowed to cooperate to form a large cavity dislocation. Barriers, such as pre-existing dislocation tangles, would hinder this cooperative motion on one glide plane. Large numbers of tangled pre-existing dislocations would then decrease the probability of dislocation coalescence which would result in an increased dynamic fracture resistance.

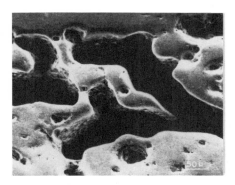

(a) well annealed

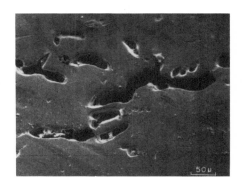

(b) 6.67% pre-strain

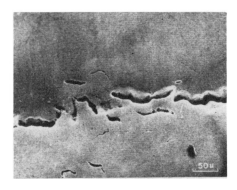

(c) 21.48% pre-strain

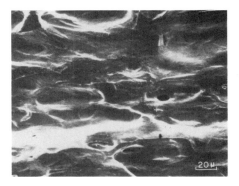

(d) typical fracture surface

Figure 5. Effect of pre-strain on dynamic fracture of pure copper. Inner boundary loading - 18 kbar.

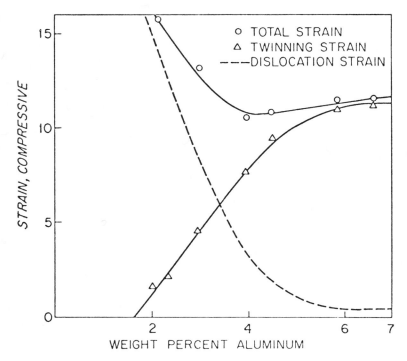

Figure 6. Dislocation and twin contribution to total strain caused by dynamic deformation in Cu-Al alloys.

2. Effects of Varying Stacking Fault Energy

The samples of increasing amounts of aluminum in copper had stacking fault energies which decreased continuously from about 80 ergs/cm^2 down to about 4 ergs/cm^2 for Al concentration up to 7% by weight (6). Twin densities measured after impact deformation increased with increasing concentrations of aluminum, in agreement with the decreasing stacking fault energy values. Figure 6 shows the total strain resulting from impact deformation together with the twin and dislocation contributions to this total strain. The twin strain was determined from the twin density measurements and the dislocation strain was taken as the difference between the total strain and the twin strain. The presence of cracks has an apparent affect on the strain of less than 0.01% and has been neglected in these estimates. Specimens with greater than 5 weight % aluminum showed no fracture while those containing less aluminum did fracture upon loading the inner specimen surface to 18 kbar.

Figure 7 displays a trend in the extent of fracture and a scanning electron micrograph showing the twin type fracture surfaces typical of the alloy specimens.

It is well established that the susceptibility to twinning increases with decreasing stacking fault energy. The stress amplitudes of both the compressive and the tensile waves are sufficient to create deformation twins. Transmission electron microscopy has shown that the twin density does not vary with distance away from the crack which suggests that the final twin density is reached on passage of the compressive wave and that, although the tensile reflection may alter the twin structure, very few new twins are created. During the passage of the tensile stress wave, the twin boundaries present in the material would hinder the cooperative motion of dislocations toward coalescing to form a cavity dislocation. Increasing twin densities would then decrease the probability of crack nucleation, so that the alloys containing more than 5% aluminum may not have spalled because the compressive wave had created a heavily twinned structure that was more resistant to fracture during the subsequent tensile stress reflection.

3. Effect of Solid State Precipitation

The study of the effect of aging treatments of 2024 aluminum on the response to impact loading shows that the solid solution condition produces the smallest degree of dynamic fracture damage, the critically aged condition displays more fracture damage and the over aged condition suffers the most severe fracture damage. Samples of 6061 aluminum age hardening alloy showed a different behavior, with specimens in the critically aged condition showing the smallest degree of fracture, the solid solution samples suffering slightly more fracture damage and the over aged specimens showed the greatest extent of dynamic fracture. The samples shown in figure 8 display the effect of aging time on fracture in both alloy systems. Scanning electron fractographs of all samples showed ductile fracture that varied little among the precipitation states in each alloy system.

2024 and 6061 aluminum alloys are basically ternary alloys containing Al-Cu-Mg, and Al-Mg-Si respectively. The commercial alloys used in this investigation contained both aluminum oxide inclusions and intermetallic inclusions, all of which are insoluble at the aging temperatures used. As a result, the distribution of these inclusions is independent on any aging treatment. These oxide and intermetallic particles will be termed inclusions, as differentiated from the metastable and equilibrium precipitates which nucleate and grow during aging.

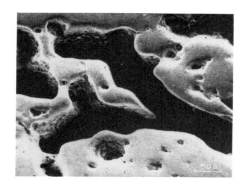

(a) pure copper

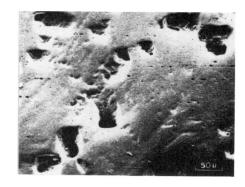

(b) Cu - 3% Al

(c) Cu - 6.62% Al

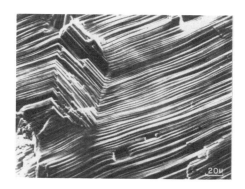

(d) typical alloy fracture surface

Figure 7. Effect of aluminum additions to copper on dynamic fracture. Inner boundary loading - 18 kbar.

METALLURGICAL PARAMETERS ON DYNAMIC FRACTURE 455

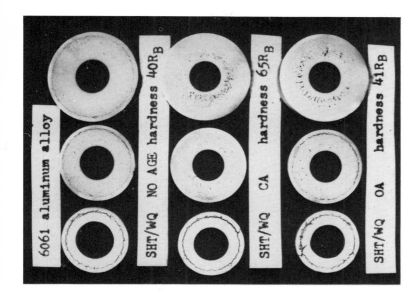

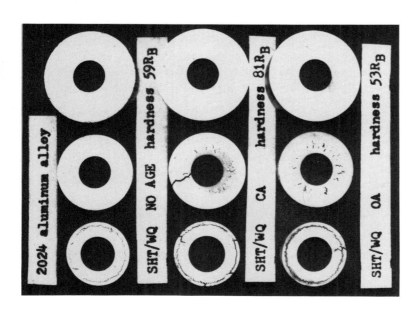

Figure 8. The effect of aging on the dynamic fracture behavior of 2024 and 6061 aluminum alloys. Inner boundary loading - 21 kbar.

After solution treatment and quenching, specimens of both alloy systems are composed of a single-phase aluminum matrix with the alloying elements in substitutional solid solution, and the insoluble inclusions also present. In this condition, both 6061 and 2024 alloys have similar microstructures and should display a relatively similar fracture behavior as has been observed.

During aging at moderate temperatures, coherent metastable zones, or phases, nucleate and grow in the supersaturated aluminum matrix. A needle shaped G-P zone precipitate in 6061 aluminum is formed and contains an ordered silicon-magnesium structure. The formation of each coherent zone results in a relatively small misfit with the parent matrix of about +2% (7). This is insufficient to cause coherency strains of a magnitude that would explain the macro-hardness that the G-P zones cause. The resistance these zones show toward the passage of dislocations through them seems to be due to the difficulty in breaking and re-ordering the magnesium-silicon bonds (8). In 2024 aluminum, the aluminum-copper-magnesium zones also have a needle shape. Their ordered structure causes a larger misfit of -6% which results in significant coherency strains (7). The hardening effect of this precipitate comes primarily from the difficulty of dislocation motion through these strain fields (8). Each G-P zone causes a similar change in the static strength of its alloy system but, as this investigation indicates, each zone has a different effect on the dynamic fracture behavior. The dynamic fracture of both 2024 and 6061 alloys is ductile in nature and nucleates and propagates along the interfaces between the aluminum matrix and some second phase particle (9,10). The lack of variation in fracture morphology with aging time as observed in this investigation and by Herring and Olson (11) indicates the following. The interfaces which are of consequence as nucleation sites and propagation paths are interfaces between the matrix and the inclusions which do not change during aging, rather than the matrix/G-P zone or matrix/precipitate interfaces which do change in both structure and distribution. Although the presence of G-P zones seems not to alter the path that a propatation crack follows, the G-P zones may change the amount of energy necessary for nucleating and moving a crack. The G-P zones in 6061 aluminum, as mentioned, have no stress field that could affect the matrix-inclusion interfaces. In contrast, the G-P zones in 2024 aluminum do have a significant strain effect, which through a difference in rigidity between the aluminum and the insoluble inclusions results in a stress along the interface. This stress would make nucleation and propagation along these interfaces easier. This could explain the observation that critically aged 2024 aluminum shows relatively greater fracture damage than critically aged 6061 aluminum.

Aging over longer times results in the loss of static hardness caused by the nucleation and growth of an equilibrium phase. In

6061 and 2024 alloys, the final precipitates are Mg_2Si and Al_2CuMg respectively, both of which are incoherent with the matrix and tend to take a spherical shape. These similar microstructures should display a relatively similar dynamic fracture as has been observed.

CONCLUSION

The preliminary results reported here have established that several metallurgical parameters do affect the dynamic fracture behavior which can be described qualitatively in terms of reasonable dislocation processes. 1) Increasing pre-strain in pure copper results in increasing dynamic fracture resistance which may be caused by an interaction of dislocations associated with the moving plastic stress wave and the dislocation networks present due to pre-strain. 2) Increasing twin densities in pure Cu-Al alloys shows increasing dynamic fracture resistance. This may be caused by an interaction between stress wave produced twins and moving dislocations. 3) Dynamic fracture in 2024 aluminum increases momotonically during aging while, in 6061 aluminum, fracture damage decreases from the solid solution condition to the critically aged condition and increases during over aging. The difference in dynamic fracture behavior may be due to the existence of coherency strains surrounding the G-P zones in critically aged 2024 aluminum and the lack of such strains associated with G-P zones in 6061 aluminum.

Each of the three sections of this investigation involved microstructures of sufficient complexity to preclude quantitative description in terms of simple dislocation models, and of sufficient heterogeneity, that microstructural explainations rather than continuum interpretations, are necessary. The qualitative and semi-quantitative results of this and similar investigations have shown that the effects of metallurgical parameters are rather specific; even between 6061 and 2024 Al, which are two similar age hardening aluminum alloys as far as their static behavior is concerned, the dynamic fracture behavior has been shown to be different. The results presented emphasize the significance of the role of metallurgical parameters in the dynamic behavior of materials, and further progress toward a better understanding of this behavior can only be made through further detailed and systematic studies.

ACKNOWLEDGMENTS

Thanks are due to Mr. D. P. Bansal for the measurement on copper and copper-aluminum, and to Professor I. M. Fyfe for many helpful suggestions and for discussions on the continuum aspects of the problem.

REFERENCES

1. Gilman, J.J., Appl. Mech. Rev., 21, 767, (1968).

2. Ensminger, R.R., Fyfe, I.M., J. Mech. Phys. Solids, 14, 231, (1966).

3. Karnes, C.H., in Mechanical Behavior of Materials Under Dynamic Loads, (New York: Springer-Verlag New York Inc.), pp. 270, 1968.

4. Gilman, J.J., Tuler, F.T., J. Fracture Mech., 6, 169, (1970).

5. Schmidt, R.M., Fyfe, I.M., University of Washington, Department of Aeronautics and Astronautics, Report 70-3, (1970).

6. Swann, P.R., in Electron Microscopy and Strength of Crystals, (New York: Interscience Publishers), pp. 131, 1963.

7. Kelly, A., Nicholson, R.B., Prog. in Materials Sci., 10, 148, (1963).

8. Van Horn, K.R., Editor, Aluminum, Vol. 1, (Metals Park, Ohio: ASM), 1967.

9. Au, R.H.C., AIAA Journal, 8, 1171, (1970).

10. Broek, D., A Study on Ductile Fracture, Ph.D. Thesis, Delft, Netherlands, (1971).

11. Herring, R.B., Olson, G.B., Army Materials and Mechanics Research Center Report AMMRCTR 71-61, (1971).

12. Bansal, D.P., Master's Thesis, University of Washington, (1972)

13. Fyfe, I.M., Schmidt, R.M., University of Washington, Department of Aeronautics and Astronautics, Report 72-10, (1972).

WAVE PROPAGATION AND SPALLATION IN TEXTURED BERYLLIUM*

A. L. Stevens and L. E. Pope

Sandia Laboratories

Albuquerque, New Mexico 87115

ABSTRACT

Plate impact experiments were conducted on polycrystalline beryllium samples of four different textures to determine the effect of texturing on the wave propagation and spall strength of beryllium. The samples were fabricated from a single billet of hot-pressed beryllium powder and characterized by chemical composition and grain size; the degree of texturing was quantified by basal plane pole figure analyses. Results indicate that (1) although the nominal spall strength is not significantly altered by the texturing, the distribution and character of the spall damage does vary markedly with texture; and (2) the elastic precursor wave is severely ramped in the as-pressed material, but appears as a sharply-rising wave in the highly textured material. A model for unrelaxed thermal microstresses is advanced to explain the suppression of the elastic precursor wave in the as-pressed beryllium.

INTRODUCTION

In the past few years several experimental studies have been made on wave propagation and spallation in polycrystalline beryllium samples fabricated by hot pressing powder to ~100% of the monocrystalline density.[1-5] This work is complemented by several other studies on cast and wrought beryllium. The results

---

*This work was supported by the U. S. Atomic Energy Commission.

of these studies have generated a number of open questions regarding the shock-loading and stress-release behavior of polycrystalline beryllium.[6] For example, compressive disturbances in both the hot-pressed and the cast beryllium exhibit in gross detail the two wave structure characteristic of most metals. That is, a compressive wave separates into a low-amplitude high-velocity elastic precursor wave followed by a lower velocity plastic wave. Unlike many other metals, however, experimental results for beryllium show that both the precursor and the plastic waves deviate substantially from idealized constant-amplitude, nondispersive disturbances, i.e., from an elastic-perfectly plastic material response. A constitutive model has been proposed[7] for shock-loaded S-200 beryllium which reasonably predicts the experimentally-observed dispersive plastic wave profiles. No explanation for the dispersive character of the elastic precursor has been proposed.

One factor which has tended to complicate and inhibit the efforts to characterize the dynamic mechanical properties of beryllium is that such parameters as the impurity content, grain size, heat treatment, and material texturing have been changed frequently in the course of improving the beryllium structural fabrication techniques. Often these parameters were modified in such a manner as to enhance the "static" mechanical properties, such as yield strength, fracture toughness and elongation to fracture, and thus their effect has been generally well characterized. The effect of changes in these process parameters on the dynamic mechanical properties and spall strength are not understood, however, and their variety tends to confound comparison of results of various investigators.

Another factor which complicates the interpretation of results is the influence of processing on the thermomechanical properties. For example, the technology has recently been developed for rolling and shear spinning structural elements from hot-pressed billets.[8] Due to the plastic anisotropy of beryllium, however, these forming processes introduce severe texturing into the resulting microstructure of the material.* The primary objective of the present investigation is to determine what effect this texturing may have on the wave propagation and spallation characteristics of the beryllium. In addition to fulfilling this primary objective, present results shed some light on the reasons for the ramped elastic precursor waves observed in previous investigations.

---

*Similar texturing results in cast beryllium when it is rolled and heat treated in order to achieve small grain size.[12]

MATERIAL CHARACTERIZATION

Plate impact samples of four different textures were fabricated from a single billet of Kawecki-Berylco hot-pressed HP-10 beryllium. The chemical composition of the material is given in Table 1. All rolling was conducted at 775°C with approximately 10% reduction per pass. All samples were annealed at 705°C for 1/2 hour subsequent to rolling. The four textures which were used are described in Table 2. These textures, achieved by rolling reduction, duplicate four textures over the range of textures which result from the shear spinning process.* The effect

---

Table 1. Chemical Composition of the Material

BeO	0.70
C	0.048
Fe	0.120
Al	0.043
Mg	0.004
Si	0.032
All other metallic impurities	0.04 max
Be	balance

Table 2. Characterization of the Four Textures of Beryllium

Set	Rolling Reduction	Grain Size (μ)	Texture Factor[a]
1	As pressed	30	~1R
2	Cross rolled 2.25:1	L-23, T-20	~3R (split poles)
3	Unidirectionally[b] rolled 3:1	L-21, T-17	~5R (split poles)
4	Unidirectionally rolled 13.5:1	L-19, T-17	~8R (split poles)

[a]Basal plane concentration in the plane of the sample disk, in multiples of the random (R) concentration in the as-pressed material.
[b]The pole figure for this material, Fig. 1c, shows some evidence that this material had undergone some cross rolling.

---

*Shear spinning results in conical frustrums; flat plates, required for the plate-impact experiments, were fabricated by rolling reduction.

of the texturing on the grain size is given in Table 2. While the
grain size is reduced with rolling reduction, the grains are
nominally equiaxed in the final annealed states. From this it is
concluded that the annealing cycle and/or the 775°C rolling
temperature was sufficient to cause recrystallization of the grains
which were elongated by the rolling process. X-ray diffraction
basal plane pole figures[9] were used as a second means of quanti-
fying the degree of texturing. This technique provides a means of
determining the orientational distribution of basal planes of
grains in the material. Pole figures, shown in Fig. 1, were made
by reflection and transmission techniques. The figures show that
rolling results in a redistribution of the (0002) poles within an
equatorial band along the cross-rolling direction (The rolling
direction is indicated as RD on the figures). There is also a
slight tendency for the (0002) poles to be split and tilted
symmetrically in the cross direction. Thus, there is a concentration of basal planes nominally parallel to the rolling plane (plane
of the sample disk) and this concentration increases with increasing
rolling reduction. Corresponding to the degree of texturing in
multiples of "random" intensity, the four textures will hereafter
be referred to as 1R, 3R, 5R and 8R. The significance of this
texturing will be examined in the discussions of results which
follow.

Sonic longitudinal wave speeds, measured through the thickness
of the plates, ranged from 13.44 mm/μsec for the as-pressed 1R
material to 13.79 mm/μsec for the most severely textured 8R
material (±1% accuracy). Density measurements indicate that
rolling did not alter the density (ρ = 1.850 gm/cc), therefore,
it is concluded that the change in sonic velocity is a result of
the texturing. This follows from longitudinal wave speed results
on single crystal beryllium:[10] 14.0 mm/μsec in the c-direction,
12.7 mm/μsec in the a-direction, and 13.3 mm/μsec in the direction
45° to the c-axis. Measured wave speeds in polycrystalline
beryllium represents some average of the wave speeds through the
individual grains which make up the polycrystalline sample. For
the nominally randomly distributed 1R material the measured wave
speed is very nearly the average of the single crystal wave speeds.
As the texturing increases, the percentage of c-axes parallel to
the wave propagation direction increases, as seen from the pole
figure analyses. This "weights" the average wave speed more
heavily toward that of the c-axis, which results in a higher
measured wave speed.

IMPACT TESTS

Sample disks of each of the four textures were subjected to
carefully controlled planar impact using a standard plate-impact

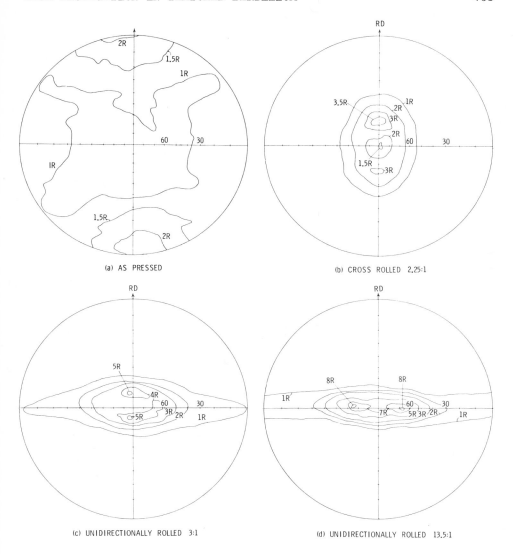

Fig. 1. Basal plane pole figures of the four textures of beryllium studied. The rolling direction is denoted by RD and the total reduction in thickness is given under each figure.

technique.[11] Each 3-mm thick 50-mm diameter disk was impacted uniformly and simultaneously over one of its faces by 1-mm thick 50-mm diameter fused silica flyer-plate disk.* As a result of the impact, plane longitudinal compressive waves propagate from the impact interface into both the target and impactor disks. These waves produce a state of global uniaxial strain† in the central region of the disks in which the only nonzero principal strain component lies in the propagation direction.[14] The initial compressive waves reflect, respectively, from the free faces of the target and flyer disks as rarefaction, or unloading waves. With the flyer disks thinner than the target disk, these rarefaction waves interact in the interior of the target disk to produce a region of tension in which the spall damage occurs. The 40-mil fused quartz flyer produced a 340 nsec square wave input into the beryllium disks. The profiles of these waves transmitted to the free back surface of the disks were recorded using the Sandia Velocity Interferometer,[15] and are shown in Fig. 2 for an

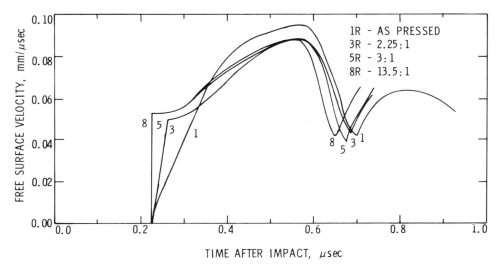

Fig. 2. Free surface velocity wave profiles resulting from 1-mm thick fused quartz impacting on 3-mm thick beryllium samples.

*For mechanical support during the rapid acceleration phase of the projectile the fused quartz flyer was backed by 0.22 gm/cc density carbon foam, which responds dynamically essentially as a void behind the quartz.
†The assumption that the motion depends only on one space variable and time must be viewed with some caution when applied to polycrystalline aggregates of crystalline solids which are strongly anisotropic. The motion is actually some average behavior of the three-dimensional motions of the individual crystallites.[10,12,13]

impact velocity of 0.12 mm/μsec (~395 fps) in each case. The corresponding impact stress is approximately 9.2 kbar* (detailed knowledge of the constitutive equation of the material is required in order to determine the exact stress amplitude).

Two observations can be made from these transmitted wave profiles: First, the structure of the elastic precursor wave changes markedly with texturing. The as-pressed 1R material exhibits a steadily rising ramp wave distributed over approximately 375 nsec. Evidence of an elastic precursor wave is almost completely suppressed in this 1R material. In contrast, a distinct elastic precursor wave is evident in the textured materials. The 3R material exhibits a ramped elastic wave with a 6.3 kbar Hugoniot elastic limit (HEL). The elastic precursors in the 5R and 8R materials rise within 4 nsec to a 6.8 kbar HEL. Such a sharply rising elastic wave has not previously been reported for polycrystalline beryllium, either for the hot pressed or cast materials. Rather, previously reported elastic precursor waves have been markedly ramped similar to that observed in the 1R material in the present investigation. In contrast, transmitted wave profiles in single crystal beryllium, for wave front normals in principal crystallographic directions[4,10] and for oblique angles,[10] all exhibit sharply rising precursor wave fronts. This strongly suggests that the change in structure of the transmitted wave in the present study is a consequence of the anisotropic character of the beryllium crystals, in aggregate, the effect of which changes with degree of texturing. This will be further discussed below.

The second observation to be made from the wave profiles of Fig. 2 is that the nominal spall strength is not significantly changed as a result of the texturing. This is deduced from the magnitude ΔU_{fs} of the pull back in the free surface velocity, i.e., the difference between the maximum free surface velocity and the velocity at the reversal point (cusp point, or lowest point on the unloading side of the wave profile). The nominal spall strength is computed using the approximate formula[4] $\sigma_s = \rho C \Delta U_{fs}/2$, where ρ is the density of the material and C is the longitudinal sound speed.

This is not the complete description of the spall damage, however. Under metallurgical examination the spall damage in the as-pressed material is seen to be transgranular cleavage cracks, somewhat randomly oriented in direction (see Fig. 3) and distributed over the center half region of the 1/8-inch thick sample. In contrast, the spall damage in the textured material is characterized by fewer cracks, which are more continuous and parallel to the plane of the plate. These observations are

*1 kbar = 14,504 psi

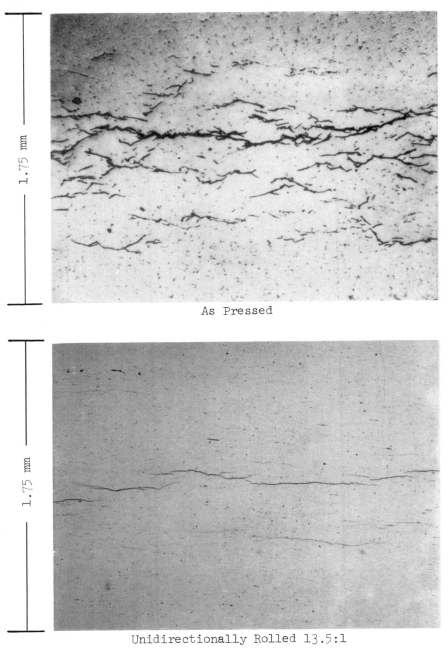

Fig. 3. Photomicrographs showing the effect of texturing on the spall damage.

consistent with the results of the pole figure analyses. In the untextured 1R material the crystals, and therefore the basal planes along which most of the cleavage fracture occurs, are randomly oriented. As a consequence, the probability of favorably oriented planes for the continuation of cleavage from one crystal to the next is lower in the 1R material than in the textured materials. Crack arrest occurs more frequently resulting in a smaller average crack length; this average crack length increases with increasing degree of texturing.[16] Residual in-plane tensile strength and bending strength also differ as a result of the texturing.

ELASTIC WAVE STRUCTURE

Let us return now to the question of structure of the elastic precursor wave. We postulate that in the as-pressed material the elastic precursor is suppressed as a consequence of residual microstresses. These are unrelaxed thermal stresses which arise because the individual grains in the polycrystalline beryllium mutually restrict their anisotropic dimensional changes in cooling from the processing temperatures down to room temperature. Armstrong and Borch[17] have estimated that average residual shear stresses of the order of a few kbar may be induced in polycrystalline beryllium as a result of this thermal anisotropy. Such residual stresses are of the order of the bulk stresses that are measured for yielding and fracture. Thus it may reasonably be assumed that these stresses should be capable of influencing the deformation and fracture behavior of this material.

The orientational dependence of the thermal stresses is such as to cause a compressive stress across basal planes and tensile stress in the basal plane. If these principal stresses are skewed from the principal crystallographic directions, as may reasonably be assumed for the polycrystalline material, a residual shear stress will exist on the basal planes, the primary dislocation glide planes of beryllium. In the absence of residual stresses, if an external load τ is applied, no plastic strain is accumulated until $\tau = \tau_0$, where τ_0 is the yield stress, and thereafter all planes are assumed to contribute equally. If residual shear stresses τ are present, the applied stress required to cause yielding can be considerably less than τ_0. In fact, if the residual stresses are distributed between $-\tau_0$ and τ_0, then macroscopic yield will begin at the onset of external load application and the usually-observed yield character will be completely suppressed. Johnson[18] has postulated the existence of a distribution function $N(\tau)$, such that $N(\tau)d\tau$ is the fractional number of slip planes with residual stresses between τ and $\tau + d\tau$. The function $N(\xi)$ is assumed to be symmetric about $\tau = 0$ with $N(\tau) = 0$ for $|\tau| > \tau_0$ and $\int_{-\tau_0}^{\tau_0} N(\xi)d\xi = 1$. In the absence of residual stresses an externally applied increment in shear stress, $d\tau$, results in a plastic strain increment $d\gamma_o$. The corresponding increment in plastic strain for the case of residual stresses is assumed to be

$$d\gamma = d\gamma_o \int_{\tau_o - \tau}^{\tau_o} N(\xi) d\xi \quad , \tag{1}$$

i.e., $d\gamma_o$ is weighted by the fraction of planes which are participating in the flow process. When $\tau \geq 2\tau_o$ then all slip planes are activated and $d\gamma = d\gamma_o$. If $N(\tau)$ and the work hardening rate $d\gamma_o/d\tau$ are known, a constitutive relation for uniaxial strain compression in a shock-loaded material can be calculated.

A function $N(\tau)$, which satisfies the conditions given above, is

$$N(\tau) = \frac{4}{35\tau_o} \left[1 + \cos \frac{\pi\tau}{\tau_o} \right]^4 \quad . \tag{2}$$

The stress-strain path computed using this function is shown in Fig. 4 using $\tau_o = 2.2$ kbar and a work hardening rate $d\gamma_o/d\tau = .0237$ kbar^{-1}, deduced from the data of reference 2. The resulting curve is seen to be a reasonable approximation to the stress-strain path for the 1R material, as constructed from the transmitted velocity wave profile of Fig. 2 using the tangent modulus method. The overshoot in the experimental curve is taken to be evidence of a strain-rate effect. Also shown for comparison in Fig. 4 is the elastic-work-hardening path with a Hugoniot elastic limit of 4.35 kbar. This value of the HEL corresponds to the value of τ_o used in the residual stress distribution function above.

This residual stress model has been incorporated into the wave propagation computer program SWAP[16] to compute the free-surface velocity loading wave profile. A wave profile computed using $\tau_o = 2.2$ is shown in Fig. 5 together with the experimentally measured free-surface wave profile and a computed wave profile assuming an elastic-perfectly-plastic material response. The method of solution used in SWAP is based on the well known method of characteristics and represents smooth wave shapes as a series of weak shocks. What is important in the present case is that all of the dispersion in the computed wave profile can be attributed entirely to the residual stress model; there is no dispersion inherent in the basic computer solution as there is in finite-difference computer program solutions where an artificial viscosity is used to maintain stable solutions.

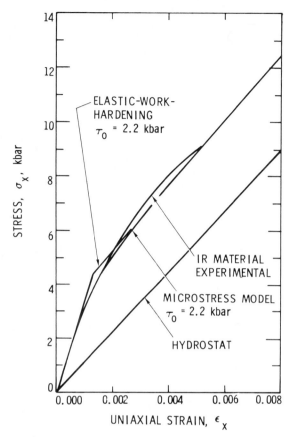

Fig. 4. Stress-strain diagram for uniaxial strain compression.

The wave profile, computed using $\tau_0 = 2.2$ kbar, follows the early part of the measured wave profile but soon falls somewhat under it in the region below 0.04 mm/µsec. The use of $\tau_0 = 3.0$ kbar (corresponding to the 6.3 kbar Hugoniot elastic limit of the 3R material) results in the computed wave rising too rapidly in the early stage but returning to follow the measured result through the middle portion of the wave profile. From this it may quickly be concluded that the distribution function of Eq. (2) is not the correct one for this material, and in the strict sense this is undoubtedly true. However, Read and Cecil[10] have shown that beryllium is strongly strain-rate dependent in the plastic region. Since SWAP uses a strain rate independent plasticity model, the computed wave profile may be expected to fall below the measured profile once yielding has occurred. On this basis we take the value $\tau_0 = 2.2$ kbar to a good representation of the 1R material.

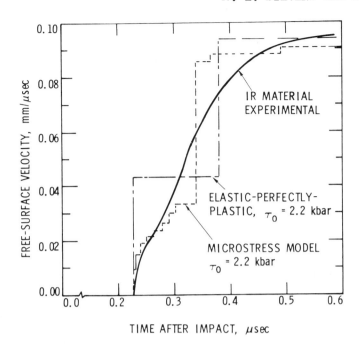

Fig. 5. Comparison of computed and experimentally measured free surface velocity loading wave profiles.

What would be needed to completely reproduce the measured wave profile is a computer program which includes the strain-rate dependent constitutive model of Read and Cecil[10] and the present residual stress distribution model but does not include the dispersive effects, such as artificial viscosity, inherent in some numerical solution techniques. No attempt has been made in the present study to construct such a program.

From comparison of the wave profile computed using the residual stress function to that using only the elastic-perfectly-plastic model, both for $\tau_0 = 2.2$ kbar, it is clear that the sharply rising elastic precursor wave can be severely suppressed by the presence of residual microstresses. In the severely textured 5R and 8R materials the crystallographic alignment which results from the rolling permits the relaxation of the thermal microstresses and the sharply rising elastic precursor, common to most metals, appears. The amount of texturing induced in the 3R material appears to be near the threshold of that required to completely relieve the microstresses.

SUMMARY

Several conclusions can be drawn from the results of this study: (1) Even though the grains in polycrystalline beryllium which have been rolled or shear spun may be equiaxed by proper heat treatment, it should not be assumed that the material is isotropic; there is a strong preference for the basal planes to be aligned nominally parallel to the plane of the rolled plate. (2) Although the nominal spall strength is not significantly altered by the severe texturing, the distribution and character of the spall damage does vary markedly with texture. (3) Unrelaxed thermal microstresses in the untextured beryllium result in the suppression of the elastic precursor wave. The crystallographic alignment which results from the rolling permits relaxation of the microstresses and the elastic precursor appears.

ACKNOWLEDGMENT

The authors wish to thank D. L. Allensworth for his assistance in performing the impact experiments and Drs. L. W. Davison and J. N. Johnson for numerous helpful discussions regarding this work.

REFERENCES

1. N. H. Froula, "The Hugoniot Equation of State of S-200 Beryllium to 1000°F," Report MSL-68-16, General Motors Corporation, Materials and Structures Laboratory, July 1968.
2. D. R. Christman and F. J. Feistman, "Dynamic Properties of S-200-E Beryllium," Report MSL-71-23, General Motors Corporation, Materials and Structures Laboratory, February 1972.
3. B. D. Jenrette and K. C. Lockhart, "Hugoniot Curves and Spall Experimental Results for Beryllium S-200 Produced by Brush and Berylco," Report B-72-67-47, Lockheed Missiles and Space Company, 20 Feb. 1968.
4. J. W. Taylor, "Stress Wave Profiles in Several Metals," Dislocation Dynamics, edited by A. R. Rosenfield, G. T. Holman, A. L. Bement, Jr., and R. I. Jaffee, McGraw-Hill Book Co., New York, 1968.
5. R. L. Warnica, "Spallation Thresholds of S-200 Beryllium, ATJ-S Graphite and Isotropic Boron Nitride at 75°F, 500°F and 1000°F, Report MSL-68-18, General Motors Corporation, Materials and Structures Laboratory, July 1968.
6. L. W. Davison and J. N. Johnson, "Elastoplastic Wave Propagation and Spallation in Beryllium: A Review," (unpublished).
7. H. E. Read and R. A. Cecil, "A Rate-Dependent Constitutive Model of Shock-Loaded S-200 Beryllium," Report DNA 2845F, Systems, Science and Software, La Jolla, California, April 1972.

8. P. B. Lindsay and R. P. Serna, "Development of a Beryllium Shear Spinning Process to Fabricate A Hot-Shaped Component," Report SCL-DR-71-0103, Sandia Laboratories, Livermore, California, July 1972.
9. M. S. Werkema, "Beryllium Recrystallization Texture," J. Appl. Cryst. $\underline{3}$, 265 (1970).
10. L. E. Pope and A. L. Stevens, "Wave Propagation in Beryllium Single Crystals," Proc. Conference on Metallurgical Effects at High Strain Rates, February 5-8, 1973 (this volume).
11. C. H. Karnes, "The Plate Impact Configuration for Determining Mechanical Properties of Materials at High Strain Rates," Mechanical Behavior of Materials Under Dynamic Loads, U. S. Lindholm, Springer Verlag, New York, 1968, pp. 270-293.
12. J. N. Johnson, "Calculation of Plane-Wave Propagation in Anisotropic Elastic-Plastic Solids," J. Appl. Phys. $\underline{43}$, 2074 (1972).
13. J. N. Johnson, "Shock Propagation Produced by Planar Impact in Linearly Elastic Anisotropic Media," J. Appl. Phys. $\underline{42}$, 5522 (1971).
14. A. L. Stevens and O. E. Jones, "Radial Stress Release Phenomena in Plate Impact Experiments: Compression-Release," J. Appl. Mech. $\underline{39}$, pp. 359-366 (1972).
15. L. M. Barker, "Fine Structure of Compressive and Release Wave Shapes in Aluminum Measured by the Velocity Interferometer Technique," Behavior of Dense Media under High Dynamic Pressures, Proc. I.U.T.A.M. Symposium, Gordon and Breach, New York, 1968, pp. 483-505.
16. A. L. Stevens, "Spall Damage Distribution and Residual Mechanical Properties of Textured Beryllium," (to be published).
17. R. W. Armstrong and N. R. Borch, "Thermal Microstresses in Beryllium and other HCP Materials," Met. Trans. $\underline{2}$, 3073 (1971).
18. J. N. Johnson, private communication.

THE INFLUENCE OF MICROSTRUCTURAL FEATURES ON DYNAMIC FRACTURE

D. A. Shockey, L. Seaman, and D. R. Curran

Stanford Research Institute

I INTRODUCTION

Depending on the application, materials are either required to resist fracture under dynamic loads, or else to break up easily and predictably under dynamic loads. Load-carrying structures, such as nuclear reactor pressure vessels, reentry vehicles, and armored vehicles, are examples of the former case; hand grenades and other fragmenting projectiles are examples of the latter.

To control and optimize the dynamic fracture behavior of materials for a given application, it is necessary to develop an understanding of the role microstructural features play in the dynamic fracture process. The purpose of this paper is to summarize our present knowledge of the effects of microstructural features on dynamic fracture and to outline an approach for assessing quantitatively the influence of microstructural variations on dynamic fracture parameters.

In the next section we describe the dynamic fracture process as involving the nucleation, growth, and coalescence of microfractures to produce spallation or fragmentation. In Sections III and IV we briefly review metallographic observations that indicate how the nucleation and growth of microfractures are influenced by such microstructural features as grain boundaries, precipitates and inclusions, inherent voids and flaws, texture, substructure, and impurity content, and in Section V we discuss the possibility of controlling dynamic fracture through microstructural control.

Section VI describes a computational model of dynamic fracture, and indicates how this model can be used to determine the effects of microstructural variations on dynamic fracture behavior.

II THE DYNAMIC FRACTURE PROCESS

Materials experience dynamic loads when they are struck by a high velocity projectile, when an explosive is detonated in close proximity, or when they are exposed to pulses of high energy radiation. When the stress level is sufficiently high and the stress duration sufficiently long, fracture can occur. Fractographic examination of dynamically loaded specimens[1-4] reveals that fracture is a gradual process consisting of several stages and requiring several microseconds to go to completion. The phenomenology is illustrated in Figure 1 with a 1-1/2 inch diameter by 1/4 inch thick polycarbonate specimen, whose transparency allows in-situ observation of the internal fracture damage resulting from flat plate impact in a gas gun. The photo shows that a large number of microfractures form under dynamic loads, and moreover that formation occurs at discrete sites in the material.

Under larger stress pulses, microfractures were observed to have grown together and begun to widen and coalesce. Under still larger stress pulses, spallation or fragmentation resulted, and the dynamic fracture process went to completion.

By more indirect methods similar observations have been made on opaque materials including metals, fiber composites, rocks, and other polymers, suggesting that this dynamic fracture phenomenology is universal. In some ductile metals, such as low alloy aluminum and copper, spherical voids form instead of planar cracks. But, regardless of the microfracture morphology the dynamic fracture process may be thought of as consisting of four basic stages:

- Rapid nucleation of microfractures at a large number of locations in the material

- Growth of the fracture nuclei in a rather symmetrical manner

- Coalescence of adjacent microfractures

- Spallation or fragmentation by the formation of one or more continuous fracture surfaces through the material.

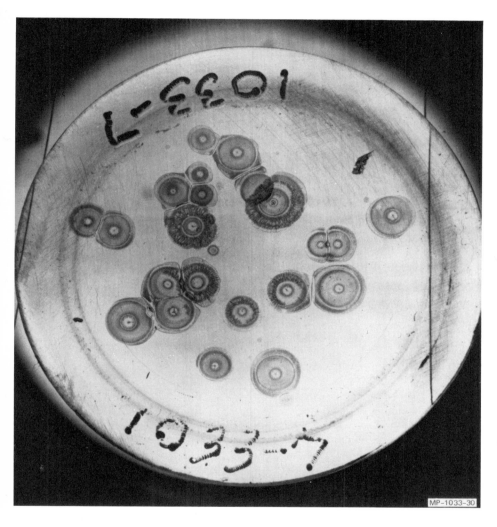

FIGURE 1 INTERNAL CRACKS IN POLYCARBONATE PRODUCED BY A SHORT-LIVED TENSILE PULSE

Thus the process by which fracture occurs under dynamic conditions is quite difference from the usual quasi-static process, whereby only one fracture forms and propagates through the material to separate it into two parts. (Fragmentation under quasi-static loads occurs by repeated branching of the original crack.)

Depending on the size and the shape of the stress pulse, the dynamic fracture process may stop before reaching the later stages. The process in the polycarbonate specimen shown in Figure 1, for example, arrested before very much coalescence had occurred. However, mechanical tests on specimens containing only minimal amounts of fracture damage have shown$_2$ that the mechanical properties may be seriously degraded. Thus in some cases the useful life of a structure may be essentially terminated the instant that the dynamic fracture process begins. This is not to say, however, that the latter stages are of no interest, for in the case of fragmentation projectiles or in rapid excavation programs the entire process must be considered. A satisfactory quantitative treatment of fragmentation, however, does not yet exist, and such a treatment should in any case be based on an understanding of the initial stages that lead up to fragmentation. For these reasons we restrict ourselves in this paper to the nucleation and growth of dynamic fracture.

III FRACTURE NUCLEATION

The nucleation process almost always involves local stress concentration due to the presence of a heterogeneity. This irregularity and nonuniformity of the stress field on a local level explains why fracture originates at discrete sites within the material and not uniformly and simultaneously throughout. Microfracture nucleation may be arbitrarily classified as heterogeneous or homogeneous according to the size of the heterogeneity involved. Microfractures that originate at gross heterogeneities generally observable by normal optical microscopy, such as inherent flaws and voids, inclusions and second phase particles, and grain boundaries, are said to have nucleated heterogeneously, whereas those originating at submicroscopic heterogeneities within the grains, such as low-angle subgrain boundaries, dislocation tangles and networks, and fine impurity or precipitate particles, are considered to have nucleated homogeneously. The latter are typically observable only by high resolution techniques such as electron microscopy and x-ray methods, if at all. In general, the smaller the heterogeneity, the smaller the volume of material that experiences the stress concentration effect, and the more difficult is microfracture nucleation. Thus higher dynamic strengths should be attainable by eliminating the larger heterogeneities, and forcing fracture nucleation to occur "homogeneously."

Inherent Flaws and Voids

A population of tiny inherent flaws exists almost invariably in materials due to the fabrication process, and we have observed many instances of heterogeneous fracture nucleation at such sites. Inherent flaws and voids may be thought of as fracture nuclei which already exist, and so the term microfracture activation may be more appropriate than microfracture nucleation.

The inherent flaw structure of injection molded polycarbonate (which is transparent and therefore easily investigated) included flaws of two types;[3] namely, a fine homogeneous dispersion of spherical voids about one micron in diameter, and a less closely spaced, random distribution of crack-like flaws having an average diameter of about 20 microns. The former are probably bubbles of trapped moisture or gases which formed as the water and gas solubility of polycarbonate decreased during the cooling phase of the molding process. The latter type likely resulted from intrusion of foreign material, such as dust particles or traces of thermally degraded resins from the mold walls. Fracture is expected to initiate preferentially at the crack-like flaws. Because of their larger size and sharper profile, the stress is concentrated more effectively and the flaws become unstable at lower nominal stresses than the smaller bubbles; hence, at stress levels near the dynamic fracture strengths, the large flaws probably control the initiation stage of the fracture process. However, at higher stresses sufficient to activate the small bubbles, these by reason of their larger number may dominate fracture initiation. Scanning electron microscope techniques were used to examine at high magnifications the center points of the penny-shaped cracks and confirmed the presence there of flaws and inclusions. Similar inherent heterogeneities were observed[3] in a polyimide. Figure 2b shows a closeup of an inherent flaw at the center of one of the fracture surfaces in Figure 2a. An instance of crack initiation at a void in rolled ingot beryllium is shown in Figure 3. The void probably was formed during the casting process by precipitation of dissolved gases and subsequently flattened during rolling. The spoke-like pattern of cleavage steps which radiate out from it identify it as the fracture initiation site.

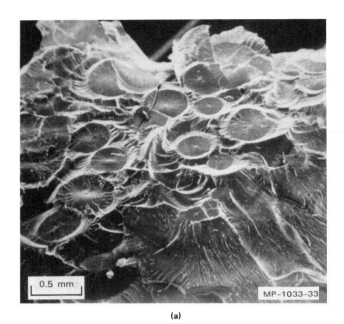

(a)

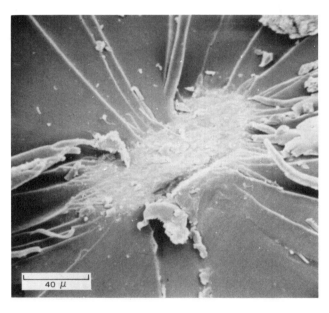

(b)

FIGURE 2 FRACTURE SURFACE OF SHOCK-LOADED POLYIMIDE (a) WITH A CLOSE-UP VIEW OF CRACK INITIATION AT AN INHERENT FLAW (b)

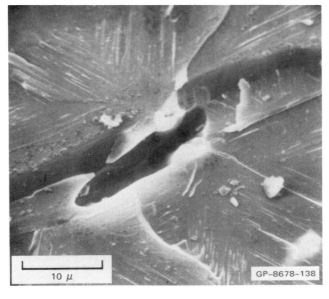

FIGURE 3 CRACK NUCLEATION IN BERYLLIUM AT A VOID

Inclusions and Second-Phase Particles

Crack nucleation at inclusions and second-phase particles may occur by (a) fracture of the inclusion, (b) fracture of the interface, or (c) fracture of the matrix under the stress concentrating effect of the inclusion or particle.

Fracture of Inclusions. An example of microfracture nucleation by fracture of brittle inclusions in 6061-T6 aluminum subjected to an underground test environment are shown in Figure 4. The grey script-like inclusions of Fe_3SiAl_{12} acquire brittle microcracks, but the fracture mode changes in the soft and ductile aluminum matrix to one of void growth. Thus both brittle cracks and ductile voids can be seen in the photomicrographs.

Separation at Interfaces. In materials containing inclusions or second phase particles that are weakly bonded to the matrix, cracks may be formed by interface separation. This mode of crack initiation may be operative in polymers at occluded dust particles or other debris. Irregular-shaped holes and depressions at apparent fracture centers may have formerly contained such foreign particles, Figure 5. The smooth-sided hydrocarbon inclusion at a fracture center in polyimide shown in Figure 6 apparently initiated a microfracture by decohesion over part of its surface, and was retained in the matrix.

Fracture of the Matrix. A third mechanism for inclusion-nucleated microfracture is fracture of the matrix material at a hard inclusion. Such a mechanism can operate when the fracture strength of the matrix is less than that of the inclusion or of the interface. Crack nucleation by this process has been observed in brittle beryllium, which has inherently weak basal cleavage planes and hard, strongly bonded BeO particles. Figure 7 shows a transgranular microcrack and twins emanating from the white BeO particle in a specimen of high-purity beryllium.

Grain Boundaries

Crack nucleation at grain boundaries is also likely a result of stress concentration there. The existence of a weak phase (or at elevated temperatures the existence of a low melting point phase) at grain boundaries can result in preferential crack nucleation in the interfaces between grains. Or in cases where grain boundaries are stronger than the grains, as for example in some precipitation hardening aluminum alloys, the stress concentrations arising from the lattice mismatch between adjacent grains in anisotropic material

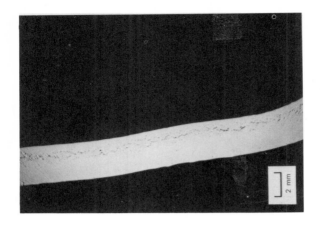
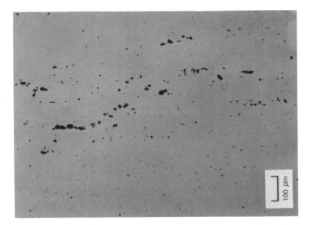
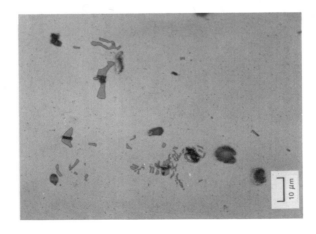

FIGURE 4 CRACK NUCLEATION AT INCLUSIONS IN 6061-T6 ALUMINUM

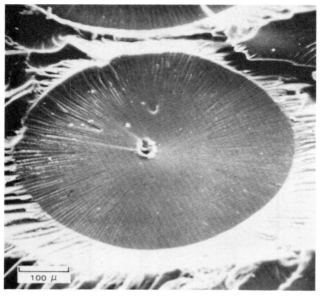

(a)

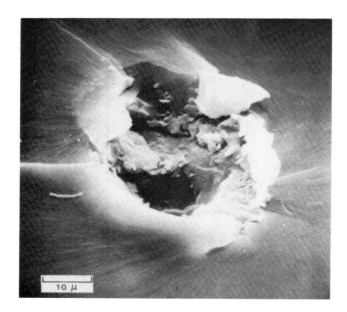

(b)

FIGURE 5   CAVITY AT A CRACK CENTER IN POLYIMIDE REMAINING AFTER PULL-OUT OF AN INCLUSION

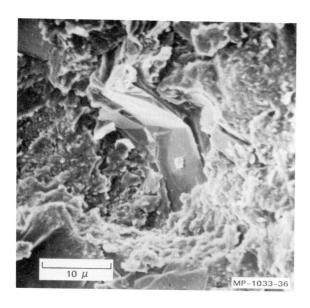

FIGURE 6 SMOOTH-SIDED INCLUSION FOUND AT A CRACK CENTER IN POLYIMIDE

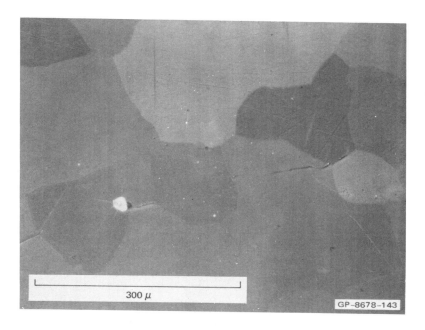

FIGURE 7 NUCLEATION OF CRACKS AND TWINS AT AN OXIDE INCLUSION IN BERYLLIUM

may be sufficient to nucleate transgranular cracking. The photomicrographs in Figure 8 show transgranular cracks in two different beryllium grades, that extend into grains after apparently nucleating at a grain boundary. Since BeO particles are densely distributed along the boundaries, however, it is not certain how much of a role they played in the fracture nucleation. The actual nucleation mechanism may have been that of dislocation pileup at the grain boundary as discussed in the next paragraph.

Subgrain Structure

Microfracture nucleation at submicroscopic heterogeneities is almost always associated with plastic flow. Large stresses can develop at tiny obstacles which impede dislocation motion and cause pileups. Many dislocation mechanisms for fracture nucleation have been hypothesized, including those of Stroh,[5] Cottrell,[6] and Gilman.[7] Low angle grain boundaries and dislocation tangles act as barriers to slip, thereby inhibiting plastic flow and raising the local stress to that required for microfracture nucleation.

In most of the fracture centers in the beryllium alloys, no heterogeneity could be found even at the high magnifications obtained with the scanning electron microscope. The site shown in Figure 9, for example, although quite obviously a nucleation site, appeared to have no heterogeneity associated with it. Our observations lead us to speculate that fracture nucleation occurred homogeneously by a plastic flow process. The fact that the nucleation threshold stress for microfracture was found to depend primarily on the shear stress and was relatively insensitive to the value of axial applied tensile stress[4] supports the speculation. However, no direct experimental evidence for dislocation pileups at subgrain heterogeneities was obtained to confirm deformation-induced crack nucleation.

Impurities

Impurity atoms in solid solution or a fine uniform dispersion of tiny precipitates can also have the effect of raising the nucleation threshold stress. The impact velocity necessary to produce incipient cracking in Armco iron was found to be nearly 50% higher than that required to produce the same amount of fracture damage in a similar iron specimen that was 99.99% pure. The effect is presumably due to the suppression of such plastic flow fracture nucleation mechanisms as large dislocation pileups and interactions, by the higher impurity content and the more strongly developed substructure of Armco

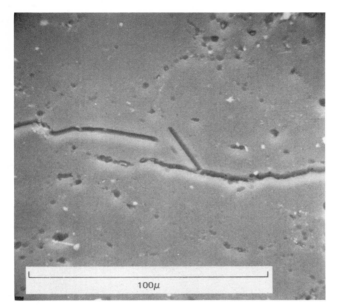

(a) N50A

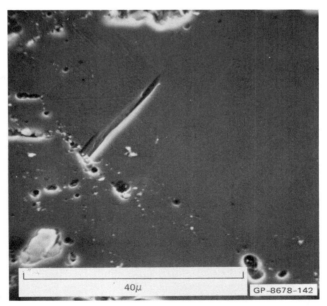

(b) S-200

FIGURE 8 APPARENT CRACK NUCLEATION AT GRAIN BOUNDARIES IN BERYLLIUM

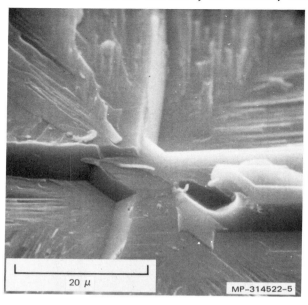

FIGURE 9   CRACK NUCLEATION SITE IN BERYLLIUM WHERE
A PLASTIC FLOW-MECHANISM POSSIBLY OPERATED

iron. Both shorten the mean free dislocation path (and hence the length of the pile up which can form) and the former gives rise to dislocation atmospheres which have a dragging effect on dislocation motion.

The impact velocity for damage in 99.999% pure aluminum, however, was found to be at least 10% higher than that in 1145 aluminum[2] which is 99.45% pure, in apparent contradictions to the iron result. This effect, however, is felt to be attributable to the presence in 1145 aluminum of Fe-Si inclusions of a size such that fracture nucleation was facilitated by occurring heterogeneously instead of homogeneously.

IV   FRACTURE GROWTH

Modes of Microfracture Growth

Growth of shock-induced microfracture usually proceeds in one of three modes; (1) viscous expansion of spherical voids, (2) transgranular propagation of planar cracks, or (3) intergranular crack growth. In soft and ductile materials, such as low alloy aluminum (Figure 4), and OFHC copper, the growth mode is by void expansion and occurs at a viscous rate controlled presumably by dislocation processes. The voids remain roughly spherical as they expand until coalescence with neighboring voids occurs. Stevens, et al.[8] observed octahedral voids in high purity aluminum single crystals and were able to explain their growth by a dislocation model.

Brittle materials fracture by the growth of planar microcracks. Materials having weak cleavage planes fracture in the transgranular mode, Figure 10b, but in cases where the grain boundaries are weaker than the grains, the cracks tend to follow the grain boundaries, Figure 10a. In general the void growth fracture mode is the most energy absorbing, so in applications where dynamic fracture resistance is important such as in armor plate, it is desirable to provide for this type of growth mode.

Growth Mode Transitions

Transitions in the fracture mode can be effected by changes in grain size, alloy content or by changes in temperature. Cracking occurs transgranularly in the finer grained beryllium alloy in Figure 10b, but proceeds along the grain boundaries in the coarse grained alloy in Figure 10a. Armco iron exhibits transgranular cleavage fracture at room temperature but the fracture mode is one of ductile void growth at $400°C$. Still another example of fracture mode transition is provided in Figure 4, which shows fracture in 6061-T6 aluminum nucleating as brittle cracks but growing as spherical voids. There is presently no way of predicting a priori the preferred fracture mode of any material, but the fact that fracture mode is sensitive to microstructure indicates the possibility of developing materials having desired growth modes.

Fracture Velocity

The velocity of fracture, although mainly of academic interest in quasi-static loading conditions, is of very practical importance under dynamic conditions. This is because the duration of a stress pulse can be smaller than the time required for microfractures to grow together to produce a continuous free surface dividing a member into two parts. Therefore, depending on the microfracture propagation speed, a given stress pulse may or may not result in spallation or fragmentation. It is generally believed that the limiting speed of crack propagation is the Rayleigh wave velocity; however, as discussed later, there is evidence that fracture velocities under dynamic loading conditions seldom attain this limiting velocity, and in fact that acceleration occurs in a viscous manner.[4]

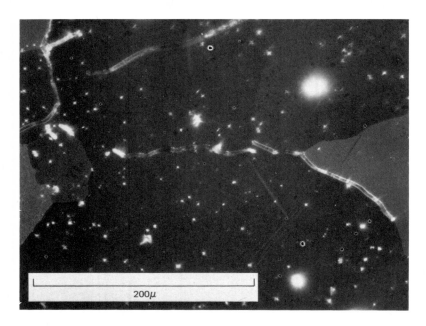

FIGURE 10(a)   INTERGRANULAR CRACKING IN HIGH PURITY BERYLLIUM

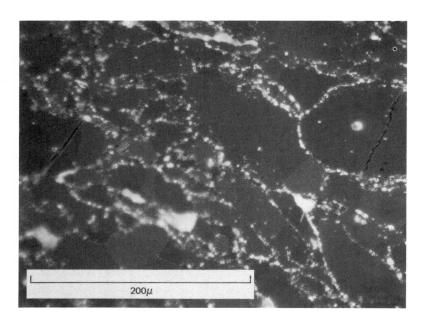

FIGURE 10(b)   TRANSGRANULAR CRACKING IN S-200 BERYLLIUM

Subsequent reverberations of the initial stress pulse can cause intermittent crack growth as illustrated in Figure 11. The second outermost concentric ring on this fracture surface in polycarbonate has been identified as an arrest line, demonstrating that the crack produced during the initial tensile pulse repropagated during the first reverberation. The suggested growth history of this crack is depicted schematically in Figure 11b. Observations of stress gage records in metals undergoing fracture, however, indicate that the reverberations are severely damped and therefore that repropagation of initial cracks is unlikely in most metals.

Crack Arrest

Once microfractures are initiated, failure resistance depends on a mechanism to arrest growth through the material. We have observed three basically different arrest mechanisms in our work; (1) plastic stretching of ligaments between interacting adjacent cracks, (2) cleavage crack arrest at grain boundaries, and (3) arrest at strong directional fiber bundles.

The first mechanism is well illustrated in polycarbonate, where the planar microcracks as they approach one another stop growing and begin to widen. The thin ligaments of the material separating the cracks undergo enormous elongation, presumably attributable to a brittle/ductile transition with decreasing cross section. This ductile-tearing-of-ligaments mechanism also operates in most metals, although to a lesser extent, as attested to by the fibrous fracture surfaces commonly observed.

Grain boundaries provide formidable barriers to transgranular crack growth, especially in materials having limited cleavage planes. Cleavage cracks in beryllium, for instance, experience serious difficulties in attempting to continue into an adjacent grain if the orientation difference between the grains is large. Note that the transgranular cleavage cracks in Figure 10b were apparently unable to continue into the adjacent grains. In such cases random texture and small grain size are most effective in limiting crack growth.

A third method of enhancing crack arrest is by providing throughout a matrix a distribution of obstacles, which are impenetrable to growing microfractures. Three-dimensional quartz phenolic composite material has reinforcing fiber bundles which stop interlaminar fracture and prevent coalescence, Figure 12.

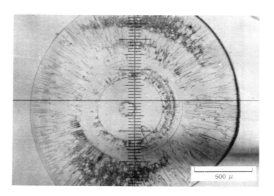

FIGURE 11(a) INTERNAL PENNY-SHAPED CRACK IN POLYCARBONATE PRODUCED UNDER SHOCK LOADING SHOWING CONCENTRIC GROWTH RANGE

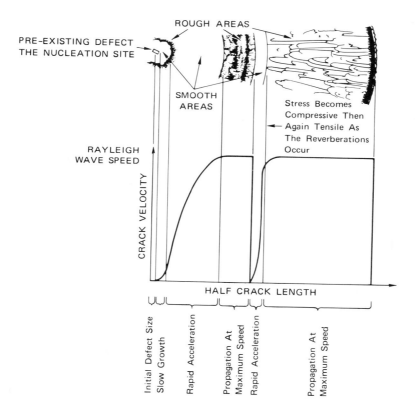

FIGURE 11(b) SCHEMATIC DEPICTION AND SUGGESTED VELOCITY HISTORY FOR THE CRACK SHOWN IN (a)

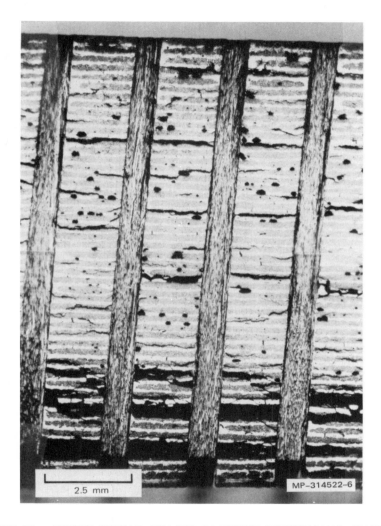

FIGURE 12   INTERLAMINAR CRACKS IN THREE-DIMENSIONAL QUARTZ PHENOLIC ARRESTED AT PERPENDICULAR FIBER BUNDLES

V CONTROL OF DYNAMIC FRACTURE

The fractographic observations of the previous sections show that microstructural features play important roles in the nucleation and growth of fracture under dynamic loads. This implies that the dynamic fracture behavior of materials should be controllable to some extent by variations in the microstructure. The following guidelines may be useful in enhancing dynamic fracture resistance in specified materials:

- A fine and uniform dispersion of impurity atoms or second-phase particles. This raises the dynamic fracture strength by strengthening the lattice and making microfracture nucleation more difficult.

- Elimination of large inclusions. These concentrate stress and act as easy nucleation sites.

- Elimination of inherent flaws and pores. These are ready-made fracture nuclei and need only be activated.

- A dense substructure. This raises the stress required for fracture nucleation and hinders fracture growth.

- A fine grain size. This limits the crack sizes to small values and inhibits homogeneous nucleation.

- A random texture. This imparts maximum crack stopping effectiveness to grain boundaries.

In addition, there appear to be ways to produce microstructures that induce desirable fracture growth modes. And based on our present knowledge of the effects of microstructural features of fracture nucleation and growth, we can infer several points pertaining to the control of fragmentation through microstructure.

Some control over the number and size of fragments should be attainable by deliberate variation of the number and spatial distribution of easy nucleation sites, such as inclusions, inherent flaws, and second-phase particles. It should be possible to attain fragment sizes proportional to grain sizes by providing for a weak phase to be present along grain boundaries and thereby inducing intergranular cracking. Some variational capability for fragment shape might be attained by changing the texture.

VI ASSESSMENT OF MICROSTRUCTURAL EFFECTS

The previous section reviewed metallographic evidence concerning the role of microstructural features in dynamic fracture response. These observations provide us with some insight into the mechanisms by which the nucleation and growth of fracture are influenced by microstructure. This approach, however, falls short of providing a quantitative measure of how continuous variations of a microstructural feature affect the dynamic fracture behavior. To attain this we need to establish dynamic fracture parameters and then measure their variation with varying microstructure. This section outlines a procedure for determining quantitatively the effects of microstructural variations on dynamic fracture behavior.

A Dynamic Fracture Model

Dynamic fracture parameters should be quantitative indices of material fracture response and should come from a realistic and quantitative model of the dynamic fracture process that is based on the sequence of events occurring during loading and leading to spallation or fragmentation. The phenomenology was deduced from fractographic examinations and is described in Section II as consisting of four stages; namely, the nucleation, growth and coalescence of many microfractures to produce spallation or fragmentation.

A dynamic fracture model which treats the initial two stages of the dynamic fracture process, nucleation and growth of microfracture, has already been developed.[1,2,4] This model allows prediction in quantitative detail of the extent of the fracture damage (the number, size, location, and orientation of the microfractures) produced by an arbitrary dynamic load. Later stages of the fracture process, coalescence and spallation or fragmentation, have not been treated. The model for a given material is developed by correlating measured size distributions of microfractures with stress histories.

In all materials thus far studied, microfractures appear at a rate well described by the following nucleation law:

$$\dot{N} = \dot{N}_o \exp\left(\frac{\sigma - \sigma_{no}}{\sigma_1}\right) \qquad (1)$$

and grow according to the viscous growth law:

$$V = \frac{\sigma - \sigma_{go}}{4\eta} R, \qquad (2)$$

where for planar microcracks

$$\sigma_{go} = \sqrt{\frac{\pi}{4R}}\, K_{Ics}, \qquad (3)$$

and where $\dot{N}$ and V are the nucleation rate and growth rate for microfractures respectivel, σ is the applied stress, and R is the radius of the microfracture. The remaining symbols are the dynamic fracture parameters characteristic of the material and are described in detail below.

To predict the extent of fracture damage produced by a known stress pulse, the material is assumed to consist of many calculational cells, each containing some portion of the total mass. SRI PUFF,[9] a finite-difference, one-dimensional wave propagation code, is used to calculate--after each increment in time--the density, internal energy, stress, and number of microfractures per unit volume and their size distribution. The calculation is then repeated for the next time increment during which the number of microfractures is increased according to the nucleation function $\dot{N}$, and their size distributions are modified by additional growth according to the viscous growth law. Degradation of the stress caused by the developing fracture damage is simultaneously computed and taken into account in each time step. This calculational procedure is repeated until the stress duration has been spanned. At the end, the crack size distributions are plotted and the stress history listed for each computational cell. Using data obtained from plate slap experiments, this model has made successful predictions of fracture damage in OFHC copper and S-200E beryllium exposed to underground test environments and in S-200E beryllium irradiated by an intense electron beam.[4]

Dynamic Fracture Parameters

The quantities $\dot{N}_o$, σ_{no}, σ_1, η, and K_{Ics}, in expressions (1) - (3) are called dynamic fracture parameters, are material dependent, and act as material properties which describe various aspects of the fracture response under dynamic loads. The first three parameters are associated with fracture nucleation, whereas the latter two determine crack growth. The values of these parameters are not determined in independent experiments but rather by an iterative procedure whereby the entire set of values is continually adjusted in the SRI PUFF wave propagation code until good agreement with experimentally measured fracture damage is attained. The values obtained by this procedure are accurate to about ±10%. The dynamic fracture parameters have the following physical meanings.

The nucleation threshold stress σ_{no} is that value of the stress below which microfractures do not appear but above which they do. It thus is the dynamic strength of the material under one-dimensional strain conditions, and in general is time and temperature dependent. For high strength brittle materials, it is a function of the dynamic fracture toughness of the material and the size of the largest inherent flaw.

The threshold nucleation rate for microfracture $\dot{N}_o$ is the number of cracks that form per cubic centimeter per second at the nucleation threshold stress. Its value is governed by the size distribution of inherent flaws and by the plastic behavior of the material.

The stress sensitivity of microfracture nucleation $1/\sigma_1$ describes the rate of increase in microfracture nucleation with increasing stress above the nucleation threshold stress.

All nucleation parameters depend upon more detailed knowledge of the actual mechanism by which microfracture nucleation occurs. For example, if inherent flaws in the material are activated by the tensile stress in the classical fracture mechanics sense, the size distribution of these flaws strongly influences the nucleation threshold stress (the same material if it has larger inherent flaws will exhibit a smaller threshold stress), the threshold nucleation rate (the same material if it has many inherent flaws near the size of the largest flaw will have a larger $\dot{N}_o$), and the stress sensitivity for nucleation (the same material with more inherent flaws will be more sensitive to stress). In this case a fourth nucleation parameter R_o characterizes the initial flaw size distribution. If, however, microfracture nucleation is dependent predominantly on plastic processes, such as any of the various hypothetical dislocation interaction mechanisms mentioned earlier, the variability of the parameters is more complicated and less well understood.

The dynamic uniaxial strain fracture toughness K_{Ics} is that value of stress intensity at tips of inherent flaws above which flaws can grow and below which flaws cannot grow, and is thus analogous to the material property defined by fracture mechanics formalism. In the dynamic fracture model, K_{Ics} is used to both start and stop microfractures.

The effective crack tip viscosity η controls the rate of crack growth, particularly in the early acceleration stages. The value represents an average obtained from the crack distributions, and as such averages nonsteady growth, i.e., hesitation at grain boundaries and precipitates, as smooth viscous growth. It is probably related to plastic deformation at crack tips, as well as the mean free crack path in the material.

Dynamic Fracture Parameters for Beryllium

Values for the dynamic fracture parameters have been determined for Armco iron,[4] 1145 aluminum,[1,2] OFHC copper,[2] Lexan polycarbonate,[3] and most recently for three grades of beryllium.[4] The parameters for beryllium, given in Table I, were obtained by analyzing the fracture damage in specimens impacted several years ago by investigators at General Motors.[10-12] The objective of the General Motors work was not to investigate microstructure effects and so the beryllium grades were not chosen to have systematic and independent variations of microstructural features. A wide range in BeO content and grain size existed between the grade known as high purity and the other two grades, as shown in Table I, but large differences also existed in other microstructural features, making it impossible to attribute changes in measured values of dynamic fracture parameters to microstructural features. For example, the high purity grade was cast and rolled, whereas the N50A and S-200 grades were hot pressed. Thus the high purity material had significantly more texture and subgrain structure than the others.

However since grain size and BeO content are considered to be the two most influential metallurgical variables on mechanical properties,[13] it was of interest to compare the values of the dynamic fracture parameters for the three beryllium grades to see if a correlation existed. Table I shows no monotonic variation with BeO content or grain size of any of the dynamic fracture parameters. (The threshold deviatoric stress σ'_{no} and the threshold axial stress σ_{no} for microfracture nucleation were chosen on the basis of the manufacturer's specifications for quasi-static tensile strength.) This is in accord with the quasi-static mechanical properties (hardness and yield strength, Table I) which show a similar indifference to BeO content and grain size. The probable explanation is that the high purity alloy, by virtue of being mechanically worked, possessed a much denser subgrain structure that resulted in a strengthening effect comparable to that derived from the smaller grains and higher BeO concentrations in the other two grades.

Table I

DYNAMIC FRACTURE PARAMETERS FOR THREE BERYLLIUM GRADES

Beryllium Alloy	High Purity	N50A	S-200
Nominal BeO Content (wt %)	<0.03	0.9	1.95
Average Grain Size (μ)	~200	25	20
Ultimate Tensile Strength (ksi)	44	35(min)	49.2
Yield Strength (ksi)	36	25(min)	33.6
Vickers Hardness No.	186	149	162
Crack Radius at Nucleation, R_o (μ)	20	45	7.5
Threshold Nucleation Rate, $\dot{N}_o$ (cm^{-3} sec^{-1})	2×10^{12}	7×10^{10}	2.5×10^{11}
Nucleation Threshold Deviator Stress, σ'_{no} (dyn/cm^2)	1.65×10^5	1.74×10^9	1.80×10^9
Nucleation Threshold Axial Stress, σ_{no} (dyn/cm^2)	2.68×10^9	2.72×10^9	2.83×10^9
Stress Sensitivity for Nucleation, $1/\sigma_1$ (cm^2/dyn)	3.92×10^{-9}	4.81×10^{-9}	1.61×10^{-9}
Effective Crack Tip Viscosity, η (dyn-sec/cm^2)	275	316	125
Dynamic Fracture Toughness, K_{Ics} (dyn cm$^{-3/2}$)	10^8	10^8	10^8

CONCLUSIONS

- Microstructural features do influence dynamic fracture behavior. Therefore it should be possible to exercise some control over the dynamic fracture process by tailoring the microstructure of materials.

- We have a qualitative understanding of the role of several microstructural features in dynamic fracture.

- New engineering parameters which describe material fracture response under dynamic loads make possible a quantitative assessment of microstructural effects on dynamic fracture behavior.

ACKNOWLEDGMENTS

It is a pleasure to thank T. W. Barbee, Jr. and R. L. Jones for critically reviewing the manuscript, and D. J. Petro for producing the photomicrographs. This work was supported by the Defense Nuclear Agency, the Air Force Weapons Laboratory, and the Ballistic Research Laboratory.

REFERENCES

1. Barbee, T., L. Seaman, and R. C. Crewdson, Dynamic Fracture Criteria of Homogeneous Materials, AFWL-TR-70-99, Air Force Weapons Laboratory, Kirtland AFB, Albuquerque, New Mexico, November 1970.

2. Seaman, L., T. Barbee, and D. R. Curran, Dynamic Fracture Criteria of Homogeneous Materials, AFWL-TR-71-156, Air Force Weapons Laboratory, Kirtland AFB, Albuquerque, New Mexico, December 1971.

3. Curran, D. R., and Shockey, D. A., Dynamic Fracture Criteria for Polycarbonate and Polyimide, DAA005-71-C-0180, Ballistic Research Laboratory, Aberdeen Proving Ground, Aberdeen, Md. April 1972.

4. Shockey, D.A., L. Seaman, and D. R. Curran, Dynamic Fracture of Beryllium Under Plate Impact, and Correlation with Electron Beam and Underground Tests, F29601-70-C-0070, Air Force Special Weapons Center, Kirtland AFB, New Mexico, January 1973.

5. Stroh, A. N., "The Cleavage of Metal Single Crystals," Phil. Mag. $\underline{3}$, 597 (1958).

6. Cottrell, A. H., "Theory of Brittle Fracture in Steel and Similar Metals," Trans. AIME 212, 192 (1958).

7. Gilman, J. J., "Fracture of Zinc-Monocrystals and Bicrystals," Trans. AIME 212, 783 (1958).

8. Stevens, A. L., L. Davison, and W. E. Warren, "Spall Fracture in Aluminum Monocrystals: A Dislocation-Dynamics Approach," J. Appl. Phys. 42 (12), 4922 (1972).

9. Seaman, L., SRI PUFF 3 Computer Code for Stress Wave Propagation, Technical Report No. AFWL-TR-70-51, Air Force Weapons Laboratory, Kirtland AFB, New Mexico, September 1970.

10. Christman, D. C., N. H. Froula, and S. G. Babcock, Dynamic Properties of Three Materials, Volume I: Beryllium, Final Report No. MSL-68-33, Materials and Structures Laboratory, General Motors Corporation, Warren, Michigan, November 1968.

11. Warnica, R. L., Spallation Thresholds of N50A Beryllium, MSL-68-1, Materials and Structures Laboratory, General Motors Corporation, Warren, Michigan, January 1968.

12. Warnica, R. L., Spallation Thresholds of S-200 Beryllium, ATJ-S Graphite and Isotropic Boron Nitride at $75°F$, $500°F$, and $1000°F$, MSL-68-18, Materials and Structures Laboratory, General Motors Corporation, Warren, Michigan, July 1968.

13. Hanes, H. D., S. W. Porembka, J. B. Melchan, and P. J. Gripshover, Physical Metallurgy of Beryllium, DMIC Report 230, Defense Metals Information Center, Battelle Memorial Institute, Columbus, Ohio, June 24, 1966.

WORK SOFTENING OF Ti-6Al-4V DUE TO ADIABATIC HEATING

A.U. Sulijoadikusumo and O.W. Dillon, Jr.

University of Kentucky

Lexington, Kentucky

I. INTRODUCTION

Many investigators have shown that at high strain rates at elevated temperature (1300-1900°F), and for large deformations, Ti-Alloys (specifically Ti-6Al-4V) exhibit considerable work softening. The mechanism of such metal softening is different from the one encountered in processes at lower strain rates (creep). In the high temperature range, Ti-6Al-4V is known to be both temperature and strain rate sensitive; i.e. an increase in strain-rate will increase the flow stress and an increase in temperature significantly decreases the flow stress. Therefore, thermal changes strongly influence the metal behavior during the deformation. When metal is undergoing large plastic deformation, approximately 95% of the plastic work is converted into heat almost instantaneously (less than 1 milliseconds). This rapid dissipation of large amounts of energy causes temperature rises in the material. Such a heating would cause (contribute) the metal softening during the process relative to what would be observed in isothermal deformations.

At high temperatures, work hardening occurs only when the deformation is still small (approximately for strains less than 10%). But as the deformation progresses, several mechanisms of work softening become dominant. In this paper we would like to study one of the softening mechanisms, namely work softening due to the temperature rise caused by plastic working itself. Thus, such a softening mechanism is rheological in nature. However, the form of the constitutive equation, that we propose in this paper, is more general than this and it provides a procedure where parameters that represent the effect of microscopic (structural) phenomena on the macroscopic (continuum) theory.

We take a particular form of a constitutive equation proposed by Hohennemser and fit it to the data obtained by Hoffmanner in experiments which were performed at both constant temperature and strain rate. Using this constitutive equation, a mathematical model for simple tension/compression is constructed. The stress-strain curves obtained from the solution of that mathematical model is compared to the experimental result obtained by Bühler. Except for a constant of multiplication, the two curves are amazingly similar. We think that the scale factor being different is probably due to the difference in initial structure of the materials.

II. CONSTITUTIVE EQUATION

Based on the experimental evidences [1-6] of Ti-6Al-4V undergoing plastic deformation at temperature in the 1300°-1900°F range, the following idealizations are drawn:

(1) The elastic part of the deformation is in the order of 10^{-5} while the plastic part is in the order of 10. Hence, the elastic-strain and the volumetric change which is associated with the elastic deformation are neglected. Then, we assume the material to be rigidly plastic.

(2) The flow of this particular metal, while it is undergoing large plastic deformations in the above temperature range, is assumed to behave as a highly viscous fluid. Such an idealization is motivated by the experimental results of Huang and DeAngelis [7] that indicate Ti-6Al-4V with fine grain structure produces superplastic behavior. Such behavior is also shown by Hoffmanner in reference [4]. The material is assumed to retain its isotropic behaviour which means that during plastic deformation the grains of the polycrystalline aggregate are assumed to have no preferred direction.

(3) The material is very temperature and strain-rate sensitive. Hence, thermal changes have a large influence on the metal behavior during deformation[1].

---

[1]Even though the constitutive equation which is used here is motivated by the thermally activated processes as used in Metallurgy [14], we avoid the use of the word "thermally activated process," since in metallurgical science this terminology is used for a process, usually simple tension, for which the strain rate is governed by Arrhenius equation [16].

In reference [8], we develop the general theory of materials of the rate type with internal state variables. From the above general theory we consider the following class of constitutive equation for a material of the rate type,

$$\underset{\sim}{D} = \Omega(\underset{\sim}{S}, \theta, \alpha_j)\underset{\sim}{S}$$

$$\dot{\alpha}_i = f_i(\underset{\sim}{D}, \underset{\sim}{S}, \theta, \alpha_j) \qquad (II.1)$$

where $\underset{\sim}{D}$ is the stretching tensor, θ is the temperature,

$$\underset{\sim}{S} = \underset{\sim}{T} - \frac{1}{3} \text{tr}(\underset{\sim}{T}) \qquad (II.2)$$

is the deviatoric stress tensor and $\underset{\sim}{T}$ is the Cauchy stress tensor. The following factors influence the choice of the constitutive equations (II.1): First, the three idealizations mentioned at the beginning of this section. Second, the form of constitutive equations proposed by Hohennemser (1937) which mainly based upon the fact that the deformation in crystalline metals tend to be stress activated.

The internal state variables (scalar parameters), α_i, can be thought of as including the effect of the fine structure (microscopic) on the larger scale properties (macroscopic) of the material. Such scalar parameters which indicate the state of a process are commonly used by metallurgist in compiling phenomenological data based on structural changes in special processes (such as dislocation densities, changing grain size, number which represents change of phase, etc)[9].[2]

Familiar equations of plasticity can also be obtained as specific examples of equation (II.1) with the choice of

$$\dot{\alpha}_1 = \sqrt{\frac{3}{2} II_{\underset{\sim}{D}}} \quad \text{or} \quad \alpha_1 = \int_{t_o}^{t} \sqrt{\frac{3}{2} II_{\underset{\sim}{D}}}\, dt + \alpha_1(t_o)$$

and $\qquad\qquad\qquad\qquad\qquad\qquad\qquad\qquad\qquad\qquad\qquad\qquad$ (II.3)

$$\dot{\alpha}_2 = \text{tr}(\underset{\sim}{S}\underset{\sim}{D}) \quad \text{or} \quad \alpha_2 = \int_{t_o}^{t} \text{tr}(\underset{\sim}{S}\underset{\sim}{D})\, dt + \alpha_2(t_o)$$

as the parameters. Where $II_{\underset{\sim}{D}}$ is the second invariant of $\underset{\sim}{D}$ and t

---

[2] We note that the variation in substructure which influences the changes in the flow stress at a given plastic strain is probably path dependent.

is time variable. From Table B.1 in reference [8], we can readily see that II_D and $\text{tr}(\underset{\sim}{S}\underset{\sim}{D})$ are members of the group of tensor invariants which in general are contained in equation (II.1)^2.

From published research [10-12], the suggested form of Ω for a material of the Von Mises type is as follows:

$$\Omega = \zeta(\theta) < \Phi\ (\mathfrak{F}) > \frac{1}{II_S^{\frac{1}{2}}} \qquad (II.4)$$

where ζ is a viscosity function, II_S is the second invariant of the deviatoric stress. Then,

$$\underset{\sim}{D} = \zeta(\theta) < \Phi\ (\mathfrak{F}) > \frac{\underset{\sim}{S}}{II_S^{\frac{1}{2}}} \qquad (II.5)$$

where

$$\mathfrak{F} = \frac{II_S^{\frac{1}{2}}}{k(\theta, \alpha_i)} - 1 \qquad (II.6)$$

$$<\Phi> = \begin{matrix} \Phi; & \mathfrak{F} > 0 \\ 0; & \mathfrak{F} \le 0 \end{matrix} \qquad (II.7)$$

A variety of possible explicit forms of $\Phi(\mathfrak{F})$ which can be fitted to an experimental result are discussed by Perzyna and Cristescu [10-12]. The experimental results of several investigators indicate that the change of the hardening parameters (α_i) has only a small effect on a material which has undergone finite plastic deformation with strains larger than 20%; i.e., the stress-strain curve becomes flat with additional isothermal plastic straining.[3] Hence, neglecting the effect of α_i's, we have

$$\underset{\sim}{D} = \Omega(\underset{\sim}{S},\theta)\underset{\sim}{S} \qquad (II.8)$$

and this is the form of constitutive equation of viscous fluid.

---

[3] Above 1/2 the melting temperature, dislocation structures may anneal out almost as fast as they form. Hence the constitutive equation (equation of state) is less influenced by internal state variables. Moreover in many instances, deformation at high temperature causes the original crystals break down into smaller subgrains or "cells". With increasing deformation the angle between the subgrain increases whereas the dislocation density in the grains stays constant hence work hardening achieves its saturation [17]. Moreover, for Ti-6Al-4V, the deformation, which produces heat, tends to change the phase of the material from α to β. The latter is known to be a softer material. In the case of extrusion at elevated temperature such a change in the grain structure as a result of deformation is observed [18].

For Ti-6Al-4V at temperatures between 1300°F - 1850°F, and at strains above 20%, the material undergoes "work softening" in which the stress decreases as the strain is increased [7]. This characteristic is not normally admitted to exist in plasticity problems (see Fig. II.1). Most experimental data is obtained for strains less than 20%; and under these conditions the work hardening phenomena are the dominant effects. However we also note that work

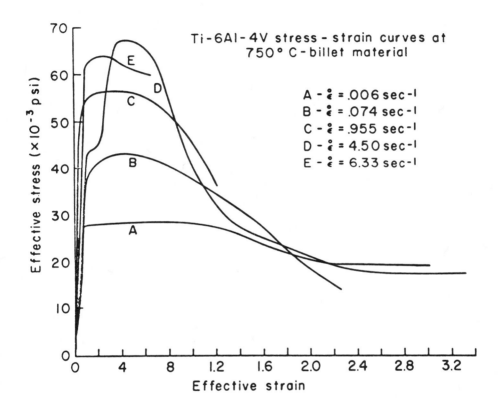

Fig. II.1 Effective stress vs. effective strain curve traced from the data obtained by Sherby and Young.

softening, when it does exist, violates the common stability postulate (Drucker's postulate) of plasticity analysis [13]. Clearly the concept of stability needs to be modified when large deformations and temperature changes are encountered. From experimental results, we know that this metal is <u>very temperature sensitive</u>, hence we assume that this decrease is mainly due to the heat produced during large plastic deformation. We proceed to present

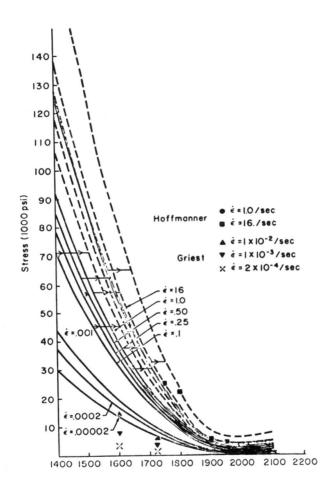

Fig. II.2 Experimental data on stress vs. temperature.

details on constructing the constitutive equation which has this character. Based on the works of Perzyna and Wierzbiecki [10-11], the following form of Φ has been successfully used for the construction of constitutive equation suitable to particular process and material,

$$\Phi(\mathfrak{F}) = \left(\frac{II_{\underset{\sim}{S}}^{\frac{1}{2}}}{k(\theta)} - 1\right)^{\beta(\theta)} \tag{II.9}$$

The parameters k, ζ, and β are material characteristics during the plastic flow. Here k is the static yield stress and β is the exponent. Hence,

$$\underset{\sim}{D} = \zeta(\theta) \left(\frac{II_{\underset{\sim}{S}}^{\frac{1}{2}}}{k(\theta)} - 1\right)^{\beta(\theta)} \frac{\underset{\sim}{S}}{II_{\underset{\sim}{S}}^{\frac{1}{2}}}$$

or (II.10)

$$\underset{\sim}{S} = k(\theta) \left[1 + \left(\frac{II_{\underset{\sim}{D}}^{\frac{1}{2}}}{\zeta(\theta)}\right)^{\frac{1}{\beta}(\theta)}\right] \frac{\underset{\sim}{D}}{II_{\underset{\sim}{D}}^{\frac{1}{2}}}$$

For simple tension or compression condition, equations (II.10) reduces to

$$D_{11} = \dot{\varepsilon} = \sqrt{\frac{2}{3}} \zeta(\theta) \left(\frac{\sqrt{\frac{2}{3}} \sigma}{k(\theta)} - 1\right)^{\beta(\theta)} \tag{II.11}$$

or

$$\sigma = \sqrt{\frac{3}{2}} k(\theta) \left(1 + \frac{\sqrt{\frac{3}{2}} \dot{\varepsilon}}{\zeta(\theta)}\right)^{\frac{1}{\beta}(\theta)} \tag{II.12}$$

where

$$S_{11} = \frac{2}{3} \sigma, \quad S_{22} = -\frac{1}{3} \sigma, \quad S_{33} = -\frac{1}{3} \sigma$$

$$D_{11} = \dot{\varepsilon}, \quad D_{22} = -\frac{1}{2} \dot{\varepsilon}, \quad D_{33} = -\frac{1}{2} \dot{\varepsilon}$$

The object is now to fit either equation (II.11) or (II.12) to the experimental data, of a simple tension or compression test with a controlled strain rate and a controlled temperature, which in this case we use the data obtained by Hoffmanner and Griest [3,4] (Fig. II.2). The result is the material characteristics k, ζ and β as functions of the temperature θ shown in Figs. II.3, II.4, and II.5. They have the following form:

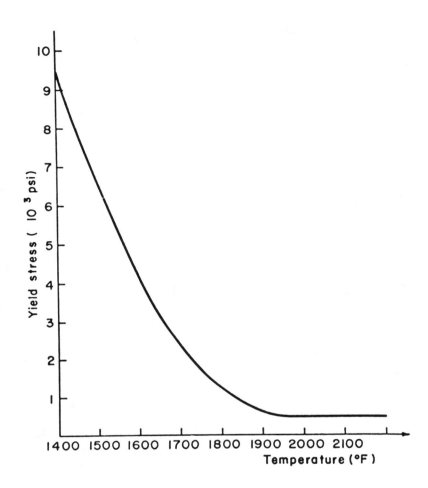

Fig. II.3 Yield stress k vs temperature, Eq. II.13.

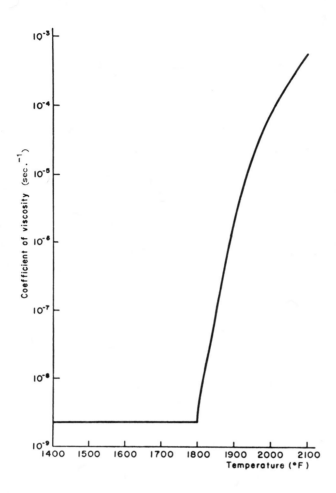

Fig. II.4 Coefficient of viscosity ζ vs. temperature Eq. II.14.

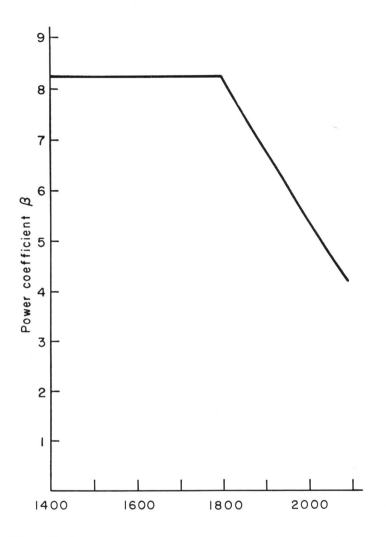

Fig. II.5 Power coefficient β vs. temperature, (°F), see Eq. II.15.

$$k(\theta) = .94 \times 10^6 + 4.28 \, (\theta - 1500)^2 + 3.852 \times 10^3 \, (\theta - 1500)$$
$$; 1300 \leq \theta \leq 1950$$

$$7.2 \times 10^4 \qquad ; \theta \geq 1950$$

(II.13)

$$\zeta(\theta) = .343485 \times 10^{-8} \qquad ; \theta \geq 1800$$
$$.343485 \times 10^{-8} + .342532 \times 10^{-5} \, (\theta - 1500)^5$$
$$; 1800 \leq \theta \leq 2100$$
$$.832365 \times 10^{-3} + .294604 \times 10^{-10} \, (\theta - 2100)^5$$
$$; 2100 \leq \theta \leq 2240$$
$$1.58535 \qquad ; \theta \leq 2240$$

(II.14)

$$\beta(\theta) = 8.2497 \qquad ; \theta \geq 1800$$
$$8.2497 - 1.37596 \times 10^{-2} \, (\theta - 1800) \qquad ; 1800 \leq \theta < 2100$$
$$4.1218 \qquad ; \theta \geq 2100$$

(II.15)

To produce a homogeneous deformation in an experiment is not simple. In the case of <u>large deformations</u> in tension, we encounter necking when the phenomena of superplasticity does not exist and bulging in the case of compression test. The stresses are found by recording the change in the cross-section area and the load during the test. A problem, which we shall also encounter in metal processing, is that the adiabatic heating effects exist in these experiments and yet they are normally ignored.

III. TENSION AND COMPRESSION PROBLEM

The governing equations for metal Ti-6Al-4V that is undergoing large and continuously plastic deformation are:

1) the constitutive equation

$$\underset{\sim}{S} = k \left\{ 1 + \left(\frac{II_{\underset{\sim}{D}}^{\frac{1}{2}}}{\zeta} \right)^{\frac{1}{\beta}} \right\} \frac{\underset{\sim}{D}}{II_{\underset{\sim}{D}}^{\frac{1}{2}}} \qquad (III.1)$$

2) the equilibrium equation

$$\nabla \underset{\sim}{T} + \rho \underset{\sim}{b} = \rho \underset{\sim}{\ddot{x}} \qquad (III.2)$$

3) the energy balance

$$\frac{1}{K} \frac{\partial K}{\partial \theta} \bar{\nabla} \theta \cdot \bar{\nabla} \theta + \bar{\nabla} \theta^2 + \frac{\gamma \dot{w}^P}{(K/L^2)} = \frac{\rho C}{(K/L^2)} \left(\frac{\partial \theta}{\partial t} + \frac{V}{L} \bar{\underset{\sim}{v}} \cdot \bar{\nabla} \theta \right) \qquad (III.3)$$

where
$$\dot{W}^P = \text{tr}(\underline{SD})$$

and K is the coefficient of conductivity, γ is the fraction of plastic work that is converted into heat ($\approx 95\%$), C is the specific heat, L is the characteristic length of the specimen, V is the characteristic speed of the process, $\nabla = \frac{\partial}{\partial x}$ is the gradient with respect to an Eulerian coordinate system $\underline{x}$, $\bar{\nabla} = \frac{\partial}{\partial \bar{x}}$ where $\bar{\underline{x}} = \frac{\underline{x}}{L}$, and $\underline{v}$ is the velocity field with dimension $\sec^{-1}$, i.e. $\bar{\underline{v}} = \underline{v}/V$. It should be noted that equations (III.2) and (III.3) are to be subjected to the appropriate boundary and initial conditions. For processes for which the following conditions are satisfied

$$\frac{\rho C L^2}{K} >> 1 \quad ; \quad \frac{\rho C V L}{K} >> 1$$

and
$$\left| \frac{1}{K} \frac{\partial K}{\partial \theta} \right| << 1 \tag{III.4}$$

equation (III.3) can be approximated by

$$\gamma \dot{W}^P = \rho C \frac{d\theta}{dt} \tag{III.5}$$

which is a locally adiabatic process [8,9].

In the temperature range mentioned in section I, the material properties of Ti-6Al-4V in equation (III.4) are nearly constant [20]. The following seem to be reasonable values to assume

$$\rho_{average} = 287 \text{ lbm/ft}^3$$

$$C_{average} = .15 \text{ btu/lb}_m {}^\circ R$$

$$K_{average} = 11 \text{ btu/hr. ft} \tag{III.6}$$

$$\left| \frac{1}{K} \frac{\partial K}{\partial \theta} \right| \leq 1.6 \times 10^{-4} << 1$$

For a specimen with 1/2" diameter and process speed $V \geq 1$ inch/sec., we find

$$\frac{\rho C L^2}{K} = 25 \text{ secs.}$$

$$\frac{\rho C V L}{K} \geq 50 \tag{III.7}$$

Hence, for high rates in the tension (or compression) test in the above temperature range, the stress-strain relation can be found by solving the following equations

$$\sigma = \sqrt{\frac{3}{2}} \, k \left(1 + \frac{\sqrt{\frac{3}{2}} \, \dot{\epsilon}}{\xi} \right)^{\frac{1}{\beta}}$$

$$\frac{d\theta}{dt} = \frac{\rho C}{J} \sigma \dot{\epsilon} \quad \text{or} \quad \theta = \theta_0 + \int_0^t \frac{\rho C}{J} \sigma \dot{\epsilon} \, dt \qquad (III.8)$$

$$\epsilon = \int_0^t \dot{\epsilon} \, d\tau \quad \text{(natural strain)}$$

where J is the Joule's constant, and k, ζ, and β are obtained through equations (II.13) up to (II.15). When we neglect the hardening, the result is shown in Fig. III.1. We can also

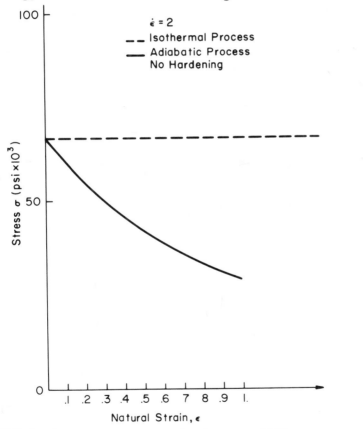

Fig. III.1 Stress versus strain for two different processes.

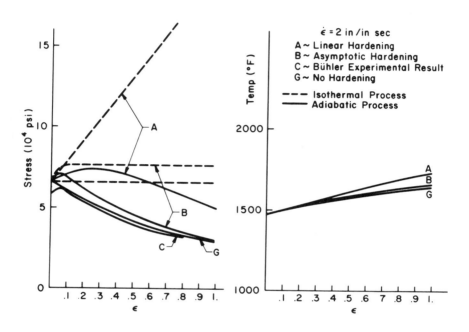

Fig.III.2^a

Fig. III.2^b

Fig.III.2^a Stress vs. strain ε for different conditions.

Fig.III.2^b Temperature vs. strain ε for several conditions.

incorporate the effect of hardening to equation (III.8), by taking

$$k(\theta,\varepsilon) = \alpha(\varepsilon) \, k(\theta)$$

In this example, the following forms of α are taken

i) $\alpha(\varepsilon) = (.63)(1 + .15(1 - e^{-34.54\varepsilon}))$ ~ asymptotic hardening
ii) $\alpha(\varepsilon) = (.63)(1 + 2\varepsilon)$ ~ linear hardening
iii) $\alpha(\varepsilon) = (.63)$ ~ no hardening

and the stress-strain curves are shown in Figs. III.2^a and III.3^a.

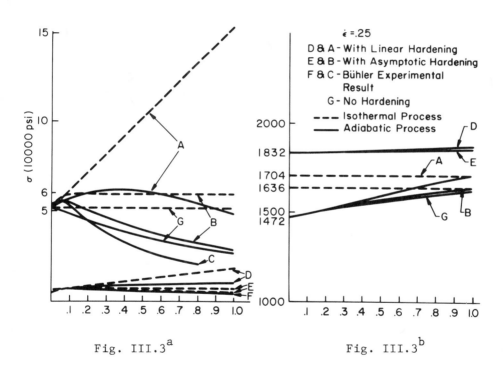

Fig. III.3^a

Fig. III.3^b

Fig. III.3^a Stress vs. strain ε for several conditions.

Fig. III.3^b Temperature vs. strain ε for several conditions.

And it is compared to the stress-strain data obtained by Bühler in his compression experiment [15]. Figures III.2^b and III.3^b show the temperature rise with respect to strain for the simple tension/compression processes with the coefficient of hardening listed in i) up to iii).

IV. DISCUSSION

As shown by Figs. III.1^a up to III.3^b the difference between the stress-strain curves, which are obtained by isothermal assumption and those that consider the temperature rise during the process are very large and by no means can the temperature effects be neglected.

For particular choices of the hardening coefficient $\alpha(\varepsilon)$, which is discussed in section III, we find that the stress-strain curves can be made to produce <u>initial</u> hardening phenomena as observed in many experiments. A constitutive equation with asymptotic hardening or no hardening (formulated in section III) can be thought of as state function model representing a process for which the dislocation density achieves its saturation value [17]. While the constitutive equation with linear hardening is model of a hypothetical metal whose dislocation density is always increasing. The latter situation would probably not exist in processes of Ti-alloys at elevated temperature and such large deformations.

When comparing our results of stress-strain curves, which is based on Hoffmanner data, with the experimental curves of Bühler, we have to use a constant of multiplication .63 in the coefficient of hardening $\alpha(\varepsilon)$. We think that such a difference is caused by the difference of the initial metal structure (or grain size). Huang and DeAngelis found in their experiment that, in Ti-6Al-4V, the grain size can either raise or decrease the yield strength by 100%. Important phenomena observed by Turbitt et.al. [21] showed that the stable condition of metal structure at elevated temperature is achieved very slowly (in term of 1-2 hrs). This means that the grain size is very much dependent on how long the specimen is in the soaking furnace.

The results of constitutive equation with asymptotic hardening and with no hardening seem to be in very good agreement with the experimental results of Bühler. We also find that in general the experimental curves are steeper than the predicted ones. This seems to indicate that other softening mechanisms such as diffusion, change of phase etc. also exists during the real process.

The main purpose of this paper is to demonstrate a temperature rise that can decrease the stress up to 50% and can raise the temperature by 200°F and therefore cannot be neglected. Such a problem is encountered in many metal forming processes such as extrusion, milling, etc. Until now, the analysis of these processes have usually been based on isothermal conditions. Further studies from both the mechanics people and the metallurgists are needed for obtaining a better constitutive equation. We also think that such a high temperature increase in a relatively short time would not be difficult to detect.

ACKNOWLEDGMENT

The authors are indebted to the Department of Defense through Project Themis under Contract F33615-69-C-1027. We are also indebted to Dr. H. Conrad, Dr. A. Adair and Mr. V. DePierre for many useful discussions.

REFERENCES

1. Young, C.M., and Sherby, O.D., "Simulation of Hot Forming Operations by Means of Torsion and Testing," Technical Report, AFML-TR-69-294, February 1970.

2. Conrad, H., "Project Themis Metal Deformation Processing," Technical Report, AFML-TR-69-284, February 1970.

3. Griest, A.J., Sabrof, A.M., and Frost, P.D., "Effect of Strain Rate and Temperature on the Compressive Flow Stresses of Three Titanium Alloys," Transaction of the A.S.M., vol. 51.

4. Hoffmanner, A.L., "Workability Testing Techniques," TRW Inc., Technical Report, AFML-TR-69-174, June 1969.

5. Wood, R.A., "Ti and Ti-Alloys DMIC Review of Recent Development", January 20, 1971.

6. Conrad, H., "Annual Report Project Themis Metal Deformation Processing," Technical Report, AFML-TR-71-18, April 1971.

7. Huang, Y., Ph.D. dissertation, Department of Metallurgy, Univ. of Ky., 1972.

8. Sulijoadikusumo, A.U., Ph.D. dissertation, Engineering Mechanics Department, University of Kentucky, 1972.

9. Dillon, O.W., Jr. "Continuum Thermo-plasticity and dislocations," Presented at The Battelle Institute Coloquim on the INELASTIC BEHAVIOUR OF SOLIDS, September 1969.

10. Perzyna, P., "The Constitutive Equations for Work Hardening and Rate Sensitive Plastic Materials," Proceeding of Vibration Problems, Warsaw 3, 4 (1963).

11. Perzyna, P., and Wierzbicki, T., "On Temperature Dependent and Strain-Rate Sensitive Plastic Materials," Bulletin De L'Academic Polonaise Des Sciences, Series des Science Techniques, Volume XII, No. 4, 1964.

12. Critescu, N., "Dynamic Plasticity, "Interscience Publishers," A Division of John Wiley & Sons, Inc., New York.

13. Hill, R., "Plasticity," Oxford Engineering Science Series.

14. Perzyna, P., "Thermodynamics of Rheological Materials with Internal Changes," Journal de Mecanique, Vol. 10, No. 3, September 1971.

15. Bühler, H., and Wagener, H.W., Bänder Bleche Rohre 6, 677(1965).

16. Brophy, J.H., Rose, R.M., and Wulff, J., "The Structure and properties of Materials," Volume II, Thermodynamic of Structure, John Wiley & Sons, Inc., New York, 1964.

17. Schoeck, G., "Theory of Creep", in "Creep and Recovery" Seminar held during the thirty eight National Metal Congress and Exposition, Cleveland, Oct. 6 to 12, 1956, sponsored by the ASM.

18. Conrad, H., "Project Themis Metal Deformation Processing," Technical Report AFML-TR-71-18, page 106-111, April 1971.

19. Tanner, R.I., and Johnson, W., "Temperature distribution in some fast metal working operations," Int. J. Mech. Sci. Pergamon Press Ltd. 1960, Vol. I., pp. 28-44.

20. Handbook of Thermophysical Properties of Solid Materials, Vol. 182, Pergamon Press, New York, 1955.

21. Turbitt, B., and Geisendorfer, R., "Effect of Elevated Temperature Exposure on the Room Temperature Properties of Ti-Alloys & Al-1M_o-1V, 6Al-4V, and 4Al-3M_o-1V," in Jaffee and Promisel (ed.), "The Science Technology and Application of Titanium," Pergamon Press.

THERMAL INSTABILITY STRAIN IN DYNAMIC PLASTIC DEFORMATION

R. S. Culver, Associate Professor
Basic Engineering Department
Colorado School of Mines
Golden, Colorado

THEORY

The Theoretical Thermal Instability Strain

When a metal is deformed slowly, geometrical and material effects combine to limit the plastic deformation. It can be shown that for a given deformation geometry the limiting, or necking, strain will be a function of the workhardening exponent, n,* in the constituitive relation.

$$\sigma = B\varepsilon^n \quad (1)$$

At high strain rates, thermal softening counteracts the stability resulting from workhardening and can cause a material instability, independent of the deformation geometry. Recht [1] showed that in orthogonal machining, once this instability condition is attained, extreme localization of the deformation occurs. Lemaire and Backofen [2] showed that this localization is accompanied by a discontinuous drop in the applied force. Because the localization occurs under conditions of pure shear, it can be concentrated on a single shear plane, resulting in far greater localization than is obtained in the conventional necking instability in tension.

Starting with the mechanical equation of state, which was originally formalized by Baron [3]

$$\left(\frac{d\sigma}{d\varepsilon}\right)_a = \left(\frac{d\sigma}{d\varepsilon}\right)_T + \left(\frac{\partial\sigma}{\partial T}\right)_\varepsilon \frac{dT}{d\varepsilon} + \left(\frac{\partial\sigma}{\partial\dot{\varepsilon}}\right)_\varepsilon \frac{d\dot{\varepsilon}}{d\varepsilon} \quad (2)$$

---

*Definition of symbols is given in Appendix 1.

Recht proposed that "catastrophic shear" would occur when the slope of the dynamic stress-strain curve, $(d\sigma/d\varepsilon)_a$ is equal to zero. Assuming the effect of strain rate to be negligible compared to thermal softening up to the point of instability, the condition for instability becomes

$$(d\sigma/d\varepsilon)_T = - (\partial\sigma/\partial T)_\varepsilon \, dT/d\varepsilon \tag{3}$$

In machining, deformation is extremely localized and it is not reasonable to expect the conditions in the deformation zone to be adiabatic. Accordingly, Recht used a model for the temperature profile based upon heat input on a single shear plane. For other metal forming operations such as forging, it is not unreasonable to assume an adiabatic temperature rise in the deforming region; the following analysis is based on this assumption.

To evaluate (3), the isothermal and adiabatic dynamic stress-strain relations will be represented by companion equations of the form shown in (1) with independent constants, B_T, n_T, and B_d, n_d. Thus,

$$(d\sigma/d\varepsilon)_T = n_T B_T \varepsilon^{n_T}/\varepsilon \tag{4}$$

and, from [3]

$$\frac{dT}{d\varepsilon} = \left(\frac{0.9}{\rho c J}\right) B_d \varepsilon^{n_d} \tag{5}$$

Substituting (4) and (5) in (3) we obtain

$$n_T \frac{B_T \varepsilon^{n_T}}{\varepsilon} = - \left(\frac{\partial\sigma}{\partial T}\right) \left(\frac{0.9}{\rho c J}\right) B_d \varepsilon^{n_d} \tag{6}$$

It is possible to simplify (6) by noting that $B_T \varepsilon^{n_T} = \sigma_T$ and $B_d \varepsilon^{n_d} = \sigma_d$. Therefore, the critical strain for thermal instability is

$$\varepsilon_i = \frac{n_T \rho c J}{0.9 |\partial\sigma/\partial T|} \frac{\sigma_T}{\sigma_d} \tag{7}$$

in which the minus sign has been eliminated because in almost all cases $\partial\sigma/\partial T$ is negative.

It has been shown experimentally [5,8] that $n_T = n_s$, the strain-hardening exponent obtained from isothermal static deformation. To calculate σ_T, the value of B_T can be computed by adding the flow stress lost by thermal softening at a strain of 100% to the dynamic strength coefficient, B_d.

For most materials the correction resulting from including the stress ratio σ_T/σ_d is less than 15% and can be ignored for a first estimate of ε_i. Values of the theoretical thermal instability strain without the stress ratio, σ_T/σ_d, are listed in Table 1 for several materials, using material property data from the open literature.

TABLE 1 - THERMAL INSTABILITY STRAIN

	Density lb/ft^3	sp. ht. Btu/°F/lb	B ksi	n	$-\partial\sigma/\partial T$ psi/°F	α in^2/sec	ε_i
Al 1100-0	169	.21	26	.32	15	.133	4.5
Al 6061-T6	169	.23	60	.075	40	.095	0.43
Copper	558	.09	75	.38	18	1.02	6.4
Armco iron	490	.12	73	.21	55	.024	1.3
1020 steel	490	.12	75	.28	51	.020	1.9
4130 steel	490	.12	120	.35	100	.013	1.2
304 S/S	501	.12	225	.51	89	.006	2.1
α-Ti(ATL-55)	283	.13	129	.17	115	.031	0.32
Ti 6Al-4V (heat treated)	276	.12	170	.08	100	.024	0.16

The instability strains given above are obviously approximate, particularly for those values above 2.0, since n and $\partial\sigma/\partial T$ were obtained for strains very much less than ε_i. The potential value of the thermal instability strain is its ability to identify the metals, such as the titanium alloys, which have low values for ε_i, for which thermal instability may be a major factor in the failure in high strain-rate forming processes. To support this argument, it should be noted that several investigators have found that titanium and its alloys have reduced ductility when deformed at high strain rates, [5,6].

Localized Plastic Strain - Conditions for Adiabaticity

When the plastic deformation is localized or the strain rate is low, the rise in temperature will be less than that given by (5) and the strain at which thermal instability would occur will be greater than that given by (7). To assess the conditions for adiabaticity, a one-dimensional model for uniform heat input, corresponding to uniform strain distribution, in a region ranging from -L < x<+L, was obtained from Carslaw and Jaeger [7].

$$\Delta T = \Delta T_a \left[1 - 2i^2\mathrm{erfc}\frac{(L+x)}{2\sqrt{\alpha t}} - 2i^2\mathrm{erfc}\frac{(L-x)}{2\sqrt{\alpha t}}\right] \quad (8)$$

This relation gives the temperature rise, ΔT, on a plane at position x as a function of the temperature rise which would be obtained under perfectly adiabatic conditions, ΔT_a. The temperature profiles in the deforming region will be controlled by the term $L/\sqrt{\alpha t}$, which relates the deformation time, t, material property, α, and the size of the deforming region, 2L. The square of this term has been defined as the thermal number

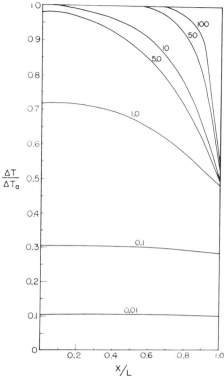

Fig. 1 - Temperature Profiles vs. R_t for uniform strain

$R_t = L^2/\alpha t$. In Fig. 1 are plotted the temperature profiles in a region of uniform deformation as a function of R_t. It can be seen that by the time R_t reaches 10, adiabatic conditions are achieved in the center of the deforming region. For $R_t = 50$, over half of the deforming region is adiabatic.

In the case of extremely localized strain, such as occurs in the machining or shearing processes, Recht's assumption of heat input on a plane is more realistic than the uniform heat input described by (8). In this case, it can be shown that the temperature increase on the shear plane, ΔT_s, is given by

$$\Delta T_s/\Delta T_a = R_t/\pi^{\frac{1}{2}} \quad (9)$$

EXPERIMENTAL INVESTIGATION OF THERMAL INSTABILITY

The Torsional Impact Test

To investigate the thermal instability condition, it is necessary to use a test configuration in which geometrical effects do not obscure the material behavior. For instance, thermal instability would only occur in the tensile test if $\varepsilon_i < n$, ie.

$$0.9\left|\frac{\partial \sigma}{\partial T}\right| > \rho c J \quad (10)$$

Baron [3] obtained this condition in a low-temperature test on steel for which $\partial\sigma/\partial T$ is large, and c is very small. Furthermore, due to the high yield strength at low temperatures, n is small. At normal forming temperatures, metals of practical interest will have an instability strain well above the necking strain, n.

In the torsional impact test, essentially pure shear deformation is obtained to very high strains. The simple stress and strain states make it convenient to investigate material behavior in this test.

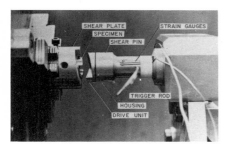

Fig. 2 - Torsional Impact Apparatus

Fig. 3 - Torsional Impact Specimen

Two torsional impact test machines have been built. The first was used to obtain detailed information on the temperature distribution in adiabatic deformation [8]. The results substantiated the conditions predicted by (5) and (8). The second was built specifically to investigate the thermal instability strain. Details of the two machines are given elsewhere [9]. In both cases, a flywheel energy source was used which was connected to a ring-shaped test section via a dog clutch. The energy source was disconnected by shearing a shear pin, thus giving independent control of strain and strain rate. Figs. 2 and 3 show the second apparatus and its specimen. The test section has a gage length of 0.2 in., O.D. = 0.75 in., I.D. = 0.69 in.

Due to the knowledge gained in the first test series, it was found that a continuous record of shear stress, plus a terminal measurement of strain distribution, would be sufficient to evaluate thermal instability. In operation, the shear plate was adjusted to terminate the test immediately after fracture, in order to protect the fractured surfaces.

Test Results

Tests were run on three materials having widely different mechanical and thermal properties: annealed mild steel, 6061-T6 aluminum, and commercially pure titanium (ATL-55). Tests were run at strain rates varying from 74 sec^{-1} to 320 sec^{-1}. A summary of the test data is given in Table 2. Plotted in Figs. 4, 5 and 6 are typical stress-time and terminal strain profiles for the three materials, as well as photomacrographs of the specimen surfaces. The terminal strain profiles were obtained by measuring the strain angles on the photomacrographs at evenly-spaced intervals along the gage length. The profiles thus obtained were averaged to give the curves in Figs. 4b, 5b, 6b.

The following results can be deduced from Figs. 4, 5, and 6.

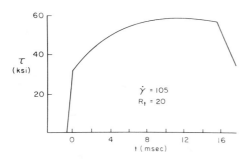

Fig.4a - Mild Steel Stress-Time Curve

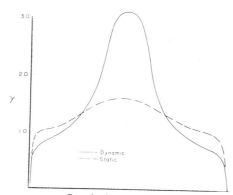

Fig.4b - Mild Steel Final Strain Distribution

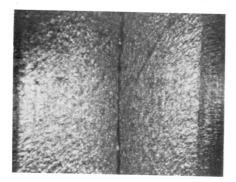

Fig.4c - Mild Steel Photomacrograph of Deformed Specimen

1. For all three materials strain localization occurred at a strain near the static fracture strain - for 6061-T6 it was slightly greater than the static value, while for mild steel and titanium it was slightly less.

2. Localization in mild steel and 6061-T6 initiated at the maximum torque, which should be the instability point since shear stress is proportional to torque. It is not clear whether a torque maximum was reached in the titanium tests.

3. Once localization began, the behavior of the three metals was quite different. The mild steel slowly localized with increasing strain and decreasing stress. Using (8) and Fig. 1 it can be shown that this localization could be attributed almost entirely to the temperature profile in the specimen. These results were confirmed by similar results on mild steel in the first series of tests in which a continuous record of strain distribution was obtained.

The 6061-T6 underwent a radical change in deformation, with very uniform deformation occurring across a narrow band of the test section up to strains 5 times as great as in the rest of the test section.

The titanium failed immediately with relatively little localized deformation in the region of the fracture. Some of the localized strain in the region of the fracture probably resulted from the rubbing of the fractured surfaces together after failure.

THERMAL INSTABILITY IN PLASTIC DEFORMATION 525

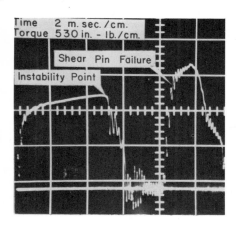

Fig. 5a - Titanium Stress Time Curve

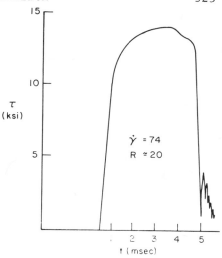

Fig. 6a - Aluminum Stress Time Curve

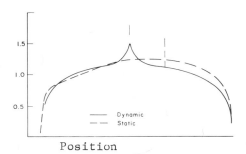

Fig. 5b - Titanium Final Strain Distribution

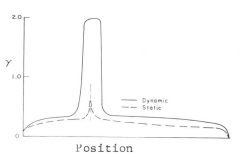

Fig. 6b - Aluminum Final Strain Distribution

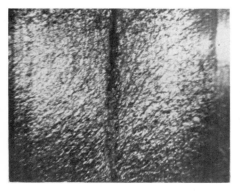

Fig. 5c - Titanium Photomacrograph of Deformed Specimen

Fig. 6c - Aluminum Photomacrograph of Deformed Specimen

TABLE 2 - TORSIONAL IMPACT TEST DATA

TEST No.	τ_y (ksi)	$\tau_{ult.}$ (ksi)	$\gamma_{inst.}$	$\gamma_{frac.}$	$\dot{\gamma}_{Sec}^{-1}$	R_t*
T1 Titanium	20.8	27.5	1.20	1.75	105	13
T2 "	20.6	29.0	1.00	1.0	320	52
T3 "	19.5	28.5	1.05	1.60	180	79
T4 "	20.0	28.8	1.20	1.40	180	64
TS2 "	18.1	25.5	1.25	1.25	Static	--
S2 Mild Steel	17.0	29.1	1.1	2.5	105	50
S3 " "	--	--	1.2	3.0	180	78
S4 " "	--	--	1.3	1.8	180	70
SS2 " "	15.7	27.0	1.65	1.65	Static	--
A1 Aluminum	11.1	14.1	.40	1.75	74	20
A2 "	11.2	14.0	.37	1.9	74	20
AS2 "	11.0	14.7	.30	.46	Static	--

*At onset of instability

This could occur during the period between the instability point and engagement of the shear pin, as illustrated in Fig. 6a.

Interpretation of Results

To assess the theoretical instability strain, it was necessary to obtain precise values of n_t, n_d, B_T, B_d, and $\partial\sigma/\partial T$ for the materials tested. (Textbook values were used for ρ and C.) A plot of the static stress-strain curve on log-log paper yielded n_s, which was taken as being equal to n_T. Values of n_d and B_d were obtained by cross-plotting stress-time data against a calculated strain-time relation based on a constant strain rate up to the point of maximum torque. In all cases, the calculated maximum uniform strain, obtained from the stress-time curve and known rate-of-twist, correlated closely with the maximum residual uniform strain estimated from the strain profiles. B_T was obtained by adding $(\partial\sigma/\partial T)\Delta T_a$, calculated for a strain of 100%, to B_d. The values of B_s, B_d, B_T, n_s, n_d, thus obtained, are given in Table 3.

TABLE 3 - EXPERIMENTAL MATERIAL CONSTANTS

	STATIC		DYNAMIC			$-\dfrac{\partial\sigma}{\partial T}$
	B_s	n_s	B_d	n_d	B_T	
Titanium	24.4	.17	28.8	.085	32.9	68
Mild Steel	25.2	.17	28.7	.13	32.4	51
Aluminum	15.9	.10	15.2	.09	17.6	40

THERMAL INSTABILITY IN PLASTIC DEFORMATION

It was found that by using values of $\partial\sigma/\partial T$ for mild steel and 6061-T6 from the open literature (see Table 1), it is possible to show that the entire decrease in the strain hardening exponent from n_s to n_d could be attributed to thermal softening. This effect was also found in the earlier test series and has been reported by other investigators. A check on the textbook value for $\partial\sigma/\partial T$ for the titanium indicated that it was too large. It would result in a negative value for n_d. Therefore, the analytical procedure was reversed and a calculated value of $\partial\sigma/\partial T$ for the titanium was obtained based upon the experimental values of n_s and n_d. The three experimentally confirmed values of $\partial\sigma/\partial T$, thus obtained are given in Table 3.

Because most material property data is presented in terms of uniaxial tensile data, the thermal instability strain was derived in terms of σ and ε. To compare the theoretical value of ε_i, with the experimental data, it is necessary to convert it to an equivalent γ_i. It can be shown that the workhardening exponent, n, will be the same for both shear and longitudinal strains, while $\tau \approx \sigma/2$, depending upon the relationship between principal stresses used. However, it has been found that the relationship, $\gamma = 2\varepsilon$, used in small strain theory, ignores second order terms which become significant at longitudinal strains above 50%. Referring to the inset in Fig. 7, $\varepsilon = \ln L_f/L_o$, while $\gamma = u/l$. Thus, from geometry, the functional relationship between γ and ε is given by

$$\varepsilon = \ln \sqrt{1 + \gamma + \gamma^2/2} \tag{11}$$

while the conventional engineering strain $\varepsilon' = (L_f-L_o)/L_o$ is related to γ by

$$\varepsilon' = \sqrt{1 + \gamma + \gamma^2/2} - 1 \tag{12}$$

These relations are plotted in Fig. 7.

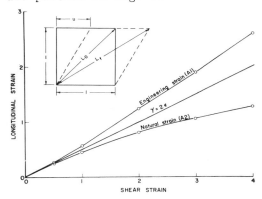

Fig. 7 - Relationship between longitudinal strain and shear strain

To obtain the thermal instability shear strain, ε_i was calculated, disregarding the term σ_T/σ_d. Using (11), an equivalent γ_i was then calculated. The correction for the stress ratio, which in this case is more correctly τ_r/τ_d since B and n were obtained from shear data, was then multiplied times the calculated value of γ_i.

The calculated values for γ_i, uncorrected and corrected, as well as the static and dynamic fracture strains and the experimental instability strain defined as the shear strain at maximum torque - are given in Table 4. Referring first to the correction for the stress ratio, it can be seen that the correction causes a 10% and 15% increase in the values of γ_i for mild steel and titanium respectively, while it increases the value for 6061-T6 by 7%.

It can be seen, from Table 4, that the theoretical and experimental values for the thermal instability strain agree quite closely for titanium. Furthermore, the catastrophic nature of the failure and the corresponding rapid drop in shear stress closely resembles the results obtained by Lemaire and Backofen [2].

TABLE 4 - INSTABILITY AND FRACTURE STRAINS

| | theory | | experimental | | |
		(corrected)	(dynamic)		(static)
	γ_i	γ_i	γ_i	γ_f	γ_f
Titanium	1.20	1.40	1.15	1.5	1.25
Mild Steel	3.50	4.00	1.20	3.0	1.60
Aluminum	1.35	1.45	.35	1.8	.30

For mild steel and 6061-T6, the results are not so clear-cut. Both materials went through a torque maximum as predicted, but at a strain well below the theoretical instability strain. Rather than the extreme localization leading directly to fracture found with titanium, a wider localized zone was obtained in which continued strain was obtained with continually falling shear stress up to strains approximately equal to the theoretical instability strains, at which point failure occurred. At this time it is not possible to state that failure resulted at these very large strains as a result of thermal instability, although it is possible that within the diffuse neck a secondary instability, similar to the thermal instability previously described, lead to failure.

Discussion of Results

From the results presented, it can be tentatively concluded that the theoretical prediction of the instability strain is valid for materials with a low instability strain, i.e. in those cases where the instability strain is approximately equal to the conventional fracture strain. For those materials for which the thermal instability strain is well above the static fracture strain, the maximum in the torque curve does not cause catastrophic failure, but rather the initiation of a diffuse neck. Further tests on other materials with high and low values for γ_i will be made to clarify these points.

In its present form, it would appear that the thermal instability strain relation could be used to predict whether or not a given metal is likely to suffer thermally induced catastrophic failure when being subjected to large-amplitude plastic deformation.

In conclusion, it should be noted that no claim is made that the thermal instability strain relation (7) is capable of accurately predicting the strain at which a thermal instability would occur. This is particularly true for those materials such as mild steel for which γ_i is over 300%, since n and possibly $\partial\sigma/\partial T$ will, in general, not be constant out to such large values of strain. However, for materials such as the titanium, in which the implications of the predicted thermal instability strain become more important, the material properties used in the analysis were obtained for strains equivalent to γ_i. Under the best circumstances, the theoretical value of γ_i would only be good to about $\pm$ 10%, so it is questionable whether the inclusion of the stress-ratio correction term is justified.

APPENDIX 1 - Nomenclature

B	strength coefficient	ΔT_s	temperature rise on a single plane
c	specific heat		
J	mech. equivalent of heat	u	displacement in x direction
L	half width of plastic deformation zone	W_p	plastic work
		x	distance
m	mass	α	thermal diffusivity
n	workhardening exponent	γ	shear strain
R_t	thermal number, $L^2/\alpha t$		
t	time	ϵ	long. natural strain
T	temperature	$\acute{\epsilon}$	long. engineering strain
ΔT	temperature rise	σ	long. true stress
ΔT_a	adiabatic temperature rise	τ	shear stress
subscript $_s$	static deformation	subscript $_d$	dynamic deformation

REFERENCES

1. Recht, R. F., "Catastrophic Thermoplastic Shear," Journal of Applied Mechanics, Trans. ASME, paper No. 63-WA-67.

2. Lemaire, J. C. and Backofen, W. A., "Adiabatic Instability in the Orthogonal Cutting of Steel," Met. Trans., Vol. 3, No. 2, 1972.

3. Baron, H. G., Journal of the Iron and Steel Institute, Vol. 182, No. 1, 1956, p. 354.

4. Manjoine, M. J., "Influence of Rate of Strain and Temperature on Yield Stresses of Mild Steel," J.A.M., Dec., 1944, p. A-211.

5. Orava, R. N., "The Effect of Dynamic Strain Rates on Room-Temperature Ductility," 1st Int. Conf. of Center for High Energy Forming, Estes Park, June, 1967.

6. Kumar, A., et.al., "Survey of Strain Rate Effects," 1st Int. Conf. of Center for High Energy Forming, Estes Park, June, 1967.

7. Carslaw, H. S., Jaeger, J. C., Conduction of Heat in Solids, Clarenden Press, 1947.

8. Culver, R. S., "Adiabatic Heating Effects in Dynamic Deformation" 3rd Int. Conf. of Center for High Energy Forming, Vail, July, 1971.

9. Culver, R. S., "Torsional Impact Apparatus," Experimental Mechanics, Vol. 12, No. 9, Sept., 1972.

THE PROPAGATION OF ADIABATIC SHEAR

Marvin E. Backman and Stephen A. Finnegan

Research Department
Naval Weapons Center
China Lake, California

Metallographic analysis of metals and alloys subjected to ballistic impact or explosive loading reveals thin bands that are associated with a displacement across the band. Figure 1 is a micrograph that shows examples of these bands found in SAE 4130 steel as the result of impact by a steel ball. The shear displacement is shown in this micrograph by the displacement of the surface. These bands occur in regions of general yielding and have been called <u>adiabatic shears</u> because there is evidence that these are inhomogeneous zones of shear in which plastic work heats the zone adiabatically and causes thermal softening within the zone and greatly reduces the shear resistance (Ref. 1-2).

There appear to be two distinct forms for these shear zones. Figure 2 illustrates these forms by examples of shears in SAE 4130 steel of four microstructures. Figures 2A and 2B are adiabatic shears in lamellar pearlite and Widmanstätten ferrite. The shear bands are evident as narrow bands of intense shearing distortion of the grain structure. Figures 2C and 2D are adiabatic shears in tempered martensite and lower bainite. The shear bands stand out as regions that are totally different in appearance from the surrounding material. These two kinds of bands can be seen to be distinct forms of a single phenomenon since the location and the patterns of adiabatic shears are essentially the same for a wide range of materials but the same kind of loading.

This paper is primarily concerned with the mechanism of initiation and propagation of adiabatic shear. It is also necessary to consider the dependence of the form of adiabatic shear on the properties of the material. In addition, it is essential to include

FIG. 1. Shear Bands in a Plate of SAE 4130 Steel With a Tempered Martensite Microstructure Caused by Ballistic Impact at 1240 m/sec.

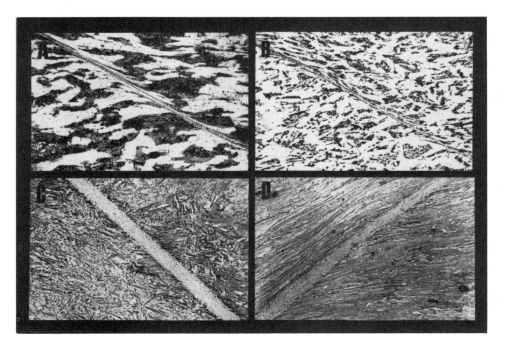

FIG. 2. Adiabatic Shears in SAE 4130 Steel of Four Microstructures: A. Lamellar Pearlite, B. Widmanstätten Ferrite, C. Tempered Martensite, D. Lower Bainite.

some results on the connection between adiabatic shear and fracture because fracture almost always accompanies adiabatic shear and certainly one of the most important aspects of adiabatic shear is its role in ultimate failure. Apparently, it is difficult to load a material so that adiabatic shears are well developed without also setting up the conditions needed for fracture; however, there are enough examples of adiabatic shear without any fracture to justify considering this process, which is strictly continuous deformation, as completely distinct from any process of fracture or rupture.

THE PROPOSED MODEL OF ADIABATIC SHEAR

Adiabatic shears initiate in regions of plastic deformation and spread outward along trajectories of consistent shape. Figure 3A is the cross-section of a 2024-T4 aluminum alloy plate that has been impacted by a steel sphere. The pattern of the adiabatic shears is shown by the pattern of fractures that have followed along the trajectories of the adiabatic shears. Figure 3B compares this to a

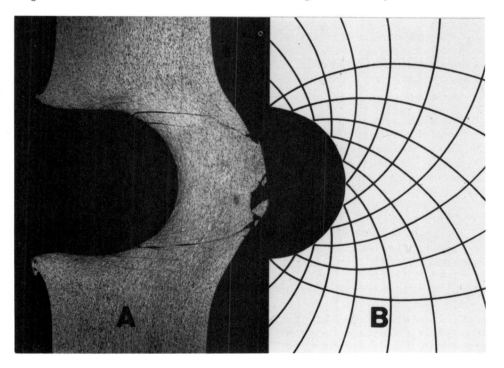

FIG. 3. Comparison of Observed Shear Patterns in Ballistic Impact With a Theoretical Slip-Line Model: A. Section of a 2024-T4 Aluminum Alloy Plate Impacted by a Rigid Sphere, B. Theoretical Pattern for Impact.

set of exponential spirals that represent the pattern of theoretical shapes for pure compressive loading around the embedded sphere. The stresses on an impacting sphere are only approximately compressive so that the agreement between the two patterns cannot be expected to be complete. There is enough similarity to show tendency for adiabatic shears to follow slip surfaces. This same pattern is repeated for many materials and over a velocity range from the onset of adiabatic shear to complete perforation, and after perforation continues to control the process of fragmentation. Adiabatic shears first develop near the locus of points on the contact surface at which the slip surfaces are in the direction of travel of the sphere. The most obvious effect of increasing the impact velocity is to increase the number of adiabatic shears on either side of the first shear and the distance that these travel away from the contact surface. The details of fragmentation due to impact have been discussed in Ref. 3.

Adiabatic shears are also formed in the explosive loading of a hollow cylinder. Figure 4A shows a fragment from an explosively loaded cylinder. Figure 4B shows theoretical slip surface patterns for compressive loading (which in this case can be expected to be much more applicable). In this particular case fractures have been initiated at the outer surface and have served as the means of initiating the adiabatic shears. The shears clearly follow the slip surface pattern and have determined the failure surfaces for fragmentation.

The examination of many fragments recovered from tests such as this have shown that adiabatic shears are initiated at flaws, pits, scratches and inhomogeneities in the material. The shear may start at the inner surface but under certain conditions the outside surface may develop tensile fractures (as in Fig. 4A in which the cylinder has been chilled to make it more brittle). The tip of the tensile fracture is the point of initiation in this case.

The common features of these observations on impact and explosive loading are that adiabatic shears have recognizable points of origin and that the shears propagate from this origin like a running crack. The thing that propagates is not a separation of the material but rather the zone of inhomogeneous deformation. It is proposed here that the zone of inhomogeneous shear travels along a slip surface as the result of the growth of a region that reverts to elastic deformation and has the slip surface as one boundary. The zone of inhomogeneous shear then develops as the consequence of the change in mode of deformation across the slip surface.

Metallographic examination of adiabatic shears often reveals significant differences in the state of strain through differences

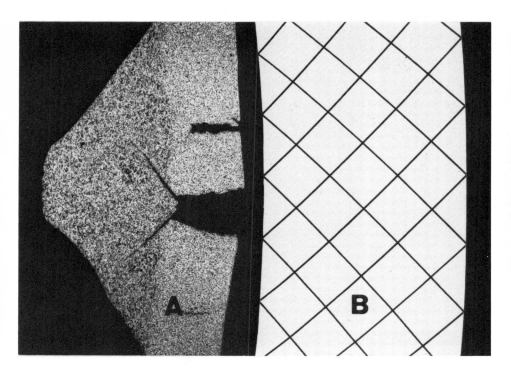

FIG. 4. Comparison of Observed Shear Patterns in an Internally, Explosively Loaded Cylinder With a Theoretical Slip-Line Model: A. Section of a SAE 1020 Steel Cylinder Fragment After Explosive Loading, B. Theoretical Slip-Line Pattern.

in the distortion of the grain texture. Figure 5 shows three examples of inhomogeneous shears induced by impact and the degrees of difference in the distortion of grain texture that can occur. These are examples in which the difference is quite obvious and therefore represent somewhat extreme conditions.

Figure 6 shows schematically the proposed structure of the adiabatic shear. The zone of inhomogeneous shear is traveling to the left into a region of general yielding. The adiabatic shear consists of the zone of inhomogeneous shear, the tip which is a region of transient conditions, and regimes of steady elastic and plastic conditions on either side of the inhomogeneous shear zone.

The tip of the adiabatic shear is the region in which the material along the slip surface makes the transition from the state of general yielding to the stable elastic and plastic regimes. The

FIG. 5. Micrographs of Three Structures of SAE 4130 Steel: A. Lower Bainite, B. Lamellar Pearlite, C. Granular Pearlite.

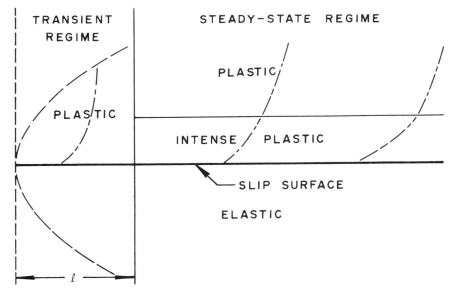

FIG. 6. A Diagram of the Zones of Elastic and Plastic Deformation in the Tip and Steady Regions.

THE PROPAGATION OF ADIABATIC SHEAR

conditions that can exist at a slip surface under quasi-static deformation processes are determined in the theory of plasticity by the properties of the characteristic surfaces of the governing equations and in this way the theory of plasticity provides an insight into the nature of the changes that have to take place at the tip.

The governing equations of plasticity are hyperbolic and therefore there exist characteristic surfaces for a given loading condition (the slip surfaces) across which discontinuities are permissible in certain components of stress, strain, and particle velocity and their derivatives. These include the tangential component of velocity and the normal components of stress that are parallel to the slip surface. The shear stress on the surface has to be continuous. References 2 and 4 describe a process by which continuity of stress can be achieved across the slip surface through an inhomogeneous shear that produces thermal softening due to the work done by the inhomogeneous shear. When the adiabatic shear is viewed as an analog of a running crack, thermal softening must be assumed to occur in the tip of the advancing shear and thus sets up the steady conditions that are required by the equations of plasticity.

In Fig. 6 the shear zone is viewed in a plane that is normal to the slip surface so that the slip surface is represented by a horizontal line. Below the slip surface the material is in an elastic state while above the surface the material is in a plastic state. There will be a difference in flow rate $\dot{\varepsilon}$ along the slip surface that generates the discontinuity in the velocity component parallel to the slip surface that produces inhomogeneous shearing deformation. The inhomogeneous shear is propagated into the surrounding material at the rate for plastic wave propagation in the deforming medium, $C_P = (E_T/\rho)^{1/2}$ where E_T is the tangent modulus for the current state of deformation and specifically includes the effect of thermal softening due to the heat generated adiabatically by plastic work, and ρ is the density of the material. Using the main elements of Recht's calculation of the condition for catastrophic shear the modulus becomes

$$E_T = E - k\int_0^\eta \tau d\eta$$

where E is the original modulus, τ is the current shear stress, η is the current shear and k is a constant that combines conversion constants and an assumed constant rate of thermal softening. The inhomogeneous shear can be estimated as

$$\eta = |v|/C_p = \dot{\varepsilon}\ell/C_p$$

where $|V|$ is the current discontinuity in the velocity component parallel to the slip surface, and $\dot{\varepsilon}$ is the deformation rate component tangential to the shear. The modulus decreases with increasing distance behind the front of the tip and at a length ℓ the modulus is zero and the condition for catastrophic shear has been achieved. This is also the point for the termination of the penetration of the shear generated by the discontinuity of velocity at the slip surface. The deformations will be qualitative as shown by the dotted lines in Fig. 6, tending to increase in magnitude of the shear and decrease in propagation rate with increasing distance behind the tip so that the penetration of shears from the slip line dies out with increasing distance behind the tip. The limiting penetration for large deformations defines the size of the shear zone. The above estimate of the rate of shear penetration implies that the distance at which a given penetration of shear has occurred will decrease with increased strain rate in the process of general yielding. The larger the strain rate the shorter the distance ℓ to achieve the limiting size.

This model assumes that at the front of the tip the stress conditions are such that there is a change across the slip surface from plastic to elastic state. This can happen if the discontinuity of stresses that exist in the steady regimes behind the tip propagates forward into the tip without too great a distortion of the discontinuity. One hypothesis is that this will happen if the zone achieves some critical size within a critical tip length.

The width of shear zones has been measured for various materials and conditions of loading. The zones are found to vary from 3μ to 10μ in width and the width appears to depend more on the type of material than on the loading conditions (e.g., the impact velocity). Thicknesses are essentially the same for impact velocities from 0.5 to 3.0 km/sec for a given material.

This model implies that the conditions necessary for the initiation of adiabatic shear are a minimum plastic strain rate and a local state of stress that produces an elastic regime of the order of the tip length and then generates a velocity discontinuity across a slip surface sufficient to generate the inhomogeneous shear. These requirements make it possible to account for some of the more obvious features of adiabatic shearing under simple forms of impact and explosive loading.

Under explosive loading the distribution of strain rate, stress and particle velocity are cylindrically symmetric everywhere and have sufficient intensity to induce adiabatic shears. Consequently, these form at notches as in Fig. 4, or at pits or scratches as shown in Fig. 7, or at any inhomogeneity or flaw that is capable of intensifying local stress and strain conditions. In the explosive

THE PROPAGATION OF ADIABATIC SHEAR

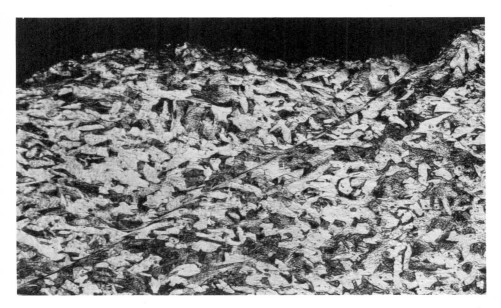

FIG. 7. Micrograph of an Adiabatic Shear in a Fragment Recovered From an Explosively Loaded SAE 1020 Steel Cylinder in Which the Shear Formed at a Pit on the Surface.

loading of a hollow cylinder the shear surfaces are all identically oriented with respect to the direction of rigid body motion that occurs in any region that has reverted to elastic strain.

For the impact of a rigid sphere against a plate the rigid body motion of the region of the plate in an elastic state will develop at the contact surface and have the same velocity as the sphere; hence, the discontinuity of velocity across a slip surface will depend in part on the orientation of the slip surface with respect to the direction of motion of the sphere and will vary over the spherical surface as

$$|V| = (\int_0^\ell \dot{\varepsilon} d\ell) \cos \theta$$

where $|V|$ is the velocity due to the local strain rate $\dot{\varepsilon}$ and ℓ the length of the current elastic region. If the strain rate is uniformly distributed the rigid body contribution would tend to maximize at 45 degrees, but if the strain rate varies as $\dot{\varepsilon}(\theta) = \dot{\varepsilon}_0 \cos \theta$ then the point of maximum $|V|$ is near 70 degrees. Metallographic observations of the point of initiation of the dominant adiabatic shears indicate that these form at angles from 45 degrees to 70 degrees. In many craters there are also other shear zones that are much shorter than the dominant shears and apparently have died out as contribution to the velocity discontinuity by rigid body motion has decreased for angles below 45 degrees.

The appearance of adiabatic shears indicates that there are two forms of shears: deformed zones which are zones of intense deformation of the original material, and transformed zones in which the material appears to have undergone a permanent change of structure. Adiabatic shears have the same patterns for both forms. With the model proposed here, the temperatures that are required for the phase change may not occur until the material is in the steady state regime of the adiabatic shear. The existence of the two forms is important in that it gives some evidence of the involvement of heat in the process. Craters in several steel, aluminum, copper, and titanium alloys have been examined to find other indications of high temperatures. Some evidence of recrystallization has been found in the areas immediately adjacent to an adiabatic shear within a mild steel sphere that had impacted against a steel plate. Figure 8 shows this region of recrystallization around the fracture that had followed the adiabatic shear.

It has been suggested (Ref. 1) that a phase change occurs by the martensite transformation, a diffusionless transformation that can take place in the time interval available during ballistic impact. The involvement of this transformation would limit the materials that can exhibit transformed zones. Ballistic impact tests

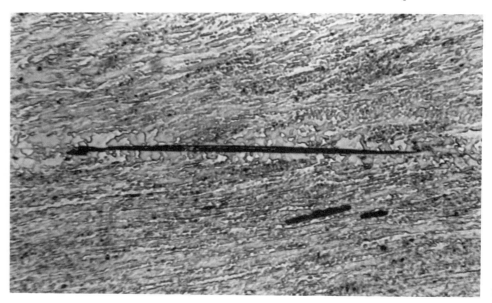

FIG. 8. Micrograph Showing Evidence of Recrystallization Around an Adiabatic Shear and Its Associated Fracture.

THE PROPAGATION OF ADIABATIC SHEAR

have been carried out on several steels, aluminum alloys, brasses, and titanium alloys with the result that only those materials that can undergo the martensite transformation produce the shear zones of the transformed type. There has been some controversy as to whether the bands of the transformed type are indeed martensite (Ref. 5). Recent work on these bands using electron microscopy has shown that these bands are martensite (Ref. 6).

Fracture is observed somewhere along the length of an adiabatic shear often enough to require some discussion of the connection between the two phenomena. The position taken here is that these are separate but interconnected effects. If adiabatic shear is viewed simply as an inhomogeneous shear, it can be seen that the deformations are extremely large and any dependence of ultimate failure on total strain would be more than satisfied if it were not for the effect of adiabatic heating. Hence, on the basis of the amount of deformation, some form of failure is expected but is suppressed by thermal effects. Furthermore, the fractures that show up in the micrographs are tensile fractures that open up and are revealed as voids in the material. The adiabatic shears occur during compressive conditions and thus the visible fractures must have formed after the compressive conditions had disappeared.

The variety of forms of fracture that accompany adiabatic shear also tends to indicate that adiabatic shear and fractures are separate phenomena. Figure 9 shows fractures that have occurred in four materials that develop deformed shear zones. These fractures are apparently a direct result of the large deformations that have occurred in the zone. These fractures are confined within the zone of intense deformation and conform to the direction of the adiabatic shear.

Figure 10 shows the fractures associated with transformed shear zones. There are many instances where the fractures are confined to the shear zone and follow its direction as is the case for the deformed zones. An example is shown in Fig. 10A. There are also fractures that cut across the shear zone diagonally, extend beyond it, or are only very near the zone but not in it. These are shown in Fig. 10B, 10C and 10D. If the shear zone is followed from the contact surface with the impacting body to its termination, the fractures often change from fractures like Fig. 10A to fractures like those of Fig. 10B. The change of phase consists of a change from austenite which is formed during application of intense pressure and temperature to martensite as the zone is quenched by the cool material adjacent to the zone. Fractures that are formed after such a change of phase are obviously only indirectly related to the formation of the shear zones.

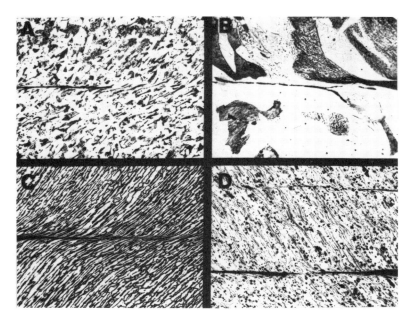

FIG. 9. Four Micrographs of Fractures Within Adiabatic Shears of the Deformed Type.

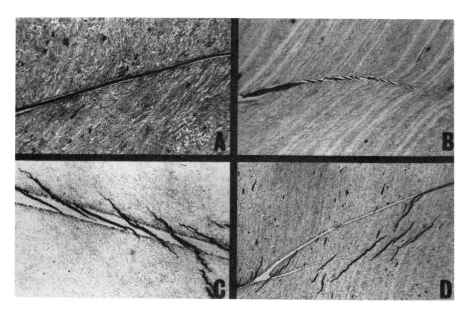

FIG. 10. Four Micrographs of Fractures Associated With Adiabatic Shears of the Transformed Type.

REFERENCES

1. C. Zener and J. H. Holloman. J APPL PHYS, 1944, Vol. 15, p. 22.

2. C. Zener. Fracturing of Metals, ASM, Cleveland, Ohio, (1948), pp. 3-31.

3. M. Backman and S. A. Finnegan. Fracture Mechanisms in Perforating Impacts, Proc., Second Canadian Congress of Applied Mechanics, 20-23 May 1969, University of Waterloo, Ontario, Canada, p. 121-2.

4. R. F. Recht. Trans, ASME, J APPL MECH, 1964, Vol. 31, p. 189.

5. S. A. Manion and T. A. C. Stock. INT J FRAC MECH, 1970, Vol. 6, No. 1, p. 106.

6. A. L. Wingrove. A Note on the Structure of Adiabatic Shear Bands in Steel, Technical Memorandum 33, Australian Defence Scientific Service, Defence Standards Lab., Department of Supply, Maribyrnong, Victoria, March 1971.

SHEAR STRENGTH OF IMPACT LOADED X-CUT QUARTZ AS INDICATED BY

ELECTRICAL RESPONSE MEASUREMENTS

R. A. Graham

Sandia Laboratories

Albuquerque, New Mexico 87115

RESUME

The present study is an experimental investigation of x-cut quartz under shock loading with particular emphasis on the determination of stress-volume states above the Hugoniot elastic limit. X-cut quartz is of particular interest since the material exhibits an unusually large Hugoniot elastic limit and previous authors[1-3] have found evidence for a substantial reduction in shear strength above the Hugoniot elastic limit; this reduction in shear strength is in sharp contrast to the behavior of metals. Because x-cut quartz is piezoelectric, it is possible to utilize electrical response measurements on impact-loaded samples to deduce mechanical properties of interest.

The present work continues previous investigations of the piezoelectric response of x-cut quartz within the elastic range[4] which led to definition of the longitudinal linear and nonlinear piezoelectric, dielectric, and elastic constants, as well as stress and electric field thresholds[5] for shock-induced conductivity. With these properties carefully defined, it appears feasible to determine the more complex behavior accompanying the inelastic behavior of x-cut quartz.

The investigation was carried out by impacting x-cut quartz disks upon x-cut quartz samples. As the mechanical disturbances produced by impacts at various velocities propagated through the samples, measurements of time-resolved current from the samples were accomplished. Experiments were performed at stresses from 26 to 90 kbar. The current-time profiles were analyzed with various electrostatic models to deduce the inelastic properties.

The current-time profiles showed increasing complexity as the impact stress was increased above 26 kbar. The changes in waveform are characterized by various relaxations in current with relaxation times which range from about 10^{-8} sec to about 10^{-5} sec. The relaxations indicate that either inelastic behavior or shock-induced conductivity has been initiated in the sample. To determine the inelastic effects, electrical relaxations must be identified. The electrical relaxations (i.e. conductivity) were identified by recognizing that shock-induced conductivity in quartz is controlled by the magnitude and sign of the electric field. Experiments at various sample thicknesses act to change the magnitude of the electric field thereby changing the conductivity. On the other hand, mechanical response is not changed by changing the sample thickness.

Detailed analysis of the data leads to the following conclusions: 1. the Hugoniot elastic limit is 60 kbar, 2. above 60 kbar, quartz is found to exhibit a substantial reduction in shear strength, 3. from 40 to 60 kbar, stress relaxation is observed, 4. shock-induced conductivity acts to change the current-time profiles if electric fields of 1.5×10^6 volt cm^{-1} are exceeded.

1. F. W. Neilson and W. B. Benedick, Bull. Am. Phys. Soc. 5, 511 (1960).
2. J. Wackerle, J. Appl. Phys. 33, 922 (1962).
3. R. Fowles, J. Geophys. Res. 72, 5729 (1967).
4. R. A. Graham, Phys. Rev. B6, 4779 (1972).
5. R. A. Graham and G. E. Ingram, J. Appl. Phys. 43, 826 (1972).

METALLURGICAL EFFECTS AT HIGH STRAIN RATES IN THE SECONDARY SHEAR ZONE OF THE MACHINING OPERATION

P.K. WRIGHT

Department of Scientific & Industrial Research
Auckland, New Zealand

The paper analyses the strain rates and conditions of deformation that occur during semi-orthogonal metal machining. Emphasis is placed upon the area termed the secondary shear zone or flow zone. It is demonstrated that when machining at high rates of metal removal conditions of 'seizure' occur at the chip-tool interface to cause a high strain rate deformation within the chip flow zone. These processes promote considerable adiabatic heat generation with consequent metallurgical changes in the chip structure and a reduction in the stress acting along the interface.

INTRODUCTION

Figure 1a illustrates the two-dimensional model of the cutting mechanism for the production of a continuous chip without a built-up-edge. This model represents a vertical section through the tool cutting edge and chip in conventional turning. The undeformed material of thickness t_1 is transformed into the chip of thickness t_2 in a shear process along the primary shear plane AB, the latter being inclined at the shear plane angle $\emptyset$ to the cutting direction. The chip shape is not controlled by the tool, unlike other working operations such as extrusion where the die profile determines the shape of the product, and the shear plane angle can vary according to material being machined and rate of metal removal. However, since the shear angle is a measure of the degree of plastic deformation and therefore controls the cutting forces in order to understand the cutting process more fully the prediction of the shear plane angle has assumed major importance. Merchant[1] proposed that the shear plane would adopt such an angle so that the minimum amount of energy is consumed during chip formation. He

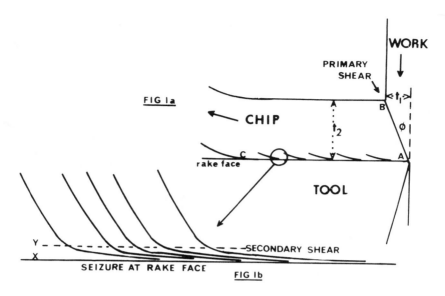

Fig.1. Two dimensional model of cutting to show chip formation.

derived the relationship $2\phi = C + a + \lambda$ where C is the machining constant, a is the rake face angle and λ is the angle of friction at the tool-chip interface proposed to describe the conditions along the contact zone (AC). The validity of this equation was tested but although reasonable agreement was obtained when machining plastics, the results from the cutting of metals were inconsistent. Many attempts have since been made to establish a more rigorous shear angle relationship and metallurgical type examinations have been made of the mechanism of primary shear, often by cinephotography[2]. However, in these analyses minimal attention has been focussed upon the mechanisms that govern the passage of the chip along the rake face of the tool. The present paper presents evidence to explain why the concept of an 'angle of friction' is unsatisfactory and proceeds to consider the deformation processes and stresses that occur at the interface.

Using a V.D.F. experimental lathe a variety of work materials was machined with M34 high-speed steel tools in a semi-orthogonal turning operation. The experiments investigated tool wear mechanisms and the mode of chip flow across the tool primarily by metallurgical sectioning of worn tools and adhering swarf. To obviate any false interpretation of the chip formation the use of a 'quick-stopping' device was essential. This consisted of a humane killer gun[3] securely mounted above the pivoted toolholder carrying the tool tip. During cutting the tool holder was supported by a shear pin. Upon firing the gun, its bolt impacted onto the toolholder, causing the shear pin to fracture and the tool to swing rapidly out of the cutting position thereby 'freezing' the forming chip in the actual cutting conditions (figure 2).

CONDITIONS AT THE CHIP-TOOL INTERFACE

The contact between two static metal surfaces is comprised of minute asperities touching at isolated points so that the real area of contact is considerably less than the apparent one. Under an applied normal and tangential load the asperities weld together and relative movement occurs by yielding in shear at the very tips of the softer material[4]. This process of adhesion and shearing gives rise to a frictional force, which is proportional to the normal force, the ratio being the coefficient of friction. In metal cutting however, the normal stresses acting on the tool-work interface are extremely high. For example, using a dynamometer to measure cutting force, it was possible to calculate that the average normal stress acting on the tool rake in figure 2 was 275 N/mm^2 (40,000 lbs/sq.in). Under the influence of these very high normal stresses the two surfaces are forced together to such an extent that the real and apparent areas of contact become equal. Relative tangential movement then occurs actually within the softer of the two materials and the concept of friction must be abandoned since the tangential stress cannot exceed the shear stress of the softer chip material, whereas the normal stress can vary according to rate of metal removal and rake angle. These have been termed conditions of 'sticking friction'[5] or 'seizure'[6], and metallographic observations on sections through cemented carbide tools and adhering swarf suggest that complete atomic bonding occurs

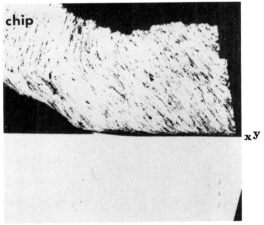

Fig.2. Section through worn tool and chip after 'quick-stopping'. Machining low carbon iron at 183 m/min (600 ft/min) and 0.5 mm/rev (0.02 ins/rev)   x 25

over much of the contact zone[6]. Such 'seized' surfaces were always observed in the present work. In the examination of unetched specimens no gaps could be observed at the interface of chip and tool: in etched specimens (e.g. figure 6) the 'seizure' process promoted tool wear by 'plucking' or deformation indicative of the strength of the bond.

Under seizure conditions the movement of the chip involves intense shear in a very narrow zone close to the tool face termed the secondary shear zone or flow zone (figures 1b, 2, 3). For example the width of the flow zone in figure 2 was measured to be 0.05 mm (0.002ins). The atoms of work material at the very interface (X) must be considered to be stationary, a very steep velocity gradient then occuring across the flow zone XY until the bulk chip speed is attained. The strain lines in this zone were often denoted by a phase of the material being machined (figure 3). Particularly at high rates of metal removal the secondary shear zone deformation was characterized by high strain rates. For example a simple calculation involving the chip dimensions and measurement from figure 2 reveals that a mean strain rate of 10^4 per second occurred across XY when machining low carbon iron at a cutting speed of 183 m/min (600 ft/min) and a feed rate of 0.50 mm/rev (0.02 ins/rev).

TEMPERATURES AND METALLURGICAL CHANGES AT THE INTERFACE

Work is continuously being done on an element of chip material as it is strained in passing through the flow zone (A to C in figure 1). Microhardness testing in the chip flow zone revealed that at comparatively low rates of metal removal considerable work-

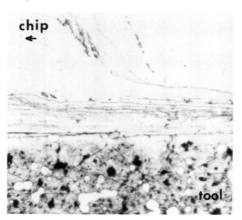

Fig.3. Flow zone when machining A.I.S.I. type 321 austenitic stainless steel at 15 m/min (50 ft/min) and 0.41 mm/rev (0.016 ins/rev) x 1300. The δ-ferrite phase denotes the strain lines across XY.

hardening takes place. However, as the rate of metal removal is
increased the amount of energy consumed in the high strain rate
deformation also causes considerable adiabatic heat generation,
which is concentrated into a very small volume of metal. The
corresponding temperature rise is therefore very high and the
secondary shear zone can constitute a continuous heat source
producing very high temperatures in the tool. In the experimental
work it was possible to develop two new methods for measuring tool
temperatures and for plotting the temperature contours that exist
in the tool. The methods[7,8], depend upon the fact that if high
speed steel is heated to temperatures of 650°C or above, upon
cooling a) its room temperature hardness is reduced and b) its
etching reaction in 2% nital varies according to the temperature
that it was previously subjected to. The temperatures at any
position in the tool that had been heated over 650°C could be
determined with an accuracy of $\pm 25°$ and a temperature contour map
drawn for the cutting condition. Figure 4 (from the tool in
figure 2) demonstrates that when machining the low carbon iron at
183 m/min (600 ft/min) and 0.5 mm/rev (0.02 ins/rev) the maximum
temperature at the rake face was approximately 1000°C. Since this
temperature was caused solely by the heat source of the chip it is
reasonable to assume that the temperatures in the actual flow zone
were also at least 1000°C at the rear of the contact length.

   The metallurgical changes that occur in the chip material of
the flow zone as a result of the heat source are complex. In
figure 5 (from figure 2) recovery, recrystallization and grain
growth processes are evident in the flow zone to produce equi-axed
grains of ferrite up to 8 microns in diameter. The structures in
figure 5 do not, however, represent those that would have existed
during actual machining. Most of the observed metallurgical
changes are more likely to have taken place after 'quick-stopping'
when the heat available in the flow zone was used to rapidly anneal
the deformed material. During machining an element of chip
material entering the flow zone at the cutting edge (A in figure 1)
is initially relatively cold but is then heated up to 1000°C as it
is strained in its passage through the zone. Assuming a linear

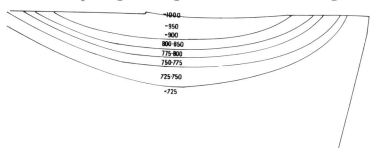

Fig.4. Temperature contour map constructed for machining low
        carbon iron as in figure 2.

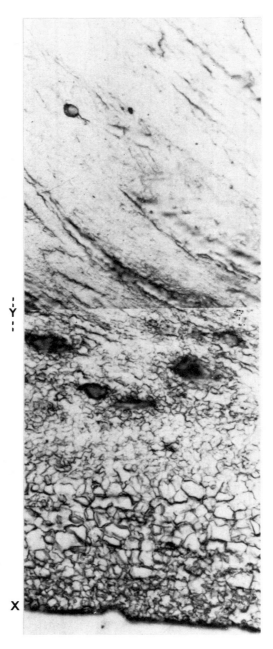

Fig.5. Rear of flow zone in figure 2   x 1700.

SECONDARY SHEAR ZONE OF THE MACHINING OPERATION

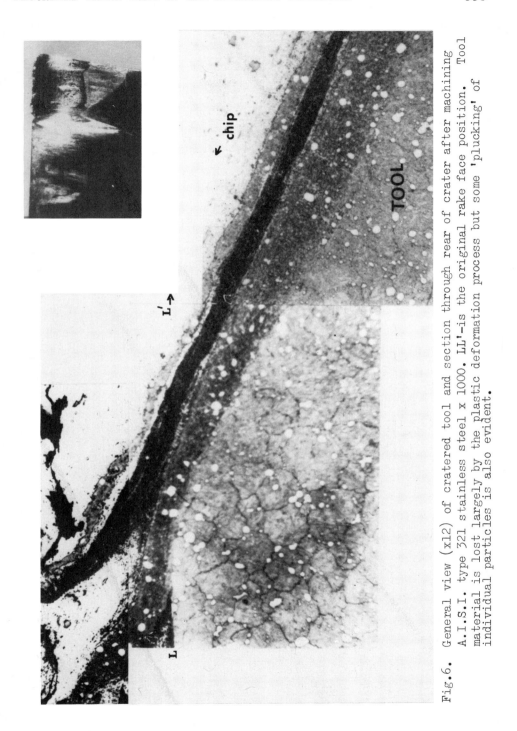

Fig.6. General view (x12) of cratered tool and section through rear of crater after machining A.I.S.I. type 321 stainless steel x 1000. LL'-is the original rake face position. Tool material is lost largely by the plastic deformation process but some 'plucking' of individual particles is also evident.

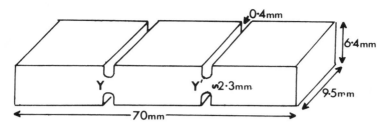

Fig. 7. Specimen for hot shear testing - M34 high speed steel.

velocity gradient across XY it can be calculated that the time taken for an element of chip material to pass through the flow zone midway between X and Y is approximately 10 milliseconds in the cutting conditions of figures 2 and 5. Due to the progressive heating mechanism the amount of time the element spends actually at $1000^{\circ}C$ is clearly much less than 10 milliseconds and therefore it is necessary to determine whether the heating cycle was sufficient to dynamically anneal the material to produce localized conditions of hot-working. This is possible by assessing the shear stress of the work material at the locality of the highest temperature region and a procedure for this follows in the next two sections.

ELEVATED TEMPERATURE YIELD STRESS IN SHEAR OF HIGH-SPEED STEEL

At a temperature of $1000^{\circ}C$ the shear strength of the high speed steel tool material is greatly reduced. Microexamination of sections through worn tools revealed that at high cutting speeds the thermally weakened tool material could be sheared away by the softer chip material to form a crater in the rake face. Figure 6 is a section through the exit end of a crater with thin tongues of the high-speed steel being sheared away by the work material to which it is welded (in this case austenitic stainless steel). This wear mechanism has been previously described[7,8] and the small lip seen in the middle of the rake face in figure 2 has also been sheared from the tool in this way. The mechanism arises because the high stresses necessary for the high strain rate deformation of the work material in secondary shear are transmitted into the thermally weakened tool material thereby deforming its surface layers at a much lower strain rate. Since the temperatures at the tool face can be found, the shear stress at the interface can be determined if the minimum flow stress of the tool steel can be measured at a low strain rate. An apparatus was thus constructed for measuring the low strain rate shear strength of the tool material at elevated temperatures. The outer portions of the test specimen (figure 7) were rigidly held in a heat resistant stainless steel jig whilst a plunger operated by a compression tester sheared the centre portion along Y and Y'. A small furnace was built around the jig to allow shear tests to be carried out at elevated temperatures, and a graph of shear yield stress versus temperature constructed for the tool material. The graph constitutes the right hand side of figure 8, and also figure 9. In examinations of partly sheared specimens the shear strain rate was calculated to be 0.16 sec^{-1} and in the following analysis it is assumed that the tool material is carried out of the crater region in a deformation of this strain rate.

ELEVATED TEMPERATURE YIELD STRESS IN SHEAR OF WORK MATERIALS

In high speed tension and shear tests other workers have demonstrated that the yield strength of metals is increased as the strain rate is increased but that this increase in strength is offset by any rise in temperature. Campbell and Ferguson[9], investigating the shear yield stress behaviour of mild steel (0.12% carbon), used the same type of specimen as shown in figure 7 heated up to various temperatures, and measured the shear yield stress in an apparatus designed to function up to very high strain rates. Their results are shown on the left hand side of figure 8 for strain rates of 10^4 to 4×10^4 sec^{-1}. The extrapolation of these curves to higher temperatures cannot be justified because the temperature interval is too great and it is not possible to estimate the influence of the ferrite/austenite transition or the influence of adiabatic heating (rather than external heating to one uniform temperature). However, for the purpose of a simple

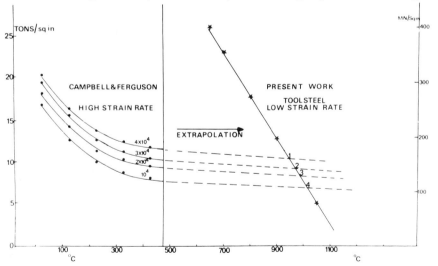

Fig. 8. Hypothetical extrapolation of results obtained[9] for high strain rate deformation of mild steel and results from hot shear testing at low strain rate of high-speed steel.

model, the general trend of the curves for high strain rate conditions suggests that they could intersect the low strain rate shear yield stress curve for the high-speed steel at temperatures between 900 and 1100°C. This model summarizes crater formation whereby the stress required to deform the material of the flow zone at a <u>high</u> strain rate is transmitted into the tool (due to 'seizure') and the latter is then deformed at a very <u>low</u> strain rate.

Figure 9 illustrates how it is possible to present actual experimental cutting data in the same terms as figure 8. Sections of worn tools used to machine the materials indicated were examined to find the conditions at which shearing of the tool surface was first detected (column A in the table). The temperature under these conditions was then determined using the techniques described[7] (Column B). From this temperature and the shear stress curve in figure 9 the shear stress acting at the interface could then be estimated. The values of stress indicate that when machining austentic stainless steel, in the conditions of the present work, the chip was not appreciably softened in the flow zone, despite any adiabatic heating, and high stresses – approximately 370 N/mm^2 (24 tons/in^2) – were required for the secondary shear zone deformation which, in turn, were transmitted into the tool to cause crater formation at a comparatively low rate

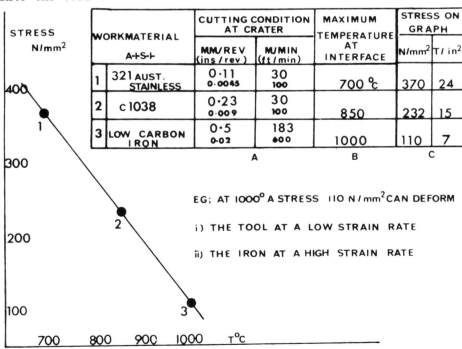

Fig.9 Actual machining data presented in same terms as Figure 8.

of metal removal. In contrast the very low figure of 110 N/mm^2 (7 tons/in^2) obtained when machining the low carbon iron indicates that at high rates of metal removal the adiabatic heat generation caused dynamic chip softening, overcoming any work hardening effect. These localized hot working conditions reduced the stress to deform the material in secondary shear, thereby causing lower stresses to act on the tool and inhibit cratering until very high rates of metal removal.

EFFECT OF SECONDARY SHEAR IN MACHINING

The mean stress on the primary shear plane is not very different from the yield stress of the work material measured in conventional tensile tests[3]. The shear plane angle and hence the cutting force, however, are dependant upon the feed force and in high speed cutting this is the force required to shear the work material in the flow zone. In this region the value of the yield stress must vary along the contact length, increasing, due to work-hardening, near the cutting edge and then decreasing rapidly as the hottest point is approached. These two interacting conditions are complex and laboratory tests to simulate them show little chance of success. Study of behaviour 'in situ' (e.g. figure 9), using actual machining operations, should provide a method of understanding the tool face force without which a comprehensive appreciation of the mechanics of cutting is impossible.

CONCLUSIONS

1. In the cutting conditions of the present work, a 'seizure' rather than a 'friction' mechanism existed at the tool-chip interface.

2. The seizure conditions caused relative chip motion to occur in a narrow band close to the tool termed the secondary shear zone or flow zone. The following conclusions were reached concerning the flow zone existing when machining a low carbon iron at 183 m/min (600 ft/min) and 0.5 mm/rev (0.02 ins/rev);-

   a) The average strain rate in the zone was 10^4 per second.

   b) This high strain rate deformation promoted considerable adiabatic heat generation and maximum temperatures of 1000°C were caused at the rake face of the tool and in the flow zone.

   c) An element of chip material passing through the contact zone was dynamically annealed resulting in lower stresses acting on the tool.

3. The soft work material deformed at a very high strain rate can transmit stresses onto the rake face of the harder tool that deform the latter at high temperatures to form a crater.

4. The complex reaction between work hardening and chip softening, that occur simultaneously as an element passes through the flow zone, governs the value of tangential stress acting along the interface which in turn effects the shear plane angle.

ACKNOWLEDGEMENT

This work was carried out at the Department of Industrial Metallurgy, University of Birmingham, England. The author wishes to sincerely thank Dr E.M. Trent who supervised the research and who has made valuable criticisms in the preparation of this paper.

REFERENCES

1. Merchant, M.E. - J. of App. Phy. 16 (1945) P.267.

2. Palmer, W.B. and Oxley P.L.B. - Proc.Inst.Mech.Engrs. 173 (1959) P.623.

3. Williams J.E. Smart, E.F. and Milner, D.R. - Metallurgia (1970) 81.

4. Bowden, F.P. and Tabor D. - Friction and Lubrication of Solids, 1954 (Oxford Univ. Press).

5. Zorev, N.N. - Conf. on Int.Res. in Prod.Eng. 1963 P.42.

6. Trent, E.M. - I.S.I. Report No. 94 Machinability P.11.

7. Wright P.K. - Ph.D. dissertation, Univ. of Birmingham 1971.

8. Wright P.K. and Trent E.M. - To be published in J.I.S.I.

9. Campbell, J.D. and Ferguson, W.G. - Phil.Mag. 1970 21, 169, P.63.

MINIATURE EXPLOSIVE BONDING WITH A PRIMARY EXPLOSIVE

J. L. Edwards, B. H. Cranston, and G. Krauss

Western Electric Company, Lisle, Ill., Western Electric Company, Princeton, N. J., and Lehigh University, Bethlehem, Pa. respectively

ABSTRACT

Lead azide, a primary explosive, was used to produce miniature gold-tanalum and copper-nickel bonds, typically 0.125" by 0.187" in area. A linear relationship between the detonation velocity and density of the explosive material was obtained experimentally, and the three types of interface that are common to secondary explosive bonding systems were produced. The system parameters, i.e., explosive and flyer plate mass, under which each type of interface formed were measured. Wave formation at the bond interface was studied by preferentially etching away the flyer plate. In addition, base plates were observed that had been abraded by the action of the jetting phenomenon that is present in the larger secondary explosive bonding systems. Mechanical properties of gold-tantalum bonds were measured by a 90° peel test and by microhardness across the bond interface. Failure occurred only in the lead of the bond and not at the bond interface. Microhardness measurements show a decrease in hardness of the cold worked tantalum, even at a substantial distance from the bond interface. Although it was possible to produce gold-tantalum bonds that had melting at the interface, no evidence of solid state diffusion across these melt zones was found. The melt region along the interface had a constant composition of approximately 20 weight percent tantalum, and did not vary from bond to bond although there was a large change in bonding parameters. No intermetallics or non equilibrium phases were detected in the melt zone by X-ray analysis, replication electron microscopy or electron microprobe techniques.

INTRODUCTION

Although secondary explosives have been used for bonding and cladding of large plates, no work has been reported outside of the Western Electric research program concerning the use of a primary explosive for miniature bonding applications. (1) Electrical connections between metal strips 1/2 to 10 mils thick and other metals in the form of thick or thin films are examples of explosive bonding application where the use of a primary explosive is necessary to reduce the scale of bonding to the requirements of the electronics industry. Primary explosives used on a small scale offer a better balance of properties than do secondary explosives because of the much lower critical mass necessary to support detonation and their capacity to be detonated by electrical, mechanical or thermal means, in contrast to secondaries that require detonator caps. Because most detonator caps are larger than the areas in thin film devices that are to be bonded, secondary explosives can not be considered for the significant reduction in bonding scale that is needed for miniature bonding application.

This report describes an investigation that evaluates the use of a primary explosive containing lead azide as a major constituent to produce miniature gold-tantalum and copper-nickel bonds. Explosive characteristics and the properties and nature of the bond interfaces were examined. In order to conserve space, an experimental procedure section has been eliminated and instead pertinent experimental techniques or comments are included as the results of the investigation are discussed.

EXPLOSIVE PROPERTIES

The formulated lead azide explosive was characterized by measurement of density, detonation velocity, and specific impulse as a function of explosive thickness.

The explosive density was measured by an indirect technique that consisted of screening the explosive charges in 5 mil steps on flyer plates* of a known mass, density and thickness. Then by measuring the explosive mass and thickness after drying and by correcting the total mass that was measured for the amount of explosive material present, it was possible to calculate the explosive density. The explosive detonation velocity was determined by measuring the travel time of an explosive train between two points with an oscilloscope. With density and detonation velocity data (for a given thickness of screened layer) it was possible to

---

*The flyer plate is the upper plate on which explosive is placed in a parallel plate explosive bonding system.

calculate[2] the explosive reaction zone pressure for a given explosive charge, as a function of the charge thickness. For a series of gold-tantalum bonds, the explosive reaction pressure varied from 900,000 to 1.3 million psi with increasing detonation velocity produced by increasing explosive thickness.

Since a fraction of the reaction zone pressure is transferred to the flyer plate in the form of an oblique pressure front, it will in turn, through a momentum change impart a velocity to the flyer plate.[2] By measuring the oblique front as a function of time for a given explosive thickness, it was possible to calculate the specific impulse of the explosive. The pressure profiles as a function of time were measured with a ballistic pressure quartz transducer that was designed to measure shock waves and explosive pressures. The peak pressure front varied from 19,000 psi for a 5.2 μsec duration with 5 mil charges to over 95,000 psi for a 3.7 μsec duration with 20 mil thick charges. These pressures produce calculated values of specific impulse that varied from 11.4×10^4 dyne-sec/gram for 5 mil charges to 12.8×10^4 dyne-sec/gram for 20 mil charges. From the specific impulse data it was possible to calculate[3] the flyer plate velocity of the parallel plate miniature bonding system. From the flyer plate velocity data the magnitude of the pressure wave created by the collision of the plates could be determined.[2] In the gold-tantalum system the pressures increased in the collision region from 1.8 to almost 19 million psi with increasing flyer plate velocity, greatly exceeding the tensile strength of the cold worked tantalum and insuring that explosive welding could occur.

The different effects of the reaction and collision pressures of a copper-nickel bond are shown in Figure 1. The almost vertical plume of explosive reaction products jetting upward from the flyer plate represents the explosive reaction zone pressure, and the horizontal expulsion of the white ionized air from between the plates is produced by the pressure wave that is created in the collision region between the plates. The copper flyer plate is 5 mils thick and the detonation was initiated by the glowing tungsten element visible in the right hand side of the photograph.

FIGURE 1: Side view of a 1/8" by 3/16" copper to nickel bond showing the ionized air being expelled from between the plates in a time elapsed photograph, 2.5X

MINIATURE BONDING

It was possible to produce the three types of interfaces found in large scale secondary explosive bonds in the gold-tantalum system. Figure 2 shows examples of straight, wavy and melt type interfaces in gold-tantalum bonds and Figure 3 shows the range of explosive conditions that are associated with each type of interface. Metallographic examination showed that in general, for the wavy type interfaces the wave height or amplitude increases as the flyer plate velocity or mass ratio increase according to Figure 3. Increases in wave amplitude were always accompanied by a decrease in frequency, i.e., the number of waves in a given length. A three dimensional view of a wavy interface in a gold-tantalum bond where the gold flyer plate has been etched away is shown in the scanning electron micrograph of Figure 4.

(a)   (b)   (c)

FIGURE 2: Examples of (a) straight, (b) wavy, and (c) melt type interface produced by primary explosive bonding of gold-tantalum sheets, 500X

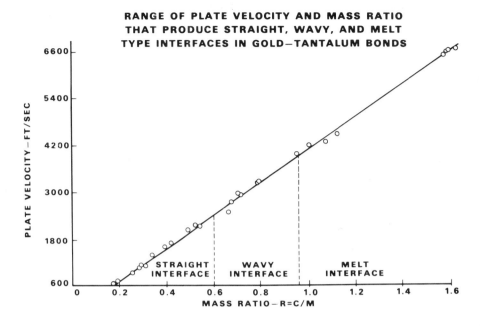

FIGURE 3

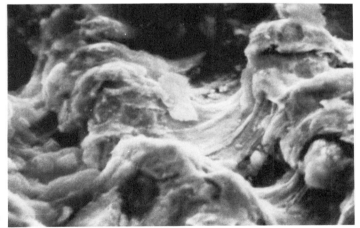

FIGURE 4: Scanning electron micrograph (45° view) of wave formations in tantalum from a gold to tantalum bond. The gold was etched away to show the wave formations in tantalum. 1200X.

In cases where the base plate extended past the flyer plate, it was possible to observe the effects of the jetting phenomenon that have been observed in secondary explosive bonding systems. The base plate acted as a witness stand similar to the experimental set-up of Bergmann, Cowan and Holtzman.(4) Figure 5 shows where the action of the jetting phenomenon has abraded the base plate adjacent to the terminal portion of the wavy interface of a gold-tantalum bond. This micrograph shows the top portion of the edge of a bond in which the gold flyer plate has been etched from the tantalum base plate to show the bond interface, as well as the lower edge of the bond where the jet abraded the base plate.

FIGURE 5: Top view of a gold to tantalum bond in which the gold has been etched away showing jet abrasion in the tantalum substrate not covered with flyer plate substrate. Scanning electron micrograph 100X.

MECHANICAL PROPERTIES OF BONDS

The mechanical properties of the gold-tantalum bonds were measured for all three types of interfaces by a 90° peel test where the load to failure was measured in grams. Failures occurred in the leads to the bonds and in no case was separation of the interface observed. Failure was due primarily to the deformation of the bond lead in the bond area. The average strength of the leads increased slightly, from 1400 to 1760 grams, as the explosive charge thickness increased. In recording the data from the peel test, it was noted that the bonds did not display an increase in strength as a function of the direction of detonation.

Microhardness was measured across gold-tantalum bonds on all three types of interfaces as well as the transition type interfaces, and was found to be independent of interface type within the sensitivity of the technique. Microhardness data is plotted in Figure 6 in increments of 25 microns from the bond interface of the initially cold worked gold and tantalum plates. The hardness of the gold is quite stable as shown in Figure 6 and its bonded value is approximately equal to the hardness value before bonding as indicated by the dashed line. On the tantalum sides of the bonds there is a distinct decrease in hardness as a result of the bonding as indicated by the dashed line. This decrease in hardness starting at a distance of 25 microns from the interface was present in all five types of interfaces.

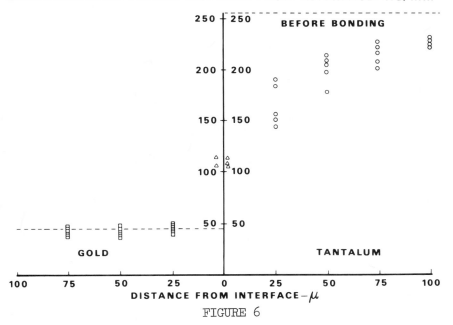

FIGURE 6

The difference in the response of the two metals to explosive bonding is quite interesting. Both the gold and tantalum were in the severely cold worked state prior to bonding and it is not likely that the pressure wave generated in the collision region increased the dislocation density of either metal substantially, as has been reported in secondary explosive bonding systems. (5-7) In view of the high dislocation density present in the cold worked metals before bonding, it is quite possible that the pressure wave was sufficient to produce in the tantalum a dynamic recovery that caused the observed microhardness decrease. Microstructural changes, typically subgrain formation, responsible for producing dynamic recovery in explosive bonds have been reported by Lucus et al., (5) Trueb, (6) and Buck and Hornbogen.(7) The mechanism of the recovery consists primarily of the cross slip of screw dislocations into subgrain configurations. In the case of the Au-Ta bonds the BCC Ta with its many slip systems would be expected to cross slip readily, while the FCC Au with its low stacking fault energy (8) and tendency to form extended dislocations would be expected to exhibit the opposite behavior, i.e., limited cross slip. The observed decrease in hardness of the Ta and the unchanged hardness of the Au are, therefore, consistent with their different expected dynamic recovery behaviors.

MELT ZONE ANALYSIS

Microprobe analysis of the melt type interfaces showed a composition of about 20 weight percent tantalum within the melt areas. The composition varied from 18.2 to 21.2 weight percent within this range from bond to bond. It should be noted that although the composition is constant from bond to bond, this means that the melt zone composition was independent of the explosive bonding parameters in which the plate velocity varied by more than 2500 ft/sec as is shown in Figure 3. X-ray scans for tantalum atoms across a melt type interface showed that the atoms are mixed uniformly within the melted region of the bond. The composition profile across these melted bond interfaces was characterized by a sharp transition at each interface as would be expected of a diffusionless type interface.

X-ray diffraction of bonds where the Au or Ta were preferentially etched away showed that the melt zone was a mixture of gold and tantalum atoms that had a very fine solidification structure with no preferred orientation. When the tantalum was etched completely from the bond, the melt zone pattern showed no evidence of tantalum lines but indicated an increase in lattice spacing for the gold pattern. This increase in lattice spacing of the gold in the melt zone, indicated that the melt zone was a substitutional solid solution of gold with tantalum atoms. Recalling from the previous section that the microhardness of the cold rolled gold was approximately 40 Kg/mm^2 as compared to the 100 Kg/mm^2 for the melt area, it was possible that this increase in hardness is in part due to solid solution hardening. Some of the hardening could also be attributed to the fine solidification structure as well as a high concentration of defects that might be expected to result from the rapid quench of the melt.

The constant chemical composition of the melt zones, may be interpreted partly with the information about the gold-tantalum system reported by Raub et al.(9) Up to about 1000°C they have reported that the solubility limit of Ta in Au is about 10 weight percent tantalum. In gold rich alloys they have reported an apparent eutectic reaction between FCC gold and a BCT intermetallic compound, TaAu within the range of 12 to 42% tantalum. Because the melt zones of the miniature bonds always formed at a constant chemical composition, it is quite possible that the melt zone formed as a eutectic alloy that would have a lower melting point than either of its components. However, the intermetallic compound TaAu, which has a distinct lattice parameter as reported by Raub et al.,(9) was not detected by X-ray analysis or from replication electron microscopy. If the melt zone was a hypoeutectic alloy that had gold as the primary phase, and since it was exposed by chemical etch for X-ray analysis, it is quite possible that the eutectic structure that contained the intermetallic was more active than either base metal and consequently was etched away and not detected in the diffraction patterns.

In an effort to determine more about the structure of the melt zones, gold-tantalum bonds were etched and replicated for viewing at high magnifications in the electron microscope. The carbon replicas were rotationally and directionally shadowed with platinum. Difficulty was encountered in removing the formvar that was used to produce the replicas because of the step that formed at the interface, which resulted from etching. The formation of this step has been encountered by others (5-7, 10) who have attempted replication of explosive bonds. To counteract the formation of steps at the interface, the samples were etched for both gold and tantalum, so that the surface of each metal would be removed at approximately the same rate. This procedure minimized the step at the interface that results in preferentially etching only one element. Success was obtained with this method although the melt zones tended to be far more reactive to the etches than either base metal. Figure 7 is a replicated electron micrograph of a sample that is considered to be a transition interface of the straight to wavy type. Although this bond showed no melt zone when it was examined with the light microscope, examination of the replica at high magnification proved that the sample contained small pockets of once molten metal, now containing a fine solidification structure, as shown in Figure 7. The white regions in the electron micrograph are the result of directional shadowing along the interface with platinum. Although the number of these areas were small in comparison to the interface length, these pockets of solidified melt do exist, even under conditions where they were not expected to form.

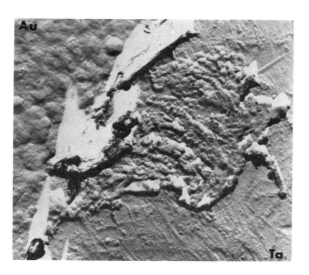

FIGURE 7: A gold-tantalum bond showing a small melt region, electron replica micrograph, 20,000X.

ACKNOWLEDGEMENTS

Thanks are extended to D. A. Machusak, R. E. Woods, J. A. Carnevale and S. J. Feltham of the Western Electric Company and to D. Bush of Lehigh University for experimental assistance. This paper is based on a thesis submitted by J. L. Edwards in partial fulfillment of the requirements of the degree of Master of Science at Lehigh University as a part of the Lehigh University-Western Electric Company Masters Program.

REFERENCES

(1) B. H. Cranston, "Explosive Bonding of Electrical Interconnections," IEEE Transactions on Parts, Hybrids and Packaging, Vol. PHP-8, No. 3, (September 1972) pp 27-32.

(2) Rinehart, John S., and Pearson, John, Explosive Working of Metals, New York, the MacMillan Co., 1963.

(3) Wright, Edward S., and Bayce, Arthur E., "Current Methods and Results in Explosive Welding," Stanford Research Institute, Menlo Park, California, March 1965.

(4) Bergmann, O. R., Cowan, G. R., and Holtzman, A. H. "Experimental Evidence of Jet Formation During Explosive Cladding," *Transactions of the Metallurgical Society of AIME,* Vol. 236, (May 1966) pp. 646-653.

(5) Lucus, W., et al., "Some Metallurgical Observations on Explosive Welding," Second International Conference on the Center for High Energy Forming, Denver, 1969.

(6) Trueb, Lucien F., "Electron Microscope Investigations of Explosive-Bonded Metals, "*Transactions of the Metallurgical Society of AIME*, Vol. 242, (June 1968) pp. 1057-1065.

(7) Buck, G. and Hornbogen, G., "Metallographic Investigation on Explosive-Welded Seams, "*Metallwissenchaft und Technik*, Vol. 20, No. 1, pp. 9-12, January 1966, Translated by Frank C. Farnham Co., Inc.

(8) Hull, Derek, *Introduction to Dislocations*, Oxford, Pergamon Press, 1969.

(9) Raub, E., Beeskow, H. and Menzel, D., "Tantalum-Gold Alloys," *Z. Metallkunde*, Vol. 52, No. 3, pp. 189-193, 1961, Translated by Frank C. Farnham Co., Inc.

(10) Trueb, Lucien F., "Microstructural Effects of Heat Treatment on the Bond Interface of Explosively Welded Metals," *Metallurgical Transactions*, Vol. 2, (January 1971) pp. 145-153.

MICROSTRUCTURAL EFFECTS OF HIGH STRAIN RATE DEFORMATION

W. C. Leslie

Department of Materials and Metallurgical Engineering

University of Michigan, Ann Arbor, Michigan

I. INTRODUCTION

This paper constitutes an attempt to present the current state of knowledge of the effects of high strain rates on the microstructures of metals and alloys. It is limited, arbitrarily, to the effects produced by the passage of planar shock waves of known amplitude; i.e., to the effects of uniaxial strain.

Fracture, spalling and explosive welding are not included, nor are the microstructural effects of high-strain-rate forming. The microstructures to be discussed are, necessarily, residual features remaining after passage of the shock wave and after the specimen has returned to ambient temperature. The task has been simplified by the existence of several prior reviews of the same topic,[2-8] written since the highly stimulating Estes Park Conference of 1960.[1] The work cited here has all been done since that conference.

The objectives of this review are several:
1. To describe the unique microstructural features produced by shock loading.
2. To point out how shock loading can be used as a metallurgical research tool.
3. To emphasize specific instances where our knowledge of structures and how they originate is still deficient.

The effects of shock loading on face-centered cubic metals and alloys, on body-centered cubic metals and alloys (including steels) and on ordered alloys will be discussed in that sequence, followed by a brief discussion of the interactions of shock waves with second-phase particles and inclusions.

II. FACE-CENTERED CUBIC METALS AND ALLOYS

A. Copper and Nickel

Because they are the simplest, the microstructures of face-centered cubic metals and alloys are considered first. With the exception of mechanical twins, the structures developed in copper and nickel resemble those generated by conventional cold deformation. The form of the dislocation arrays is dictated by the stacking fault energy; when this is comparatively high, as in copper and nickel, cross-slip of dislocations is comparatively easy and dislocation cells are formed.[9-24]

In copper, the hardness, dislocation density, stored energy, yield stress and ultimate tensile strength increase with increasing shock pressure, then reach maxima.[12,13,17,19,22] It is important to note that these quantities also increase with increased shock duration.[13] The dislocation cell size decreases with increasing shock pressure.[12] The energy stored in copper during shock loading at ambient temperatures is approximately twice as great as in specimens deformed slowly to the same strain.[17] After exposure to very high shock pressures (1000 kbars)* the structure of copper changes to ill-defined cells, often elongated.[25]

The dislocation substructures developed by the shock loading of polycrystalline nickel are similar to those in copper.[11,18,20,23,24] As the shock pressure increases from 80 to 460 kbars, the average dislocation cell diameter decreases from 0.8 to 0.1µ (8×10^{-4} to 1×10^{-4} mm).[24] In one respect, however, nickel differs from copper. When shock loaded to 1000 kbars, the dislocation density in polycrystalline nickel is very high (10^{12} to 10^{13} per cm^2), but no cell structure forms.[23]

Mechanical twins can be generated in both copper and nickel by shock loading. There is poor agreement as to the shock pressure required to produce twins in copper; this ranges from 16 to 28 kbars,[14] to 75 kbars,[17] to 112 kbars[19] to 345 kbars.[22] Twins do not appear after exposure to a pressure of 1000 kbars.[25] Some of this disparity may be due to variations in the duration of loading. The lowest pressure cited may be in error because of a non-planar shock front.[3] A reasonable estimate is that mechanical twins can be detected in polycrystalline copper after shock loading to a pressure of 75 kbars; they are present in every grain after exposure to a pressure of 435 kbars,[26] but are no longer present after shock loading to 1000 kbars.[25]

---

*1 kbar = 100 M Pa = 14,503 psi.

In copper single crystals, twins are formed more readily when the shock wave travels in the [001] direction (145 kbar) than in the [111] direction (200 kbar).[9] Twins can form during the passages of both the initial compressive wave and the subsequent tensile rarefaction wave.[9,27]

The formation of twins in shock-loaded copper made possible the first experimental determination of the twinning shear in a face-centered cubic metal.[28] This was estimated to be between 0.620 and 0.701, as compared to the theoretical 0.707. Some changes in crystallographic texture occur during the shock loading of copper.[28,29] When the shock waves were planar, the resulting textures varied with shock pressure.[29] Low shock pressures were comparable in effect to conventional slow compression, increasing the number of {110} planes parallel to the surface. At higher pressures the texture generated resembles that produced by conventional tensile straining, in that the density of {111} and {100} planes in the surface increases. These textural changes are not related to the formation of twins.

Because of the higher stacking fault energy of nickel, deformation twins are more difficult to form than in copper. As in the case of copper, there is rather poor agreement on the shock pressure required to form twins. The determinations are: less than 350 kbars,[11,33] 370 kbars,[31] less than 460 kbars[24] and none at pressures near 1000 kbars.[23] An estimate of 370 kbars seems reasonable for the minimum required pressure. Because of their very low concentration, it is obvious that deformation twins in nickel contribute very little to strengthening.

The energy stored in shock-loaded copper or nickel is greater, probably by a factor of about 2, than in specimens strained slowly to the same strain;[17,20] therefore, the annealing characteristics also differ. In copper, the first stage of annealing (temperatures up to 100°C) is characterized by a decrease in density and an increase in electrical resistivity.[32] These effects are opposite to those seen after conventional cold deformation. The authors speculate that these changes are due to stress relief and microcrack growth. In the second stage, (75° to 200°C) there is a slight release of stored energy, a slight decrease in resistivity and a slight increase in density. In the third stage (above 200°C) the major softening occurs. This softening is accomplished by recovery rather than by recrystallization. No recrystallization was observed even after shock loading to 435 kbars.

The annealing of nickel has been studied less thoroughly, but it is known that recrystallization does not occur during annealing after loading up to 330 kbars,[20,23] but does occur after loading to 460 kbars and above.[23,34]

B. Aluminum

As compared to copper and nickel, comparatively little attention has been paid to the structure of shock-loaded aluminum. The observations that have been made indicate that, surprisingly, no dislocation cell structure is formed, at least at pressures as high as 150 kbars.[35] On the basis of stacking fault energy and ease of cross-slip, a cell structure would be expected, and indeed, is formed during conventional cold deformation. Shock loading to 150 kbars pressure produces a high density ($\sim 1.5 \times 10^{10}$ cm^2) of randomly distributed, heavily jogged dislocations, plus a high density of point defect clusters and dislocation loops. This degree of defect clustering does not seem to occur in copper or nickel, although even in those metals the point defect concentration is much higher after shock loading than after conventional deformation.[17,20] The reasonable explanation of the structure is that dislocations are pinned by point defect clusters and dislocation loops.[35] Although these observations on the defect structure of shock-hardened aluminum are indisputable, it would be comforting to have supporting data, particularly if such were obtained on more massive specimens than 0.001-inch thick foil.[35]

C. Face-Centered Cubic Alloys

Several studies of shock-loaded copper-base alloys have been conducted to determine the effect of stacking fault energy on the residual microstructure. These alloys include Cu-Al,[14,25] silicon bronze[15] and 70/30 (alpha) brass.[33,36-38] In such alloys, the stacking fault energy can be reduced from approximately 80 ergs/cm^2 for copper to as little as about 3 ergs/cm^2.[38] In every instance, as the stacking fault energy is reduced, planar arrays of dislocations, stacking faults and twins replace the dislocation cell structure of shock-loaded copper.

The same effect is seen in nickel-base alloys, such as Chromel-A (80Ni, 20Cr) and Inconel 600 (16Cr, 7Fe, 1Mn).[24,31,46] These have stacking fault energies of 30 to 40 ergs/cm^2, as compared to about 300 ergs/cm^2 for nickel.

The austenitic (fcc) iron-base alloys, Fe-30Ni,[39] Fe-32Ni[52] and Fe-23Ni-0.6C[52] have been shock loaded. A ferromagnetic-to-paramagnetic transition occurs in Fe-30Ni at 25 kbars.[39] The other two alloys, when subjected to pressures of 170 or 270 kbars, display the usual structures produced in fcc alloys of low stacking fault energy—deformation twins on {111} and planar arrays of dislocations on the same planes. No phase change was detected. The hardening of the Fe-23Ni-0.6C alloy by a given shock pressure was much greater than the hardening of the Fe-32Ni alloy, even though the defect structures were similar. There is, apparently, an interaction of the carbon atoms with shock-produced defects.

The structures of at least four different austenitic stainless steels have been examined after shock loading; these are Types 302,[41] 304,[36,37,40,42-44,48] 316,[49] and A286.[41] Again, the predominant microstructural features are arrays of dislocations on {111} planes, accompanied by stacking faults and twins at pressures above about 120 kbars. The density of twins increases with increasing shock pressure, passes through a maximum, then decreases when pressures exceed about 750 kbars. At this and higher pressures, recovery phenomena complicate the interpretation of the residual structures.[42]

There is one point of apparent disagreement, however, regarding the structures of shock-loaded austenitic stainless steels. Koepke, et al,[41] report the formation of α martensite in Type 302 shocked at ambient temperature and both α and ε martensite in the same steel shocked at -196°C. Kangilaski, et al,[44] report 20% by volume of α martensite in Type 304 shocked at a pressure of 320 kbars. Gelles[45] states that there was evidence of an unspecified shock-induced allotropic phase transformation in Type 316. On the other hand, Murr and coworkers[36,37,40,42,43,47,48] report no such transformations. We must conclude that transformation to α martensite, and perhaps to ε martensite, can occur when metastable austenitic stainless steels are subjected to planar shock waves. The transformation may occur in the rarefaction stage. It is not known why such transformations have not been observed in all instances. Because of its composition, the austenite of A286 is stable against strain-induced transformation.[41]

Because of the high density of defects generated in austenite by shock loading, this procedure can be used, prior to an aging treatment, to provide more nucleation sites for the precipitation of carbides[44] or intermetallic compounds.[41] The resulting finer dispersion can reduce the subsequent steady-state creep rate and increase the stress-rupture life. The effects are analogous to those obtained earlier when the defects were generated by conventional cold work.[53]

Austenitic Hadfield manganese steel (∿12%Mn, 1%C) was one of the first alloys to be shock loaded. Its hardening by this technique remains one of the few commercial applications of shock waves to metallurgy. The hardness can be increased from about 200 to about 450 DPH.[49-51] The most recent studies agree that the resulting residual structures are typical of fcc alloys with comparatively low stacking-fault energies, i.e., planar arrays of dislocations and twins on {111}.[49,51] Twins form at about 125 kbars, but again, duration of the stress is important.[51] No ε martensite was detected.[49] The mechanism of the phenomenal work-hardening rate of Hadfield's steel has been a matter for conjecture since development of this steel in 1882. Most attention has been focussed on the microstructure after conventional deformation, and, particularly in recent years, on the formation of twins and ε martensite during cold work. However, we have seen that there is little difference in the residual defect structures in shock-hardened Fe-32Ni, Fe-23Ni-0.6C or Hadfield manganese steel, but the hardening is much the greatest in the Hadfield steel. I conclude, therefore, that the critical factor in the high work-hardening rate of this steel is not the defect structure as such, but the interaction between dislocations and Mn-C complexes in solid solution in the austenite. This type of interaction, at much lower concentrations of manganese and carbon or nitrogen in alpha iron, is known to control the strength of low-carbon steels at slightly elevated temperatures.[54]

The very apparent relation between deformation twinning and stacking fault energy in fcc metals and alloys seems to favor the twinning mechanism of Cohen and Weertman[55] over that of Cowan[56] as the principal mode of twin formation during shock deformation. In the former, an a/2 [011] dislocation splits into an a/6 [$\bar{2}$11] mobile Shockley partial and an a/3 [111] Frank sessile partial, creating a stacking fault. Cowan suggested that twins might form profusely when the theoretical shear strength of the crystal is exceeded. Although this may indeed occur, the observations indicate that profuse twinning occurs in several alloys long before the theoretical shear strength is exceeded.

As mentioned previously, the annealing of shock-loaded stainless steels can be complicated by precipitation of carbides on defects.[44] In general, recovery and recrystallization proceed more rapidly in cold-rolled than in shock-loaded stainless,[47] but this is not true for specimens loaded to 750 or 1200 kbars. Apparently, a high density of point defects is then available to drive recovery and recrystallization.[47]

III. BODY-CENTERED CUBIC METALS AND ALLOYS

A. Iron

<u>Dislocation structures and twinning</u>. When alpha iron is shock-loaded at pressures under 130 kbars, the residual dislocation configuration resembles that developed by conventional deformation at low temperatures,[57,58] comparatively straight screw dislocations form on {110} planes in <111> directions. In contrast, in face-centered cubic metals, no one type of dislocation predominates. In body-centered cubic metals strained at low temperatures or at high strain rates, the edge components of dislocations can move at

higher rates than the screw components, hence elongated segments of
the latter remain in the structure. These screw components are
unable to cross-slip at low temperatures or high strain rates.[59-62]
Because of this lack of cross-slip, no cell structure forms.

These uniform dislocation distributions are always accompanied
by deformation twins, for twinning begins in moderately pure iron
at shock pressures of about 2.5 kbars, corresponding to a maximum
resolved shear stress of about 0.75 kbars.[63,64] Only a minor amount
of study has been devoted to the nucleation of such twins; twins
are generated at the intersection of two slip planes, at least one
of which is of the {112} type, and a dislocation dissociation
mechanism has been proposed.[65]

Studies of twinning in shock-loaded single crystals of 3%Si
iron[52,66] indicated that mechanical twins formed only on twinning
systems that are expected to operate in compression. Twins did
not form on systems with the highest resolved shear stress when
operation of these systems involved an elongation of the specimen
against the acting compressive stress. No "conjugate" twins of
the type reported to form in shock-loaded single crystals of molybdenum[67] were found.

By alloying iron to eliminate the pressure-induced $\alpha \rightarrow \varepsilon \rightarrow \alpha$
phase transformations, it is possible to determine the effect of
shock pressure on twin density; this increases with increasing
shock pressure, at least up to 330 kbars.[52] It would be interesting
to determine whether this effect is reversed at higher pressures,
as it is in face-centered cubic metals.

Twinning in shock-loaded iron can be prevented by a pre-
existing dislocation structure[52,68,69] which can be generated by
prior deformation[68,69] or by a prior phase transformation.[52] To
be effective, however, the dislocation structure should be cellular; planar arrays, such as are generated by the subsequent shock
wave, would not prevent twinning. The multitude of pre-existing
slip dislocations in the cell structure allows slip on a number
of systems which would not normally operate during shock deformation, obviating the need for twinning as a deformation mechanism.
Obviously, if the pre-existing dislocations are pinned by strain
aging, they will be ineffective in preventing subsequent twinning.[70]

The yielding and flow of iron is highly sensitive to strain
rate; the dynamic yield strength is much larger than the quasi-
static yield stress. The Hugoniot elastic limit (HEL), which is
the amplitude of the elastic precursor wave, is related to the
dynamic yield strength. It now seems fairly clear[64,69,71] that
the commonly observed yield drop in iron during shock deformation
is caused by deformation twinning and that the magnitude of the

HEL is determined by dislocation motion. If twinning is eliminated, the yield drop disappears.

Pressure-induced transformation structures. The factor that makes shock-loaded iron so interesting, is, of course, the pressure-induced phase transformation, bcc (α) $\rightarrow$ hcp (ε). In unalloyed iron, this occurs at a dynamic pressure of 129 $\pm$ 1 kbars.[72] The transformation pressure is raised by additions of vanadium, cobalt, molybdenum or silicon.[66,72] Among the common alloying elements, only nickel and manganese lower the transformation pressure.[73-77] Sufficient concentrations of the last two elements can change the transformation to $\alpha \rightarrow \gamma$ (fcc). In most instances, the specimen reverts to bcc by a second martensitic transformation in the rarefaction portion of the wave.

The residual structure in iron, following the two successive pressure-induced transformations, usually resembles the lath martensite formed in low-carbon steels quenched from austenite.[52] Occasionally, the laths may be twin related, as in the martensite formed by cooling 18-8 stainless steel to -195°C.[78] The heterogeneous component of the transformation strain is provided by a high density of slip dislocations within the laths and at their boundaries. These pressure-induced transformations, occurring at or near ambient temperature, allow us to develop martensitic structures in high-purity iron. Such structures are impossible to obtain by quenching in any but very thin specimens, because of the loss of defects by recovery during cooling.

With the exception of a fine paper by Bowden and Kelly,[79] the difficult problem of the crystallography of the pressure-induced phase transformation in iron has been largely ignored. These authors state that the $\alpha \rightarrow \varepsilon$ transformation occurs by way of an invariant plane strain on $\{112\}_\alpha$ (that is, $\{10\bar{1}0\}_\varepsilon$), plus a small uniform dilatation. Usually, a small inhomogeneous shear on $\{0001\}_\varepsilon$ must be included, apparently in the form of slip. The second martensitic transformation upon unloading, $\varepsilon \rightarrow \alpha$, seems to be a reversal of the prior $\alpha \rightarrow \varepsilon$ transformation.

With the proper alloy additions to iron, the close-packed phase, whether fcc γ[52,79-81] or hcp ε,[76,77] can be retained at ambient pressure and temperature. These "reverse martensites" have unusual structures. The fcc structure in Fe-32%Ni, formed by shock loading bcc martensite, is characterized by profuse fine twinning on $\{111\}_\gamma$.[52,79] The hardness is about 100 DPH units higher than in "reverse martensites" formed by rapid reheating[82] because of the greater density of defects.

The occurrence of partial transformation of α to γ in Fe-Ni alloys and retention of γ after release of shock pressure makes permits determination of the crystallography of the transformation.[79] The two phases have approximately the Kurdjumov-Sachs relationship, $(111)_\gamma \parallel (101)_\alpha$, $[1\bar{1}0]_\gamma \parallel [11\bar{1}]_\alpha$, which is the same as found when the $\gamma \to \alpha$ transformation occurs by temperature change.[52] At comparatively low shock pressures (100 kbars) the $\alpha \to \gamma$ martensitic transformation is a direct reversal of the conventional $\gamma \to \alpha$ transformation, with a habit plane $(225)_\gamma$. At higher pressures the habit plane becomes $(112)_\gamma$ and the pressure-induced γ twins on $\{111\}_\gamma$.[79]

It is unfortunate that no attempt has been made to determine the fine structure of ε martensite in shock-loaded Fe-Mn alloys, nor to determine the orientation relationship between α and ε.

The pressure-induced martensitic transformation in iron produces a large concentration of lattice defects. One attempt has been made to obtain a comparative measure of this concentration.[83] Ingot iron previously shock loaded to 350 kbars pressure was nitrogenated at 350°C. It absorbed a total of 0.051%N, 0.014% in "normal" lattice sites and 0.037% at defects. The pressure-induced transformation was more effective in generating defects than cold rolling to 88% reduction in thickness.

Annealing of shock-loaded iron. When iron is shock loaded at pressures below 130 kbars, the rate and extent of recrystallization during subsequent annealing are limited. However, when a phase change occurs, the rate of subsequent recrystallization increases markedly[57,84] because of the discontinuous increase of stored energy and the high density of nucleation sites for recrystallized grains. Perhaps the most important consequence is that shock loading allows us to develop a structure of equi-axed, randomly oriented, fine grains in high-purity iron; which we could not do before.[57,85] Measurements of stored energy in shock-transformed pure iron would be useful as a base for quantitative studies of energy release during the tempering of martensitic steels.

B. Ferritic Steels

Several commercial steels have been shock loaded in the normalized (ferrite-pearlite), annealed (spheroidite) and quenched and tempered (martensite) conditions.[86-91] Steels containing free ferrite are much more susceptible to hardening than are entirely pearlitic or martensitic steels.[89,90] The free ferrite undergoes the pressure-induced phase transformations at or near the usual pressures, but much higher pressures (>200 kbars) are required to transform ferrite lamellae in pearlite.[90] Bowden and Kelly[90] propose that this increase is due to the mechanical resistance of the carbide lamellae to propagation of the transformation shear.

There is disagreement as to the hardening of high-carbon pearlitic steels by shock loading. Bowden and Kelly[90] show a sharp increase in hardness of a 0.95%C steel after loading to 250 kbars; Koepke, et al,[89] show only a slight increase of hardness in a 0.95%C steel shock loaded to 285 kbars.

Steels that are initially martensitic show only minor increases in hardness after shock loading and little apparent change in structure.[89] After one martensitic transformation has occurred, there is not much to be gained by following it with another.[86]

In ferritic steels, the weight of evidence indicates that in strengthening by shock loading a penalty of embrittlement is exacted. This has been a barrier to acceptance of shock loading as a strengthening mechanism. Watters, et al,[91] however, claim that there is no shift in the Charpy V-notch transition temperature (15 ft-lbs) of an AISI 1009 steel after shock loading in the range 14-95 kbars. This contrasts with Dieter,[92] who found an upward shift of about 80°F in the transition temperature of a similar steel after loading to 95 kbars. Loading to pressures above that required for the phase transformation leads to severe embrittlement.[92]

When 18 and 20% nickel maraged steels are shock loaded in the hardened condition, the yield and tensile strengths increase slightly and the elongation, reduction of area and impact energy decrease slightly.[50] When an H11 tool steel in the martensitic condition is shock loaded, the subsequent tempering process is accelerated.[86]

Shock loading has been employed as a tool for the study of the ausforming process.[52] It was found that the presence of a high density of lattice defects in austenite was not sufficient, in itself, to strengthen the subsequently formed martensite. Strengthening by ausforming must come from precipitation in the austenite, changes in tempering of the austenite, or from directional refining of the martensite plates.[87]

The suggestion has been made[93] that shock hardening of stainless and of ferritic steels may simulate radiation damage in these materials, but this proposal does not seem to have been pursued.

The studies of the various structures developed in shock-loaded iron have advanced our understanding of the stress and thermal history of iron meteorites.[94]

C. Other Body-Centered Cubic Metals

To date, no other shock-loaded bcc metal has undergone an allotropic phase transformation. Tantalum,[1] molybdenum,[67] niobium[1,95] and tungsten[95] all exhibit mechanical twinning when exposed to shock, the twin density increasing with shock pressure. No extensive examinations of the fine structure by transmission electron microscopy have been done. No overall improvement of mechanical properties after shock loading has been noted.

IV. ORDERED ALLOYS

The effects of explosive shock on the ordered structures Cu_3Au (cubic, $L1_2$),[96,97] beta brass (bcc, B2),[98] 49%Fe-49%Co-2%V (bcc,B2)[99] and $Fe_3Al(DO_3)$,[52,59] have been investigated. In Cu_3Au, there was a sharp increase in stored energy at pressures between 290 and 370 kbars and at 475 kbars the alloy was disordered.[96,97] This is in accord with Cowan's suggestion[56] that shock pressures which exceed the theoretical shear strength can create dislocations as needed to form the boundary between the shocked and the unaltered portion of the specimen, thus destroying the order. For Fe_3Al, the highest pressure used (370 kbars) was insufficient to destroy the order, although many twins were generated.[52] It would be interesting to repeat this experiment at the pressure predicted to destroy the order.[56]

In the ordered Fe-Co-V alloy[99] the pressure was also insufficient to destroy completely the order. Anti-phase boundaries and a few twins were created at 260 kbars. In the shocked, ordered beta brass the dislocations produced at 300 kbars were mostly screws, which were removed by annealing above 300°C.[98]

V. EFFECTS OF INCLUSIONS AND SECOND PHASES

Probably because of the inherent experimental difficulties, little work has been done on the structural effects of the interactions between inclusions or second-phase particles and shock waves. The dense array of fine thoria particles in TD-Ni or in TD-NiCr prevents the formation of twins or of a dislocation cell structure; only a dense continuum of dislocations is generated by shock pressures in the range 80-460 kbars.[24] Coherent precipitates have less drastic effects. Such particles in Inconel 600 do not prevent twinning.[31] Coherency is lost at pressures above about 200 kbars.

The presence of larger inclusions or second-phase spheroidal particles can have marked effects on the surrounding matrix during passage of a shock wave. On the assumptions of an elastic matrix and a rigid insert. Pao and Mow[100] concluded that if no boundary displacement were allowed, stresses became infinitely large at certain points in the matrix. Although metals are not elastic under high shock pressures inclusions certainly serve to increase local stresses, to the extent that inclusions in shock-loaded, <u>untransformed</u> iron can be surrounded by an envelope of martensite.[52] This effect will certainly not benefit the mechanical properties.

VI. CONCLUSIONS AND RECOMMENDATIONS

1. Insufficient attention has been paid in the past to the effect of shock duration on residual microstructures.
2. In most face-centered cubic metals and alloys, the stacking fault energy is the dominant factor controlling the residual microstructure. When the stacking fault energy exceeds about 30 ergs/cm^2, cross-slip of dislocations allows the formation of a cellular structure and mechanical twinning is difficult. When the stacking fault energy is lower, planar arrays of dislocations on $\{111\}_\gamma$ are formed, and twinning is common.
3. After very high shock pressure, ~1000 kbars, twins are not left in fcc metals, and dense clouds of dislocations remain. This change in structure needs further study.
4. Aluminum is an exception to the rule that cellular dislocation structures form in fcc metals with high stacking fault energies. A high density of point defects pins dislocations, producing a random distribution of heavily jogged dislocations.
5. There is apparent disagreement as to whether the shock loading of metastable austenitic stainless steels (302 and 304) leads to the martensitic transformation $\gamma \rightarrow \alpha$.
6. The most likely cause of the phenomenal rate of work hardening of Hadfield manganese steel is the interaction between dislocations and Mn-C couples in solution in austenite.
7. The fine structure of ε iron and its crystallographic relationship with α should be determined by direct observation. This could be done in shock-loaded Fe-Mn alloys.
8. The energy stored in shock-transformed high-purity iron should be determined.
9. Cowan's prediction of the shock pressure required to exceed the theoretical shear stress of an alloy should be tested with several ordered alloys to determine whether the order is destroyed at the predicted pressure.
10. The stresses and structures generated when shock waves encounter an inclusion in a metal warrant further study.

References

1. *Response of Metals to High-Velocity Deformation*, Interscience, N.Y. 1961.
2. E. Hornbogen, *High Energy Rate Working of Metals*, Sandefjord-Lillehammer, Norway, Sept. 1964, p.345.
3. A.S. Appleton, Appl. Matl. Res., 1965, vol.4, p.195.
4. D.G. Doran and R.K. Linde, Solid State Phys., 1966, vol.19, p.229.
5. E.G. Zukas, Metals Engr. Quart., May, 1966, vol.6, p.1.
6. H.E. Otto and R. Mikesell, *First Int. Conf. Center for High Energy Rate Forming*, Estes Park, Col. June, 1967.
7. A.H. Jones, C.J. Maiden and W.M. Isbell, *Mechanical Behavior of Materials Under Pressure*, H. Ll. D. Pugh, ed. p.680, Elsevier, London, 1970.
8. D.R. Curran, *Shock Waves and the Mechanical Properties of Solids*, J.J. Burke and V. Weiss, ed. p. 121, Syracuse University Press, 1971.
9. R.J. DeAngelis and J.B. Cohen, J. Metals, 1963, vol.15, p.681.
10. C.A. Verbraak, Metal Progress, 1963, vol.83, p.109.
11. R.L. Nolder and G. Thomas, Acta Met., 1964, vol.12, p.227.
12. O. Johari and G. Thomas, Ibid, 1964, vol.12, p.679.
13. A.S. Appleton and J.S. Waddington, Ibid, 1964, vol.12, p.956.
14. O. Johari and G. Thomas, Ibid, 1964, vol.12, p.1153.
15. J.R. Holland and H.M. Otte, Bull. Am. Phys. Soc. [2], 1965, vol.10, p.710.
16. A.S. Appleton and J.S. Waddington, Phil. Mag., 1965, vol.12, p.273.
17. D.C. Brillhart, R.J. DeAngelis, A.G. Preban, J.B. Cohen and P. Gordon, Trans. AIME, 1967, vol.239, p.836.
18. M.F. Rose, T. Berger and M.C. Inman, Ibid, 1967, vol.239, p.1998.
19. J. George, Phil. Mag., 1967, vol.15, p.497.
20. H. Kressel and N. Brown, J. Appl. Phys., 1967, vol.38, p.1618.
21. S.H. Carpenter, Phil. Mag., 1968, vol.17, p.855.
22. F.I. Grace, J. Appl. Phys., 1969, vol.40, p.2649.
23. L.F. Trueb, Ibid, 1969, vol.40, p.2976.
24. L.E. Murr, H.R. Vydyanath and J.V. Foltz, Met. Trans., 1970, vol.1, p.3215.
25. S.V. Hooker, J.V. Foltz and F.I. Grace, Proc. EMSA, 1970, p.422.
26. R.J. DeAngelis and J.B. Cohen, *Deformation Twinning*, R.E. Reed-Hill, J.P. Hirth and H. Rogers, eds., p.430, Gordon and Breach, N.Y., 1964.
27. E.G. Zukas and J.W. Taylor, Trans. AIME, 1965, vol.233, p.828.
28. J.B. Cohen, A. Nelson and R.J. DeAngelis, Ibid, 1966, vol.236, p.133.
29. R.J. DeAngelis and J.B. Cohen, Trans. ASM, 1965, vol.58, p.700.
30. J. Desoyer and J. Jacquesson, *High Energy Rate Working of Metals*, Sandefjord-Lillehammer, Norway, Sept. 1964.

31. L.E. Murr and J.V. Foltz, J. Appl. Phys., 1969, vol.40, p.3796.
32. D.C. Brillhart, A.G. Preban and P. Gordon, Met. Trans., 1970, vol.1, p.969.
33. M.F. Rose and F.I. Grace, Brit. J. Appl. Phys.,1967, vol.18, p.671.
34. L.E. Murr and H.R. Vydyanath, Acta Met., 1970, vol.18, p.1047.
35. M.F. Rose and T.L. Berger, Phil. Mag., 1968, vol.17, p.1121.
36. L.E. Murr and F.I. Grace, Trans. AIME, 1969, vol.245, p.2225.
37. L.E. Murr and F.I. Grace, Exp. Mech., 1969, vol.5, p.145.
38. F.I. Grace, M.C. Inman and L.E. Murr, Brit. J. Appl. Phys., 1968, vol.1, p.1437.
39. R.A. Graham, D.H. Anderson and J.R. Holland, J. Appl. Phys., 1967, vol.38, p.223.
40. M.C. Inman, L.E. Murr and M.F. Rose, Advances in Electron Metallography, ASTM, STP396, 1966, vol.6, p.39.
41. B.G. Koepke, R.P. Jewett and W.T. Chandler, Tech. Doc. Report No.ML TDR 64-282, Air Force Materials Laboratory, Oct. 1964.
42. L.E. Murr and M.F. Rose, Proc. EMSA, 1968, p.254.
43. J.V. Foltz and L.E. Murr, Ibid, 1969, p.196.
44. M. Kangilaski, J.S. Perrin, R.A. Wullaert and A.A. Bauer, Met. Trans., 1971, vol.2, p.2607.
45. S.H. Gelles, Proc. EMSA, 1971, p.106.
46. L.E. Murr and J.V. Foltz, Ibid, 1969, p.194.
47. L.E. Murr and M.F. Rose, Phil. Mag., 1968, vol.18, p.281.
48. J.A. Korbonski and L.E. Murr, Proc. EMSA, 1970, p.428.
49. W.N. Roberts, Trans. AIME, 1964, vol.230, p.372.
50. L.A. Potteiger, NWL Report No.1934, 9 Sept. 1964.
51. A.R. Champion and R.W. Rohde, J. Appl. Phys., 1970, vol.41, p.2213.
52. W.C. Leslie, D.W. Stevens and M. Cohen, High Strength Materials, V.F. Zackay, ed., p.382, Wiley and Sons, N.Y., 1965.
53. F. Garofalo, F. vonGemmingen and W.F. Domis, Trans. ASM, 1961, vol.54, p.430.
54. J.D. Baird, Metal. Rev., 1971, vol.16, No.149.
55. J.B. Cohen and J. Weertman, Acta Met., 1963, vol.11, p.996.
56. G.R. Cowan, Trans. AIME, 1965, vol.233, p.1120.
57. W.C. Leslie, E. Hornbogen and G.E. Dieter, J. Iron and Steel Inst., 1962, vol.200, p.622.
58. E. Hornbogen, Acta Met., 1962, vol.10, p.978.
59. H. Kressel and N. Brown, Ibid, 1966, vol.14, p.1860.
60. H.D. Solomon, C.J. McMahon and W.C. Leslie, Trans. ASM, 1969, vol.62, p.886.
61. T. Imura, H. Saka and N. Yukawa, J. Phys. Soc. Japan, 1969, vol.26, p.1327.
62. H.D. Solomon and C.J. McMahon, Acta Met., 1971, vol.19, p.291.
63. R.W. Rohde and J.N. Johnson, Proc. 2nd Int. Conf. Strength Metals and Alloys, p.985, ASM, Cleveland, O., 1970.
64. J.N. Johnson and R.W. Rohde, J. Appl. Phys., 1971, vol.42, p.4171.
65. R. Priestner and W.C. Leslie, Phil. Mag., 1965, vol.11, p.895.

66. E.G. Zukas, C.M. Fowler, F.S. Minshall and J. O'Rourke, Trans. AIME, 1963, vol.227, p.746.
67. C.A. Verbraak, Science and Technology of W, Ta, Mo, Nb and Their Alloys, p.219, Pergamon Press, N.Y., 1964.
68. S. Mahajan, First Int. Conf. Center for High Energy Rate Forming, Estes Park, June, 1967.
69. R.W. Rohde, W.C. Leslie and R.C. Glenn, Met. Trans., 1972, vol.3, p.323.
70. J.R. Holland, Acta Met., 1967, vol.15, p.691.
71. R.W. Rohde, Ibid, 1969, vol.17, p.353.
72. T.R. Loree, C.M. Fowler, E.G. Zukas and F.S. Minshall, J. Appl. Phys., 1966, vol.37, p.1918.
73. T.R. Loree, R.H. Warnes, E.G. Zukas and C.M. Fowler, Science, 1966, vol.153, p.1277.
74. A. Christou, Scripta Met., 1970, vol.4, p.437.
75. A. Christou and J.V. Foltz, Phil. Mag., 1971, vol.23, p.1551.
76. A. Christou and N. Brown, J. Appl. Phys., 1971, vol.42, p.4160.
77. A. Christou and N. Brown, Met. Trans., 1972, vol.3, p.867.
78. P.M. Kelly and J. Nutting, J. Iron and Steel Inst., 1961, vol.197, p.199.
79. H.G. Bowden and P.M. Kelly, Acta Met., 1967, vol.15, p.1489.
80. R.W. Rohde, J.R. Holland and R.A. Graham, Trans. AIME, 1968, vol.242, p.2017.
81. R.W. Rohde, Acta Met., 1970, vol.18, p.903.
82. G. Krauss, Jr. and M. Cohen, Trans. AIME, 1962, vol.224, p.1212.
83. H.A. Wriedt, R.J. Sober and W.C. Leslie, Met. Trans., 1970, vol.1, p.3351.
84. A. Christou, Phil. Mag., 1972, vol.26, p.97.
85. E.G. Zukas and R.G. McQueen, Trans. AIME, 1962, vol.221, p.412.
86. S.M. Silverman, L. Godfrey, H.A. Hauser, E.T. Seaward, Tech. Doc. Rept. ASD-TDR-62-442, Aug. 1962.
87. B.A. Stein and P.C. Johnson, Trans. AIME, 1963, vol.227, p.1188.
88. O.E. Jones and J.R. Holland, J. Appl. Phys., 1964, vol.35, p.1771.
89. B.G. Koepke, R.P. Jewett and W.T. Chandler, Trans. ASM, 1965, vol.58, p.510.
90. H.G. Bowden and P.M. Kelly, Metal Science J., 1967, vol.1, p.75.
91. J.L. Watters, T.R. Wilshaw and A.S. Tetelman, Met. Trans., 1970, vol.1, p.2849.
92. G.E. Dieter, Strengthening Mechanisms in Solids, p.279, ASM, Cleveland, O., 1962.
93. D.R. Ireland, M. Kangilaski and R.A. Wullaert, Trans. Am. Nucl. Soc., 1970, vol.13, p.140.
94. A.V. Jain and M.E. Lipschutz, Nature, 1968, vol.220, p.139.
95. M.J. Klein and F.A. Rough, Trans. ASM, 1964, vol.57, p.86.
96. P. Beardmore, A.H. Holtzman and M.B. Bever, Trans. AIME, 1964, vol.230, p.725.
97. D.E. Mikkola and J.B. Cohen, Acta Met., 1966, vol.14, p.105
98. J.V. Rinnovatore and N. Brown, Trans. TMS-AIME, 1967, vol.239, p.225.

99. J.C. Desoyer, J.P. Eymery and P. Moine, Acta Met., 1972, vol.20, p.959.
100. Y.-H. Pao and C.C. Mow, J. Appl. Phys., 1963, vol.34, p.493.

THE SUBSTRUCTURE AND PROPERTIES OF EXPLOSIVELY LOADED Cu-8.6 % Ge

D. J. Borich
General Motors Corporation, Milford, MI 48042

D. E. Mikkola
Michigan Technological University, Houghton, MI 49931

ABSTRACT

The shock-induced defect microstructure in Cu-8.6 at. pct. Ge has been studied with x-ray diffraction. Samples of the alloy were explosively loaded at 100, 180, 350, and 475 kbars with planar shock waves using the driver plate technique. Complete analysis of the x-ray diffraction effects was made on each sample along with measurements of the changes in microhardness, yield strength, and electrical resistivity. The effective particle size decreased in a regular manner with shock pressure. Similarly, the rms microstrains, stacking fault probability, and twin fault probability increased with shock pressure. The twin fault probability was four times as large as the stacking fault probability. Generally, the substructure was more isotropic at higher shock pressures. Semiquantitative relationships have been developed relating the yield strength to the dislocation density and to the various substructure parameters.

INTRODUCTION

The response of metals to the passage of high velocity shock waves has received considerable attention.[1-4] Early work established the general microstructural behavior including the occurrence of slip lines and twin-like surface markings.[1-5] More recently, shock-induced defect structures have been examined with transmission electron microscopy[6-10] and x-ray diffraction.[10-12] These studies have confirmed the presence of deformation twins and have established that the substructure is characterized by high dislocation densities. Also, it has been noted that at very high shock pressures (> ~600 kbars) transient heating effects can cause thermal recovery.[2,3,13-15]

The purpose of the current study has been to examine the substructure of Cu-8.6 at. pct. Ge deformed by shock loading over the pressure range 100-475 kbars. X-ray diffraction was used as the primary tool with some supporting observations being made with transmission electron microscopy. The emphasis of the study has been to provide as quantitative information as possible about the as-shocked substructure and to relate the changes in substructure to changes in properties. Cu-8.6 at. pct. Ge has a very low stacking fault energy (~5 ergs/cm^2) and therefore provides the opportunity to study in detail the creation of stacking faults and microtwins as important parts of the deformed substructure. In particular, it was of interest in the study to establish whether the numbers and varieties of substructural defects change in a uniform manner with shock pressure or whether there are abrupt changes at certain critical pressures.

EXPERIMENTAL PROCEDURE

The Cu-Ge alloy used in this study was prepared by Engelhard Industries from 99.999 pct. pure Cu and Ge. Approximately 400 g of a mixture of Cu with nominally 8.5 at. pct. Ge was melted in a graphite crucible and poured into a graphite mold. Melting was done under vacuum and the chamber was backfilled with helium while pouring into the mold. The ingot was then rolled to 50 pct. R.A. and samples were cut from various points for chemical analysis. Results of the chemical analysis showed 8.6 at. pct. Ge with 0.001 at. pct. Pt the principal impurity.

The ingot as received from Engelhard was homogenized in an argon atmosphere by heating from 600°C to 825°C over a period of six days, holding at 825°C for five days, and then furnace cooling to room temperature. The material was then cross-rolled to a thickness of 0.045 in. in steps of 50 pct. R.A. with intermediate anneals at 375°C for one hour. The resulting material had a fine grain size (~5μ) with a minimum of texture.

Twelve 3 x 1 x 0.045 in. specimens of the alloy were machined from the as-rolled sheet and lapped flat with 1μ Alundum powder. These specimens were given a final annealing treatment and then subjected to explosive loading in groups of three arranged to make a 3 x 3 in. square at pressures of 100, 180, 350, and 475 kbars at the Eastern Laboratories of E. I. du Pont de Nemours and Co. The shock loading procedure has been described in detail elsewhere.[3] Briefly, the shock waves were generated by the driver-plate technique. A constant driver-plate thickness was used and the shock loading pressures were controlled by varying both the position of the driver plate assembly relative to the specimen assembly and the amount of explosive. Close-fitting spall plates were coupled to all free surfaces of the specimen group to minimize reflections and all specimens were shot into water to minimize thermal recovery effects. All shock-loaded specimens were stored at -25°C until measurements were made.

Each of the shock-loaded specimens was cut into four pieces with a jewelers saw and the surfaces metallographically prepared to provide specimens for subsequent measurements--three specimens 1 x 0.875 x 0.045 in. for x-ray diffraction and hardness measurements and one specimen 3 x 0.125 x 0.045 in. for electrical resistance measurements.

X-ray measurements were made with a Picker x-ray diffractometer automated with a Humphrey Electronics control unit. All reflections available with filtered Cu K_α radiation were recorded by point counting. The intensity, angle, and counting time were recorded on punched paper tape and then analyzed with an IBM 360 computer. The data analysis programs have been designed to perform the necessary data corrections, conduct a Fourier analysis of the diffraction peak profiles, and output the necessary data to evaluate the particle size, rms microstrain, stacking fault probability, multilayer fault probability, residual stress, and change in lattice constant. The programs include plotting routines which provide plots of the raw data, data after correction, corrected peak profiles, etc.

Changes in electrical resistance were measured with a precision Kelvin bridge coupled to a dc null detector. The specimen resistances were ~0.001 ohm and the sensitivity of the system was less than 0.01 pct. with a reproducibility for the entire system of less than 0.03 pct. Five measurements were made on each specimen and the resistance of an annealed standard was measured prior to and after each measurement on a specimen. All data were then analyzed by computer to correct for temperature variations and to determine the mean and standard deviation.

Microhardness measurements were made with a Tukon tester using a 100 g load and a 136° diamond pyramid indenter. Ten measurements were made on each specimen and these were subjected to the same statistical analysis as the electrical resistance data. Yield strengths were determined with an Instron tensile tester on tensile specimens (0.5 x 0.115 x 0.045 in. gage section) cut from the shock-loaded specimens.

EXPERIMENTAL RESULTS

The variation of the measured substructure and property parameters with shock loading pressure is illustrated in Figs. 1-5. The effective particle size measured in the <111> and <100> directions decreases with increasing shock loading pressure as shown in Fig. 1. The decrease is quite regular with no abrupt changes and there is some indication that the particle size becomes crystallographically more isotropic as the shock loading pressure increases. For the pressures 100, 180, 350, and 475 kbars the ratio $D_e(111)/D_e(100)$ has the values 1.61, 1.51, 1.34, and 1.24, respectively.

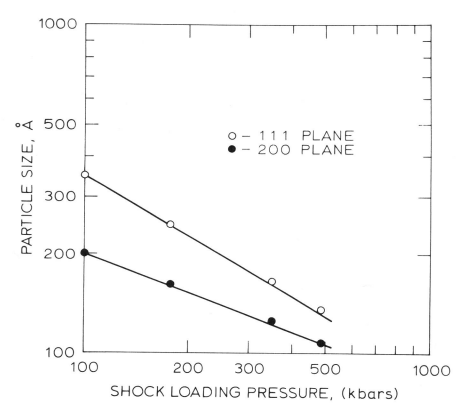

Fig. 1 Effective particle size $D_e(hk\ell)$ vs. shock loading pressure for Cu-8.6 at. pct. Ge.

Similar plots for the rms microstrains in the two crystallographic directions are shown in Fig. 2. The microstrains increase with pressure, but the relative change with pressure is not as great as for the particle size. As with the particle size the

PROPERTIES OF EXPLOSIVELY LOADED Cu-8.6% Ge

strains become more isotropic as the pressure increases--the ratio for the <111>/<100> directions has a value of 0.91 at 100 kbars, 0.96 at 180 kbars, and ~1.0 at both 350 and 475 kbars.

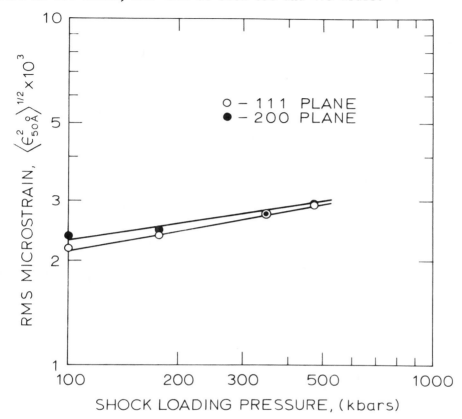

Fig. 2 Rms microstrain $\langle \epsilon^2_{50\text{Å}} \rangle_{hk\ell}$ vs. shock loading pressure for Cu-8.6 at. pct. Ge.

Figure 3 shows the behavior of the stacking fault and twin fault probabilities as a function of shock pressure. The stacking fault probability α was determined from least squares analysis of the peak shifts of the first eight peaks. The twin fault probability was calculated from the asymmetry of the 111 and 200 peaks. Generally, α and β have the same variation with shock pressure with $\beta \simeq 4\alpha$.

The least squares analysis of the changes in peak separations of the first eight peaks also provided values for the residual stress (σ) and the fractional change in lattice constant ($\Delta a/a$).

In all cases the major part of the peak shifts was caused by stacking faults. The residual stresses were compressive and less than 12,000 psi for all shock pressures. The fractional lattice constant change was also small with values less than 0.0003.

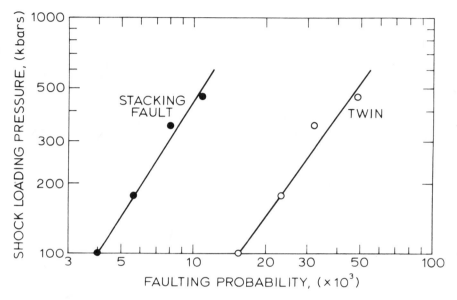

Fig. 3 Variation of stacking fault probability α and twin fault probability β with shock loading pressure for Cu-8.6 at. pct. Ge.

The basic features of the substructure; i.e., high dislocation density, occurrence of stacking faults and twin faults, were verified with transmission electron microscopy. Although the values of the various x-ray parameters could be substantiated with the transmission electron micrographs, it was not possible to make exact quantitative comparisons because of difficulties in resolving details of the cold-worked structure.

The microhardness values for the same specimens along with that for an annealed specimen are plotted in Fig. 4. A rapid increase occurs at low shock pressures followed by a smooth increase over the range 100-475 kbars. This behavior is similar to that reported for other metals deformed by shock loading.[16,17]

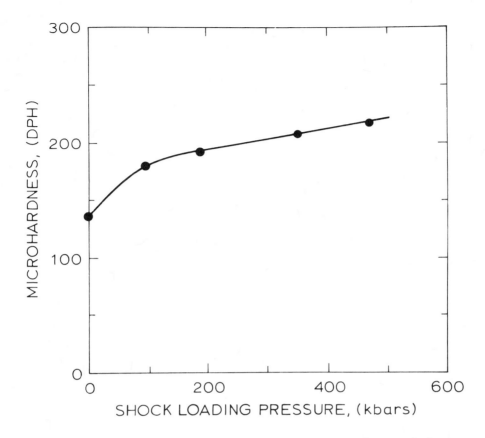

Fig. 4 Microhardness vs. shock loading pressure for Cu-8.6 at. pct. Ge.

Figure 5 shows the variation of the percent increase in electrical resistance with shock pressure for the same specimens. As with the hardness there is a rapid change at low pressures followed by a smooth increase.

DISCUSSION

Explosive loading in the range 100-475 kbars introduces a large amount of cold work into the Cu-8.6 at. pct. Ge alloy. As the pressure is increased, the substructure changes in a uniform way with no indication of abrupt changes corresponding to critical pressures for defect creation and/or multiplication. There is some evidence that the dislocation distribution becomes more isotropic as the pressure is increased; however, this might be

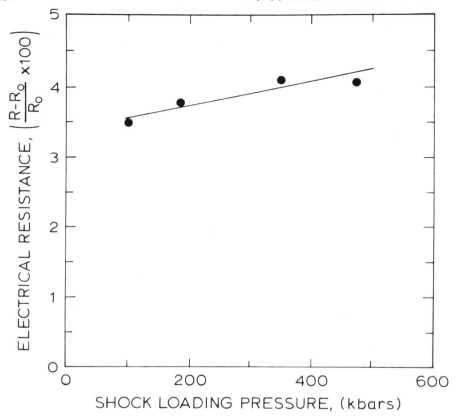

Fig. 5 Electrical resistance vs. shock loading pressure for Cu-8.6 at. pct. Ge.

expected because Cu-8.6 at. pct. Ge has a low stacking fault energy and characteristically the substructure involves planar defect arrays. At high pressures, the higher dislocation densities and multiple slip conditions may make the substructure more isotropic.

The changes in effective particle size with shock loading strain are similar to those found for explosively loaded Cu[18] as shown in Fig. 6. The basic differences are that the particle size for Cu is considerably larger and there is no pronounced anisotropy at low pressures. Similar data for the rms microstrains in Fig. 7 show that the behavior for the Cu and Cu-8.6 at. pct. Ge is almost exactly the same except that the microstrains in the Cu-8.6 at. pct. Ge at any level of strain are about twice those in Cu.

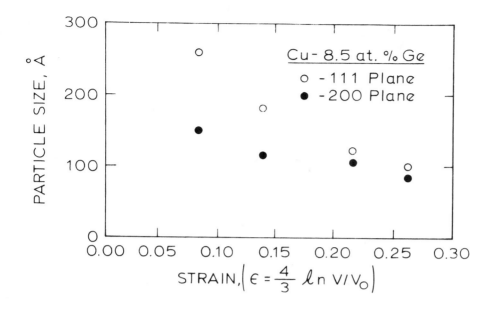
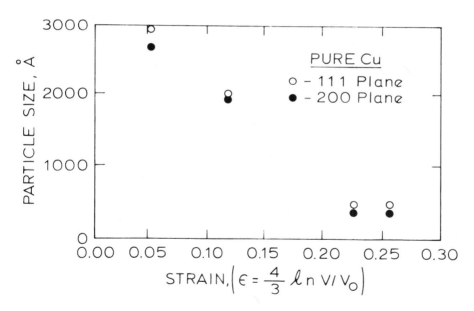

Fig. 6 Effective particle size $D_e(hk\ell)$ vs. shock loading strain for Cu-8.6 at. pct. Ge and pure Cu. Data for Cu from Ref. 18.

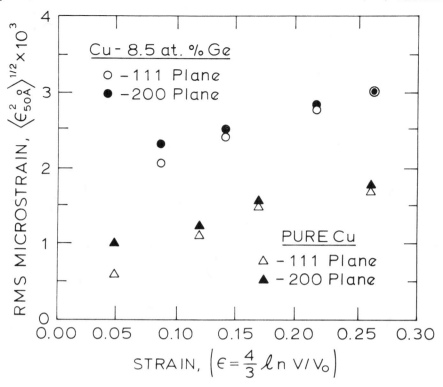

Fig. 7 Rms microstrain $\langle \varepsilon^2_{50\text{Å}} \rangle^{1/2}_{hk\ell}$ vs. shock loading strain for Cu-8.6 at. pct. Ge and pure Cu. Data for Cu from Ref. 18.

The rms microstrains determined from x-ray diffraction line broadening represent the non-uniform strains associated with the strain fields of the dislocation substructure.[19] The values reported above are those obtained as averages over a column length of 50Å which has become a standard for reporting these data. Table I compares the rms microstrains at column lengths of 50, 100, and 200Å for the various shock pressures. The ratios of the strains are the same for all shock pressures although the dislocation density increases with increases in shock loading pressure. This suggests that there is no major change in the nature of the dislocation arrangement with shock pressure and that the increased dislocation density simply results in a smaller particle size and a proportionate increase in the rms microstrains at all column lengths.

Both the stacking fault probability and twin fault probability increase in a uniform manner with increases in shock pressure (Fig. 3). As with the particle size and microstrain, there is no evidence for an abrupt change in these parameters with shock pressure. The relative number of twin faults is considerably greater than found for filings of the same alloy[20]; i.e., $\beta/\alpha \sim 4$ for shock loading vs. $\beta/\alpha \sim 0.5$ for filing. This difference is most likely a result of the higher strain rates associated with shock loading. It might be noted that the fact that both fault probabilities have the same pressure dependence is in agreement with those models of twinning based on the formation of stacking faults.[21-23]

TABLE I.

Rms Microstrains for Various Averaging Distances (L) for Cu-8.6 at. pct. Ge Shock Loaded at Different Pressures.

Shock Pressure (kbars)	Averaging Distance (Å)	Rms Microstrains ($\times 10^5$) $<\varepsilon_L^2>^{1/2}_{111}$	$<\varepsilon_L^2>^{1/2}_{100}$
100	50	205	230
	100	137	168
	200	95	110
180	50	240	248
	100	163	172
	200	116	122
350	50	275	280
	100	165	216
	200	125	122
475	50	295	300
	100	216	220
	200	146	140

Figure 8 shows a comparison of the current twin fault data with that for 70/30 α-brass.[11] The stacking fault energy of 70/30 α-brass is ~ 15 ergs/cm^2 and that for Cu-8.6 at. pct. Ge is ~ 5 ergs/cm^2. The increase with shock pressure is similar for the two materials. Although it is difficult to draw conclusions from the absolute magnitudes of twin fault probabilities measured in different laboratories, it appears that there is no large difference between the number of twin faults in the two materials at any given shock pressure.

It has been reported on the basis of transmission electron microscopy studies that there exists a critical shock loading pressure $\sim 200-300$ kbars above which there is an abrupt change in mode of deformation from slip to twinning.[6,24] It appears that this is not the case and that the number of twin faults increases quite uniformly with shock pressure. If a critical pressure

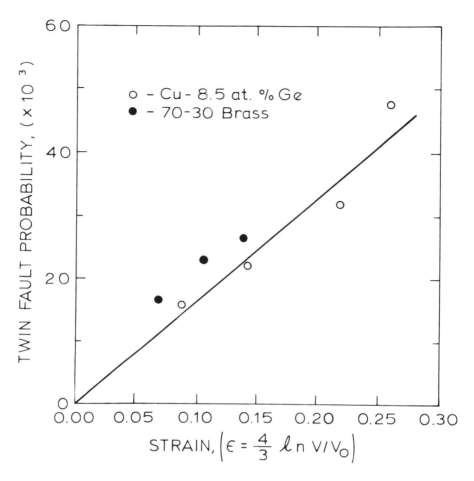

Fig. 8 Twin fault probability β vs. shock loading strain for Cu-8.6 at. pct. Ge and 70/30 α-brass. Data for brass from Ref. 11.

exists it must occur at lower pressures; e.g., in the range reported by Johari and Thomas.[7] It must be recognized that the shock loading conditions can affect the incidence of twinning;[7,25] however, it seems that the critical pressures which have been reported to be in the range 200-300 kbars may also be the result of two other factors: 1) the difficulty in detecting very thin and/or segmented twins in a high dislocation density structure with transmission electron microscopy, and 2) the loss of twins caused by the relaxation resulting from preparation of thin foils. This latter point may be very important since annealing studies on the samples used in this study have shown that for low shock pressures a large fraction of the twin faults disappear in a recovery or relaxation type process.[26]

Average dislocation densities (ρ) were calculated from the effective particle sizes ($D_e(hk\ell)$) and rms microstrains ($<\varepsilon^2>^{1/2}$) in the usual manner with the relation $\rho = 2\sqrt{3} <\varepsilon^2>^{1/2}/bD$;[27,28] b is the Burgers vector. The values for the 100, 180, 350, and 475 kbar specimens were 1.2×10^{11}, 1.7×10^{11}, 2.5×10^{11}, and $3.4 \times 10^{11}/cm^2$, respectively. For comparison, similar calculations give a dislocation density of $\sim 15 \times 10^{11}/cm^2$ for the same alloy deformed by filing[20] and $\sim 0.6 \times 10^{11}/cm^2$ for Cu deformed by explosive loading at 435 kbars.[18] Comparing dislocation densities calculated from the microstrains to those calculated from the particle sizes[27,28] shows that the dislocation density based on the microstrains is larger by about a factor of four. This suggests a tendency toward grouped dislocations or pile-ups rather than single dislocations and is an expected result for an alloy of this type.

It is generally accepted that the hardening caused by cold work is mostly due to the elastic interactions of mobile dislocations with those accumulated in the materials by the deformation. One of the limitations in describing these processes in detail is a lack of quantitative descriptions of the nature of the substructure in the cold worked state. Because diffraction data of the type generated here provides a quantitative description of the substructure, an attempt was made to relate the changes in structure to changes in microhardness or yield strength measured on the same specimens. (Yield strength measurements were made on several specimens which showed that the yield strength could be approximated by (1000/3)(DPH) where DPH is the diamond pyramid hardness value.)

As a first attempt to relate the structural parameters to the mechanical properties, a best fit expression relating the average dislocation densities (calculated from the particle sizes and microstrains) to the yield strength was determined with a least squares analysis of the form:

$$\sigma = \sigma_o + K\rho^m$$

Figure 9 shows the variation of the change in yield strength $\sigma - \sigma_o$ with the dislocation density for the four shock loading pressures. The best fit values of the constants m and K were found to be 0.5 and 5,100 dynes/cm, respectively. The resulting expression given below is of the form most commonly used to relate dislocation density to flow stress: (c.f. Ref. 29)

$$\sigma = \sigma_o + 0.45 \; Gb\sqrt{\rho}$$

(here G is the shear modulus and b the Burgers vector.) The

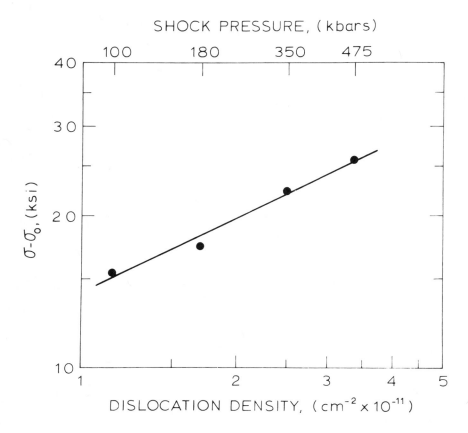

Fig. 9 Change in yield strength ($\sigma-\sigma_0$) vs. dislocation density for shock loaded Cu-8.6 at. pct. Ge.

constant 0.45 found in this study is in agreement with those values given in the literature.[29]

As a second step the relation between each of the substructural parameters and the yield strength was explored with expressions of the form:

$$\sigma = \sigma_0 + K_1 X^m$$

where X is the substructural parameter and K_1 and m are constants. As an example of these data, the variation of $\sigma-\sigma_0$ with the particle size $D_e(111)$ is shown in Fig. 10. The best fit for each of the parameters as determined by least squares analysis gave the values shown in Table II for the exponent m. These data show that the dependence of $\sigma-\sigma_0$ on α^{-1}, β^{-1}, and $D_e(111)$ is the same; i.e., $m \simeq -0.5$. This is expected because the reciprocals of the two

PROPERTIES OF EXPLOSIVELY LOADED Cu-8.6% Ge 601

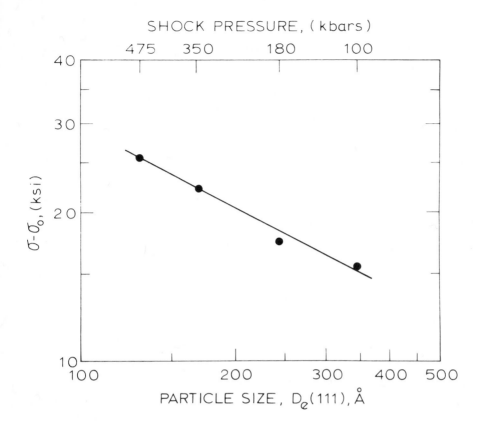

Fig. 10 Change in yield strength ($\sigma-\sigma_o$) vs. effective particle size $D_e(111)$ for shock loaded Cu-8.6 at. pct. Ge.

TABLE II.
Calculated Values of the Exponent m in the Expression $\sigma = \sigma_o + KX^m$ Relating the Yield Strength (σ) to Various Substructural Parameters (X).

Parameter	X	m
Stacking fault probability	α	0.55
Twin fault probability	β	0.46
Effective particle size	$D_e(111)$	-0.53
	$D_e(100)$	-0.88
Rms microstrain	$\langle\epsilon^2_{50\text{Å}}\rangle^{1/2}_{111}$	1.4
	$\langle\epsilon^2_{50\text{Å}}\rangle^{1/2}_{100}$	1.9

fault probabilities, α and β, represent the number of {111} planes between faults and therefore contribute to $D_e(hk\ell)$. The value of m for $D_e(100)$ is slightly larger in magnitude at -0.88 and reflects the difference in the behavior of $D_e(100)$ with shock pressure (Fig. 1). Although this difference is most likely related to the crystallographic asymmetry of both the substructure and the deformation process, the exact reason for the difference is not clear at present. A value of -0.5 for the exponent relating $\sigma-\sigma_0$ to particle size is in agreement with a Hall-Petch type relation:[30,31]

$$\sigma = \sigma_o + K_2 D^{-1/2}$$

This implies that the barriers to dislocation motion (which occur at an average spacing defined by the particle size) are such that a stress build-up occurs within a "particle" which causes nucleation of slip in an adjacent "particle". A similar behavior has been noted between subgrain diameter and yield strength in Al by Ball[32] and in Fe by Warrington.[33]

The exponent relating the rms microstrain in the <111> direction to the yield strength is near 1.5 as shown in Table II. (As with the particle size, changing from the <111> to the <100> crystallographic direction gives a slightly larger exponent.) Although the significance of the value of ∿1.5 for this exponent is not clear, it should be noted that the rms microstrain represents the non-uniform strain existing over a given averaging distance and can be related to the average stored energy in the cold-worked structure.[19,27,28] The primary source of these non-uniform strains is the dislocation network itself with some contribution from other defects such as faults. Therefore, whereas the particle size represents the mean free path between obstacles, the rms microstrain is directly related to the long-range stress fields associated with the dislocations and dislocation boundaries making up the obstacles.

The above functional relationships have been arrived at by treating each substructural parameter separately, primarily because the as-shocked data are not extensive enough to allow a least squares fit of two or more parameters simultaneously. However, as part of an annealing study we have gathered sufficient data points over a large number of isotherms to verify that the exponents given above are essentially unchanged when the yield strength is related to all four substructural parameters simultaneously.[26]

We have concluded therefore that the yield strength of explosively loaded Cu-8.6 at. pct. Ge can be related to the four substructural parameters with an expression of the form:

$$\sigma = \sigma_o + K_1 D_e^{-1/2} + K_2 [<\varepsilon^2_{50\text{Å}}>^{1/2}]^{3/2} + K_3 \alpha^{1/2} + K_4 \beta^{1/2}$$

As mentioned previously, the terms for α and β can be included as part of the particle size (D_e) term. Based on this expression it is apparent that the same dislocation density can give different values for the yield strength depending on the nature of the dislocation distribution. For example, in a highly polygonized structure the microstrain term might be expected to make a much smaller contribution so that the D_e term would be dominant.

The dislocation distribution developed by shock loading is quite uniform and therefore in order to establish the generality of these observations we are currently investigating the behavior of the same alloy deformed by rolling. In addition, we are examining the behavior of other materials, such as Cu, that form a cell structure on deformation.

At the present time we have made no attempt to establish relationships between the electrical resistance changes and the structural changes. Preliminary examination in connection with the annealing studies indicates that point defects play an important role in determining the electrical resistance changes. Considering this point, the smooth increase in the electrical resistance with shock pressure (Fig. 5) indicates that there is no evidence for thermal recovery of the residual substructure during the shock loading and specimen recovery processes for pressures up to 475 kbars.

ACKNOWLEDGEMENTS

The authors gratefully acknowledge the financial support of the U. S. Atomic Energy Commission and Michigan Technological University. We also thank Mr. E. T. Puuri for invaluable technical assistance, Dr. H. Albert of Engelhard Industries who arranged for the alloy, and Dr. A. Holtzman of E. I. du Pont de Nemours who arranged for the explosive loading.

REFERENCES

1. C. S. Smith: Trans. TMS-AIME 212, (1958) p. 574.
2. G. E. Dieter: in Strengthening Mechanisms in Solids, ASM, Metals Park, Ohio, 1962, p. 279.
3. G. E. Dieter: in Response of Metals to High Velocity Deformation, P. G. Shewmon and V. F. Zackay, eds., Interscience, New York, 1961, p. 409.
4. E. G. Zukas: Metals Engr. Quart. 6, (1966) p. 1.
5. C. S. Smith and C. M. Fowler: in Response of Metals to High Velocity Deformation, P. G. Shewmon and V. F. Zackay, eds., Interscience, New York, 1961, p. 309.

6. R. L. Nolder and G. Thomas: Acta Met. 11 (1963) p. 994 and Acta Met. 12 (1964) p. 227.
7. O. Johari and G. Thomas: Acta Met. 12 (1964) p. 1153.
8. L. E. Murr and F. I. Grace: Trans. TMS-AIME 245 (1969) p. 2225.
9. G. T. Higgins: Met. Trans. 2 (1971) p. 1277.
10. R. J. DeAngelis and J. B. Cohen: Deformation Twinning, Gordon and Breach, New York, 1964, p. 430.
11. H. S. Yu, Ph.D. Thesis, University of Denver, 1967.
12. D. E. Mikkola and J. B. Cohen: Acta Met. 14 (1966) p. 105.
13. R. G. McQueen and S. P. Marsh: J. Appl. Phys. 31 (1960) p. 1253.
14. L. E. Murr and M. F. Rose: Phil. Mag. 18 (1968) p. 281.
15. M. F. Rose and M. C. Inman: Phil. Mag. 19 (1969) p. 925.
16. D. C. Brillhart, A. G. Preban and P. Gordon: Met. Trans. 1 (1970) p. 969.
17. A. S. Appleton, G. E. Dieter and M. B. Bever: Trans. TMS-AIME 221 (1961) p. 90.
18. D. C. Brillhart, R. J. DeAngelis, A. G. Preban, J. B. Cohen and P. Gordon: Trans. TMS-AIME 239 (1967) p. 836.
19. B. E. Warren: Progress in Metal Physics, Pergamon Press, New York, 8 (1959) p. 147.
20. W. G. Truckner and D. E. Mikkola: J. Appl. Phys. 40 (1969) p. 5021.
21. H. Suzuki and C. S. Barrett: Acta Met. 6 (1958) p. 156.
22. J. A. Venables: Phil. Mag. 6 (1961) p. 379.
23. J. B. Cohen and J. Weertman: Acta Met. 11 (1963) p. 997.
24. F. I. Grace, M. C. Inman and L. E. Murr: Brit. J. Appl. Phys. 1 (1968) p. 1437.
25. A. R. Champion and R. W. Rohde: J. Appl. Phys. 41 (1970) p. 2213.
26. D. J. Borich: Ph.D. Thesis, Michigan Technological University, 1973.
27. R. E. Smallman and K. H. Westmacott: Phil. Mag. 2 (1957) p. 669.
28. G. K. Williamson and R. E. Smallman: Phil. Mag. 1 (1956) p. 34.
29. H. Conrad, S. Feuerstein and L. Rice: Mat. Sci. Eng. 2 (1967) p. 157.
30. E. O. Hall: Proc. Phys. Soc. (London) B64 (1951) p. 742, 747; N. J. Petch: J. Iron Steel Inst. 173 (1953) p. 25.
31. J. P. Hirth: Met. Trans. 3 (1972) p. 3047.
32. C. J. Ball: Phil. Mag. 7 (1957) p. 1011.
33. D. H. Warrington: J. Iron Steel Inst. 201 (1963) p. 610.

FRAGMENTATION, STRUCTURE AND MECHANICAL PROPERTIES OF SOME STEELS AND PURE ALUMINUM AFTER SHOCK LOADING

Torkel Arvidsson and Lars Eriksson

Senior Scientists, Res. Inst. of National Defense

Box 98   S-147 00   TUMBA   Sweden

INTRODUCTION

The background of the studies presented here is our interest to understand the fragmentation behavior of High Explosive Shells. Ten to twenty years of experimental and theoretical work has given us a broad basis of data and semiempirical expressions. The present work represents an approach to the study of one part of the process separately, namely the shock loading of the case material. The paper is divided into three parts: 1. Fragmentation experiments, 2. Mild steel subjected to long, high explosive generated shock waves, 3. Aluminum subjected to flyer plate generated shocks with different durations.

FRAGMENTATION EXPERIMENTS

Among the previously performed experimental series we choose one in which the fragment size distribution from high explosive (Comp B) filled steel cylinders was determined (Johnsson 1972). Five different steels with carbon content varying from .16 to 3.45 wt% were used (see Table I). The cylinders were normalized prior to the experiment. Fig 1 shows the microstructre for some of the steels. The fragment mass distribution was described by the relation (Lundberg and Kemgren 1962).

$$M = M_o \exp(-\gamma m_1) \qquad (1)$$

where M = total weight of all fragments with masses > m_1
M_o = total weight of the shell case

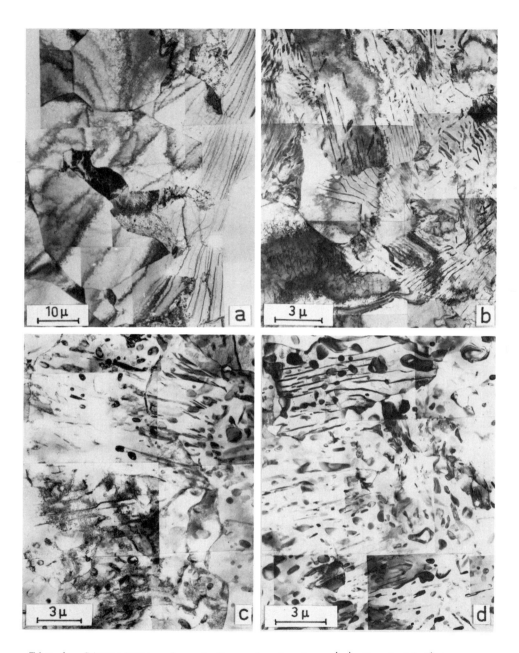

Fig 1. Structure of undeformed steels. (a) Fe-0.16 % C, (b) Fe-0.57 % C, (c) Fe-1.17 % C and (d) Fe-1.36 % C.

STRUCTURE AND MECHANICAL PROPERTIES OF STEELS AND ALUMINUM

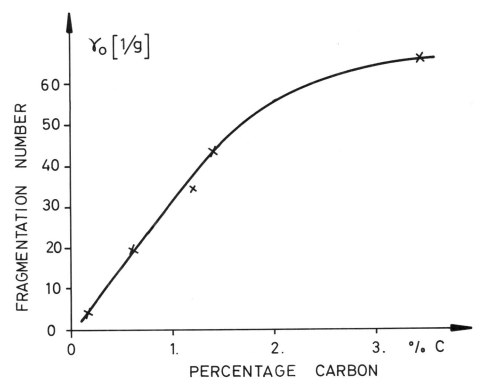

Fig 2. Fragmentation number vs percentage carbon from fragmentation experiments with explosive filled cylinders.

Element present %	Steel				
	A	B	C	D	E
Carbon	.16	.57	1.17	1.36	3.45
Silicon	.23	.56	.59	.66	2.35
Manganese	.71	.81	.72	.74	.53
Phosphorous	.009	.020	.021	.020	.030
Sulphur	.023	.020	.022	.020	.006
Chromium					.07
Nickel					1.31
Magnesium					.08

Table I

The geometry dependence of the parameter γ is found experimentally to be

$$\gamma = \gamma_o \exp(-26.4\, \phi - 15.8\, d/\phi) \quad (2)$$

where ϕ and d is the outer diameter and wall thickness of the shell case (expressed in meters).

The "fragmentation number" γ_o (dimension 1/mass) depends on the properties of the high explosive and the shell case material. The relation (1) is in good agreement with experimental data for $m_1 > .5$ g. Fig 2 shows that the fragmentation number γ_o increases (which corresponds to smaller fragments) with increasing carbon content which is a well known effect of the more brittle behavior at higher carbon contents. The fragments from the .16 % C steel showed only shear fracture surfaces, inclined approximately $45°$ to the outer and inner surfaces of the cylinder while the other four steels all had both radial tension fracture surfaces (adjacent the outer surface) and $45°$ shear surfaces (at the inner surface) (see Fig 3 a and b and Fig 4). This agrees with a model (Hoggatt & Recht 1968) of the fracture process starting with inward growing radial cracks and ending with shear fracture of the inner part of the cylinder, subjected to compressive stresses by the detonation products.

The hardness examination of the unshocked steels and of the fragments (Fig 5) showed that the hardening due to shock loading and plastic deformation decreases with increasing carbon content. This is in agreement with the results of Koepke et al (1965) showing that plane wave shock hardening of carbon steels decreases with increasing carbon content. Moreover we found that the absolute value of hardness after fragmentation decreases with increasing carbon content above 1.2 % C. This must be due to the increasing brittleness which reduces the amount of plastic deformation before fragmentation. The last result is a good illustration to why we are interested to isolate the effect of flowing deformation from the plastic deformation.

MILD STEEL SUBJECTED TO LONG, HIGH EXPLOSIVE GENERATED SHOCK WAVES

For these experiments we used a very simple shock loading technique employing a plane wave lens initiated high explosive (Baratol) to generate a plane shock wave in the steel sample. The peak pressure was varied by placing attenuator plates (lucite) of different thickness between the high explosive and the sample. The sample was softly recovered in foam rubber and water. The pressure-time history in the sample produced by this method has a more or less triangular shape and a relatively long duration (in the order of 10 µs). Our acceptance of the method despite its clear disadvantages (varying peak pressures and pulse durations within the sample, different pulse durations for different peak pressures, spall fractures in the center of the sample etc) was based on the assumption that the dominant

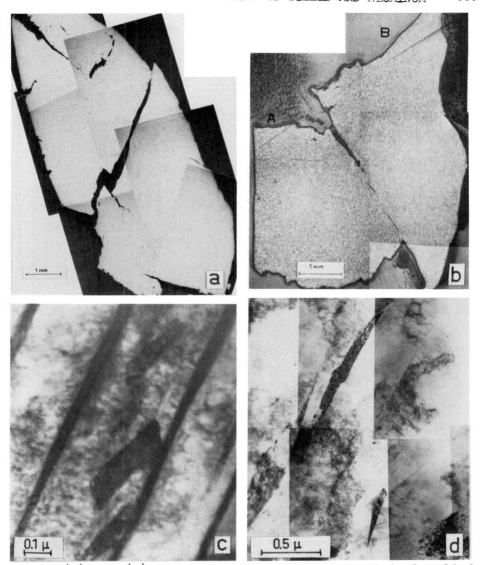

Fig 3. (a) and (b) sections perpendicular to original cylinder axis through fragments of Fe-0.16 % C and Fe-0.57 % C respectively. Radial surface (A in (b)) is nearly perpendicular to the outer cylinder surface (to the left in (b)). Shear surface (B in (b)) is inclined approximately 45° to the inner cylinder surface (to the right in (b)). Note absence of radial surfaces in (a). - (c) and (d) are 1 000 keV TEM images from a Fe-0.57 % C fragment. (c) shear plates contained in the ferrite between two cementite lamellae in a pearlite grain, and (d) a ferrite structure typical pressures exceeding 130 kbar.

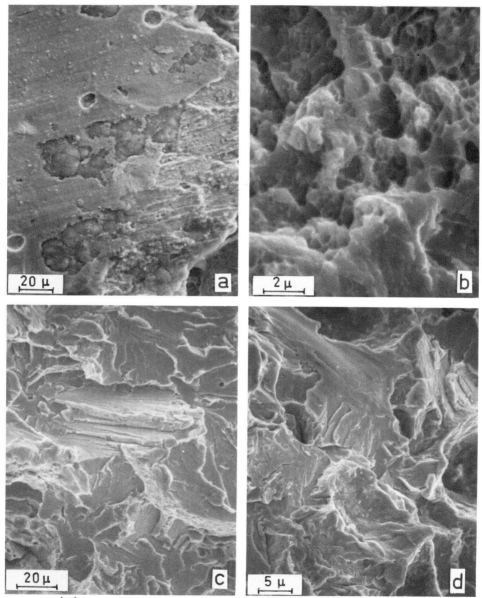

Fig 4. (a) and (b) show shear surfaces of Fe-0.16 % C and Fe-0.57 % C fragments respectively. Upper left hand part of (a) is typical of regions close to the inner surface. (b) dimples in the region close to the radial surface. (c) and (d) cleavage facets on the radial surface, close to the outer surface, on fragments of Fe-0.57 % C and Fe-3.45 % C respectively

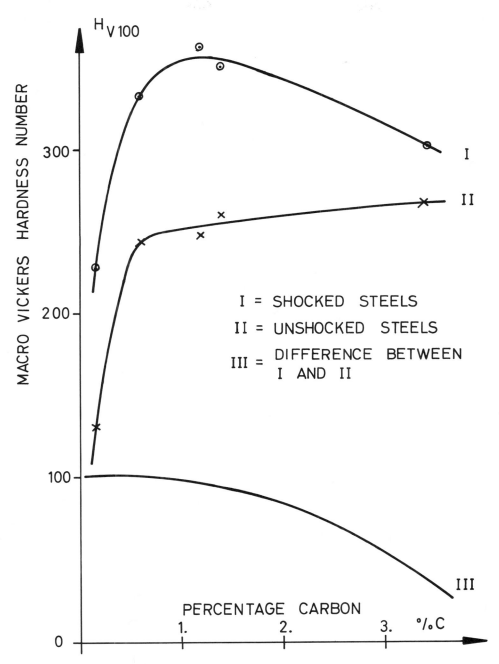

Fig 5. Vickers macro hardness vs percentage carbon for five steels. The curve III illustrates the increase of hardness.

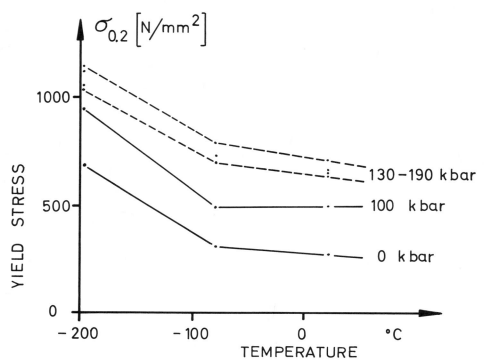

Fig 6. Yield stress $\sigma_{0.2}$ vs temperature for various shock pressures. Fe-0.16 % C.

characteristic parameter of the loading was its peak pressure. This implies that the relaxation time for relevant physical phenomena (dislocation movement, twin formation, phase transition) should be short compared to the characteristic time constant for the shock pulse. This first series of tests were made using as sample material mild steel with a carbon content of 0.16 wt%, normalized at 900°C for 35 minutes. Its average grain size was 23 μm. The peak pressures close to the interface between the attenuator plate and the sample, where all the examinations after the shot were performed, were 100,130,150,170 and 190 kbars. This pressure interval was chosen because the rate of change of the shock hardening is reported in the literature to be especially large around the α-ϵ phase transition pressure (130 kbars). An examination of the macro Vickers hardness after the shot showed an increase from 127 H_v for the unloaded steel to 185 H_v after 100 kbars and 223-239 after 130-190 kbars. Small tensile test pieces (1.5 mm diameter, 15 mm length) were cut from a 5 mm thick disk at the attenuator sample interface and tested at room temperature -78°C and -196°C (strain rate 0.001 sec^{-1}). Fig 6 shows the yield stress vs temperature for samples subjected to different shock pressures. All curves have

STRUCTURE AND MECHANICAL PROPERTIES OF STEELS AND ALUMINUM 613

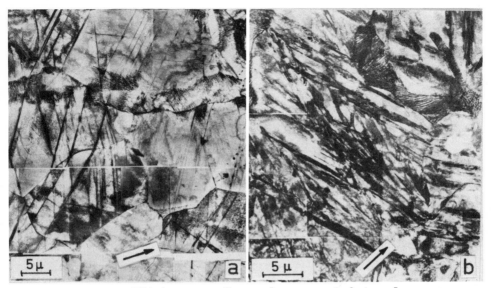

Fig 7. 1000 keV TEM images of samples exposed to planar shock waves. (a) and (b) show the structures obtained in Fe-0.16 % C steel after shock loading with the explosive lens method to 100 and 150 kbars respectively.

approximately the same temperature dependence which shows that only the temperature independent part of the yield stress is affected by the shock loading. Fig 6 also shows that the yield stress is increased rapidly when shock pressure is varied from 0 to 100 kbars and from 100 to 130 kbars. Above 130 kbars, however, a much slower rate of change is achieved. This may be explained by assuming the pressure pulse duration being long enough to allow the phase transition to go to completion for a shock peak pressure only slightly above the transition pressure. The hardening due to change of structure induced by the phase transition is then fully developed as soon as the transition pressure is passed and further increase of the pressure will only have a small effect on the hardening. Electron microscope examination of the 100 and 150 kbars shots (Fig 7) also showed clear signs of the phase transition. This behavior is consistent with the hardness results reported above. It is also in accordance with the result reported by Koepke et al 1971, who found a phase transition relaxation time at 158 kbars of the order of 0.8 μs for a mild steel similar to ours, which is short compared to our pulse durations. Because there has been a discussion in the literature about the role of the deformation twins in the increase of yield stress we tried to use our experimental results to test the hypothesis that the twin boundaries are responsible for the whole yield stress rise. From optical micrographs of the shocked material we could calculate an equivalent

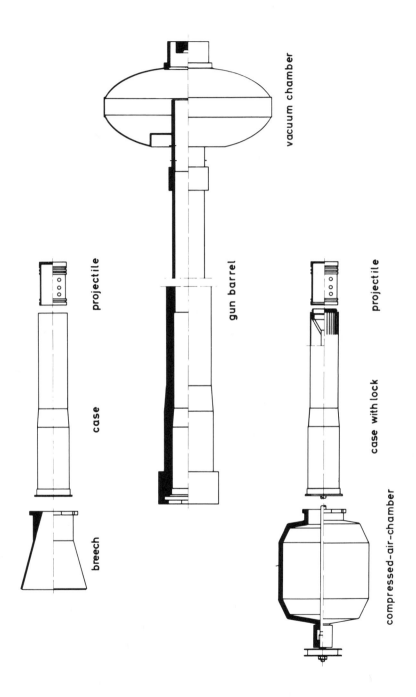

Fig 8. Over-all view of gun. A 90 mm smooth bored rifle of 2.70 meters length is used.

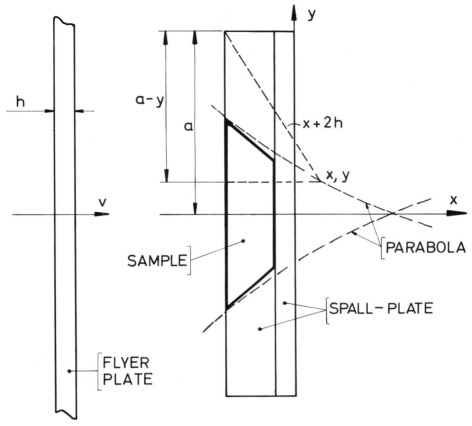

Fig 9. The maximum size of the sample is determined by the thickness of the flyer-plate and the outer diameter of the spall-plate. The sample must be situated inside the parabola $(a-y)^2 = 4hx + 4h^2$ because the entire pulse must pass before the arrival of the edge unloading. The spall-plate to the right takes care of the reflected wave.

new grain size, counting the twins as new grains. The scatter in the grain size figures was large and we could only get an approximate mean value for all shocked samples. Using this value in the Hall-Petch equation we calculated the yield stress increase which we should expect if the twin boundaries were as effective as grain boundaries in stopping dislocation movement. This yield stress increase was about half the value we found experimentally. This shows that the increase of the dislocation density is an important and may be the dominant factor in the shock hardening. Our conclusion of the above presented results is that the pressure pulse duration is a most important parameter in the shock hardening phenomena. Therefore, we are

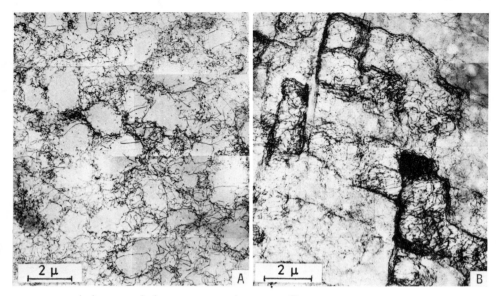

Fig 10. (a) and (b) show Al (99.99 %) after flyer plate impacts at a velocity of 100 and 750 m/sec respectively.

setting up experimental arrangements, using a gun accelerated flyer plate technique, where we intend to study the influence of shock pulse durations and shock pressure systematically. The first results are presented below.

ALUMINUM SUBJECTED TO FLYER PLATE GENERATED SHOCKS WITH DIFFERENT DURATIONS

A gun technique (Fig 8) has many advantages in an investigation of the response of a material to shock loading. One of the most important is the wide range of predetermined controlled pulse shapes, durations and amplitudes than can be obtained by varying the velocity, thickness and composition of the flyer plate positioned on the head of a projectile. A 90 mm smooth bored recoilless rifle is used. The flat-faced projectile is launched under conditions of intermediate vacuum of 10^{-2} torr. to eliminate air-cushion effects at impact. The projectile was glued to the case preventing it from being launched by the air pressure as vacuum exists in the gun. When only air pressure was used the projectile received a velocity of 0.07 mm/μs. The maximum velocity was 1 μs/mm when using gun-powder. The geometry of flyer plate and target is schematically shown in Fig 9. From the beginning we only used gun-powder as driving medium. Today, however, we have switched to compressed-air. In the first series of experiments we shock loaded aluminum (99.99 % Al, a pure material with no phasetransformation). The material was

STRUCTURE AND MECHANICAL PROPERTIES OF STEELS AND ALUMINUM 617

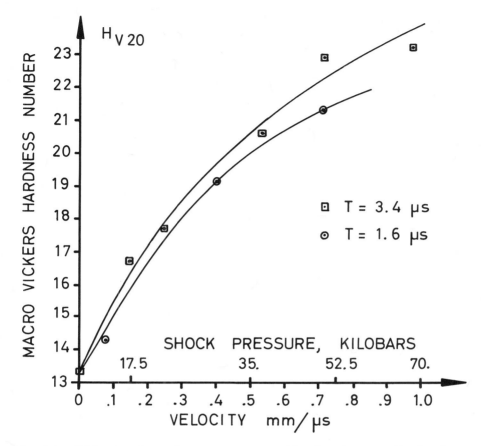

Fig 11. Effect of peak pressure on the hardness of aluminum Al (99.99 %) at two different pulse durations.

annealed 1 hr at 600°C. The grain size was 0.9 ± 0.1 mm. The dislocation structure of two shocked samples are shown in Fig 10. The hardness examinations are given in Fig 11. The hardness increases about 74 % in the pressure interval 0 to 70 kbars while the 3.4 µs pulse gives 3-4 % higher hardness than the 1.6 µs pulse.

CONCLUSIONS

Fragmentation experiments reveal increasing number of fragment with increasing carbon content. Examinations show that both duration and peak pressure are important to the hardening. Possibilities are given to separate the flowing deformation from plastic deformation.

REFERENCES

1. H. G. Bowden and P. M. Kelly, Acta Met. 15, 1489 (1967).
2. C. R. Hoggatt and R. F. Recht, J. Appl. Phys. 39, 1856 (1968).
3. AEA Johnsson, FOA 2 Report C 2529-D4 (1972).
4. B. G. Koepke, R. P. Jewett, and W. T. Chandler, Trans. ASM 58, 510 (1965).
5. B. G. Koepke, R. P. Jewett, W. T. Chandler, and T. E. Scott, Met. Trans. 2, 2043 (1971).
6. L. Lundberg and E. Kemgren, FOA 2 Report AH 2167-253 (1962).

TWINNING IN SHOCK LOADED Fe-Al ALLOYS

M. Bouchard* and F. Claisse**

* Université du Québec, Chicoutimi, Québec, Canada

** Centre de Recherche de l'Etat Solide,
Université Laval, Québec, Canada.

Most metals and alloys, specially the b.c.c. ones, are highly susceptible to twinning when the deformation rate is high or when the temperature is low. However, those which are ordered are not twinned after a fast deformation as observed by Elam (1), Kramer and Madin (2), Barrett (3), and Bouchard et al (4).

Laves (5) has suggested that long range order should impede twinning because the product of the twinning transformation is a crystal with a different ordered arrangement and the internal energy would prevent the "pseudo-twin" to form.

The experiments of Bolling and Richman (6) on ordered and disordered Fe_3Be alloys indicate that both alloys "twin" under compression and that ordered Fe_3Be detwins when the load is removed. That particular behavior of ordered b.c.c. alloys has been predicted by Claisse (7) who showed that atomic forces between next neighbor atoms in the "pseudo-twin" are in such directions that the latter should detwin spontaneously and elastically.

The influence of order on twinning can be studied by changing either the composition of the alloy, the thermal treatment, or both of them. The present paper is the result of such a research made on Fe-Al alloys. One purpose of the research was to find some indication that these alloys in the ordered state, whether do not twin under explosive shock or twin and detwin successively.

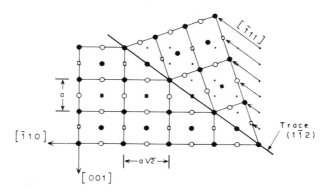

Fig. 1: Atomic movements during formation of a "twin" in the DO3 structure. Black symbols represent atoms in the projection plane, open symbols are atoms above or below the projection plane; round symbols are Fe atoms, square symbols are Al atoms.

ALLOYS

Iron-rich alloys in the Fe-Al system form an extensive solid solution which includes long range ordered phases based on both the DO3 and B2 structures (8-13). According to Rimlinger (8), alloys containing less than 19 at % Al have "α" bcc disordered structure; those between 19 to 23 at % Al are ordered particles of Fe_3Al (DO3) surrounded by disordered "α" phase. Alloys in the range 23 to 29 at % Al exibit homogeneous DO3 type long range order (LRO). Between 29 and 33 at % Al, both the DO3 and B2 ordered phases exist while the B2 structure only is found above 33 at % Al.

When the influence of a twinning shear on an alloy with a DO3 superlattice is examined graphically (Fig. 1), it turns out that the net fraction of A-B bonds destroyed by shear is smaller than in the case of the B2 superlattice. In this case, only the second nearest neighbours bonds are modified by twinning. In fact, the number of A-B bonds destroyed by shear increase gradually with aluminium content between 19 and 50 at % Al.

EXPERIMENTAL

Iron-aluminium alloys containing between 13 and 38 at % Al have been prepared from iron of 99.9% purity and aluminium 99.99% purity in an induction furnace under vacuum.

The ingots were hot-rolled and annealed for three days

at 1100°C to promote homogeneisation and to coarsen the grains. A coarse grain structure was desired since this, as a rule, promotes twinning. The grain size after annealing was 0.5 - 1.0 mm. A few single crystals of each alloy were also grown from the melt by the Bridgman technique. Cylindrical specimens 2.54 cm in lenght and 0.63 cm in diameter were machined from the annealed stock.

Tests were performed on specimens in three heat treated conditions:

a) to produce maximum DO_3 type LRO and anti-phase domain (APD) size, specimens were heated at 1000°C and then slowly cooled to room temperature at an average rate of 10°C/hour.

b) to maximize the B_2 type LRO and APD size, specimens were heated to 600°C for six hours, then quenched into water at room temperature.

c) to minimize both long range order and APD size, specimens were heated to 1000°C for two hours and quenched into water at room temperature. It is known however that a high degree of LRO is still present in those alloys, even after rapid quenching from high temperature (14-16).

The equipment used for shock loading experiments is shown in Fig. 2. The specimens were closely fitted into a steel cylinder and placed over a steel disk to prevent scabbing at the lower end of the specimen. Tetryl was used as the explosive material. The charges were cylinders one inch in diameter and one inch long. The pressure generated by the explosive charge was estimated to be 200 kilobars. For low temperature experiments, the specimen and its assembly were kept in liquid nitrogen; when required, they were removed from the bath, the charge was rapidly placed in position and ignited as in room temperature experiments. The heat capacity of the assembly was large enough to prevent any significant heating of the specimen before shock loading occurred.

After shock loading, each specimen was cut axially and the surface was mechanically polished. Final polishing and etching were made electrolytically.

The quasi-static compression tests at liquid nitrogen temperature were made with an Instron testing machine, Model TTC, at a constant crosshead velocity of 0.02 inch/minute. Specimens were deformed between two hardened steel blocks.

RESULTS

A. Deformation by Explosive Shock

Polycrystalline Fe-Al alloys with different heat treatment were shock loaded at -195°C and then were examined for the fraction of grains in which twins were observed. The results are given in Fig. 3. It is obvious that the alloys can be classified in three groups depending on their susceptibility for twinning: alloys with less than 20 at % Al twin easily while those with more than 30 at % seem not to twin at all; in the intermediate group, the susceptibility for twinning decreases rapidly between 23 and 29 at % Al. Each of these three groups will now be analysed separately and results with single crystals will be included also.

1. <u>Alloys with less than 20 at % Al</u>. Numerous twins have been found in single crystals and polycrystals of these alloys after deformation by explosive shock. For example, up to 8 different twinning systems have been observed in a single crystal containing 13 at % Al which has been explosively loaded at liquid nitrogen temperature (Fig. 4). All the traces are compatible with (112) twinning planes associated with twinning systems operating in compression.

In this composition range, the alloys are disordered, whatever their thermal treatment was; consequently, they are expected to twin under impact as most bcc metals do.

2. <u>Alloys with 20 to 30 at % Al</u>. The twinning susceptibility of iron-aluminium alloys decreases very rapidly when the aluminium content increases from 23 to 30 at % Al. This is observed as a net decrease in the percentage of twinned grains in polycrystals as well as a decrease in the number of twinning systems in each grain. For example, in the alloy Fe-30 at % Al, twins were observed on a few grains only, usually located near grain intersections (Fig. 5a).

Specimens in this range of composition, specially near the 30 at % Al provided the first evidence of detwinning. In circled area of Fig. 5a, for instance, we observe one large twin and two thin traces above it which are difficult to identify as twin traces. However, at a larger magnification (Fig. 5b), each trace appears as two lines which remind of a former twin; the central part of this twin seems to have reverted to the original matrix while the interface has remained nearly unchanged. Most of the other faint lines in Fig. 5a are also debris of former twins.

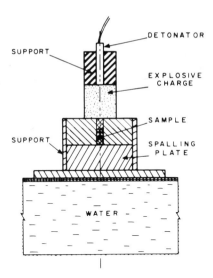

Fig. 2: Specimen assembly for shock loading at room and below room temperature.

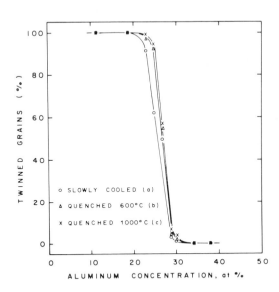

Fig. 3: Effect of aluminium content on percentage of twinned grains for specimens shock loaded at -195°C.

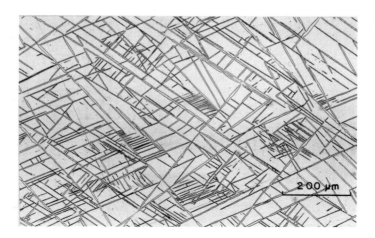

Fig. 4: Fe-13 at % Al single crystal deformed by explosive shock at -195°C. Eight twinning systems are observed.

Debris are usually associated with regions of stress which are thought to prevent detwinning; grain intersections are such regions. In other regions where the crystals are more perfect, detwinning should occur more readily; this is consistant with the fact that debris are in lower concentration in such regions. The same observation is made when a polycrystal and a single crystal of the same composition are compared. Twins in a polycrystal containing 25 at % Al are easily seen (Fig. 6a) at a magnification of 500X; practically no twins are observed in the single crystal but if examination is made at a magnification of 1500X (Fig. 6b), fine traces are observed usually grouped in pairs, as Fig. 5, which we interpret as interfaces of twins (produced by the shock wave) which have detwinned.

3. Alloys with more than 30 at % Al. These alloys are very brittle and many specimens were fractured after the explosive shock. Twins have not been observed in these alloys but numerous fine lines have been seen under the microscope, using dark field illumination. These lines seem to be residus or debris left after detwinning; their orientation is always compatible with traces of (112) planes that belong to twinning systems working in compression and having high Schmidt's factors.

4. Effect of degree of order on twinning. Although Fe-Al alloys order very rapidly (14, 15, 16), we hoped that quenching from above the critical temperature would restrict the degree of order reached and thus make twinning more likely. Results of shock loading experiments performed on polycrystal-

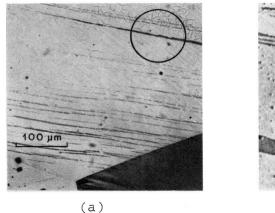 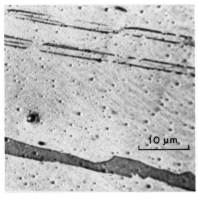

                (a)   (b)

Fig. 5: Ordered Fe-30 at % Al alloy explosively loaded at
        -195°C.

   a) Twins and debris near grain intersection.
   b) Large twin with debris delineating former twins.
      (enlargement of circled area in "a")

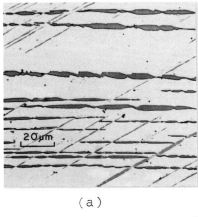 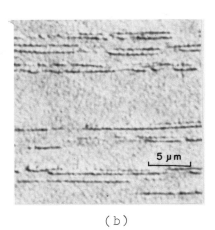

                (a)   (b)

Fig. 6: Ordered Fe-25 at % Al alloy explosively loaded at
        -195°C.

   a) Twins in a polycrystal.
   b) Debris in a single crystal.

line specimens in the three heat treated conditions described above are shown in Fig. 3. Specimens, partially disordered by quenching from high temperature (treatment "c") were in general slightly more twinned than well annealed and ordered specimens (treatment "a").

The deformation behavior of single crystals in the composition range 23 to 30 at % Al was more sensitive to heat treatment. The permanent deformation of ordered crystals was significantly smaller than for those with a lower degree of order when submitted to the same conditions of deformation by shock loading, as shown in Table I. This behavior cannot be attributed to the hardening of ordered crystals since quasi-static compression tests (17, 18) have shown that quenched specimens are slightly harder than slowly cooled specimens. It is believed that both well ordered and partially ordered crystals have been deformed by about the same quantity under the explosive shock, but that ordered crystals have detwinned more than the others. This would explain the difference in permanent deformation.

B. Quasi-Static Compression Tests

A few quasi-static compression tests were made on single crystals at liquid nitrogen temperature to see if detwinning would occur during unloading.

Twinning and detwinning have been observed on well ordered single crystals containing between 23 and 27 at % Al; crystals with higher aluminium content did not twin under our conditions of deformation. Some of the other specimens have twinned on loading, and detwinned almost completely during unloading. An interesting example is given in Fig. 7: an ordered Fe_3Al single crystal has been submitted to three cycles of deformation. During each cycle, the specimen was deformed by 2% and then slowly unloaded. The total permanent deformation after the three cycles was 0.5% only. Metallographic observations of the specimen after the compression test have shown small residus or debris similar to those observed in shock loaded crystals of the same composition. In polycrystalline specimens, detwinning was sometimes observed, but considerably less frequently than in single crystals.

DISCUSSION

The absence of twinning in specimens containing more than 30 at % Al may be attributed to two effects:

TABLE I

Effect of order on permanent deformation of single crystals deformed by explosive shock.

Crystal no.	Composition % at Al	Heat treatment *	Expected relative degree of order	Perman. deformation %
890-1	18	(a)	disordered	7.8
890-2	18	(c)	disordered	8.0
491-1	23	(a)	high	5.8
491-2	23	(c)	lower	7.5
391-1	25	(a)	high	4.3
391-2	25	(c)	lower	7.2
590-1	27	(a)	high	3.8
590-2	27	(c)	lower	7.7
691-1	30	(a)	high	6.8
691-2	30	(c)	lower	8.2
990-1	34	(a)	high	7.8
990-2	34	(c)	lower ?	8.0

* (a): slow cooled - (c): quenched from $1000°C$.

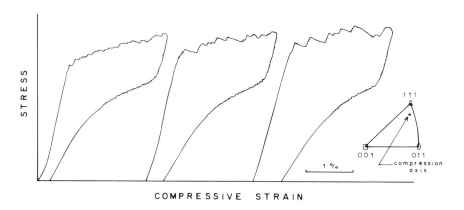

Fig. 7: Stress-strain curves for three cycles of deformation for an ordered single crystal, showing detwinning during unloading.

i) the shock wave energy is sufficient to produce twins in the alloys but twins spontaneously detwin after the wave has passed. Several observations tend to confirm this mechanism.

ii) the energy of the shock wave is too low to overcome the critical energy for twin formation.

In Fe-Al alloys which are normally ordered when composition permits, the twinning energy is a function of both composition and degree of order. An estimate of the APB energy, associated with the motion of $1/6\, a_o$ [111] (112) twinning dislocations has been made by Leamy and Kayser (17). The APB energy (γ_T) is described in terms of the quasichemical bond energies and site occupation probabilities:

$$\gamma_T = \frac{2\sqrt{2}}{3\, a_o^2}[E_{NN}(4S_{NN}^2 - S_{NNN}^2) + E_{NNN}(2S_{NNN}^2 - 5S_{NN}^2 - 4x^2 + 4xS_{NN})] \quad (1)$$

Where: E_{NN} = quasichemical energy difference for first neighbor pair energies.
E_{NNN} = quasichemical energy difference for second neighbor pair energies.
S_{NN} = first neighbor LRO parameter
S_{NNN} = second " " "
x = aluminium concentration of the alloy
a_o = lattice parameter of the DO_3 unit cell.

Since the motion of twinning dislocations leads to the formation of APB, the magnitude of the external stress must be sufficient to allow for the production of APB. The stress $\tau APB_{(T)}$ required for this process is determined largely by the APB energy, through the relation (17)

$$\tau APB_{(T)} = \gamma_T / b_T \quad (2)$$

where b_T is the magnitude of the Burger's vector of the twinning dislocation. Curves representing calculated γ_T and $\tau APB_{(T)}$ for ordered alloys are shown in Fig. 8. The stress $\tau APB_{(T)}$ increases rapidly with aluminium content; this indicates, in agreement with experiments, that twinning should become more difficult in alloys of high aluminium content. However, it is believed that an explosive shock is strong enough to overcome the internal resistance offered by order to the movement of twinning dislocations even in alloys with high aluminium content. But detwinning probably takes place immediately after the passage of the shock wave, because the twin energy (APB energy) also increases.

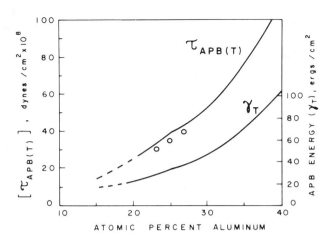

Fig. 8: Anti-phase boundary energy and the stress required for motion of twinning dislocations in ordered Fe-Al alloys (calculated with eq. 1 and 2, parameters taken from Leamy and Kayser, 1969). O = τ_T, experimental values of critical stresses for twinning.

REFERENCES

1) ELAM, C.F.: Proc. Soc. A153: 273, 1936.
2) KRAMER, I. & MADDIN, R.: Trans. AIME, 194: 197, 1952.
3) BARRETT, C.S.: Trans. AIME, 200: 1003, 1954.
4) BOUCHARD, M., CLAISSE, F., TARDIF, H.P. & DUBE, A.:
    Can. Metal. Quart., 9: 395, 1970.
5) LAVES, F.: Naturwissenschaften, 39: 546, 1952.
6) BOLLING, G.F. & RICHMAN, R.H.: Acta Met., 13: 723, 1965.
7) CLAISSE, F.: Proc. Sandefjord & Lillehammer Conference
    High Energy Rate Working of Metals, 1: 194, 1964.
8) RIMLINGER, L.: Thesis, Nancy, France (1969).
9) GUTTMAN, L, SCHNYDERS, H.C. & ARAI, G.J.:
    Phys. Rev. Letters, 22: 520, 1969.
10) LÜTJERING, G. & WARLIMONT, H.: Acta Met., 12: 1460, 1964.
11) OKAMOTO, H. & BECK, P.A.: Met. Trans., 2: 569, 1971.
12) EPPERSON, J.E. & SPRUIELL, J.E.:
    J. Phys. Chem. Solids, 30: 1721, 1969.
13) WARLIMONT, H. & THOMAS, G.: Metal Sci. J., 4: 47, 1970.
14) LAWLEY, A. & CAHN, R.W.:
    J. Phys. Chem. Solids, 20: 204, 1961.
15) DAVIES, R.G.: J. Phys. Chem. Solids, 24: 985, 1963.
16) SELISSKII, Y.P.: Fiz. Metal. Metalloved, 10, no 5, 714, 1961.
17) LEAMY, H.J. & KAYSER, F.X.: Phys. Stat. Sol. 34: 765, 1969.
18) LEAMY, H.J., KAYSER, F.X. & MARCINKOWSKI, M.J.:
    Phyl. Mag., Vol. 20, no 166, p. 763, 1969.
19) FISHER, J.C.: Acta Met., 2: 9, 1954.

Discussion by R. J. Wasilewski
           National Science Foundation

I would like to suggest that the phenomenon investigated here is one of stress-assisted martensitic transformation, not one of reversible mechanical twinning.

In spite of the geometry clearly reminiscent of the b.c.c. twin formation, the distinction is not merely a trivial one of semantics. Firstly, as already pointed out by Laves, it is meaningless to speak of twinning when the "twin" differs in structure from the "matrix". Secondly, it seems that the question whether the mechanism involved is that of twin formation--widely accepted as one of plastic (i.e. irreversible) deformation--or one of stress-assisted phase transformation, is a fundamental one. There is no reason whatever why any plastic deformation should be inherently necessary in a phase transformation. In this connection it may be noted that Laves questioned the interpretation of Bolling and Richman (6) of their observations on Fe_3Be in terms of twin formation (20), and the authors in fact agreed with his objections (21).

The results presented here suggest that the mechanism is one of stress-assisted martensite formation. It would seem that this takes place above the A_f temperature in the ordered alloys, and below it in the disordered ones. In the former case the reversion back to austenite occurs on removal of the stress; in the latter the martensite remains stable. The ordering thus affects the transformation temperature, or more accurately the A_f temperature. The geometry of this transformation in itself should not be interpreted as indicating twin formation, and the behavior predicted by Claisse (7) remains valid <u>above</u> the transformation temperature range, where the austenitic structure atom arrangement is clearly energetically favorable.

The stress effects on the martensitic transformation in TiNi discussed previously (22) provide further information interpreted as indicating that no twinning need be involved.

References:
20. F. Laves, Acta Met. <u>14</u>, 58, 1966
21. G.F. Bolling and R.<u>H</u>.Richman, Acta Met. <u>14</u>, 58, 1966
22. R.J. Wasilewski, Met. Trans. <u>2</u>, 2973, 1971

ANNEALING OF SHOCK-DEFORMED COPPER

E. A. Chojnowski* and R. W. Cahn

Materials Science Division, School of Applied Sciences

University of Sussex, Brighton, U.K.

ABSTRACT

A systematic study was made of the changes resulting from progressive annealing of foils of OFHC copper which had been deformed by planar shock-waves to a peak pressure of 155 or 410 kbar. Mechanical properties were assessed by microhardness histograms (or proof-stress), the resistivity ratio (77K/293K) was measured, and pole-figures were determined. Morphological changes were examined by TEM.

The microhardness histograms showed that substantial recovery preceded true recrystallization; in this respect, shock-deformed copper behaves differently from the cold-worked material. The fractional recovery is greater for the less severely deformed metal. The more heavily deformed foils, strongly textured before annealing, had an almost random texture after recrystallization, but the more lightly deformed foil had an almost unchanged texture after annealing.

No significant refinement of grain-size was achieved by annealing, not even by a special flash-annealing procedure using a ruby-laser. This was put down to a paucity of nucleating sites, especially in the more lightly deformed material, which appeared to recrystallize by strain-induced boundary migration.

---

* Now at the Instituto de Pesquisas e Desenvolvimento, Centro Técnico Aeroespacial, São José dos Campos, São Paulo, Brazil.

A few experiments were done with iron shock-deformed to 350 kbar. The most interesting observation was that — in contradiction with earlier findings by Zukas and McQueen — it was not possible to randomise the texture or refine the grain-size by annealing this material.

INTRODUCTION

A number of detailed studies (1-8) of the annealing of shock-deformed copper have been published since Smith's pioneer paper (1). These studies have been variously concerned with some or all of: hardness, density, resistivity, stored energy, microstructure, the peak pressures up to 1 mbar have been used. The results have shown various mutual contradictions. Most investigators found one or more recovery stages before recrystallization set in, revealed by large fractional changes in resistivity, stored energy or density. Fractional changes in hardness in this recovery stage were smaller or, in recent work (7), absent. Recrystallization of shock-deformed copper was found to be more sluggish than that of copper conventionally deformed to the same hardness (4a, 7); this sluggishness was especially pronounced when shock-deformed single crystals were examined (6), and this was attributed to the observed smallness of the misorientations in the deformed structure. Scattergood et al. (9) similarly found that shock-deformed silver-gold alloy recrystallized more sluggishly than the same material wire-drawn to the same hardness or level of retained energy. — It was one aim of the work reported here to obtain further information on some of the disputed aspects of the behaviour of shock-annealed copper on annealing.

The second aim of the work arose from findings by Zukas and McQueen (10), confirmed by Leslie (11), that when iron was shock-deformed to 400 kbar (10) or 220 kbar (11) and annealed 2 hours at $650^{\circ}C$ (10) or 20 minutes at $500^{\circ}C$ (11), the resultant grain-size was reduced to about one tenth of the original diameter and moreover these grains were randomly oriented. Other investigators found a lesser refinement of the grain size. One of us earlier found extreme grain-refinement in uranium after rapid (not shock) adiabatic deformation which heated the sample transiently to its recrystallization temperature (12). We wished to reproduce these findings and also to establish whether grain refinement and texture randomization is feasible in copper, especially since randomly oriented fine polycrystalline samples are very difficult to prepare and are of interest for some problems in physical metallurgy.

EXPERIMENTAL

Square foils of OFHC copper 0.012 in thick, in the rolled and annealed condition, and pure Swedish iron 0.010 in thick, in the rolled and annealed condition, were stacked 25 high, bolted between cover plates and protected by guard rings in the usual way, and deformed by plane shock waves from a driver plate. The assembly was driven into water to provide a prompt quench. This treatment was undertaken by the late Dr. I.C.Skidmore at A.W.R.E., Aldermaston; he also calibrated the peak pressures. We are most grateful for his collaboration. Peak pressures of 155 and 410 kbar for copper and 350 kbar for iron were used. The duration of the peak pressures was kept constant at a nominal time of 2.0 μsec. — The foils were kept at $0^{\circ}C$ until required.

The principal measure for assessing the microstructural state of annealed samples was microhardness, using a Reichert instrument. Isochronally annealed series of copper specimens were examined by 25 microhardness indentations of 30-gram load; the error bars on the corresponding plot (fig. 7) correspond to three standard deviations, i.e. a 99% confidence range. Some checks in which microhardness was measured at various depths below the original surface showed that there was no significant variation across the section. Isothermally annealed series of specimens were examined by 100 microhardness indentations arrayed uniformly in a square pattern over a square centimetre of surface, and the results displayed in the form of histograms, in the manner originated by Gordon (13) and later used by Thornton and Cahn (14) to examine the recovery of copper and aluminium annealed under stress. The advantage of this approach is that the microhardness values cluster in two separate groups associated with unrecrystallized and recrystallized components. Not only can the fraction of recrystallized material (which is impossible to assess reliably for copper by metallographic methods) be determined from such histograms, but any shift of the "deformed peak" in the histogram is incontrovertible evidence of recovery as distinct from recrystallization. All microhardness tests were performed on electrolytically polished surfaces.

0.5% proof stresses were determined from tensile tests on miniature specimens, deformed at a strain-rate of 0.001/sec. 5 specimens were used for each condition. Resistivities of isochronally annealed samples were measured on spark-machined strip specimens; five specimens were measured for each condition. Resistivities were measured at 293K and 77K, and the results plotted as R_{77K}/R_{293K}, which obviates the problem of calibrating absolute resistivities.

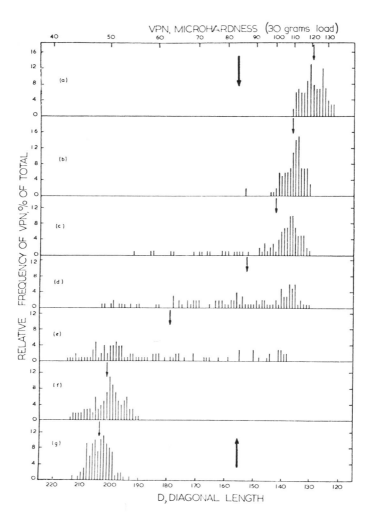

Fig. 1. Cu, 410 kbar. Annealed 250°C. (a) As shock-deformed. (b) 60 min. (c) 120 min. (d) 180 min. (e) 300 min. (f) 600 min. (g) Before shock-deformation.

Electron microscope specimens were jet-thinned by standard techniques, and (111) pole figures were recorded by the standard Schulz diffractometric technique, in reflection.

Some copper samples were subjected to flash-anneals by means of a ruby laser beam focused down to a 3 mm circle; the specimen was polished and spark-machined so that a 3 mm disc, about 0.08mm thick, was retained only by 4 slender radial spokes, with a view to isolating the

ANNEALING OF SHOCK-DEFORMED COPPER 635

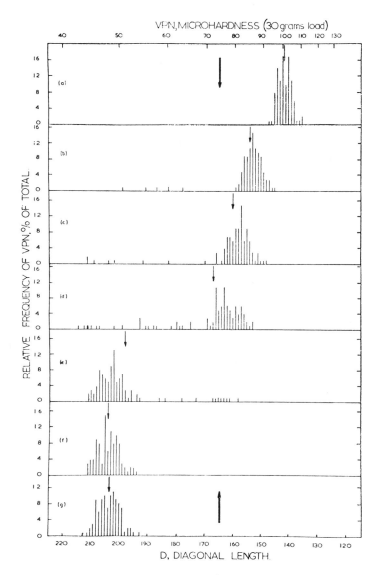

Fig. 2. Cu, 155 kbar. Annealed 450°C. (a) As shock-deformed.
(b) 1 min. (c) 2 min. (d) 4 min. (e) 6 min. (f) 8 min.
(g) Before shock-deformation.

disc thermally. The disc was coated with carbon paint. An ultrafine thermocouple linked to an oscilloscope provided a rough estimate of the peak temperature attained during the 0.9 msec pulse from a 30-joule laser; apparent temperatures in the range 650-760°C were used. A servo-circuit was arranged to spray the disc with water within about 35 msec after the laser pulse. The procedure was based upon that described by Speich and Fisher (15).

RESULTS

Isothermal annealing runs were made with copper specimens as follows: Cu 410 kbar, 200°C, 250°C, 300°C, 350°C; Cu 155 kbar, 300°C, 350°C, 400°C, 450°C. Figs. 1 and 2 show typical microhardness histograms for Cu (410) at 250°C and Cu (155) at 450°C. In each histogram, the thin arrow shows for each condition the mean of the 100 individual microhardness values. When recrystallization begins, a few microhardness values are necessarily in error, when the corresponding indentation straddles both recrystallized and unrecrystallized regions. This was also found by Gordon in his original use of this technique. We believe that these intermediate hardness values do not measure any physically distinct condition. The thick arrows represent our estimate of the best dividing line between hardness values to be aggregated with the "shock-deformed" peak and those to be aggregated with the "recrystallized peak". This dividing line was the same for all samples deformed at a constant peak pressure, irrespective of annealing temperature. — The absolute hardness values obtained for the two states of shock-deformation fit neatly on the assembled plot of hardness vs. peak pressure recently published by Hooker et al. (8).

The specimen histograms show unmistakably that the "shock-deformed" peak moves towards lower hardness values with progressive annealing; the deformed material <u>recovers</u>, particularly in the 155 kbar series. Mostly this happens before recrystallization begins, but continues even after the "recrystallized" peak puts in its appearance. This form of recovery is scarcely detectable when copper normally cold-worked is annealed (12, 16).

Figs. 3 and 4 show plots of fraction of specimen recrystallized as a function of isothermal annealing time; the fractions recrystallized were determined from the relative numbers of hardness values above and below the dividing line on the histograms.

Fig. 5 shows the activation plots constructed from the times, $t_{0.5}$, required to reach 0.5 fraction recrystallized at each temperature.

Fig. 6 represents an alternative way of indicating the extent of recovery of the shock-deformed component of the structure. Here the <u>mean</u>

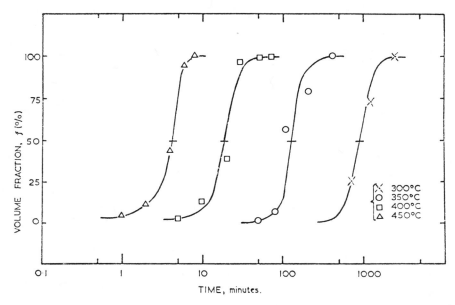

Fig. 3. Cu, 155 kbar. Fraction recrystallized as function of annealing time.

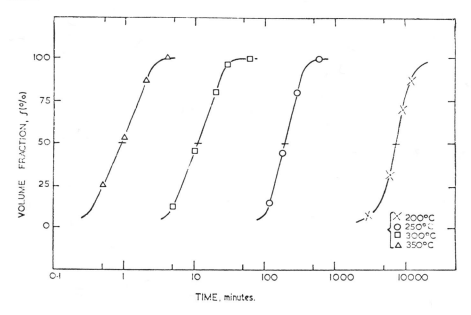

Fig. 4. Cu, 410 kbar. Fraction recrystallized as function of annealing time.

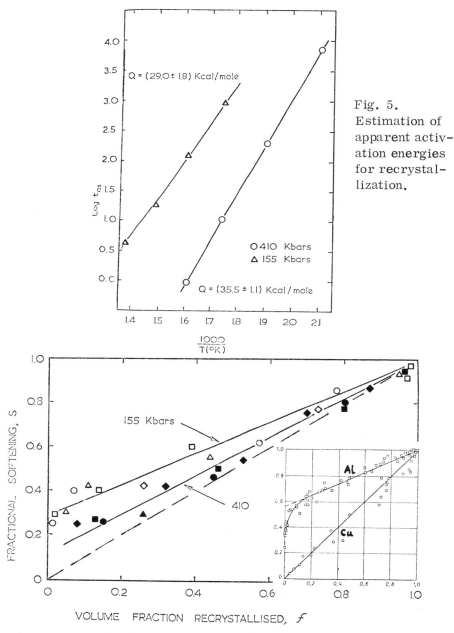

Fig. 5. Estimation of apparent activation energies for recrystallization.

Fig. 6. Fractional softening, $(H_s - H_t)/(H_s - H_0)$, as function of volume fraction recrystallized. H_s, H_t, H_0 are hardness values as shock-deformed, annealed for time t and before shock-deformation, respectively. (Inset) Curve similar to fig. 6 for cold-worked Cu and Al. From Lücke and Rixen (16). - Increasing temperatures in sequence: ◆●■▲ ◇○□△

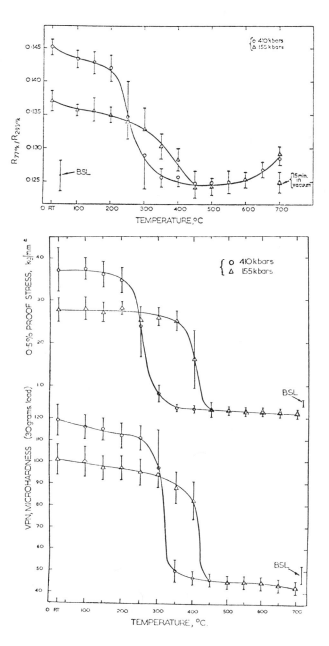

Fig. 7. Cu 410 and Cu 155 isochronally annealed for 15 min. Changes in resistivity ratio, 0.5% proof stress and microhardness.

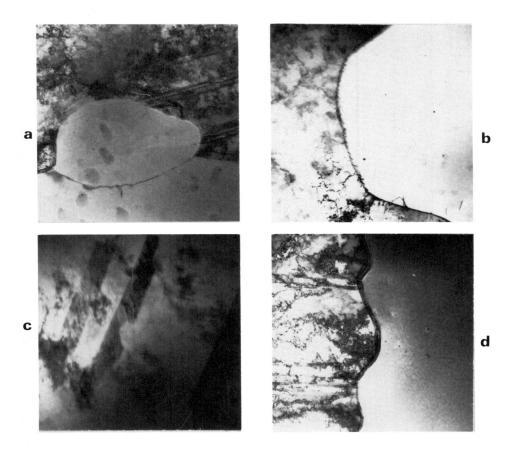

Fig. 8. (a) Cu 410, ann. 15 min at 200°C. 18K. (b) Cu 155, ann. 15 min at 400°C. 18K. (c) Cu 155, ann. 15 min at 350°C. 27K. (d) Cu 410, ann. 15 min at 250°C. 18K.

---

hardness obtained from each histogram is plotted as a function of fraction recrystallized. If there were no recovery at all, the points should fall along the dashed line, as indeed they do for copper deformed by normal cold-work (the inset to fig. 6, taken from Lücke and Rixen's paper, which also shows how extensively cold-worked aluminium can recover). The trend is quite clear, and in particular the more lightly shock-deformed copper recovers substantially.

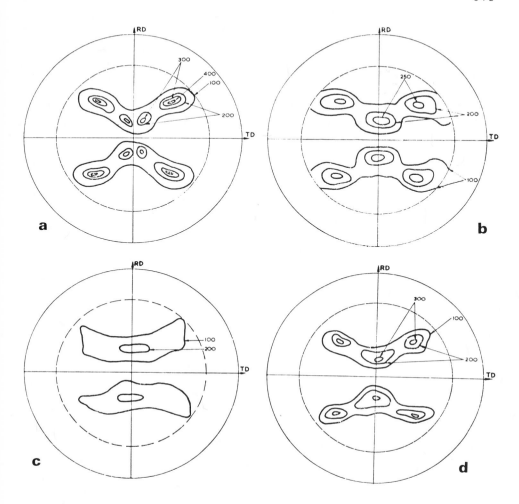

Fig. 9. (111) pole figures. (a) Copper before shock-deformation. (b) Cu 410. (c) Cu 410, annealed 10 min at 350°C. (d) Cu 155 annealed 10 min at 450°C.

Fig. 7 refers to a series of isochronal 15-minute anneals and juxtaposes resistivity ratios, proof stress and microhardness. (The anomaly in resistivity ratio for the highest temperature was traced to hydrogen absorption; specimens were normally annealed in that gas). It is clear from all curves that there is only a single continuous recovery stage until recrystallization intervenes; this is inconsistent with earlier reports, especially from Gordon's laboratory, that two recovery stages exist.

TEM showed, in summary, that cell structures were poorly defined in the shock-deformed state and did not sharpen much on annealing (no cells were seen as clear as those in Hooker's recent paper (8)). There was copious deformation twinning for both peak pressures. The annealed 410 kbar samples showed numerous small new grains, some emerging from the intersection of twin lamellae with grain-boundaries (fig. 8a). The 155 kbar samples showed no small new grains, and all indications were that recrystallization was initiated exclusively by strain-induced boundary migration (17, 18). Fig. 8(b) shows an example of a grain initiated in this way. Contrary to some earlier reports, little nucleation at twin intersections or by strain-induced boundary migration of twin interfaces was seen: fig. 8(c) is one of the few instances seen of the latter. New grains often grow preferentially into twin lamellae (fig. 8d). So far as could be judged, misorientations within as-deformed grains were small, especially in Cu 155.

Fig. 9(a) shows a (111) pole figure of the annealed copper foil before shock-deformation, and fig. 9(b) shows that changes after shock-deformation to 410 kbar were slight. The strong cube-texture component in particular is preserved; this is consistent with the smallness of the misorientations observed by TEM. Fig. 9(d) shows annealed Cu 155, and again there is little change, which is entirely consistent with the postulated preponderance of strain-induced boundary migration. Fig. 9(c) shows, however, that annealed Cu 410 shows a substantially reduced texture, though it is by no means random. Increased annealing time led to a further slight reduction of texture.

Recrystallization of the shock-deformed iron left the sharp initial texture quite unaltered.

The initial grain-size of the copper was 15.8 ± 1.2 μm. Isothermal annealing of the Cu 410 at 200-350°C gave final grain-sizes in the range 15.0 ± 1.0 μm at all these temperatures. For Cu 155 the recrystallized grain-sizes at 300-450°C ranged from 16.4 to 21.5 μm — again little change. Numerous flash-annealing experiments were done because it was thought that rapid primary growth of new grains, combined with sluggish nucleation, might be responsible for the failure to refine the grain-size. For a range of nominal annealing temperatures (650-760°C), recrystallized grain-sizes (shown to be recrystallized by microhardness tests) were all in the range 14.4 — 18.2 μm. The initial hypothesis which prompted the flash-annealing experiments was thus shown to be false.

The iron had an initial grain-size of $10.2 \pm 1.8\,\mu$m, whereas after recrystallization at 500°C for 1 hour, the mean grain-size was $12.7 \pm 1.5\,\mu$m. The earlier reports of grain refinement in iron were thus not confirmed.

INTERPRETATION

Most of the essential interpretations of our individual results have been put forward in the preceding section. Taking the various lines of evidence together, the following interpretation is proposed:

The copper shock-deformed to 155 kbar peak pressure recrystallizes quite sluggishly. Nucleation is impeded by the smallness of the misorientations, itself an intrinsic consequence of the mode of deformation. Nucleation of new, differently oriented grains is thus impossible, and only strain-induced boundary migration initiates recrystallization. The slowness of recrystallization gives plenty of time for effective recovery, even though copper, as a metal with low stacking-fault energy, normally is resistant to dislocation climb and therefore to cell formation and recovery.

Cu 410 recrystallizes more rapidly, partly by true nucleation (e.g. fig. 8a); this is consistent with the partial reduction of the sharpness of texture, if we postulate that in shock-deformed metal nuclei are more or less random and subject to no selective growth condition. The readier recrystallization gives less time for recovery.

The activation energies deduced from fig. 5 are unusually sharply defined but difficult to interpret, as is usually the case with activation energies from recrystallization kinetics. For Cu 155, the value of 29.0 ± 1.8 kcal/mole is due to Gordon's figure, and also identical to Burton and Greenwood's (19) value for grain-boundary self-diffusion in copper, 29.0 ± 2.0 kcal/mole. The activation energy for bulk vacancy migration (20) is 26.2 kcal/mole, but that for bulk self-diffusion is much higher (47.1 kcal/mole). If recrystallization kinetics in Cu 155 are indeed limited by strain-induced boundary migration, the match with Burton and Greenwood's value can readily be interpreted. — The difference in activation energies between Cu 410 and Cu 155 is consistent with the proposed difference in recrystallization mechanisms.

It must be pointed out that the various published experiments on copper were done with metal of different purities and allowance must be made for this in comparing the results. For example, Gordon's experiments on cold-worked metal (13) were done with 99.999% grade, whereas our metal was probably somewhat below 99.99% grade. It is known that dislocation climb (21) and thus polygonization are more difficult in purer copper. Subject to this caveat, the distinction between the recovery capacity of cold-worked and shock-deformed copper is the most substantial feature established by our experiments.

Our failure to repeat Zukas and McQueen's, and Leslie's reports of grain refinement and texture loss in iron cannot be interpreted on the basis of the available information. The only obvious difference (peak pressure apart) is the lesser purity of our iron. The disagreement can only be resolved by further, more systematic experiments.

REFERENCES

1. C.S.Smith, Trans.AIME., 212 (1958) 574.
2. C.S.Smith and C.M.Fowler, in "Response of Metals to High Velocity Deformation", P.G.Shewmon and V.F.Zackay, eds., (AIME, Interscience, 1961), p. 309.
3. A.S.Iyer and P.Gordon, Trans.AIME., 224 (1962) 1077.
4. J.B.Cohen, A. Nelson and R.J. de Angelis, Trans.AIME., 236, (1966) 133.
4a. G.T.Higgins, Met. Trans., 2(1971)1277.
5. D.C.Brillhart, R.J. de Angelis, A. Preban, J.B.Cohen and P.Gordon, Trans.AIME., 239 (1967) 836.
6. J. George, Phil.Mag., 15 (1967) 497.
7. J.R.Till, Ph.D. thesis, University of Liverpool, U.K. (1968).
8. S.V.Hooker, J.V.Foltz and F.I.Grace, Met.Trans., 2 (1971) 2290.
9. R.O.Scattergood, P.Beardmore and M.B.Bever, Trans.AIME., 227 (1963) 1468.
10. E.G.Zukas and R.G.McQueen, Trans.AIME., 221 (1961) 412.
11. W.C.Leslie, E.Hornbogen and E.Dieter, J.Iron & Steel Inst., 200 (1962) 622.
12. J.A.Sabato and R.W.Cahn, J.Nucl.Mater., 3 (1961) 115.
13. P.Gordon, Trans.AIME., 203 (1955) 1043.
14. P.H.Thornton and R.W.Cahn, J.Inst.Metals, 89 (1960-61) 455.
15. G.R.Speich and R.M.Fisher, in "Recrystallization, Grain Growth and Textures", H.Margolin et al. eds., (ASM, 1966) p.563.
16. K. Lücke and R.Rixen, Z. Metallkde., 59 (1968) 321.
17. J.E.Bailey, Phil.Mag., 5 (1960) 833.
18. R.D.Doherty and R.W.Cahn, J.Less-Common Metals, 28 (1972) 279.
19. B.Burton and G.W.Greenwood, in "Quantitative Relation between Properties and Microstructure", D.G.Brandon and A.Rosen, eds. (Israel Universities Press, 1969) p.373.
20. C.J.Meechan and R.R.Eggleston, Acta Met., 2 (1954) 680.
21. F.W.Young, J.Appl.Physics, 29 (1958) 760.

ENERGY ABSORPTION AND SUBSTRUCTURE IN SHOCK LOADED COPPER SINGLE CRYSTALS

A. M. Dietrich
Ballistic Research Laboratories
and
V. A. Greenhut
Rutgers University

ABSTRACT

The stored energy of shock loaded copper single crystals with cube axis and non symmetric orientations relative to the explosive shock wave direction is measured by differential power analysis. Recrystallization and energy release on annealing are strongly affected by a reflected tensile wave from the free surface and are sensitive to initial crystallographic orientation. Lattice defects produced by the shock wave are correlated to the recrystallization behavior. An unexpected endothermic effect is observed in <100> crystals.

INTRODUCTION

When a metal is plastically deformed, the mechanical energy expended in doing plastic work is partitioned into sensible heat generated within the material and latent energy stored in the material. In contrast to elastic deformation, plastic stored energy is not released on removal of loads. Mechanisms of the change of plastic work into other forms of energy (heat, stored energy) have been treated analytically[1,2,3,4]; and generally, these theoretical studies predict that the stored energy of plasticity is associated with the production of lattice defects during plastic flow. Two types of experiment for studying stored energy phenomena in plasticity are defined by Bever and Titchener[5]; of these, quantitative annealing experiments on deformed metals provide, in addition to the amount of energy stored, information concerning the kinetics of release of stored energy[6]. Release of this energy by cold worked metals is accompanied by changes in density, electrical resistivity, hardness, and microstructure, confirming a relation between energy storage during plastic deformation and the defect structures responsible for property changes[7]. Further, specific reactions are associated with recovery of specific structure sensitive properties[6,7]. Thus, energy may be stored in several defect modes and its reappearance as heat on subsequent annealing will be controlled by the relative stability of the different modes of storage with respect to thermal activation.

Investigation of stored energy due to shock loading is of interest because of the contrasts provided by comparing shock deformation of metals with conventional cold working. In shock loading, a metal is subjected to stresses much in excess of its static yield strength for periods of time on the order of microseconds; large amounts of energy are available over this time period; but, as a consequence of the stress state developed, macroscopic deformations tend to be small. A shock loaded gold-silver alloy is observed to store more energy after shock loading than after cold working to equivalent macroscopic strains[8]. Qualitative differences in recovery/recrystallization rates between cold worked and shock loaded copper are reported by Iyer and Gordon[9].

The purpose of this paper is to explore relations between lattice defects, stored energy, and the release of stored energy during annealing of shock loaded metals. The results shown in following sections of this paper indicate the importance of relating stored energy to deformation substructure as opposed to macroscopically defined variables such as strain or flow stress. Experiments are reported which examine the stored energy response of explosively shock loaded copper single crystals of two different orientations relative to the direction of shock propagation. A pair of crystals for each orientation is described. One crystal from each pair is subjected to a compression pulse only (through the use of momentum traps) while the other from the pair is spalled by allowing a tensile wave to reflect from its free surface. In addition, cold rolled polycrystalline copper samples examined by the same techniques are discussed to establish continuity with other work on stored energy.

EXPERIMENTAL PROCEDURE

High purity copper single crystals grown from the melt by a Bridgman technique, one inch long and one-half inch in diameter, were subjected to an explosive shock wave using techniques described by Glass, et. al.,[10], and recovered in water. Approximate peak pressures are determined for these shots from the data of Golaski[11]. Peak pressures of 100 kbars at the buffer-crystal interface were applied. After shock loading, samples (3mm in diameter, 1-2mm thick) were spark machined from various locations in the crystal specimens, and prepared for subsequent operations by electropolishing. The following analyses are made on these samples: observation of lattice defect structures [by transmission electron microscopy] after shock loading; measurement of energy changes in shocked samples during annealing by differential power analysis (DPA); and observation of structure after annealing using both optical and electron microscopy.

Uses of DPA in the investigation of plastically stored energy are discussed more completely elsewhere[6,7]. Two samples, one deformed-one undeformed, are heated at a constant (linear) rate and the power difference required to maintain their heating rates equal is measured in successive runs. Differences between the first experimental run and subsequent runs to establish 'baseline' (no-reaction) behavior can be interpreted in terms of reactions occuring in the deformed sample as it is heated. Plots of power difference versus time may be readily integrated to give energy released or absorbed by a reaction. Our apparatus is calibrated between room temperature and 773°K by measuring the melting points and heats of fusion of indium, tin and lead. Melting points are all within 0.5°K of literature values[12]. Assuming the heat of fusion of indium is 6.80 cal-gm^{-1}, the DPA determined heats of fusion for tin and lead are 14.67 and 5.44 cal-gm^{-1} respectively and compare well to literature values[12] of 14.50 and 5.50 cal-gm^{-1} (same order).

Some caveats in this part of the experiment (DPA) are necessary. First, in an anisothermal annealing cycle the reaction kinetics depend on the heating rate. The results presented here are all for a heating rate of 10° K-min^{-1}, and relative comparisons are valid. Secondly, interpretation of power difference curves relies on the assumption that the dummy sample is thermally stable. Annealed single crystal dummies are employed, and no evidence of thermal instability has been found. It is important to emphasize that DPA experiments compare the behavior of a sample in its <u>deformed</u> state to the sample in its <u>annealed</u> condition, not to the dummy sample.

Samples from cold rolled polycrystalline copper, referred to in the preceeding section, were prepared for analysis in the same way as the shocked samples after rolling deformations of 92, 95.4, and 97.5% nominal reduction of thickness. Since these samples provide a convenient place to touch base with classical studies, the next section begins with a description of their annealing response.

EXPERIMENTAL RESULTS

Stored Energy and Recrystallization

Cold rolled polycrystalline samples were examined in order to evaluate the experimental procedure in terms of classical results[6,7,13]. After rolling, the samples exhibit the energy release behavior on annealing shown in Figure 1. Peak temperatures reproduced ±2°K and energies ±5%. The curves shown are characteristic of results obtained for compression of copper in uniaxial stress[13]. In general, stored energy will increase and the release curve will shift to lower temperatures with increasing plastic deformation[13]; however, for the large reductions here, stored energy has essentially saturated; i.e., increasing deformation stores only small increments of energy[5,14].

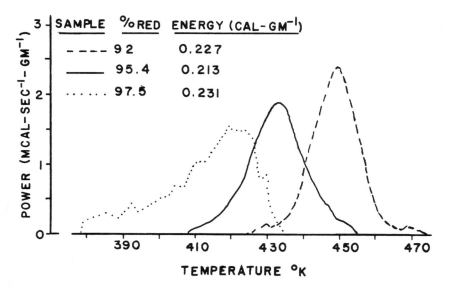

Figure 1. Energy release curves for cold rolled polycrystalline copper annealed at 10 K-min^{-1} after deformations of 92, 95.4 and 97.5% nominal reduction. Typical reaction curves are shown. The energy released is shown in the legend and these values are averaged from six tests in each instance.

In spite of the apparent saturation, a shift in release kinetics toward lower temperatures with increasing deformation is still noted. The samples are recrystallized as verified by optical microscopy.

Single crystals with a non-symmetric orientation (NSO) relative to the shock axis were explosively shock loaded as described

in the experimental procedure section: one with and one without
momentum traps. Quantitative annealing of samples from different
locations in the spalled (no momentum traps) NSO crystal shows a
release of energy like that in Figure 2. Energy output and peak
power are of the same order as the cold rolled case; however, shock
loaded samples have a broader response with the suggestion that
more than one reaction peak is present. Both energy release and
temperature range of release depend upon sample location in the
crystal as summarized in Figure 3. Stored energy decreases with
increasing distance from the explosive metal interface while
release temperature range decreases sharply in the vicinity of the
spall fracture plane. Recrystallized grain size is also sensitive
to location in the shocked crystal, and the final grain structure
is much finer immediately adjacent to the spall fracture. This
refinement of grain size is directly related to the region of shock
induced porosity surrounding the spall fracture and correlates well
with the sharp reduction in reaction temperature. See Figure 4.
Similar studies of the unspalled NSO crystal reveal that recrystallization initiates at 750°K at the high pressure end of this
crystal. Other samples farther from the explosive metal interface
did not show activity on annealing or recrystallized structure.

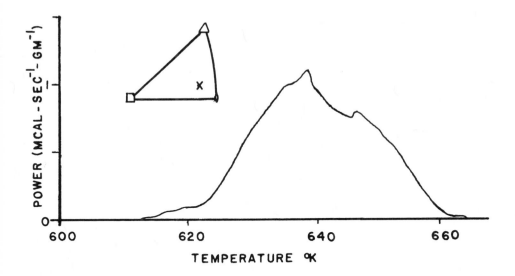

Figure 2. Energy release curve for a sample from shock loaded
copper single crystal of orientation (NSO) shown. The sample was
taken from an area above the spall fracture close to the explosive-
metal interface. Heating rate 10°K-min^{-1}.

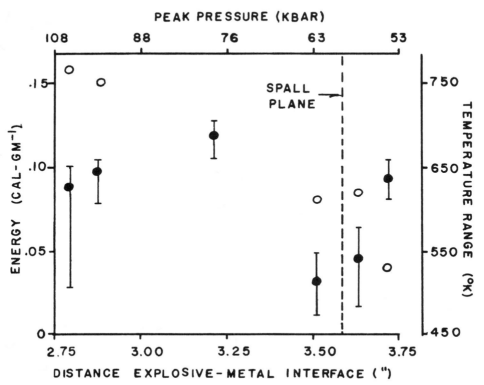

Figure 3. Summary of results of annealing samples from different locations in the spalled non-symmetric orientation crystal. Energy release (o) and the reaction temperature ranges of that release (I) are shown. Temperatures of maximum power are indicated (·). Approximate peak compressive pressures are shown on the upper abscissa. Position of major spall fracture is indicated.

In the <100> orientation crystals samples exhibit more complex (and somewhat surprising) behavior. This phenomenon is described in the next section prior to evaluating their recrystallization and stored energy response.

ENERGY ABSORPTION IN SHOCK LOADED COPPER

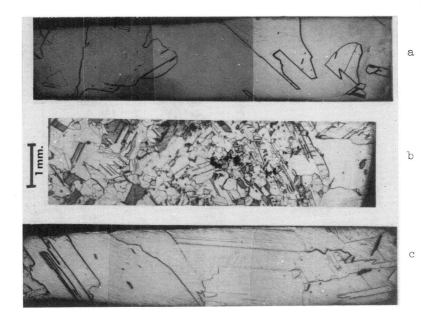

Figure 4. Recrystallized grain structure of samples from spalled NSO crystal, a. 3.24 inches from the explosive metal interface, b. 3.51 inches from the explosive metal interface, immediately adjacent to spall (note shock porosity), c. 3.71 inches from the explosive metal interface, below spall fracture.

Endothermic Effects

Quantitative annealing of samples from spalled and unspalled <100> crystals shows power difference curves like those in Figure 5. Between room temperature and 575°K a strong endothermic displacement is seen; the sample absorbs more energy during the initial annealing run than it does during subsequent baseline runs. For all samples from the unspalled <100> crystal this power difference diminishes gradually between 530° and 575°K. On the other hand, it terminates sharply in two of the samples from the spalled crystal taken above the spall. These two samples also exhibit small exothermic reactions at higher temperatures (688-710 K and 658-706 K) and have recrystallized. Two other samples from intermediate sites exhibit no reactivity and are not recrystallized. Results related to the endothermic effect are summarized in Figure 6 and indicate that this effect is sensitive to sample location.

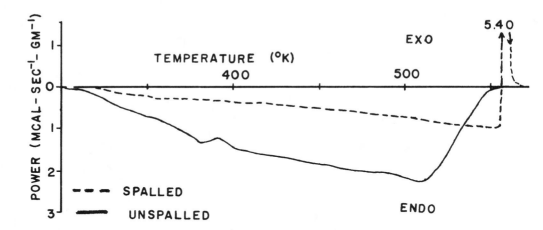

Figure 5. Examples of the endothermic effect in samples from spalled and unspalled <100> copper crystals. Power absorbtion is plotted versus temperature for a heating rate of 10 K-min^{-1}.

It is interesting to note that the unspalled sample shows a maximum endothermic power response in the region where spall fracture occurs in the other <100> crystal. Studies of structural changes uniquely associated with this endothermic effect are continuing.

Qualitative similarities in the recrystallization response between the spalled <100> crystal and the spalled NSO crystal are noted although the energy release behavior is quantitatively different. Figure 7 is a plot of these results. Release temperature ranges tend to be higher in samples from the spalled <100> crystal. Amounts of energy released in the spall regions of these two crystals are comparable. Refinement of grain size in the spall region also occurs in the <100> crystal.

Defect Structure-Shocked Crystals

Transmission electron microscopy of samples from the same crystals used in the annealing studies show that all the shocked crystals have the following structural features - very small dislocation cells (a few tenths of a micron in radius) and regions of micro-twin formation separated from the cell structure by narrow areas of high lattice distortion (rotation). Micro-twin

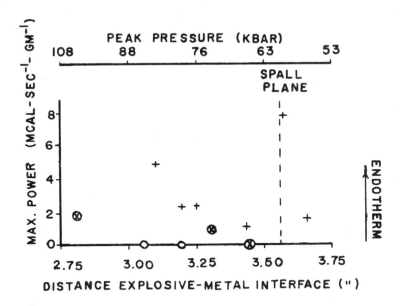

Figure 6. The magnitude of the endothermic effect plotted as peak power difference is shown for different sample locations in un-spalled (+) and spalled (o) <100> crystals. A value of zero indicates that no endothermic effect is observed. An (x) indicates that a separate exothermic reaction is observed and that the sample is recrystallized.

regions tend to be 10-20 microns in diameter. Individual twins have {111} habit planes[15]. Figure 8 shows examples of these structures. Relative misorientation of dislocation cells is small throughout most of the crystals except in the vicinity of spall fracture. In the region of porosity surrounding the spall x-ray macrotopographs show that more severe lattice rotations occur (Kingman[16]). The porosity itself is an additional structural feature in the spalled crystals.

DISCUSSION

Energy stored in conventionally cold worked metals is released on annealing, and several structural changes may be associated with this release: recovery of point defects, rearrangement of dislocations and recrystallization. With respect to recrystallization, theoretical considerations due to Cahn[17] indicate that energy stored by cold working is insufficient to drive homogeneous

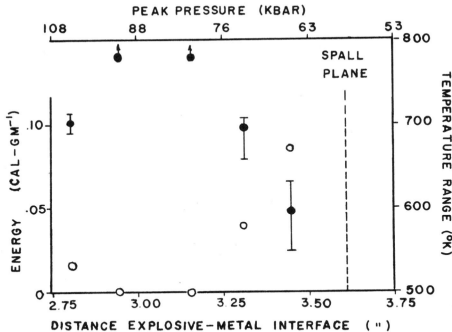

Figure 7. Summary of results of annealing samples from different locations in the spalled <100> orientation crystal. Energy released (o) and the reaction temperature ranges (I) of that release are shown. Temperatures of maximum power output are indicated (•). Approximate peak pressures of the initial compressive pulse are shown on the upper abscissa. Exothermic energies given here consider only reactions separated from the endothermic effect. Note that two of the samples had no reaction up to 773°K.

nucleation of new grains (by stastical fluctuations). As an alternative, he proposes a process whereby nucleation sites are "preformed" to obviate surface energy restraints. Heterogeneous nucleation at internal interfaces is also possible. Implied in these theoretical approaches is the necessity to separate nucleation phenomena from the amount of energy stored. From an experimental point of view it is important to realize that stored energy of plastic flow is an averaged quantity. It is averaged over a sample as if this energy were homogeneously distributed throughout the sample. It is patently obvious that this is not true, but by paying attention to defect structures and the kinetics of energy release on annealing, indications of the degree of localization of stored energy are obtained.

Severe reductions in cold rolling produce a saturated level of stored energy in polycrystalline copper; however, the shift of

Figure 8. Electron micrographs showing structures characteristic of the as-shocked sample prior to annealing. Observation plane is (110). Twins (left) lie on {111} habit planes[15].

release to lower temperatures with increasing deformation after saturation demonstrates that nuclei of increasing effectiveness are being formed. The kinetics are very sensitive to the presence of nuclei, but the (averaged) stored energy is not.

Recrystallization behavior of shock loaded crystals is strongly influenced by the occurence of a reflected tensile wave. Comparing the two NSO crystals shows the effect of the reflected wave is to lower the recrystallization reaction temperature. Results from the spalled NSO crystal indicate that different nuclei predominate depending on distance from the explosive-metal interface. Near the high pressure end energy storage is largest, but nucleation is difficult. Final grain size in this region also indicates a paucity of nuclei, when compared to the spall zone. Small misorientations between cells in these areas away from the spall explain this behavior relative to the easier nucleation in the spall region where pore surfaces, localization of flow between pores and higher cell misorientations may all contribute to nucleating new grains. The critical mechanism which lowers the recrystallization temperatures in the spalled NSO samples (away from the spall region) relative to the unspalled case is not isolated but perhaps it depends on the distribution of microtwin regions in the samples. Recrystallization is promoted in the vicinity of larger twins in shock loaded copper[15]. The spalled <100> crystal shows a similar recrystallization pattern and this pattern is also rationalized in terms of the same changes in nucleation sites.

Another consideration is the defect structure(s) responsible for storing energy. If it is assumed that all of the energy is stored in dislocation cell boundaries and further assumed that a unique energy per unit boundary area can be ascribed to the boundaries, then a variation in average cell radius by a factor, f, infers an (averaged) energy variation by a factor, f^2. Limited studies to date neither confirm nor eliminate this approach. Cell sizes change by approximately a factor of two from the high pressure end (smallest cells) to the low pressure end (largest cells) although this size change is complicated in the spalled samples by increased cell misorientation around the spall which may violate the assumption of constant boundary energy. Micro-twin regions are not felt to contribute significantly to the amount of stored energy although they may provide nucleation sites for recrystallization. Excess point defects created by the shock have probably disappeared prior to annealing[4,18].

The endothermic effect observed in <100> orientation crystals is not understood. Structural changes occuring with this phenomenon are still being investigated. Two broad interpretations seem to be open. On one hand it may be a heat capacity change arising from a structure change at the high temperature end., i.e., between 530° and 575°K. On the other hand, it may represent a true reaction, (an athermal growth process in reverse?). At present the heat capacity model does not seem tenable in view of the fact that the heat capacity change would have to be 10% or more. Existing theories cannot account for a change of this magnitude in a cold worked structure[19]. Easily rationalized changes would be on the order of a few tenths of a percent, which fact is consistent with observations in the other samples tested.

CONCLUSIONS

Complete constitutive equations for plastic deformation which include energy partition behavior (heat plus stored energy) are necessary. Presuming that well defined thermodynamic functions exist for plasticity, such functions (e.g., entropy or energy) must contain parameters related to the defect structure. Macroscopic parameters, alone, appear insufficient to define these functions. Although the amount of energy stored in a metal after passage of a shock wave is small relative to the amount delivered by the shock wave, its relation to crystal lattice defects which produce changes in material properties makes it another tool to improve understanding of these lattice defects.

More study is needed to isolate structure(s) associated with energy storage although cell boundaries are likely candidates. The endothermic effect in <100> crystals is not understood, but is sufficiently documented and related to the shock wave that the authors are not disposed to dismiss it as an artifact.

Clearly, because severe gradients in response exist in these shocked crystals, it is desirable to use small samples from specific locations in annealing experiments. At the same time one would like to avoid surface nucleation which may cause a lower recrystallization temperature than that characteristic of the bulk structure. Different loading configurations to suppress gradients so that larger samples can be used might be tried. The following conclusions are proposed from these results:

1. It is important to relate the stored energy of plasticity to structural defects as opposed to macroscopic parameters of deformation.

2. The recrystallization nuclei present in shock loaded samples do not have a simple relationship with the amount of energy stored.

3. A reflected tensile wave significantly influences the number and type of nuclei present. Nucleation in the region of spall is very efficient.

4. Nucleation away from the spall fracture or in unspalled samples probably occurs by a 'preform' mechanism; in the spall region nucleation at pore surfaces, higher cell misorientations and localized flow combine to give more effective nuclei.

5. Quantitative aspects of recrystallization after shocking are sensitive to initial crystallographic orientation relative to the shock axis. A phenomenon observed by Smith and Fowler[15].

The authors wish to thank Dr. C. M. Glass, Mr. S. K. Golaski, Dr. W. Oldfield and Ms. P. W. Kingman for many helpful discussions and Mr. Ming Guang Chen for his work on electron microscopy of samples. One of the authors (VAG) was supported on AROD contract number DA-72-G109.

1. Nicholas, J.F., Acta Met v. 7, p544-548, 1959.
2. Koehler, J.S., Phys. Rev. v.60, p. 397-410, 1941.
3. Moore, J.T. and Kuhlmann-Wilsdorf, D., Crys. Lat. Def. v. 1, p. 201-210, 1970.
4. van den Berkel, A., Acta. Met., v. 11, p 97-105, 1963.
5. Bever, M.B., and Titchener, A.L., Prog. in Metal Phys, v. 7, p. 247-338, 1958.
6. Clarebrough, L.M., Hargreaves, M.E., Michell, D. and West, G.W., Proc. Roy. Soc. v. 230(A), p. 507-524, 1953.
7. Clarebrough, L.M., Hargreaves, M.E. and West, C. W., Proc. Roy. Soc., v. 232(A), p. 252-270, 1955.
8. Appleton, A.S., Dieter, G.E. and Bever, M.B., Trans. Met. Soc. AIME, v. 221, p. 90-94, 1961.

9. Iyer, A.S. and Gordon, P., Trans. Met. Soc., AIME, v. 224, p. 107-109, 1962.
10. Glass, C.M., Golaski, S.K., Misey, J.J., and Moss, G.L., Symposium on Dynamic Behavior of Materials, ASTM Tech Pub. 338, p. 282-305, 1962.
11. Golaski, S.K., unpublished data.
12. Hultgren, R., Orr, R.L., Anderson, P.D, and Kelly, K.K., Selected Values of Thermodynamic Properties of Metals and Alloys, Wiley, 1963.
13. Clarebrough, L.M., Hargreaves, M.E. and Lorretto, M.H., Acta. Met., v. 6, p. 725-735, 1958.
14. Taylor, G.I. and Quinney, M.A., Roc. Roy. Soc., v. 143(A), p. 307-326, 1937.
15. Smith, C.S. and Fowler, C.M., Response of Metals to High Velocity Deformation, p. 309-335, Intersci Pub., 1961.
16. Kingman, P.W., private communication.
17. Cahn, R.W., Recrystallization, Grain Growth and Textures, p. 99-128, ASM, 1965.
18. Henderson, J.W. and Koehler, J.S., Phys Rev., v. 104, p. 626-633, 1956.
19. Nabarro, F.R.N., Theory of Crystal Dislocations, p. 683-704, Oxford, Clarendon Press, 1967.

X-RAY TOPOGRAPHY OF SHOCK LOADED COPPER CRYSTALS

P. W. Kingman

Ballistic Research Laboratories

Aberdeen Proving Ground, Maryland 21005

ABSTRACT

Single crystals of high purity copper subjected to shock loading have been examined by X-ray diffraction topography. The topographs reveal a basic substructure consisting of narrow, close packed kinks normal to highly stressed slip directions and extending rather homogeneously throughout the sample. In the vicinity of fractures, where gross macroscopic deformation occurs, complex interactions are observed. The substructure has been examined as a function of orientation and pressure.

INTRODUCTION

The detailed substructures of shock-loaded metals have been extensively examined in the past by transmission electron microscopy (TEM). Because of the limited field of view at very high magnification, these studies do not reveal the long-range macrostructure. In the present work, X-ray diffraction topography (XRT) has been applied to shock-loaded copper crystals, utilizing a simple macro-topographic arrangement which permits large areas (up to 1" x 1/2") to be imaged. Since X-ray topography is characterized by limited spatial resolution and high angular resolution, the information provided is complimentary to that derived from TEM and optical microscopy. It is also possible to observe the internal structure of the crystal without resorting to the preparation of thin foils, thus minimizing the possibility of internal rearrangement. In the case of shock loading, an added advantage is that the preservation of pre-polished surfaces is unnecessary since the substructal indications are not effaced by surface preparation.

It might be noted that XRT of highly deformed crystals differs from that of nearly perfect crystals since there is no possibility of resolving individual dislocations or detailed dislocation interactions; rather one sees lattice rotations and inhomogeneities due to large arrays of dislocations over an extended area. Because the angular resolution is very high even in the presence of very large dislocation densities, one is still able to derive information about the substructure and dislocation arrangement, particularly when the topographs are correlated with TEM and optical observations.

EXPERIMENTAL METHOD

All of the observations to be described were made on single crystals of high purity copper, 1" x 1/2" diameter, explosively loaded by a plane wave generator. All samples were shocked in polycrystalline copper surrounds without spall plates. The crystallographic orientations of the stress axes are given in Figure 1 and the peak pressures in Table I; further experimental details are given by Glass, et. al.[1]. The samples were sectioned to provide an axial plane surface for observation. The principal X-ray technique was a simple Berg-Barrett X-ray topographic arrangement[2,3], using a broad, filtered beam. Under such conditions, the resolution of fine grained X-ray film (Type R) was adequate. Once the overall substructural pattern was established however, the use of a parallel monochromator permitted the delineation of more limited regions at higher angular resolution on nuclear track plates.

EXPERIMENTAL RESULTS

Detailed topographic studies were made on two overlapping groups of specimens. The first group, Y-1, DD1, and S1, were all shocked at low pressures and were examined to evaluate the effect of orientation on the deformation modes. The second group, DD-1 thru DD-5, were chosen to evaluate the substructure over a range of pressures for a single orientation. (A limited study was also made of S-3).

Figure 2 is a topograph of crystal Y-1, showing nearly the entire sample cross section. The diffracting plane is the cross slip plane ($1\bar{1}\bar{1}$). The principal features of the topograph may be categorized as follows:

1. Gross macroscopic deformation: Except for a few areas in the fracture region nearly the entire area below the spall is in reflecting position (the lowest part of the sample is outside the

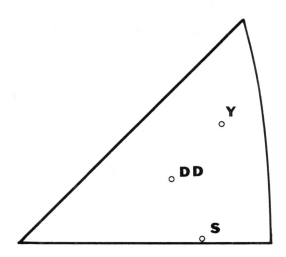

Figure 1. Crystallographic Orientation of sample axes.

Sample	Peak pressure	
DD-1	31	Kb
DD-2	48	
DD-3	66	
DD-5	142	
Y-1	31	
S-1	31	
S-3	66	

Table I. Pressure Data for Copper Crystals.

Figure 2. Axial Section through Sample Y-1, Filtered Cu Radiation.

beam). This implies that the $(1\bar{1}\bar{1})$ diffracting plane remains macroscopically nearly plane, i.e., the net long range lattice rotation is negligible. This sample did not fracture completely; and as might be expected, the area above the spall was misoriented from the remainder of the sample.

2. Sub-boundary structure: Curved low-angle sub-boundaries are seen in several regions. It appears that these boundaries may be growth misorientations; their relationship with the deformation substructure is not entirely clear.

3. Deformation modes: Numerous deformation markings are observed, with wide variations from one region of the crystal to another. The most prominent markings appear to have a well defined crystallographic direction, but in other cases, particularly near the fracture area, the directionality becomes indistinct. Since only a single surface is observed, assignment of indices to these traces is not necessarily unique; however, the dense and rather homogeneous closely-packed traces which extend throughout the sample are traces of the plane normal to the primary slip direction, (101). In general, {111} traces do not predominate in the topographs, although they can be discerned in some regions, particularly near the sample surfaces; it is the masses of fine kinks associated with the {011} planes which are commonly seen. The identity of the coarse banded structure extending down one side of the sample is less certain. The direction of the bands corresponds

with the (110) trace normal to the conjugate slip direction, but the assymetrical localization of the bands suggests that a growth structure may also have been involved.

Figure 3 shows a limited region of the crystal at higher magnification. It is seen that the individual striations are themselves inhomogeneous and discontinuous, with a tendency to bifurcate. The spacing of these kinks is of the order of 100μ, while the lengths of the individual segments are an order of magnitude larger.

Figure 3. Details of Kink Bands in Sample Y-1, Filtered Cu Radiation.

Within these segments second-order entities can be just resolved. The broad "bands" so prominent in Figure 2 now appear as systematic perturbations of the basic narrow primary kinks.

The specimen axis of sample Y-1 was a single-slip orientation. Sample S-1 was selected because the stress axis lay on a symmetry line; thus two slip systems were stressed simultaneously. Figure 4 is a (200) reflection from S-1, showing a dual kink structure running throughout the sample. Again, subgrain boundaries are seen, but their influence on the overall structure appears minimal. The long range lattice rotation is slight, and the kink structure away from the partial fracture is quite homogeneous. Near the fracture region the local rotations appear more pronounced, and in

Figure 4. Axial Section of Sample S-1, Filtered Cu Radiation.

the center of the sample adjacent to the fracture area there appears to be strong interaction between the deformation modes.

The two principal sets of bands are identified as the {011} traces normal to the two equally stressed "primary" systems. To further elucidate the interaction region, topographs were taken using monochromatic radiation. Figure 5 shows that in addition to the primary kinks a horizontal banding appears prominent in the interaction region. These striations are nearly washed out in the filtered radiation topographs taken with similar diffraction geometry. A microhardness traverse of the sample gave essentially uniform hardnesses through the primary kink region, but increased by 10 to 15% in the striated "interaction" region.

The series DD-1 - DD-5 was investigated primarily to ascertain the effect of pressure on the deformation structure. Since all samples were taken originally from the same crystal rod and had subsequently been sectioned on the same plane, it was possible to take topographs under identical diffraction conditions. Figures 6 and 7 illustrate the two extreme pressures. The general appearance of DD-1 is similar to Y-1, except that complete fracture has occurred and the inhomogeneity near the fracture is much greater. A substantial net lattice rotation is present, particularly near the fracture area, and it was necessary to rotate the sample progressively to bring successive regions into diffracting position.

X-RAY TOPOGRAPHY OF SHOCK LOADED COPPER CRYSTALS 665

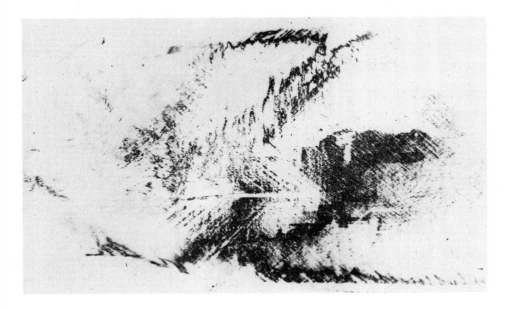

Figure 5. Axial Section of Sample S-1, Cu Kα Radiation.

Figure 6   Axial Sections of Sample DD-1, 31 Kb, Filtered Cu
    Radiation (111) Reflection. Montage of Topographs at Two
    Angular Settings.

Figure 7. Axial Section of Sample DD-5, 142 Kb, Filtered Cu Radiation, (111) Reflection.

Near the fracture only a narrow strip appears at each setting. In addition to the primary kinks, numerous other traces are seen, many poorly defined. In contrast, DD-5 shows very little net rotation right up to the fracture region, and the deformation traces, although thicker in texture, appear very homogeneously distributed, with the primary kink structure strongly predominating. In general, the intermediate samples show corresponding behavior, with progressive localization of the gross lattice rotation and increasing homogeneity of the general kink structure. Topographs at S-3 also follow the same trend.

DISCUSSION

The most salient feature of the deformation substructure revealed by XRT thus far is the homogeneous alignment of substructural entities along crystallographic planes, primarily the (011) planes, throughout the sample, combined with the virtual absence of long-range net lattice rotation except in the fracture region. Since the region of gross rotation diminishes and the deformation mode becomes more uniform as the pressure increases, it appears that the uniform short range structure is most closely related to the shock loading process, while the additional inhomogeneities probably derive from subsequent loading conditions.

A possible model for this structure is the piling up of large groups of dislocations on the primary slip planes throughout the crystal, each group having an excess of dislocations on one sign. These groups then tend to align themselves on successive slip planes in vertical arrays such that successive groups have opposite signs. The resulting structure consists of two-dimensional "kink walls" perpendicular to the slip direction. Extended "super braids" of dislocations having a net excess of one sign alternate with "super braids" of opposite sign to produce local lattice rotations while the net long range rotation remains small. Analogous kink walls have been observed by Wilkens[4] in lightly deformed copper crystals. Although the imperfection densities differ by orders of magnitude and the substructural entities occur on a finer scale, the lattice rotation distributions are strikingly similar, suggesting that the morphology of the imperfection distribution may be quite similar even though the detailed arrangement cannot be the same.

The homogeneity of this structure throughout the sample suggests that the component lattice imperfections are generated quite uniformly throughout the sample and that dislocation movement is blocked with equal uniformity. In samples shocked at slightly higher pressures, twins are observed on a very fine scale. Such twin assemblages could serve as blocks. Sessile dislocations may also be formed by interaction with intersecting systems, and the uniaxial strain conditions existing during shock loading make this an attractive possibility. Because the structure is so uniformly distributed thoughout the specimen, it is suggested that while the ultimately observed macroscopic structure may not be achieved until rather late in the total deformation history, its nature is determined by the microscopic blocking mechanisms set up during the early stages of shock loading. The correlation of XRT and TEM should provide a broader insight into the total shock deformation process.

CONCLUSION

It has been shown that X-ray diffraction topography of shocked crystals provides a unique insight into the nature of the final deformation substructure. The present study has established the existence of a homogeneous lattice rotation distribution in shock loaded spall-fracture samples. This kink band structure extending through the crystal is believed to be uniquely related to the shock loading process. The information provided by XRT compliments that obtained by other methods such as optical and election microscopy, and should be a useful tool in the total delineation of deformation structures.

I would like to acknowledge the patience and encouragement of Dr. C. Glass throughout this work, and also the many helpful discussions with Dr. A. Dietrich and Mr. S. Golaski at BRL and Drs. V. Greenhut and S. Weissmann at Rutgers University.

REFERENCES

1. Glass, C.M., S.K. Golaski, J.J. Misey, and G.L. Moss, Symposium on Dynamic Behavior of Materials, Publication No. 336, American Soc. For Testing and Materials, 1962, pp 282-305.
2. Berg, W. F., Z. F. Kristallographic, Bd 89, 1934, pp 286-294.
3. Barrett, C. S., Transactions AIME, Vol. 161, 1945, p. 15.
4. Wilkens, M., Canadian J. of Physics, Vol. 45, No. 2, Part 2, 1966, p. 567.

THE EFFECT OF HEAT TREATMENT ON THE MECHANICAL PROPERTIES AND MICROSTRUCTURE OF EXPLOSION WELDED 6061 ALUMINUM ALLOY

Robert H. Wittman

Denver Research Institute
University of Denver
Denver, Colorado 80210

INTRODUCTION AND BACKGROUND

The unique microstructure generated by explosion welding and the mechanics of weld formation has been the subject of numerous research programs at laboratories in many countries[1,2,3,4]. The results of these investigations seem to develop a clear argument for explosion welding occuring as an essentially solid state bonding mechanism based on the disruption of surface contaminants by "jetting" from oblique high velocity collision while simultaneously establishing atomic contact of newly formed surfaces. Some welding by melting and fusion does occur; however, this is regarded as a secondary phenomenon which should be minimized in order to achieve maximum weld quality. The result is a weld interface that is usually characterized by a wavy flow pattern of direct solid phase bond alternating with regions of solidified melt. The extent of flow deformation and melting can be altered by the impact conditions, i.e. the explosion welding parameters.

The conditions which lead to welding are produced by the oblique collision of explosively driven metal plates as illustrated in Figure 1. When the angle and velocity of impact are within certain critical limits, "jetting" will occur[1]. The term "jetting" is used here and in the explosion welding literature to describe the spray of metal in liquid or vapor state that emanates from the apex of the angled collision surfaces. A coherent slug of metal, as in the Munroe effect, is not observed in a welding collision. "Jetting" in explosion welding is thought to result from an unbalanced triaxial state of stress in the collision region. High velocity flow in the direction of propagation would occur when the hydrostatic stress component exceeds the Hugoniot elastic limit, and a net shear stress

of sufficient magnitude is developed in the propagation direction
by the oblique collision. This condition leads to shearing flow
just ahead of the imaginary collision point and produces atomically
clean surfaces. As the atomically clean surfaces produced by jetting
come together, the force normal to the plane of the weld is of sufficient magnitude to produce cohesive interaction at the atomic level.
Nothing more is required to form a metallurgical weld.

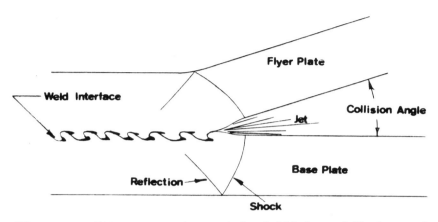

Figure 1. Illustrated Asymmetric Collision of Moving and
Stationary Plates Leading to Jetting and Welding.

When viewed in the electron microscope, the weld interface in
the region of solid phase bond closely resembles an oversize grain
boundary between highly elongated grains of the two welded metals[6,7].
Solid state diffusion effects are observable adjacent to melted areas,
but generally occur only to a limited depth since the heat generated
at the interface by impacting and jetting is rapidly dissipated in
the adjacent bulk of cold metal. Explosion welds are thus sometimes
called diffusionless, but this is true only on a macroscopic scale.
Furthermore, effects of dynamic recovery and recrystallization have
been observed near the weld interface as a result of heat generated
by the collision[6]. The weld zone is known to contain higher densities of dislocations and presumably higher concentrations of point
defects than the parent metals[5,6]. The concentration of defects
generated and the extent of welding-initiated diffusion, recovery
and recrystallization are thought to be directly related to the kinetic energy of the impacting flyer plate, the time duration of impact generated stresses, and the heat generation and dissipative
properties of the metals combination being joined.

Although there has been extensive investigation of explosion weld microstructures and the mechanics of weld formation, little work has been reported on the response of explosion weld mechanical properties to post weld heat treatment. It has generally been assumed that the annealing kinetics would be increased due to the higher concentration of defect structure introduced by explosion welding.

Precipitation hardened aluminum alloys are important structural materials that are difficult to fusion weld. They can, however, be explosion welded to produce high weld strengths, and characterization of their explosion weldability should contribute to applications development. Aluminum alloy 6061 is representative of the precipitation strengthened aluminum alloys, and has characteristics that make it an interesting candidate for the study of explosion welding and annealing behavior of explosion welds. The alloy in the precipitation hardened (T6) condition has a high tensile strength relative to the annealed (O) condition. The mechanical properties of the solution treated (T4) condition, the T6 condition and the cold worked state can be sharply modified at temperatures above 350°F, a temperature attainable during explosion welding. In general the mechanism of 6061 alloy strengthening involves precipitate-dislocation and to some extent point defect interactions, all of which are subject to modification by shock waves, high strain rate flow, and elevated temperatures that accompany explosion welding. It is these interactions and the resulting property modifications that are of interest in this investigation.

EXPLOSION WELDING PROCEDURE

Aluminum alloy 6061 was explosion welded in the T6, T4, and O conditions using a wide range of explosive loading parameters at a constant standoff using the parallel plate geometry. The explosion welding configuration is illustrated in Figure 2. In all instances during this investigation the flyer and base plates were 1/4 inch thick by 6 inches by 12 inches, separated by a gap of 1/8 inch. The base plate was resting on a 4 inch thick steel anvil supported by a concrete pad. The explosive was a granular dynamite composition, Red Cross 40% Extra, produced by du Pont. The detonation velocity of the dynamite varied as a function of the explosive loading as shown in Figure 3, although the packing density was maintained constant at 1.3 g/cm^3. Surfaces were prepared for welding by dipping the sheared plates in hot NaOH solution, washing with water and air drying to produce a smooth bright metal finish.

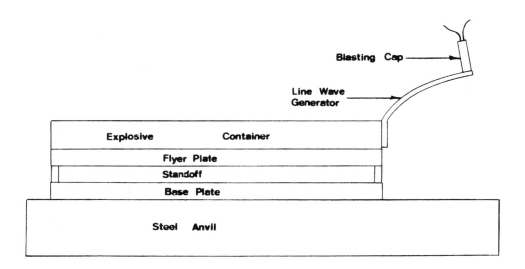

Figure 2. Parallel Plate Explosion Welding Configuration

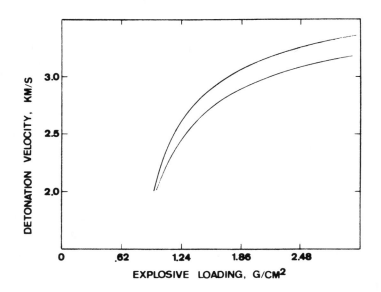

Figure 3. Detonation Velocity of 40% Extra Dynamite as a Function of Explosive Loading at a Density of 1.3 g/cm^3.

MATERIALS AND WELD STRENGTH TEST METHOD

The 6061 alloy aluminum plate was purchased commercially in the T651 condition and where required heat treated to the T4 and O condition. Typical tensile properties parallel to the rolling direction along with Brinnel hardness for the T6, T4, and O starting stock are given in Table I and conform to the Aluminum Association, Aluminum Standards and Data[8], typical properties.

Table I. Typical Properties of 6061 Alloy Prior to Explosion Welding

Condition	.2% Yield Strength	Ultimate Strength	Elongation Percent in 1 inch	BHN
O	9,000 psi	18,200 psi	24	32
T4	20,400 psi	34,000 psi	23	58
T6	39,600 psi	44,500 psi	12	94

The tensile strength of the resulting explosion welds was measured using a "zero gauge length" test specimen. The test configuration illustrated in Figure 4 has been used to measure the strength of a homogeneous aluminum plate in the thickness direction and found to yield a fracture strength that is equal to the ultimate tensile strength measured using a standard sheet or plate tensile specimen oriented in the rolling direction. When the weld zone is properly located, the test specimen accurately reflects the strength of the weld and material immediately adjacent as failure is made to occur in that plane. The weld tensile strength is determined by dividing the load at fracture by the initial test area.

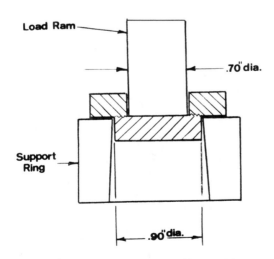

Figure 4. Weld Test Configuration.

DETERMINATION OF MAXIMUM WELD TENSILE STRENGTH

Using the procedure described above, 6061 aluminum alloy plates were welded in the T6, T4, and 0 condition at 5 explosive load variations. Four weld tensile strength values were determined for each welded plate using the test configuration in Figure 4. The average weld tensile strength as a function of explosive loading for the T6, T4, and 0 condition is shown in Figure 5. Optimum explosive loadings for maximum weld tensile strength occur at 1.24 g/cm^2, 1.55 g/cm^2 and 1.24 g/cm^2 for the 0, T4, and T6 condition in that order.

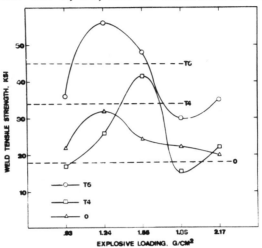

Figure 5. Average Weld Tensile Strength as a Function of Explosive Loading at a Constant Standoff of 1/8 inch for 1/4 inch thick 6061-0, T4, and T6 Aluminum Alloy Plates. Horizontal Dashed Lines are the Parent Metal Strengths.

The collision conditions and stress wave intensity at the weld zone were characterized for the 6061-T6 optimum weld condition. The collision velocity normal to the flyer plate and collision angle was measured using pulsed x-ray and velocity probe techniques and found to be 300 m/sec and 6.5°. From Hugoniot data a 300 m/sec planar impact between 6061 aluminum alloy plates would be expected to produce a peak pressure of approximately 25 kbar[9]. Confirming this expectation, pressure at the explosion welded interface was measured using a manganin piezo-resistive gauge and found to have a peak value ranging between 18 and 25 kbar for a 300 m/sec oblique impact. Similar flyer plate impact velocities and collision pressures would be expected for the 6061-0 and T4 conditions welded using the 1.24 g/cm^2 and 1.55 g/cm^2 explosive loadings. Measurement of residual temperature at the surface of the welded plates indicates that a temperature of 200°F is attained several minutes after welding and remains at that level for 10 to 15 minutes.

ANNEALING BEHAVIOR OF EXPLOSION WELDED ALUMINUM ALLOY

Explosion welds of maximum weld tensile strength for 6061 aluminum alloy in the O, T4, and T6 condition were subjected to isochronal anneals at temperatures up to 775°F. The weld tensile strength of heat treated samples was determined and compared to unshocked parent aluminum alloy and to cold rolled aluminum alloy subjected to the same isochronal annealing schedule.

Using the optimum explosive loading and the configuration described previously for maximum weld tensile strength, 5 sets of 6061 aluminum plates were explosion welded at each of the 3 initial heat treatment conditions to produce a supply of specimens for post weld annealing treatments. Two weld tensile strength measurements were made on each plate prior to annealing to verify that at least 90% of the expected maximum weld tensile strength was attained. Plates having a weld tensile strength less than 90% of expected optimum were rejected from annealing and property characterization.

Interface and surface Brinell hardness was determined for the explosion welded O, T4, and T6 initial heat treatment conditions. Then 6061 alloy plates in O, T4, and T6 initial condition were cold rolled to the approximate hardness of the corresponding explosion welded alloy. These samples and samples of undeformed 6061 alloy in the O, T4, and T6 initial were prepared for inclusion in the isochronal heat treatment schedule.

Heat treatment of the explosion welded, cold rolled, and undeformed aluminum alloy samples was conducted in an air atmosphere, electric muffle furnace. The heat treatment time was 3 hours at 350°F, 450°F to 500°F, 650°F, and 775°F, followed by air cooling to room temperature. Samples annealed at 650°F and 775°F were furnace cooled to 500°F before air cooling to preclude any solutionizing and aging effect. Tensile samples of the type shown in Figure 4 from explosion welded plates and undeformed samples from 1/4 inch thick plates were tension tested after heat treatment. However, because of the large rolling reductions required to produce equivalent hardness levels, and the resulting thin sheet, the cold rolled material could not be tension tested using the weld tensile strength specimen. For comparison to weld tensile strength, the UTS of cold rolled plates was determined using standard flat sheet tensile samples oriented parallel to the rolling direction. The comparison is apparently accurate as indicated previously.

The weld tensile strength is graphically compared to the unde formed and cold rolled tensile strength for the O, T4, and T6 conditions, each appearing as a function of isochronal annealing temperature in Figures 6, 7, and 8.

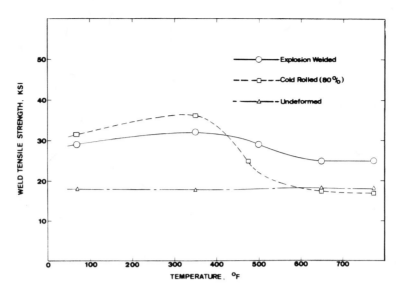

Figure 6. Average Weld Tensile Strength, Undeformed Parent Metal Strength and Ultimate Tensile Strength of Cold Reduced 6061-0 Aluminum Alloy as a Function of 3 Hour Annealing Temperatures.

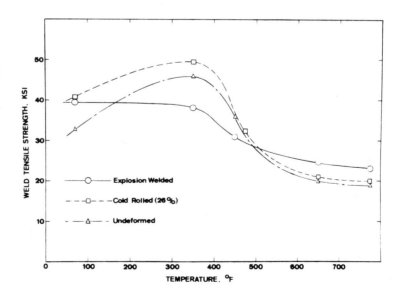

Figure 7. Average Weld Tensile Strength, Undeformed Parent Metal Strength and Ultimate Tensile Strength of Cold Reduced 6061-T4 Aluminum Alloy as a Function of 3 Hour Annealing Temperatures.

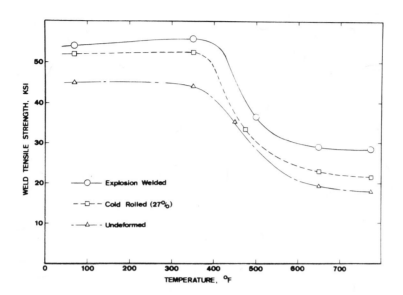

Figure 8. Average Weld Tensile Strength, Undeformed Parent Metal Strength and Ultimate Tensile Strength of Cold Reduced 6061-T6 Aluminum Alloy as a Function of 3 Hour Annealing Temperatures.

Metallurgical examination of the weld interface was conducted for the as-welded and annealed samples representing the O, T4, and T6 initial conditions. In each case the optical microstructure in the as-welded condition was the same as the 350°F annealed condition, and the microstructure after both 650°F and 775°F were the same. The optical microstructure is therefore adequately represented for the O, T4, and T6 initial conditions by the as-welded and the 650°F annealed microstructures. The as-welded and annealed microstructures are compared in Figure 9.

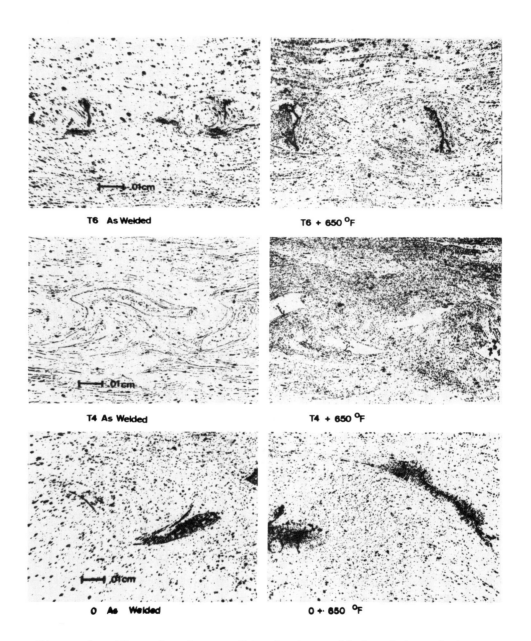

Figure 9. Microstructures of Explosion Welded and Annealed 6061-T6, T4, and O Aluminum Alloy.

DISCUSSION

Explosion Welding of 6061 Aluminum

Variations in explosive loading at a constant standoff will produce variations in the impact velocity and peak pressure at the weld interface. As noted previously the peak pressure at an explosive loading of 1.24 g/cm^2 is approximately 25 kbar. Estimates of impact velocity based on the Gurney method[10] of predicting the velocity imparted to a rigid plate by an explosion would put the impact pressure of the highest explosive loading at 50 kbar. As might be expected the weld tensile strength (Figure 5) exhibited a wide fluctuation as the explosive loading was varied for the O, T4, and T6 condition. Significantly the maximum weld tensile strength was greater than the tensile strength of the unshocked parent metal by up to 67% for O condition and 20% for the T6 condition. Hardness measurements at the welded plate interface and surface indicate a similar increase for the O and T6 condition. Since the O condition is stable with respect to temperature, the increase in strength and hardness must be attributed to the combined effects of shock and strain hardening and unrelated to temperatures produced during welding. The effect of elevated temperature on the T6 condition would be to lower the strength and hardness, not to increase as was observed. Therefore the increase in weld tensile strength and hardness of the T6 condition above the level of unshocked parent metal must also be due to the combined effects of shock and strain hardening. The situation for the T4 alloy was somewhat different. The maximum weld tensile strength was 25% greater than the unshocked parent metal strength while the hardness increased 50% at the same explosive loading condition. The explanation for the difference in behavior of the T4 alloy develops a clearer picture of the importance of thermal-mechanical stability on weldability.

Metallographic examination (Figure 9) shows that wavy flow deformation is limited to the weld zone, extending only 0.020 to 0.030 inches maximum to either side of the weld interface for any of the three initial alloy conditions. The surface and interface hardness difference, compared in the following table, for the T6 and O alloy conditions reflect this non-uniform distribution of strain hardening.

Table II. Brinell Hardness of Explosion Welded 6061 Aluminum Alloy

Initial Weld Condition	Initial Hardness	Hardness After Welding	
		Surface	Interface
Annealed (O)	32	48	60
Solution Treated (T4)	58	94	90
Precipitation Hardened (T6)	94	110	118

However, the surface and interface hardness values for the T4 alloy condition are nearly equal. For the temperature insensitive 0 condition, the difference between the initial hardness and the surface hardness should reflect the contribution of shock wave deformation, while the difference between the interface and surface hardness should reflect the contribution of flow deformation. A similar relation would be expected for the T6 condition. The T4 alloy would be sensitive not only to deformation hardening but to hardening by temperatures below 350°F. Because the surface hardness of the T4 condition has reached that of the peak aged condition, it is believed that not only shock wave deformation but temperature induced aging effects are responsible for the final strength and hardness. The interface hardness of the T4 condition is thought to be less than the surface hardness because of overaging. The microstructure indicates some melting at the weld interface had occurred and it is certain, because the heat flux originates at the weld interface, that temperatures above 200°F and below the melting point (1200°F) were present in the weld zone for a period of time not less than 3 to 5 minutes. In fact, surface temperatures of 200°F have been measured 15 minutes after welding.

The weld tensile strength is observed to fluctuate widely with explosive load even though the surface hardness was nearly the same at each explosive loading for a particular T6, T4, or 0 initial alloy condition. This would indicate that features at the weld interface must be responsible for controlling weld tensile strength, not the residual properties of the welded plates. Metallographic examination of the weld interface at each explosive loading clearly shows that wave amplitude and weld defects increase as explosive loading increases. Micrographs of the weld interface for 6061-T6 alloy as a function of explosive loading are shown in Figure 10. The major weld defects are associated with interfacial melting. The solidified melt pockets contain numerous solidification voids and hot tears resulting from thermal contraction. The strength of the solidified melt would approach the annealed (0) condition in any of the 3 initial alloy heat treatments, and tensile fracture could be expected to occur by link-up of weld defects if they are present in sufficient size and number. At explosive loadings of 1.86 and 2.17 g/cm for the 0 and T6 alloy condition, tensile fracture does occur at less than parent metal strength in the weld plane by ductile tears linking periodically spaced defects. At the lower explosive loadings where weld defects are minimal, tensile failure occurs at high strength levels by slant fractures at 45° to the weld plane for the T6 and 0 conditions. In the T4 condition fracture occurred along the wavy interface even at the peak strength condition which contains few voids but does exhibit a substantial amount of interfacial melting.

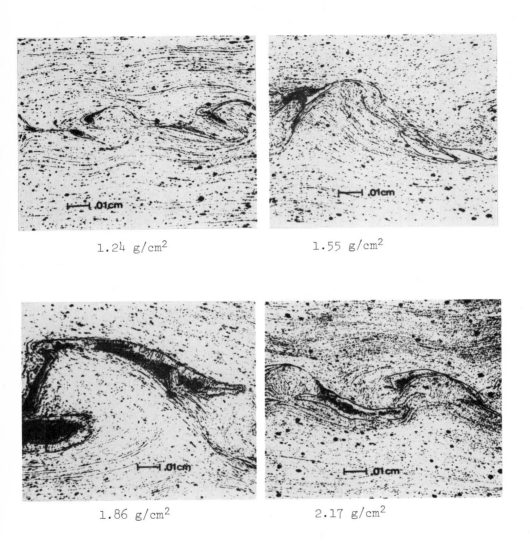

Figure 10. Microstructures of Explosion Welded 6061-T6 Aluminum Alloy as a Function of Explosive Loading.

It is therefore concluded that weld defects in the form of voids control weld strength when the plate collision velocity was higher than optimum, and work hardening at the weld zone controls weld strength at or near optimum collision parameters. The T4 alloy is a special case where the generation of heat during welding is sufficient at any impact pressure level to produce voids and/or melt that reflects the strength of annealed 6061 alloy.

A measured collision pressure of 25Kb during welding is sufficient to account for not more than a 30°C temperature rise in the compressed state[11] and less of a residual temperature increase after relaxation to atmospheric pressure. The temperature increase to produce melting must therefore be due to heat generated by the irreversible work of plastic deformation during adiabatic conditions. This seems plausible when one considers that effective strains of 2.3 have been measured in the zone of wavy deformation[12]. If heat generation by flow is responsible for melting and associated weld defects then it may be possible to establish a weldability criterion base on a knowledge of some flow stress parameter as a function of temperature and strain rate as well as the specific heat.

On the basis of weld defects, type of fracture, and explosive loading range that will produce parent metal tensile strength or greater, the T4 initial condition is rated the least weldable and the O condition the most weldable.

Response to Heat Treatment

Fracture of the as-welded T6 and O alloy in the maximum weld strength condition occurs at 45° to the tensile axis (along the plane of maximum shear) without appreciable plastic flow. Loss of weld ductility would be expected due to the high degree of work hardening that has occurred during explosion welding. Post weld heat treatments were conducted in an effort to restore all or part of the weld ductility without reducing the weld tensile strength below the undeformed parent metal strength level.

Examination of weld tensile fractures after various heat treatments do reveal a change in fracture appearance above a particular temperature. For the 6061-T6 and O alloy welds, fracture transitions from a slant fracture with little ductility after 350°F to a ductile, fiberous fracture after 500°F. The 6061-T4 welds continue to fracture in the weld plane with little ductility after all heat treatments. It was, however, established previously that the T4 weld strength is being controlled by weld defects and not the metallurgical condition of the matrix.

The weld tensile strength after the 500°F heat treatment for the explosion welded T6 alloy was approximately 38,000 psi or 6,000 psi below the tensile strength of the undeformed T6 condition, and 17,000 psi below the weld tensile strength after 350°F heat treatment. Further heat treatment experiments involving both time and temperatures between 350°F and 500°F may make it possible to produce a fiberous fracture and maintain the tensile strength above 40,000 psi.

The weld tensile strength of the explosion welded 6061-0 alloy after 500°F heat treatment was 28,000 psi or 10,000 psi higher than the tensile strength of the undeformed 6061-0 alloy and approximately 3,000 psi higher than 6061-0 alloy cold rolled 80% and annealed for 3 hours at 475°F.

Therefore it is not only possible to recover a significant amount of weld ductility, but the maximum short time exposure temperature possible before the weld strength is reduced below parent metal strength has been increased substantially.

Perhaps the most significant observation from this data is the higher strength level, compared to cold rolled and undeformed 0, T4, and T6 alloy, maintained after annealing for 3 hours at temperatures up to 775°F. Not only are the weld tensile strengths higher, reflecting properties at the weld interface, but hardness measurement indicates the increase occurs throughout the welded plates. An explanation for the high comparative strength levels maintained after high temperature anneals is not possible by examination of the optical micrographs. However, examination of the wavy interface region of 6061-T6 alloy by transmission electron microscopy after 650°F-3 hour heat treatment reveals a structure that is consistent with the higher strength. A transmission micrograph is shown in Figure 11.

After annealing at 650°F for 3 hours, the structure of undeformed 6061 alloy would consist of rod-shaped Mg_2Si precipitates with a spacing of 0.1 to 0.5 microns[13]. The structure seen in Figure 11 appears to consist of dislocations and possibly solute atoms precipitated on a stable dislocation network. Determination of the structure and the mechanism of formation were unfortunately beyond the scope of the investigation. It is important to note that high rate deformation via explosion welding has been more effective in extending the elevated temperature properties than conventional deformation. However, raising elevated temperature properties in precipitation hardened aluminum by thermal-mechanical processing has been observed previously[14,15].

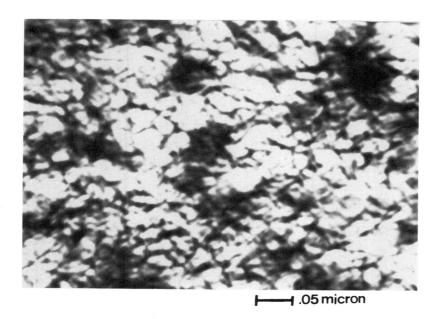

|—————| .05 micron

Figure 11. Transmission Electron Microstructure of Explosion Welded 6061-T6 Aluminum Alloy After Annealing 3 Hours at 650°F.

CONCLUSIONS

Aluminum alloy 6061 can be explosion welded to produce joint strengths greater than the tensile strength of unshocked parent metal. The reduction of weld tensile strength as explosive loading increases above the optimum value is primarily due to interfacial melting and the associated solidification voids. Weld tensile strength above parent metal tensile strength is due to strain hardening. Melting at the interface is due to the adiabatic temperature rise resulting from the irreversible energy component of work in plastic deformation. Weldability of the 6061 alloy increases as the thermal stability of the initial heat treatment condition increases. The O condition is most weldable and the T4 condition is least weldable. Heat treatment after explosion welding can restore weld ductility without significantly reducing weld tensile strength. A high strain rate thermal-mechanical strengthening phenomenon increases the strength of 6061 alloy relative to undeformed material after exposure to temperatures up to 775°F.

REFERENCES

1. G. R. Cowan and A. H. Holtzman, J. Appl. Phys. 34 (1963) 2, p. 928-939.

2. A. Burkhardt, E. Hornbogen and K. Keller, Z. Metallkde, 58(1967) 6, p. 410-415.

3. A. S. Bahrani, T. J. Black and B. Crossland, Proc. Royal Soc. 296 A (1967).

4. A. A. Deribas, Physics of Strengthening and Welding by Explosion, Science Publishers, Novosibirsk, 1972.

5. L. F. Trueb, Trans. AIME 262 (1968), p. 1057-1065.

6. W. Lucas, J. Inst. Metals 99 (1971), p. 335-340.

7. L. F. Trueb, Met. Trans. 2 (1971), p. 145-153.

8. Aluminum Standard and Data - 1972-73, The Aluminum Association, New York.

9. C. D. Lundergen and W. Herrmann, J. Appl. Phys. 34 (1963), p. 2046-2052.

10. J. E. Kennedy, Proc. Symp., Behavior and Utilization of Explosives in Engineering Design, New Mexico Section ASME, March 1972, Albuquerque, N.M. 87110.

11. J. M. Walsh, M. H. Rice, R. G. McQueen and F. L. Yarger, Phy. Review 108 (1957) 2, p. 196-216.

12. R. H. Wittman, Unpublished Data.

13. J. C. Swearengen, Mater. Sci. Eng. 10 (1972), p. 103-117.

14. H. A. Lipsitt and C. M. Sargent, Proc. Vol. III, 2nd International Conference on Strength of Metals and Alloys, ASM, Asilomar, Calif. 1970.

15. S. Pattanaik, V. Srinivasan and M. L. Bhatia, Scripta Met. 6 (1972), p. 191-196.

Acknowledgment

The results in this paper are primarily from work conducted under AMMRC contract DA 19-066-AMC-266(X) reported in more detail in AMMRC Report CR 66-05/31(F). Portions of the work reported were conducted under an AFWL contract in conjunction with the Air Force Academy. A more complete discussion of the collision mechanics and welding mechanism can be found in AFWL Report, "Explosive Welding Development for Civil Engineering Construction Applications," edited by Majors D. Merkle and C. Lindbergh, to be published in 1973.

Discussion by J. Lipkin
        Sandia Laboratories, Mechanical Response Division

    The unusually high weld tensile strengths reported for explosion welded 6061 in the -T6 condition may be an artifact of the "zero gage length" test specimen used by the author. The stress state in the vicinity of the fracture surface is complicated by the geometrical constraints of two 90° notches in close proximity. In addition, the material adjacent to the fracture surface provides additional constraint, leading to some degree of tri-axiality in the resulting stress field. It is therefore not likely that the actual stress state that is associated with an observed mode of fracture can be suitably approximated as one-dimensional. The test sample configuration and the test procedure may be useful for proof testing weld strengths, but unless the observed fracture characteristics are identical for all test samples it is not realistic to associate fracture <u>stresses</u> in the material with fracture <u>loads</u> obtained in the test.

    For example, the author states that fracture at the optimum weld strength for 6061 in the -T6 and -O conditions occurs on 45° planes with apparently little ductility, while fracture for the homogeneous material is fibrous and ductile appearing. If a cleavage type fracture occurs on an inclined plane, the stress normal to this plane is the fracture stress. This normal stress is somewhat lower than the P/A value given by the author (where A is the projected area of the "zero gage length" sample) simply because of the orientation of the fracture plane. The appearance of different fracture modes seems to be clear evidence that the actual stress state and the macro properties of the sample are interacting to such an extent that the failure "stress" determined in a simplistic way from the test configuration is not necessarily a good measure of the actual fracture strength of the material.

Author's Reply

The author's choice of words regarding slant fractures is probably misleading. Fractures do occur at angles of about 45° to the load axis with little ductility, but these are shear ruptures, characteristic of a plane stress state and not cleavage fractures. If cleavage had occurred it would be necessary to use the area of the included fracture plane to determine a normal stress, as you indicate; however, cleavage fracture would not be expected from a face centered cubic aluminum alloy. For plane stress fractures it is appropriate to use the fracture load and the projected fracture area normal to the load axis to determine a fracture stress.

Although the stress state in the vicinity of the fracture plane is complicated as indicated, there may be offsetting factors that cause the fracture stress of the "zero gage length" tensile specimen to approximate the tensile stress measured in finite gage length tensile samples, which is observed experimentally. The 90° notches would act as stress concentrators and thus tend to reduce the apparent strength, while radial constraint leading to some degree of triaxiality would increase the apparent strength. The author believes the unusually high strength levels are real based not only on the test results but on the fact that the measured strength increases correlate well with the hardness increase measured in the vicinity of the weld interface (i.e. the fracture plane).

Admittedly aluminum may present special circumstances (low strain hardening, no cleavage, high toughness) which cause a complex stress state to produce results approximately equal to a simple tension test. With other alloys the results may be less favorable. My experience, however, indicates the "zero gage length" test configuration does provide a useful comparison of weld effects in many alloys, and the fracture stress correlates well with tensile strength and hardness variations. The test is preferred to shear and chisel tests for quality of information about the weld zone.

INDEX

Acceleration wave equation, 71
Activation energy, 388, 643
Adiabatic shear, 520
  fracture, 533
  fragmentation, 534
  heat generation, 511, 547
  initiation, 531, 534
  propagation, 531, 537
  recrystallization, 540
  slip surfaces, 534
  temperature rise, 416, 540, 551, 684
Age hardening, 443
Alloys
  disordered, 620
  face centered cubic, 571
  iron-aluminum, 620
  ordered, 581
Aluminum, 39, 150, 251, 429
  99.999% pure, 486
  mechanical properties, 417
  melting, 247
  physical properties, 417
  shock-loaded, 616
Aluminum alloy, 151, 533
  Al-1145, 486
  Al-2024, 443
  Al-2024-T4, 533
  Al-6061, 443, 671
  Al-6061-T6, 480, 523
  inclusions, 453
  intermetallic particles, 453
  oxide, 453
  precipitation state, 448
Anisotropy plastic, 360
Annealing, 573, 653
  of shock loaded metals, 646
Anti-phase domain, 581, 621
Attenuation
  effects, 429
  stress wave, 343, 431
  ultrasonic, 255, 263, 301
Austenite, 575
Austenitic stainless steels, 575
  See steel

Barium fluoride, 195
  single crystals, 188
Bauschinger effect, 98
Beryllium, 45, 340
  high purity, 480
  S-200, 460
  single crystals, 349
Biaxial state of strain, 445
Bismuth
  melting, 247
Body-centered cubic metals, 576
Bonds
  gold tantalum, 559
  copper nickel, 559
  mechanical properties of, 559
Brinell hardness, 675
Brittle fracture mode, 438
Bulk modulus, 212

Capacitor, 114
  bank system, 429
  gauge, 429
Carbon replicas, 567
Cells
  formation, 642, 643
  size, 147
Cerium, 179
Chapman-Jouguet pressure, 110
Cleavage
  transgranular, 489
  steps, 477
Coercive force, 225
Cohesive energy, 212
Cold work, 599, 644, 646, 653
Compressibility data, 171, 288
Compression
  plane wave, 40
  quasi-static, 626
  shock, 52
  test, 511
Compressive wave, 573
Computer codes, 494
Conjugate force, 98
Conjugate slip, 663

Conservation law, 57
  momentum, 196, 433
Constitutive equation, 42, 57, 101, 293, 335, 501, 656
Coordinates
  Lagrangian, 58
  material, 19
Copper, 41, 158, 572
  annealing of shock-deformed, 632
  cold rolled polycrystalline, 646
  dynamic fracture strength, 443
  grain size, 642
  mechanical properties, 417
  physical properties, 417
  single crystals, 646, 659
Copper alloy, 443
  Cu-Ge alloy, 588
  hard beryllium, 419
Coulomb, 205
  d-d ... interaction, 207
  interaction energy, 208
Crack
  arrest mechanisms, 489
  nucleation, 415, 437, 450
  nucleation at grain boundaries, 480
  propagation of, 437
  separation process, 416
  transgranular growth, 489
Crack tip viscosity, 496
Creep
  behavior, 149
  diffusion controlled, 386
Cross-slip, 572
Crystal (single), 34, 96, 624
  beryllium, 45, 349
  copper, 573, 646, 659
  Fe-3%Si, 577
  zinc, 379
Crystallographic directions, 351
Curie point, 240

Debye temperature, 163, 418
Debye theory, 25
Defects point, 143
  structure, 656
Deformation
  history, 105
  inhomogeneous, 534
  isothermal, 98
  low temperature, 311
  mechanisms, 444
  permanent, 626
  plastic, 311, 608, 648
  shock, 226, 333, 646
  state, 647
  substructure, 138, 646
  twinning, 229, 642
  twins, 573, 613
  wavy, 679, 682
  zone, 520
Demagnetization, 202
  thermal, 238
Detonation velocity, 559, 671
Dielectric loss, 369
Differential power analysis (DPA), 647
Diffusion, 670
Dislocation, 660
  average ... densities, 599
  cells, 572, 652
  climb, 643
  coalescence of, 450
  critical stress for depinning, 314
  density, 42, 140, 361, 381, 443, 565, 572, 596, 615
  distribution, 147, 603
  dynamics, 34, 101
  forest, 388
  generation rates, 4, 255
  interactions with electrons, 311
  interface, 327
  jogs, 574
  line, 96
  loops, 574

Dislocation (continued)
  mechanics, 319
  microstructure, 47
  mobile ... density, 270, 336
  multiplication, 44, 289
  networks, 450, 476
  nucleation, 44
  phonon drag, 418
  pinning points, 420
  rearrangement of, 653
  saturation density, 504
  sessile, 667
  substructures, 102, 140, 572
  supersonic, 319
  tangles, 476
  vibrational frequency of the, 418
  viscosities, 418
Dislocation components
  edge, 576
  screw, 577
Dislocation damping mechanisms,
  phonon, 321
  electron, 321
Dislocation drag, 287
  coefficient, 256, 277
  electron, 261
  mechanisms, 311
  phonon, 256
  viscous, 260
Dislocation mobility, 41, 45, 93
  athermal, 385
  diffusion-controlled, 385
  dislocation-drag, 385
  non-dissipative forces, 278
  thermal activation, 298, 338, 385
  viscous drag mechanism, 391
Dislocation scattering
  radiation, 257
  reradiation, 266
  strain-field, 257
Dislocations, partial
  Frank, 576
  Shockley, 576

Dislocation velocity, 4, 41, 98, 277, 336
  inertial effects, 419
  in lead, aluminum, potassium, iron, 321
  limiting, 287
  mean, 419
  mobile, 419
  obstacle controlled, 281
  relativistic effects, 264, 303, 338
  viscous drag controlled, 281
Drucker's postulate, 505
Ductile-brittle transition, 139
Ductility, 147
Dynamic(s), 40, 379
  recovery, 565
  unloading, 49
  yielding, 34, 367
Dynamic fracture
  effect of pre-strain on, 451
  microstructural features, 473
  strength, 443
Dynamic stress-strain curve, 520
Dynamic yield behavior
  influence of heat treatment, 368
  influence of point defects, 368
  strength, 577

Effective particle size, 587
Elastic-perfectly plastic
  material response, 460
Elastic-plastic constitutive
  relation, 83
Elastic-plastic transition, 431
Elastic precursor, 4, 431, 460
Elasticity
  non-linear, 294
Electrical resistance, 587, 589

Electron
  micrograph, 562
  microprobe, 559
  replicated ... micrograph, 567
Electron beam experiment, 251
Electronic structure, 178
Electronic transition, 179
Electromagnetic transducer, 116
Embrittlement, 580
Endothermic effects, 651
Energy
  anti-phase boundary (APB), 629
  d band, 209
  latent, 645
  partition, 656
  plastic stored, 645
  release, 646, 648
  shock wave, 628
  stored, 573, 648, 655
  twinning, 628
Entropy, 23, 59, 173
Equation of state, 3, 23, 157, 251
  caloric, 110
  mechanical, 138, 519
  Mie-Grüneisen, 65
Erbium, 177
Etch pit, 264
Europium, 176
Eutectic, 567
Exothermic reactions, 651
Exploding wire technique, 447
Explosive
  bonding, 559
  lens, 108, 608
  reaction pressure, 561
  reaction products, 561
  shock, 619
  specific impulse of the, 561
  welding, 669
Explosive loading, 531, 534, 587, 593, 660, 679

Explosives, 132
  primary, 559
  secondary, 560

Face-centered cubic, 41
Fatigue, 139
Faults
  stacking, 587, 591
  twin, 587, 591
Ferromagnetic, 213, 225
Flash-annealing, 642
Flaw size distribution, 495
Flow equations, 19
Flow potential, 93, 99
Flow stress, 109, 278, 646
Fluorite structure, 186
Flyer plates, 560, 669
Forming
  beryllium, 460
  electrohydraulic, 136
  electromagnetic (EM), 134
  explosive, 131
  high energy rate (HERF), 130
  high-velocity, 129
  impact extrusion, 129
  post-forming properties, 139
  pressure, 134
Fourier analysis, 589
Fracture, 7, 143, 229, 443, 477, 608, 662, 682
  brittle, 412
  criteria, 416
  ductile-brittle, 439
  fractographs, 474
  incipient, 436, 438
  mechanism, 436
  morphology, 450
  radial cracks, 447
  rate of, 440, 447
  spall, 415
  static, 415
  strength, 436
  toughness, 149, 495
  velocity, 487

INDEX

Fracture nucleation
  coalescence, 474
  dislocation pileup, 484
  effect of impurities, 484
  heterogeneous, 476
  homogeneous, 477
  mechanisms, 484
Fragmentation, 473, 605
  "number", 608
Free atomic d-functions, 207
Free energy, 94, 215
Free-surface velocity measurements, 429

Gadolinium, 179
Gages, 122
  manganin, 123
  piezoelectric, 122
  quartz, 2
Gibbs free energy, 157
Glide plane, 41
Grain, boundaries, 642
  low-angle subgrain boundaries, 476
  refinement, 632
  size, 462
Grüneisen
  equations of state, 65, 157, 181
  parameter, 25, 158, 174, 252
Guns, 112, 616
  gas, 107
  recovery experiments with, 126
Gurney method, 679

Hadfield manganese steel, 575
  See steel
Hall-Petch equation, 602, 615
Hardening, 144, 575, 599, 613, 679
Hardness
  micro, 237, 550, 565, 587, 633, 636
  vickers, 612
Heat capacity, 656

Heterogeneous nucleation, 654
Hole growth
  mechanical criteria, 420
  nucleation, 415
  temperature rises during, 420
  viscous models, 421
Hoop stress, 447
Hugoniot, 37, 171, 181
  equation of state, 23, 173
  jump relations, 37
  pressure, 159
Hugoniot elastic limit (HEL), 109, 468, 545, 577
  for beryllium, 468

Impact
  rigid sphere, 531
  welding, 9
Impedance matching, 111
Impulse, 560
Inelastic behavior, 545
Interface
  bond, 560
  melting, 559, 562
Interferometer
  laser, 118
  Michelson, 121
  velocity, 118
  VISAR, 121
Internal energy, 26
Internal friction, 281
Internal state variable, 83, 94
Iron (Fe), 49, 172, 201, 235
  alpha, 212, 576
  annealing of shock-loaded, 579
  Armco, 484
  Fe-Al, 620
  Fe_3Al, 620
  Fe_3Be ordered and disordered, 619
  flow of, 577
  low carbon, 549
  mechanical properties, 417
  nickel, 225

Iron (continued)
  phase transformation, 223
  physical properties, 417
  shock-loaded, 577
  shock-loaded single crystals of 3%Si, 577
Isentrope, 28
Isochronal anneals, 675
Isotherm, 25, 158
Isothermal behavior, 94

Jetting phenomenon, 564, 669
Jump conditions, 17, 71

Kinetics
  phase transformations, 183
  relations, 97
  release, 648
Kinks, 663
  walls, 667
Kurdjumov-Sachs relationship, 579

Lagrangian coordinate, 58
Lattice
  defects, 645
  fractional change constant, 591
  rotations, 660
Lax-Wendroff difference equations, 77
Lead, 180
Lead azide, 559
Liquid state, 178
Lithium fluoride, 367
Long range order, 619
Low-angle sub-boundaries, 662
Lunar samples, 238

Machining, 547
Magnetic
  hysteresis, 230
  moment, 212
Magnetization
  remanent, 225, 232
  thermal remanent, 236

Manganin piezo-resistive gauge, 674
Martensite, 232
  α, 575
  ϵ, 575
  lath, 578
  martensitic transformations, 236
  reverse, 578
Material instability, 519
Melting, 172, 251
  phase line, 176
  rates, 247
Meteorite impact, 225
Microfracture, 144
  coalescence, 473
  growth, 473
  nucleation, 473
  nucleation stress sensitivity, 495
  nucleation threshold stress, 484
  viscous growth law, 494
Microprobe analysis, 566
Microstrains, rms, 587
Microscopy
  optical and electron, 443
Microstructure, 94, 140, 444, 571, 677
  of face centered cubic metals and alloys, 572
  shock-induced defect, 587
Mie-Grüneisen equation, 175
Miniature bonding, 560
Modulus
  instantaneous, 20
Momentum traps, 649
Mossbauer, 202
Munroe effect, 669
Murnaghan equations, 65

Nickel, 572
Normality condition, 93, 97
Nucleation, 643, 655
  sites, 575, 656

INDEX

Numerical
  algorithms, 61
Numerical analysis
  algorithms, 61
  convergence, 64
  shock fitting, 71
  solution method, 57
  stability, 64

Oblique collision, 669
Orthorhombic, 186

Paramagnetic, 201, 225
Particles
  second phase, 476
Peel test, 559
Phase
  boundaries, 29, 173
  changes, 157, 179
  diagram, 180
  line, 181
  mixed, 29, 172
  ordered, 620
Phase transformation, 7, 28, 143, 171, 575, 613
  iron, 125, 223, 612
  melting, 247
  rate, 251
Phonon drag, 419
Plane wave, 34, 108, 288
  compression, 40
  deformation, 35
  explosive lens, 108, 608
  generator, 660
  longitudinal, 36
  propagation, 42
  shock compression, 34, 587
  shock hardening, 608
  See shock hardening
Plastic deformation, 648
  flow, 255, 645
  incompressible, 100
  increment, 99
  macroscopic, 94
  straining, 94, 382
  strain rate, 99

Plastic wave, 349, 460
  multiple, 360
Plastic work, 52, 101, 295
Plasticity
  continuum, 93
  equations, 503, 537
  homogeneous, 420
Plate-impact experiments, 459, 462
Point defects
  clusters, 574
  recovery of, 653
Pole figures, 459, 634
Polycarbonate, 474
Polygonization, 644
Polyimide, 477
Porous materials, 177
Potential, 102
  dual, 100
  flow, 93, 99
Precipitation
  of carbides, 576
Precursor, 42
  amplitude, 40, 373
  decay, 305, 373, 380
  elastic, 43, 288, 349
Pressure-volume relationships, 157
Prestraining, 147, 443
Primary slip direction, 662
Pulse duration, 46, 615
Pure shear, 519

Quantum mechanical
  tunneling, 418
Quartz, 545
  crystals, 113
  gage, 188
  transducer, 561
Quartz phenolic
  three-dimensional, 489
Quasi-harmonic solid, 159
Quasi-static loading, 355

Radial release process, 49, 445

Rankine Hugoniot equations, 111, 160, 173
Rarefaction, 157, 229, 575
Rate law, 101
Rayleigh line, 23, 37
Rayleigh wave velocity, 487
Recovery, 573, 632, 670
Recrystallization, 462, 573, 632, 648, 670
  behavior of shock loaded crystals, 655
  grain size, 649
  nuclei, 657
  temperatures, 655
Reflected waves, 157, 655
Release wave profile, 223, 248
Remagnetization shock, 225, 227
Residual
  microstresses, 46, 467
  shear stresses, 467, 591
    thermal anisotropy, 467
Resistivities, 633
Richtmyer, 72

Scab thickness, 434
  See spall
Scanning electron microscopy, 415, 477
Screw dislocations, 566, 576
  cross slip of, 566
Schrödinger, 204
Secondary explosive bonding systems, 559
Shear, 229, 537
  critical resolved, 40
  inhomogeneous, 537
  resolved, 355
  strength, 545
  wave velocity, 319
Shear zones
  martensite, 541
  secondary, 547
  width of, 538
Shock, 108, 287, 331, 633, 646
  adiabat, 25
  compression, 34, 185

Shock (continued)
  demagnetization, 5
  duration, 572
  elastic, 373
  energy, 628
  hardening, 149, 237, 615, 608
  hydrostat, 161
  induced porosity, 649
  lithification, 225
  melting, 225
  metamorphism, 227
  phase transition, 4
  plastic, 228, 373
  plane wave, 633
  polarization, 8
  propagation, 34, 247, 349
  remagnetization, 227
  stability, 26, 29
  steady, 189, 290
  structure, 2
  techniques, 587
  temperature, 24
  thickness, 292
  velocity, 22
Shock loading, 33, 125, 181, 270, 349, 571, 580, 587, 590, 605, 608, 621
Shock-velocity-particle velocity relationship, 158
Shocked liquids, 125
Single crystals
  See crystals
Slip bands, 420
Slip systems, 96, 349
  basal, 355
  primary, 662
  prismatic, 355
  pyramidal, 355
  theory, 103
Sound speed, 20, 171, 462
Spall, 110, 349, 429, 473, 662
  fracture plane, 363, 649
  strength, 459, 465
Stacking fault, 142
  energy, 47, 443, 572, 588, 643

Steady wave, 72, 189, 290
Steel, 39
　annealed mild, 523
　austenitic stainless, 575
　carbon, 608
　ferritic, 579
　Hadfield manganese, 575
　martensitic, 579
　mild, 612
　pearlitic, 579
　SAE 4130, 531
　stainless, 143, 550
Strain, 235, 295, 528, 646
　boundary migration, 642
　dynamic fracture, 528
　induced transformation, 140
　plastic, 295, 336
　uniaxial, 34, 336, 571
Strain aging, 235, 577
Strain-hardening, 99
　exponent, 520
Stress
　Cauchy tensor, 503
　combined behavior, 105
　corrosion, 139
　deviatoric tensor, 503
　effective, 389
　fracture, 404
　macroscopic, 96
　micro-slip, 404
　proof, 633
　quasistatic yield, 577
　relaxation, 39, 312, 362, 367, 546
　residual, 101
　thermodynamic, 97
　torsion, 278
　triaxial, 447
Stress rupture behavior, 149
Stress-strain relations, 223, 520
Stress waves
　compressive, 573
　planar shock waves, 571
　plane, 109
　propagation, 57, 252, 335

Stress waves (continued)
　rarefaction, 573
　release, 342
Stretching tensor, 503
Subgrain
　boundaries, 663
　formation, 565
Substructures, 590, 660
　Cu-8.6 at. pct. Ge, 588
　parameters, 600
Superconductors
　plastic deformation, 255
　temperature dependence of dislocation motion, 314
Superplasticity, 502

Tantalum, 39
Taylor hardening, 105
TD-Ni, TD-NiCr, 581
Tensile rarefaction wave
　See stress waves
Tensile stress, 354, 502
Tensor
　Cauchy, 503
　deviatoric, 503
　invariants, 504
　macroscopic plastic strain, 93
　plastic strain-rate, 41
　stretching, 503
Terminal strain profiles, 523
Texture, 459, 492, 573, 632
Thermal, 519
　conductivity, 265
　energy, 110
　expansion, 158
　instability strain, 519
　metamorphism, 230
　number, 521
　softening, 519, 537
Thermodynamic force, 93
Thermomechanical processing, 150
Thin films, 560
Titanium, 140
　commercially pure, 523

Titanium (continued)
  Ti-6Al-4V, 501
Torsional impact test, 522
Transformation, 232
  $\alpha \longrightarrow \epsilon$, 201, 236
  allotropic, 575
  kinetics, 172
  martensitic, 236
  phase, 186
  polymorphic, 185
  pressure-induced phase, 578
  rate of ___, 190
  shock induced, 185
Transformation matrix, 41
Transgranular cleavage, 465
Transition, 180
  Curie temperature, 227
  ferromagnetic-to-paramagnetic 574
  point, 202
  pressure, 183, 215
Transmission electron microscopy, 450, 659
Triaxial stress state, 447
Twin, 667
  density, 443
  "detwins", 619
  faults, 597
  former, 622
  lamellae, 642
  loss of ___, 598
  mechanical, 572
  micro, 655
  morphology, 443
  "pseudo-twins", 619
Twinning, 49, 142, 230, 619
  activation energy, 403
  deformation, 229, 401, 642
  detwinning, 4
  dislocations, 628
  energy, 622
  nucleation rate, 401
  shear, 573
Twinning Stress
  grain size dependence, 401
  strain rate dependence, 401

Twinning Stress (continued)
  Temperature dependence, 401

Uniaxial strain, 34, 336, 571

Velocity
  free surface, 110
  particle, 108, 353
  Rayleigh, 487
  shock, 37, 108, 357
Viscosity, 72, 502
Von Mises material, 504
Von Neumann, 72

Wave
  acceleration waves, 17
  amplitude, 680
  longitudinal, 15, 325
  plane compression, 17
  propagation, 459
  radial release, 51
  ramped elastic, 465
  rarefaction, 17, 38
  shear, 329
  steady compression, 17
Welds, welding
  defects, 680
  ductility, 682
  explosive, 669
  microstructures, 671
  post heat treatment, 682
  tensile strength, 673
Wigner-Seitz (W-S), 203
Work hardening, 501, 682
  exponent, 519
  rate, 575
Work softening, 501

X-ray analysis, 559
X-ray diffraction, 186, 587
  pole figures, 462
  topography, 659

Yield criterion, 36, 99
Yielding, 39
  process, 7

Yielding (continued)
  shock-induced, 39
Yield strength, 237, 587
  dynamic, 353
Yield stress, 613
  elevated temperature, 555, 612
  quasi-static, 370

Zinc, 279
  single crystal, 379
Zirconium, 183